THE WILEY
ENGINEER'S
DESK REFERENCE

THE WILEY ENGINEER'S DESK REFERENCE

A CONCISE GUIDE FOR THE PROFESSIONAL ENGINEER

SANFORD I. HEISLER, P.E.
Bechtel Power Corporation
San Francisco, California

A Wiley-Interscience Publication
JOHN WILEY & SONS
New York · Chichester · Brisbane · Toronto · Singapore

Library of Congress Cataloging in Publication Data:

Heisler, Sanford I.
  The Wiley engineer's desk reference.

  "A Wiley-Interscience publication."
  Includes index.
  1. Engineering—Handbooks, manuals, etc.   I. Title.

TA151.H424 1984   620   83-21690
ISBN 0-471-86632-6

Printed in the United States of America

10 9 8 7 6 5 4 3 2 1

PREFACE

The Wiley Engineer's Desk Reference is oriented toward the working engineer for use as a desk reference to aid in his or her day-to-day work. As a result, this volume includes a significant amount of material, including widely used equipment information, oriented toward the practical problems daily faced by the engineer. Theoretical or derivative data of limited interest is thus minimized. To keep the size of the book convenient, much engineering data has been omitted; but to assist engineers in handling work of a more complex or theoretical nature, references for this type of information are included as well. In addition, SI as well as English units have been shown where appropriate.

The Engineers' Manual by Ralph Hudson and *The Engineer's Companion* by Mott Souders, although valuable references, have tended to become less useful over the years, due to the growth of and developments in engineering. This *Engineer's Desk Reference* has been developed as a successor volume with several purposes in mind: update the fundamental information; expand the coverage to newer techniques; and perhaps most important, provide a source of data to assist in practical problem solving. Both Hudson and Souders have been widely drawn upon for data and we wish to acknowledge their contribution.

Numerous discussions with professors in academia over the past two decades have also indicated a growing need for more practical application information for use as a teaching aid in our universities. This has been further confirmed by my colleagues in industry, where engineers frequently must work across the traditional discipline lines. This volume is intended to fulfill both these needs.

In addition, most engineers during their careers find themselves in a discipline different from their major field of study and/or experience. This volume is intended to assist the engineer in these circumstances by providing overviews of the technology in these less-familiar areas.

The growth of codes and standards has been rapid and far reaching, and they affect not only the manner of solution for many engineering problems but also load assumptions, permitted stress levels, design details, and so forth; thus references to significant codes have also been included.

This reference contains sections on Controls, Economics, and Energy Sources not commonly found in a single volume today. With the shifting emphasis and the changing economic values of energy and energy-related factors, their inclusion is particularly necessary. In addition, they emphasize the area of application and thus represent practical design data. We have tried to structure these sections to avoid premature obsolescence, but where cost is discussed, it is necessarily time qualified.

To permit the maximum utilization of this volume, the use of advanced mathematics has been limited. This is particularly the case in Section 6, Controls. Most

To make this book easier to use,

English

v

control references tend either to treat control systems in a detailed, complex mathematical way, with heavy theoretical emphasis, or to discuss only the hardware available, such as valves, instruments, and so on. We have tried to achieve a middle ground and present a balanced view of the subject of particular value to the process or applications engineer, while recognizing that a particular user may wish to work in greater depth, utilizing the references included.

The section on Engineering Practice, Section 9, is new and presents data derived from several decades of engineering practice. While not all-inclusive, it covers many areas of activity. Here again the limitations of space require that much information be presented in summary or abbreviated form.

Given the development and widespread use of electronic calculators and their powerful mathematical and arithmetic capability, we have limited the use of tabular data in this reference to only those items most repetitively used. For example, the trigonometric functions of various angles, radians, and so on, are given for the principal or most commonly used values only. We anticipate that, where necessary, the user would interpolate or refer to data stored either in the memory of his/her calculator or to tables of reference that would have those values shown to a large number of significant places.

Throughout the book, superscript numbers (except in exponents) are used to denote references at the end of each section.

Needless words

I should like to acknowledge the efforts and assistance of Beverly Duke and Rose Huber for manuscript preparation, coordination, and overall editorial assistance. In addition, to the encouragement by Dean Robert Steidel of the University of Califonia at Berkeley, considerable assistance and professional comments by Distinguished Professor Donald Othmer of Polytechnic Institute of New York, and Mr. John Hefler of Bechtel Power Corporation have been most useful. Without the active assistance of Doris Lanctot and Carolyn Planakis for librarian and illustrative assistance, the preparation of this volume would not have been possible. The early assistance of Professor Jerald Henderson of the University of California at Davis was particularly helpful also in determining the need and scope of the book. Lastly, I should like to acknowledge and thank my wife Lois Heisler for her active support and assistance in all phases of this work.

As with all professional activities, nothing can take the place of common sense and an inquiring mind. Thus, this reference cannot substitute for true professional judgment and care, but hopefully will provide some guidance and points of reference to the engineer.

We welcome suggestions and comments on the contents of this reference with a view toward improving future editions and making the volume more immediately useful and of interest to the practicing engineer.

SANFORD I. HEISLER

San Francisco, California
February, 1984

CONTENTS

4   Thermodynamics/Heat Transfer   207

5   Electricity and Electronics   241

J. M. Shulman
Fellow Engineer (Retired)
Westinghouse Electric Corporation

THE WILEY
ENGINEER'S
DESK REFERENCE

SECTION 1
MATHEMATICS

1.1 SYMBOLS AND ABBREVIATIONS

Symbol	Meaning	Symbol	Meaning
$+$	Plus / Positive	$\angle$	Angle
$-$	Minus / Negative	$\perp$	Perpendicular to
$\pm$	Plus or minus / Positive or negative	$\parallel$	Parallel to
		$(\ )$	Parentheses
		$[\ ]$	Brackets
		$\{\ \}$	Braces
$\mp$	Minus or plus / Negative or positive	$\overline{}$	Vinculum
		$a°$	a degrees (angle)
$\times$ or $\cdot$	Multiplied by	a'	a minutes (angle) / a prime
$\div$ or $:$	Divided by		a seconds (angle)
$=$, or $::$	Equals, as	a''	a second
$\neq$	Does not equal		a double-prime
$\approx$	Equals approximately	a'''	a third
$>$	Greater than		a triple-prime
$<$	Less than	a_n	a sub n
$\geqq$	Greater than or equal to	sin	Sine
		cos	Cosine
$\leqq$	Less than or equal to	tan	Tangent
$\equiv$	Is identical to	cot	Cotangent
$\rightarrow$ or $\doteq$	Approaches as a limit	sec	Secant
$\propto$	Varies directly as	csc	Cosecant
$\therefore$	Therefore	vers	Versed sine
$\sqrt{}$	Square root	covers	Coversed sine
$\sqrt[n]{}$	nth root	exsec	Exsecant
a^n	nth power of a		Anti-sine a
$n!$	$1 \cdot 2 \cdot 3 \cdots n$	$\sin^{-1} a$	Angle whose inverse sine is a
log	Common logarithm / Briggsian logarithm	sinh	Hyperbolic sine
		cosh	Hyperbolic cosine
	Natural logarithm	tanh	Hyperbolic tangent
ln or $\log_e$	Hyberbolic " / Napierian "	$\sinh^{-1} a$	Anti-hyperbolic sine a / Angle whose hyperbolic sine is a
e or ϵ	Base (2.718) of natural system of logarithms	$P(x, y)$	Rect. coord. of point P
π	Pi (3.1416)		

$P(r, \theta)$	Polar coord. of point P	$\dfrac{\partial z}{\partial x}$	Partial derivative of z with respect to x
$f(x), F(x),$ or $\phi(x)$	Function of x	$\dfrac{\partial^2 z}{\partial x\, \partial y}$	Second partial deriv. of z with respect to y and x
Δy	Increment of y		
Σ	Summation of		
∞	Infinity	$\displaystyle\int$	Integral of
dy	Differential of y		
$\dfrac{dy}{dx}$ or $f'(x)$	Derivative of $y = f(x)$ with respect to x	$\displaystyle\int_a^b$	Integral between the limits a and b
$\dfrac{d^2 y}{dx^2}$ or $f''(x)$	Second deriv. of $y = f(x)$ with respect to x	j	Imaginary quantity $(\sqrt{-1})$
$\dfrac{d^n y}{dx^n}$ or $f^{(n)}(x)$	nth deriv. of $y = f(x)$ with respect to x	$x = a + jb$	Symbolic vector notation

1.2   ALGEBRA

Powers and Roots (1)

$$a^n = a \cdot a \cdot a \cdots \text{ to } n \text{ factors.} \qquad a^{-n} = \frac{1}{a^n}.$$

$$a^m \cdot a^n = a^{m+n}; \quad \frac{a^m}{a^n} = a^{m-n}. \qquad (ab)^n = a^n b^n; \qquad \left(\frac{a}{b}\right)^n = \frac{a^n}{b^n}.$$

$$(a^m)^n = (a^n)^m = a^{mn}. \qquad (\sqrt[n]{a})^n = a.$$

$$a^{1/n} = \sqrt[n]{a}; \qquad a^{m/n} = \sqrt[n]{a^m}. \qquad \sqrt[n]{ab} = \sqrt[n]{a}\sqrt[n]{b}; \qquad \sqrt[n]{\frac{a}{b}} = \frac{\sqrt[n]{a}}{\sqrt[n]{b}}.$$

$$\sqrt[n]{\sqrt[m]{a}} = \sqrt[mn]{a}.$$

Operations with Zero and Infinity (2)

$$a \cdot 0 = 0; \ a \cdot \infty = \infty; \ 0 \cdot \infty \text{ is indeterminate}$$

$$\frac{0}{a} = 0; \qquad \frac{a}{0} = \infty; \qquad \frac{0}{0} \quad \text{is indeterminate}$$

$$\frac{\infty}{a} = \infty; \qquad \frac{a}{\infty} = 0; \qquad \frac{\infty}{\infty} \quad \text{is indeterminate}$$

$$a^0 = 1; \qquad 0^a = 0; \qquad 0^0 \quad \text{is indeterminate}$$

$$\infty^a = \infty; \qquad\qquad\qquad \infty^0 \quad \text{is indeterminate}$$

$$\left.\right\}\ \text{see Section 1.15.}$$

$$a^\infty = \infty, \text{ if } a^2 > 1; \ a^\infty = 0, \text{ if } a^2 < 1; \ a^\infty = 1, \text{ if } a^2 = 1$$
$$a^{-\infty} = 0, \text{ if } a^2 > 1; \ a^{-\infty} = \infty, \text{ if } a^2 < 1; \ a^{-\infty} = 1, \text{ if } a^2 = \ \ \left.\right\}\ \text{see Section 1.15.}$$
$$a - a = 0; \ \infty - a = \infty; \ \infty - \infty \text{ is indeterminate}$$

Binomial Expansions (3)

$(a \pm b)^2 = a^2 \pm 2ab + b^2.$
$(a \pm b)^3 = a^3 \pm 3a^2b + 3ab^2 \pm b^3.$
$(a \pm b)^4 = a^4 \pm 4a^3b + 6a^2b^2 \pm 4ab^3 + b^4.$
$(a \pm b)^n = a^n \pm \dfrac{n}{1} a^{n-1}b + \dfrac{n(n-1)}{1 \cdot 2} a^{n-2}b^2 \pm \dfrac{n(n-1)(n-2)}{1 \cdot 2 \cdot 3} a^{n-3}b^3 + \cdots$

NOTE. n may be positive or negative, integral or fractional. When n is a positive integer, the series has $(n + 1)$ terms; otherwise the number of terms is infinite.

Logarithms (4)

Definition.   If b is a finite positive number, other than 1, and $b^x = N$, then x is the logarithm of N to the base b, or $\log_b N = x$. If $\log_b N = x$, then $b^x = N$.

Properties of Logarithms.

$$\log_b b = 1; \log_b 1 = 0; \log_b 0 = \begin{cases} +\infty, \text{ when } b \text{ lies between 0 and 1,} \\ -\infty, \text{ when } b \text{ lies between 1 and } \infty. \end{cases}$$

$\log_b M \cdot N = \log_b M + \log_b N.$ $\log_b \dfrac{M}{N} = \log_b M - \log_b N.$

$\log_b N^p = p \log_b N.$ $\log_b \sqrt[r]{N^p} = \dfrac{p}{r} \log_b N.$

$\log_b N = \dfrac{\log_a N}{\log_a b}.$ $\log_b b^N = N; b^{\log_b N} = N.$

Systems of Logarithms.   Common (Briggsian)—base 10.
Natural (Napierian or hyperbolic)—base 2.7183 (designated by e or ϵ).

NOTE.   The abbreviation of "common logarithm" is "log" and the abbreviation of "natural logarithm" is "ln."

Characteristic or Integral Part (c) of the Common Logarithm of a Number (N).   If N is not less than one, c equals the number of integral figures in N, minus one.
If N is less than one, c equals 9 minus the number of zeros between the decimal point and the first significant figure, minus 10 (the -10 being written after the mantissa).

Mantissa or Decimal Part (m) of the Common Logarithm of a Number N.   If N has not more than three figures, find mantissa directly in Section 10.
If N has four figures, $m = m_1 + (f/10)(m_2 - m_1)$, where m_1 is the mantissa corresponding to the first three figures of N, m_2 is the next larger mantissa in the table and f is the fourth figure of N.

Number (N) Corresponding to a Common Logarithm that Has a Characteristic (c) and a Mantissa (m).   If N is desired to three figures, find the mantissa nearest to m in the table, Section 10, and the corresponding number is N.

If N is desired to four figures, find the next smaller mantissa, m_1, and the next larger mantissa, m_2, in the table. The first three figures of N correspond to m_1 and the fourth figure equals the nearest whole number to $10[(m - m_1)/(m_2 - m_1)]$.

NOTE. If c is positive, the number of integral figures in N equals c plus one. If c is negative (e.g., $9 - 10$, or -1), write numeric c minus one zeros between the decimal point and the first significant figure of N.

Natural Logarithm (ln) of a Number (N). Any number, N, can be written $N = N_1 \times 10^{\pm p}$, where N_1 lies between 1 and 1000. Then $\ln N = \ln N_1 \pm p \ln 10$.
If N_1 has not more than three figures, find $\ln N_1$ directly in Section 10.
If N_1 has four figures, N_2 is the number composed of the first three figures of N_1, and f is the fourth figure of N_1, then

$$\ln N_1 = \ln N_2 + \frac{f}{10}[\ln(N_2 + 1) - \ln N_2].$$

Number (N) Corresponding to a Natural Logarithm, ln N. Any logarithm, $\ln N$, can be written $\ln N = \ln N_1 \pm p \ln 10$, where $\ln N_1$ lies between $4.6052 = \ln 100$ and $6.9078 = \ln 1000$. Then $N = N_1 \times 10^{\pm p}$.
The first three figures of N_1 correspond to the next smaller logarithm, $\ln N_2$, in the table, and the fourth figure, f, of N_1 equals the nearest whole number to $10\{(\ln N_1 - \ln N_2)/[\ln(N_2 + 1) - \ln N_2]\}$.

The Solution of Algebraic Equations (5)

The Quadratic Equation.

If

$$ax^2 + bx + c = 0,$$

then
$$x = \frac{-b \pm \sqrt{b^2 - 4ac}}{2a} = \frac{2c}{-b \mp \sqrt{b^2 - 4ac}}.$$

If $b^2 - 4ac = 0$ $\begin{cases} > & \text{the roots are real and unequal,} \\ & \text{the roots are real and equal,} \\ < & \text{the roots are imaginary.} \end{cases}$ The second equation serves best when the two values of x are nearly equal.

The Cubic Equation. Any cubic equation, $y^3 + py^2 + qy + r = 0$ may be reduced to the form $x^3 + ax + b = 0$ by substituting for y the value $[x - (p/3)]$. Here $a = \frac{1}{3}(3q - p^2)$, $b = \frac{1}{27}(2p^3 - 9pq + 27r)$.

Algebraic Solution of $x^3 + ax + b = 0$.

Let

$$A = \sqrt[3]{-\frac{b}{2} + \sqrt{\frac{b^2}{4} + \frac{a^3}{27}}}, \quad B = \sqrt[3]{-\frac{b}{2} - \sqrt{\frac{b^2}{4} + \frac{a^3}{27}}},$$

then

$$x = A + B, \quad -\frac{A + B}{2} + \frac{A - B}{2}\sqrt{-3}, \quad -\frac{A + B}{2} - \frac{A - B}{2}\sqrt{-3}.$$

$$\text{If } \frac{b^2}{4} + \frac{a^3}{27} \begin{matrix} > \\ = \\ < \end{matrix} 0 \left\{ \begin{matrix} \text{1 real root, 2 conjugate imaginary roots,} \\ \text{3 real roots of which 2 are equal,} \\ \text{3 real and unequal roots.} \end{matrix} \right.$$

Trigonometric Solution of $x^3 + ax + b = 0$. In the case where $(b^2/4) + (a^3/27) < 0$, the above formulas give the roots in a form impractical for numerical computation. In this case, a is negative. Compute the value of the angle ϕ from $\cos \phi =$

$$\sqrt{\frac{b^2}{4} \div \left(-\frac{a^3}{27}\right)}, \text{ then}$$

$$x = \mp 2\sqrt{-\frac{a}{3}} \cos \frac{\phi}{3}, \quad \mp 2\sqrt{-\frac{a}{3}} \cos \left(\frac{\phi}{3} + 120°\right), \quad \mp 2\sqrt{-\frac{a}{3}} \cos \left(\frac{\phi}{3} + 240°\right),$$

where the upper or lower signs are to be used according as b is positive or negative.

In the case where $(b^2/4) + (a^3/27) > 0$, compute the values of the angles ψ and ϕ from $\cot 2\psi = [(b^2/4) \div (a^3/27)]^{1/2}$, $\tan \phi = (\tan \psi)^{1/3}$; then the real root of the equation is

$$x = \pm 2\sqrt{\frac{a}{3} \cot 2\phi},$$

where the upper or lower sign is to be used according as b is positive or negative.

In the case where $(b^2/4) + (a^3/27) = 0$, the roots are

$$x = \mp 2\sqrt{-\frac{a}{3}}, \pm\sqrt{-\frac{a}{3}}, \pm\sqrt{-\frac{a}{3}},$$

where the upper or lower signs are to be used according as b is positive or negative.

Graphical Solution of the Cubic Equations. To find the real roots of the cubic equation

$$x^3 + ax + b = 0,$$

draw the parabola (Section 1.5) $y^2 = 2x$, and the circle (Section 1.5), the coordinates of whose center are $x = (4 - a)/4$, $y = -b/8$, and which passes through the vertex of the parabola. Measure the ordinates of the points of intersection; these give the real roots of the equation.

The Binomial Equation.   If $x^n = a$, the n roots of this equation are: if a is *positive*,

$$x = \sqrt[n]{a}\left(\cos\frac{2k\pi}{n} + \sqrt{-1}\,\sin\frac{2k\pi}{n}\right);$$

if a is *negative*,

$$x = \sqrt[n]{-a}\left[\cos\frac{(2k + 1)\pi}{n} + \sqrt{-1}\,\sin\frac{(2k + 1)\pi}{n}\right],$$

where k takes in succession the values 0, 1, 2, 3, $\cdots$, $n - 1$.

1.3   TRIGONOMETRY

Definition of Angle (6)

An angle is the amount of rotation (in a fixed plane) by which a straight line may be changed from one direction to any other direction. If the rotation is counterclockwise the angle is said to be positive, if clockwise, negative.

Measure of Angle (7)

A degree is $\frac{1}{360}$ of the plane angle about a point.

   A radian is the angle subtended at the center of a circle by an arc equal in length to the radius.

Trigonometric Functions of an Angle (8)

sine (sin) α $= \dfrac{y}{r}$.

cosine (cos) α $= \dfrac{x}{r}$.

tangent (tan) α $= \dfrac{y}{x}$.

cotangent (cot) α $= \dfrac{x}{y}$.

secant (sec) α $= \dfrac{r}{x}$.

cosecant (csc) α $= \dfrac{r}{y}$.

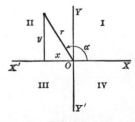

Fig. 1.1

exsecant (exsec) $\alpha = \sec \alpha - 1$.
versine (vers) $\alpha = 1 - \cos \alpha$. coversine (covers) $\alpha = 1 - \sin \alpha$.

NOTE. x is positive when measured along OX, and negative along OX'; y is positive when measured parallel to OY, and negative parallel to OY'.

Signs of the Functions (9)

Quadrant	sin	cos	tan	cot	sec	csc
I	+	+	+	+	+	+
II	+	−	−	−	−	+
III	−	−	+	+	−	−
IV	−	+	−	−	+	−

Functions of 0°, 30°, 45°, 60°, 90°, 180°, 170°, 360° (10)

	0°	30°	45°	60°	90°	180°	270°	360°
sin	0	$\dfrac{1}{2}$	$\dfrac{\sqrt{2}}{2}$	$\dfrac{\sqrt{3}}{2}$	1	0	−1	0
cos	1	$\dfrac{\sqrt{3}}{2}$	$\dfrac{\sqrt{2}}{2}$	$\dfrac{1}{2}$	0	−1	0	1
tan	0	$\dfrac{\sqrt{3}}{3}$	1	$\sqrt{3}$	∞	0	∞	0
cot	∞	$\sqrt{3}$	1	$\dfrac{\sqrt{3}}{3}$	0	∞	0	∞
sec	1	$\dfrac{2\sqrt{3}}{3}$	$\sqrt{2}$	2	∞	−1	∞	1
csc	∞	2	$\sqrt{2}$	$\dfrac{2\sqrt{3}}{3}$	1	∞	−1	∞

Fundamental Relations among the Functions (11)

$$\sin \alpha = \frac{1}{\csc \alpha}; \qquad \cos \alpha = \frac{1}{\sec \alpha}; \qquad \tan \alpha = \frac{1}{\cot \alpha} = \frac{\sin \alpha}{\cos \alpha}.$$

$$\csc \alpha = \frac{1}{\sin \alpha}; \qquad \sec \alpha = \frac{1}{\cos \alpha}; \qquad \cot \alpha = \frac{1}{\tan \alpha} = \frac{\cos \alpha}{\sin \alpha}.$$

$$\sin^2 \alpha + \cos^2 \alpha = 1; \qquad \sec^2 \alpha - \tan^2 \alpha = 1; \qquad \csc^2 \alpha - \cot^2 \alpha = 1.$$

Functions of Multiple Angles (12)

$$\sin 2\alpha = 2 \sin \alpha \cos \alpha;$$
$$\cos 2\alpha = 2 \cos^2 \alpha - 1 = 1 - 2 \sin^2 \alpha = \cos^2 \alpha - \sin^2 \alpha.$$
$$\sin 3\alpha = 3 \sin \alpha - 4 \sin^3 \alpha;$$

$\cos 3\,\alpha = 4 \cos^3 \alpha - 3 \cos \alpha.$
$\sin 4\,\alpha = 4 \sin \alpha \cos \alpha - 8 \sin^3 \alpha \cos \alpha;$
$\cos 4\,\alpha = 8 \cos^4 \alpha - 8 \cos^2 \alpha + 1.$
$\sin n\,\alpha = 2 \sin (n - 1)\alpha \cos \alpha - \sin (n - 2)\alpha;$
$\cos n\,\alpha = 2 \cos (n - 1)\alpha \cos \alpha - \cos (n - 2)\alpha.$

Functions of Half Angles (13)

$$\sin \frac{\alpha}{2} = \sqrt{\frac{1 - \cos \alpha}{2}}; \quad \cos \frac{1}{2}\alpha = \sqrt{\frac{1 + \cos \alpha}{2}}.$$

$$\tan \frac{1}{2}\alpha = \frac{1 - \cos \alpha}{\sin \alpha} = \frac{\sin \alpha}{1 + \cos \alpha} = \sqrt{\frac{1 - \cos \alpha}{1 + \cos \alpha}}.$$

Powers of Functions (14)

$\sin^2 \alpha = \frac{1}{2} (1 - \cos 2\alpha);$ $\cos^2 \alpha = \frac{1}{2} (1 + \cos 2\alpha).$

$\sin^3 \alpha = \frac{1}{4} (3 \sin \alpha - \sin 3\alpha);$ $\cos^3 \alpha = \frac{1}{4} (\cos 3\alpha + 3 \cos \alpha).$

$\sin^4 \alpha = \frac{1}{8} (\cos 4\alpha - 4 \cos 2\alpha + 3);$ $\cos^4 \alpha = \frac{1}{8} (\cos 4\alpha + 4 \cos 2\alpha + 3).$

$\sin^n \alpha = \dfrac{1}{(2\sqrt{-1})^n} \left(y - \dfrac{1}{y} \right)^n;$ $\cos^n \alpha = \dfrac{1}{(2)^n} \left(y + \dfrac{1}{y} \right)^n.$

Functions of Sum or Difference of Two Angles (15)

$\sin (\alpha \pm \beta) = \sin \alpha \cos \beta \pm \cos \alpha \sin \beta.$
$\cos (\alpha \pm \beta) = \cos \alpha \cos \beta \mp \sin \alpha \sin \beta.$

$$\tan (\alpha \pm \beta) = \frac{\tan \alpha \pm \tan \beta}{1 \mp \tan \alpha \tan \beta}.$$

Sums, Differences, and Products of Two Functions (16)

$\sin \alpha \pm \sin \beta = 2 \sin \frac{1}{2} (\alpha \pm \beta) \cos \frac{1}{2} (\alpha \mp \beta).$
$\cos \alpha + \cos \beta = 2 \cos \frac{1}{2} (\alpha + \beta) \cos \frac{1}{2} (\alpha - \beta).$
$\cos \alpha - \cos \beta = -2 \sin \frac{1}{2} (\alpha + \beta) \sin \frac{1}{2} (\alpha - \beta).$

$$\tan \alpha \pm \tan \beta = \frac{\sin (\alpha \pm \beta)}{\cos \alpha \cos \beta}.$$

$\sin^2 \alpha - \sin^2 \beta = \sin (\alpha + \beta) \sin (\alpha - \beta).$
$\cos^2 \alpha - \cos^2 \beta = -\sin (\alpha + \beta) \sin (\alpha - \beta).$
$\cos^2 \alpha - \sin^2 \beta = \cos (\alpha + \beta) \cos (\alpha - \beta).$
$\sin \alpha \sin \beta = \frac{1}{2} \cos (\alpha - \beta) - \frac{1}{2} \cos (\alpha + \beta).$
$\cos \alpha \cos \beta = \frac{1}{2} \cos (\alpha - \beta) + \frac{1}{2} \cos (\alpha + \beta).$
$\sin \alpha \cos \beta = \frac{1}{2} \sin (\alpha + \beta) + \frac{1}{2} \sin (\alpha - \beta).$

Equivalent Expressions for sin α, cos α, and tan α (17)

$$\sin \alpha = \sqrt{1 - \cos^2 \alpha} = \frac{\tan \alpha}{\sqrt{1 + \tan^2 \alpha}} = \frac{1}{\sqrt{1 + \cot^2 \alpha}} = \frac{\sqrt{\sec^2 \alpha - 1}}{\sec \alpha} = \frac{1}{\csc \alpha}$$

$$= \cos \alpha \tan \alpha = \frac{\cos \alpha}{\cot \alpha} = \frac{\tan \alpha}{\sec \alpha} = \frac{\sin 2\alpha}{2 \cos \alpha} = \sqrt{\frac{1}{2}(1 - \cos 2\alpha)}$$

$$= 2 \sin \frac{\alpha}{2} \cos \frac{\alpha}{2}.$$

$$\cos \alpha = \sqrt{1 - \sin^2 \alpha} = \frac{1}{\sqrt{1 + \tan^2 \alpha}} = \frac{\cot \alpha}{\sqrt{1 + \cot^2 \alpha}} = \frac{1}{\sec \alpha} = \frac{\sqrt{\sec^2 \alpha - 1}}{\csc \alpha}$$

$$= \sin \alpha \cot \alpha = \frac{\sin \alpha}{\tan \alpha} = \frac{\cot \alpha}{\csc \alpha} = \frac{\sin 2\alpha}{2 \sin \alpha} = \sqrt{\frac{1}{2}(1 + \cos 2\alpha)}$$

$$= \cos^2 \frac{\alpha}{2} - \sin^2 \frac{\alpha}{2} = 1 - 2 \sin^2 \frac{\alpha}{2} = 2 \cos^2 \frac{\alpha}{2} - 1.$$

$$\tan \alpha = \frac{\sin \alpha}{\sqrt{1 - \sin^2 \alpha}} = \frac{\sqrt{1 - \cos^2 \alpha}}{\cos \alpha} = \frac{1}{\cot \alpha} = \sqrt{\sec^2 \alpha - 1} = \frac{1}{\sqrt{\csc^2 \alpha - 1}}$$

$$= \frac{\sin \alpha}{\cos \alpha} = \frac{\sec \alpha}{\csc \alpha} = \frac{\sin 2\alpha}{1 + \cos 2\alpha} = \frac{1 - \cos 2\alpha}{\sin 2\alpha} = \frac{2 \tan \frac{\alpha}{2}}{1 - \tan^2 \frac{\alpha}{2}}.$$

Inverse or Anti-functions (18)

$\text{Sin}^{-1} a$ is defined as the angle whose sine is a. $\text{Sin}^{-1} a$ has an infinite number of values. If α is the value of $\sin^{-1} a$ which lies between $-90°$ and $+90°$ ($-\pi/2$ and $+\pi/2$ radians), and if n is any integer,

$$\sin^{-1} a = (-1)^n \alpha + n \cdot 180° = (-1)^n \alpha + n\pi \text{ (similarly for } \csc^{-1} a).$$

$\text{Cos}^{-1} a$ is defined as the angle whose cosine is a. $\text{Cos}^{-1} a$ has an infinite number of values. If α is the value of $\cos^{-1} a$ which lies between $0°$ and $180°$ (0 and π radians), and if n is any integer,

$$\cos^{-1} a = \pm \alpha + n \cdot 360° = \pm \alpha + 2n\pi \text{ (similarly for } \sec^{-1} a).$$

$\text{Tan}^{-1} a$ is defined as the angle whose tangent is a. $\text{Tan}^{-1} a$ has an infinite number of values. If α is the value of $\tan^{-1} a$ which lies between $0°$ and $180°$ (0 and π radians), and if n is any integer,

$$\tan^{-1} a = \alpha + n \cdot 180° = \alpha + n\pi \text{ (similarly for } \cot^{-1} a).$$

Some Relations among Inverse Functions (19)

$$\sin^{-1} a = \cos^{-1}\sqrt{1 - a^2} = \tan^{-1}\frac{a}{\sqrt{1 - a^2}} = \cot^{-1}\frac{\sqrt{1 - a^2}}{a}$$

$$= \sec^{-1}\frac{1}{\sqrt{1 - a^2}} = \csc^{-1}\frac{1}{a}.$$

$$\cos^{-1} a = \sin^{-1}\sqrt{1 - a^2} = \tan^{-1}\frac{\sqrt{1 - a^2}}{a} = \cot^{-1}\frac{a}{\sqrt{1 - a^2}}$$

$$= \sec^{-1}\frac{1}{a} = \csc^{-1}\frac{1}{\sqrt{1 - a^2}}.$$

$$\tan^{-1} a = \sin^{-1}\frac{a}{\sqrt{1 + a^2}} = \cos^{-1}\frac{1}{\sqrt{1 + a^2}} = \cot^{-1}\frac{1}{a} = \sec^{-1}\sqrt{1 + a^2}$$

$$= \csc^{-1}\frac{\sqrt{1 + a^2}}{a}.$$

$$\cot^{-1} a = \tan^{-1}\frac{1}{a}\,; \sec^{-1} a = \cos^{-1}\frac{1}{a}\,; \csc^{-1} a = \sin^{-1}\frac{1}{a}.$$

$$\text{vers}^{-1} a = \cos^{-1}(1 - a)\,; \text{covers}^{-1} a = \sin^{-1}(1 - a)\,; \text{exsec}^{-1} a = \sec^{-1}(1 + a).$$

$$\sin^{-1} a \pm \sin^{-1} b = \sin^{-1}(a\sqrt{1 - b^2} \pm b\sqrt{1 - a^2}).$$

$$\cos^{-1} a \pm \cos^{-1} b = \cos^{-1}(ab \mp \sqrt{1 - a^2}\sqrt{1 - b^2}).$$

$$\tan^{-1} a \pm \tan^{-1} b = \tan^{-1}\frac{a \pm b}{1 \mp ab}.$$

$\sin^{-1} a + \cos^{-1} a = 90°; \tan^{-1} a + \cot^{-1} a = 90°; \sec^{-1} a + \csc^{-1} a = 90°,$
 if $\sin^{-1} a, \tan^{-1} a, \csc^{-1} a$ lie between $-90°$ and $+90°$
 and $\cos^{-1} a, \cot^{-1} a, \sec^{-1} a$ lie between $0°$ and $180°$.

Properties of Plane Triangles (20)

Notation: α, β, γ = angles; a, b, c = sides.
 A = area; h_b = altitude on b; $s = \frac{1}{2}(a + b + c)$.
 r = radius of inscribed circle; R = radius of circumscribed circle.
 $\alpha + \beta + \gamma = 180° = \pi$ radians

$$\frac{a}{\sin \alpha} = \frac{b}{\sin \beta} = \frac{c}{\sin \gamma}.$$

$$\frac{a + b}{a - b} = \frac{\tan \frac{1}{2}(\alpha + \beta)}{\tan \frac{1}{2}(\alpha - \beta)}.$$

$$a^2 = b^2 + c^2 - 2bc \cos \alpha, \; a = b \cos \gamma + c \cos \beta.$$

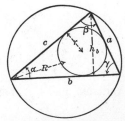

Fig. 1.2

$$\cos \alpha = \frac{b^2 + c^2 - a^2}{2bc}, \quad \sin \alpha = \frac{2}{bc} \sqrt{s(s-a)(s-b)(s-c)}.$$

$$\sin \frac{\alpha}{2} \sqrt{\frac{(s-b)(s-c)}{bc}}, \quad \cos \frac{\alpha}{2} = \sqrt{\frac{s(s-a)}{bc}},$$

$$\tan \frac{\alpha}{2} = \sqrt{\frac{(s-b)(s-c)}{s(s-a)}} = \frac{r}{s-a}.$$

$$h_b = c \sin \alpha = a \sin \gamma = \frac{2}{b} \sqrt{s(s-a)(s-b)(s-c)}.$$

$$r = \sqrt{\frac{(s-a)(s-b)(s-c)}{s}} = (s-a) \tan \frac{\alpha}{2}.$$

$$R = \frac{a}{2 \sin \alpha} = \frac{abc}{4A}.$$

$$A = \frac{1}{2} bh_b = \frac{1}{2} ab \sin \gamma = \frac{a^2 \sin \beta \sin \gamma}{2 \sin \alpha} = \sqrt{s(s-a)(s-b)(s-c)} = rs.$$

Solution of the Right Triangle (21)

Given any two sides, or one side and any acute angle α, to find the remaining parts.

$$\sin \alpha = \frac{a}{c}, \quad \cos \alpha = \frac{b}{c}, \quad \tan \alpha = \frac{a}{b}, \quad \beta = 90° - \alpha.$$
$$a = \sqrt{(c+b)(c-b)} = c \sin \alpha = b \tan \alpha.$$
$$b = \sqrt{(c+a)(c-a)} = c \cos \alpha = \frac{a}{\tan \alpha}.$$
$$c = \frac{a}{\sin \alpha} = \frac{b}{\cos \alpha} = \sqrt{a^2 + b^2}.$$
$$A = \frac{1}{2} ab = \frac{a^2}{2 \tan \alpha} = \frac{b^2 \tan \alpha}{2} = \frac{c^2 \sin 2\alpha}{4}.$$

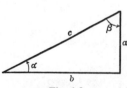

Fig. 1.3

Solution of Oblique Triangles (22)

For numerical work, use tables in Section 10.

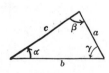

Fig. 1.4

CASE I. Given any two angles α and β, and any side c.

$$\gamma = 180° - (\alpha + \beta); \quad a = \frac{c \sin \alpha}{\sin \gamma}; \quad b = \frac{c \sin \beta}{\sin \gamma}.$$

CASE II.   Given any two sides a and c, and an angle opposite one of these, say α.

$$\sin \gamma = \frac{c \sin \alpha}{a}, \quad \beta = 180° - (\alpha + \gamma), \quad b = \frac{a \sin \beta}{\sin \alpha}.$$

NOTE. γ may have two values, $\gamma_1 < 90°$ and $\gamma_2 = 180° - \gamma_1 > 90°$. If $\alpha + \gamma_2 > 180°$, use only γ_1.

CASE III.   Given any two sides b and c and their included angle α. Use any one of the following sets of formulas:

1. $\frac{1}{2}(\beta + \gamma) = 90° - \frac{1}{2}\alpha; \quad \tan \frac{1}{2}(\beta - \gamma) = \frac{b - c}{b + c} \tan \frac{1}{2}(\beta + \gamma);$

$\beta = \frac{1}{2}(\beta + \gamma) + \frac{1}{2}(\beta - \gamma); \quad \gamma = \frac{1}{2}(\beta + \gamma) - \frac{1}{2}(\beta - \gamma); \quad a = \frac{b \sin \alpha}{\sin \beta}.$

2. $a = \sqrt{b^2 + c^2 - 2bc \cos \alpha}; \quad \sin \beta = \frac{b \sin \alpha}{a}; \quad \gamma = 180° - (\alpha + \beta).$

3. $\tan \gamma = \frac{c \sin \alpha}{b - c \cos \alpha}; \quad \beta = 180° - (\alpha + \gamma); \quad a = \frac{c \sin \alpha}{\sin \gamma}.$

CASE IV.   Given the three sides a, b, and c. Use either of the following sets of formulas.

1. $s = \frac{1}{2}(a + b + c); \quad r = \sqrt{\frac{(s - a)(s - b)(s - c)}{s}}.$

$\tan \frac{1}{2}\alpha = \frac{r}{s - a}; \quad \tan \frac{1}{2}\beta = \frac{r}{s - b}; \quad \tan \frac{1}{2}\gamma = \frac{r}{s - c}.$

2. $\cos \alpha = \frac{b^2 + c^2 - a^2}{2bc}; \quad \cos \beta = \frac{c^2 + a^2 - b^2}{2ca}; \quad \gamma = 180° - (\alpha + \beta).$

1.4 MENSURATION: LENGTHS, AREAS, VOLUMES

Notation: a, b, c, d, s denote lengths; A denotes area; V denotes volume.

Right Triangle (23) (Fig. 1.5)

$A = \frac{1}{2}ab$. [For other formulas, see (21).]

$c = \sqrt{a^2 + b^2}, \quad a = \sqrt{c^2 - b^2}, \quad b = \sqrt{c^2 - a^2}.$

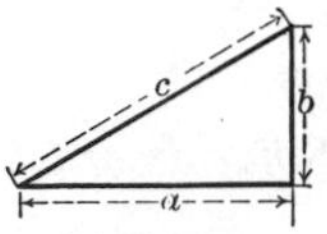

Fig. 1.5

Oblique Triangle (24) (Fig. 1.6)

$A = \frac{1}{2}bh$. [For other formulas, see (22).]

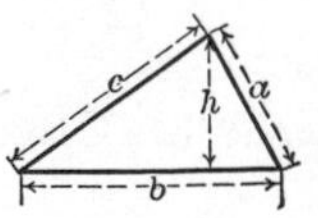

Fig. 1.6

Equilateral Triangle (25) (Fig. 1.7)

$A = \frac{1}{2}ah = \frac{1}{4}a^2\sqrt{3}. \quad r_1 = \frac{a}{2\sqrt{3}}$

| Fig. 1.7 | Fig. 1.8 | Fig. 1.9 | Fig. 1.10 |

$$h = \frac{1}{2} a\sqrt{3}. \qquad\qquad r_2 = \frac{a}{\sqrt{3}}$$

Square (26) (Fig. 1.8)

$$A = a^2; \; d = a\sqrt{2}.$$

Rectangle (27) (Fig. 1.9)

$$A = ab; \; d = \sqrt{a^2 + b^2}.$$

Parallelogram (Opposite Sides Parallel) (28) (Fig. 1.10)

$$A = ah = ab \sin \alpha.$$
$$d_1 = \sqrt{a^2 + b^2 - 2ab \cos \alpha};$$
$$d_2 = \sqrt{a^2 + b^2 + 2ab \cos \alpha}.$$

Trapezoid (One Pair of Opposite Sides Parallel) (29) (Fig. 1.11)

$$A = \tfrac{1}{2}h(a + b).$$

Trapezium (No Sides Parallel) (30) (Fig. 1.12)

$$A = \tfrac{1}{2}(ah_1 + bh_2) = \text{sum of areas of two triangles.}$$

Regular Polygon of n Sides (All Sides Equal, All Angles Equal) (31) (Fig. 1.13)

$$\beta = \frac{n-2}{n}180° = \frac{n-2}{n}\pi \text{ radians.}$$
$$\alpha = \frac{360°}{n} = \frac{2\pi}{n} \text{ radians.}$$

| Fig. 1.11 | Fig. 1.12 | Fig. 1.13 |

n	a	r	R	A
3	$2r\sqrt{3} = R\sqrt{3}$	$\frac{1}{6}\,a\sqrt{3}$	$\frac{1}{3}\,a\sqrt{3}$	$\frac{1}{4}\,a^2\sqrt{3} = 3r^2\sqrt{3} = \frac{3}{4}\,R^2\sqrt{3}$
4	$2r = R\sqrt{2}$	$\frac{1}{2}\,a$	$\frac{1}{2}\,a\sqrt{2}$	$a^2 = 4r^2 = 2R^2$
6	$\frac{2}{3}\,r\sqrt{3} = R$	$\frac{1}{2}\,a\sqrt{3}$	a	$\frac{3}{2}\,a^2\sqrt{3} = 2r^2\sqrt{3} = \frac{3}{2}\,R^2\sqrt{3}$
8	$2r(\sqrt{2}-1) = R\sqrt{2-\sqrt{2}}$	$\frac{1}{2}\,a(\sqrt{2}+1)$	$\frac{1}{2}\,a\sqrt{4+2\sqrt{2}}$	$2a^2(\sqrt{2}+1) = 8r^2(\sqrt{2}-1) = 2R^2\sqrt{2}$
n	$2r\tan\frac{\alpha}{2} = 2R\sin\frac{\alpha}{2}$	$\frac{a}{2}\cot\frac{\alpha}{2}$	$\frac{a}{2}\csc\frac{\alpha}{2}$	$\frac{na^2}{4}\cot\frac{\alpha}{2} = nr^2\tan\frac{\alpha}{2} = \frac{nR^2}{2}\sin\alpha$

Circle (32)

Notation: C = circumference, α = central angle in radians.

$C = \pi D = 2\pi R.$

$c = R\alpha = \dfrac{1}{2} D\alpha = D \cos^{-1} \dfrac{d}{R} = D \tan^{-1} \dfrac{1}{2d}.$

Fig. 1.14

$l = 2\sqrt{R^2 - d^2} = 2R \sin \dfrac{\alpha}{2} = 2d \tan \dfrac{\alpha}{2} = 2d \tan \dfrac{c}{D}.$

$d = \dfrac{1}{2}\sqrt{4R^2 - l^2} = \dfrac{1}{2}\sqrt{D^2 - l^2} = R \cos \dfrac{\alpha}{2} = \dfrac{1}{2} l \cot \dfrac{\alpha}{2} = \dfrac{1}{2} l \cot \dfrac{c}{D}.$

$h = R - d.$

$\alpha = \dfrac{c}{R} = \dfrac{2c}{D} = 2 \cos^{-1} \dfrac{d}{R} = 2 \tan^{-1} \dfrac{l}{2d} = 2 \sin^{-1} \dfrac{l}{D}.$

$A(\text{circle}) = \pi R^2 = \tfrac{1}{4} \pi D^2 = \tfrac{1}{2} RC = \tfrac{1}{4} DC.$
$A(\text{sector}) = \tfrac{1}{2} Rc = \tfrac{1}{2} R^2\alpha = \tfrac{1}{8} D^2\alpha.$

$A(\text{segment}) = A(\text{sector}) - A(\text{triangle}) = \dfrac{1}{2} R^2(\alpha - \sin \alpha) = \dfrac{1}{2} R\left(c - R \sin \dfrac{c}{R}\right)$

$\quad = R^2 \sin^{-1} \dfrac{l}{2R} - \dfrac{1}{4} l\sqrt{4R^2 - l^2} = R^2 \cos^{-1} \dfrac{d}{R} - d\sqrt{R^2 - d^2}$

$\quad = R^2 \cos^{-1} \dfrac{R - h}{R} - (R - h)\sqrt{2Rh - h^2}.$

Ellipse* (33)

$$A = \pi ab.$$

Fig. 1.15

Perimeter $(s) = \pi(a + b)\left[1 + \dfrac{1}{4}\left(\dfrac{a - b}{a + b}\right)^2 + \dfrac{1}{64}\left(\dfrac{a - b}{a + b}\right)^4 + \dfrac{1}{256}\left(\dfrac{a - b}{a + b}\right)^6 + \cdots\right]$

$\quad \approx \pi \dfrac{a + b}{4}\left[3(1 + \lambda) + \dfrac{1}{1 - \lambda}\right], \lambda = \left[\dfrac{a - b}{2(a + b)}\right]^2$

Parabola* (34)

$$A = \tfrac{2}{3} ld.$$

Length of arc $(s) = \dfrac{1}{2}\sqrt{16d^2 + l^2} + \dfrac{l^2}{8d} \ln\left(\dfrac{4d + \sqrt{16d^2 + l^2}}{l}\right)$

$\quad = l\left[1 + \dfrac{2}{3}\left(\dfrac{2d}{l}\right)^2 - \dfrac{2}{5}\left(\dfrac{2d}{l}\right)^4 + \cdots\right].$

Height of segment $(d_1) = \dfrac{d}{l^2}(l^2 - l_1^2).$

Fig. 1.16

*For definition and equation, see Analytic Geometry, Section 1.5.

Width of segment $(l_1) = l \sqrt{\dfrac{d - d_1}{d}}$.

Cycloid* (35)

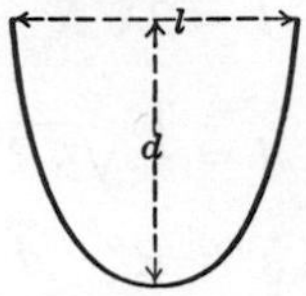

Notation: r = radius of generating circle.

Fig. 1.17

$A = 3\pi r^2$.
Length of arc $(s) = 8r$.

Catenary* (36)

Fig. 1.18

Length of arc $(s) = 1\left[1 + \dfrac{2}{3}\left(\dfrac{2d}{l}\right)^2\right]$ approximately, if d is small in comparison with l.

Area by Approximation (37) (Fig. 1.19)

Let $y_0, y_1, y_2, \ldots, y_n$ be the measured lengths of a series of equidistant parallel chords, and let h be their distance apart, then the area enclosed by any boundary is given approximately by one of the following rules.

$A_T = h[\frac{1}{2}(y_0 + y_n) + y_1 + y_2 + \cdots + y_{n-1}]$
             (Trapezoidal Rule).
$A_D = h[0.4(y_0 + y_n) + 1.1(y_1 + y_{n-1}) + y_2 + y_3 + \cdots + y_{n-2}]$
             (Durand's Rule).
$A_s = \frac{1}{3}h[(y_0 + y_n) + 4(y_1 + y_3 + \cdots + y_{n-1}) + 2(y_2 + y_4 + \cdots + y_{n-2})]$
             (Simpson's Rule, where n is even).

The larger the value of n, the greater is the accuracy of approximation. In general, for the same number of chords, A_s gives the most accurate, A_T the least accurate approximation.

Cube (38) (Fig. 1.20)

$V = a^3; \ d = a\sqrt{3}$.
Total surface $= 6a^2$.

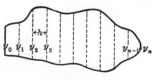

Fig. 1.19

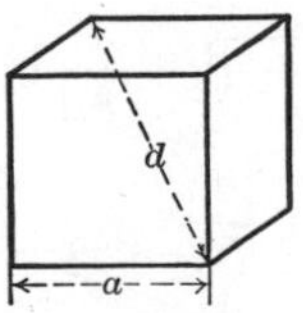

Fig. 1.20

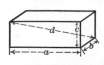

Fig. 1.21

*For definition and equation, see Analytic Geometry, Section 1.5.

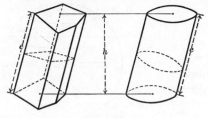

Fig. 1.22

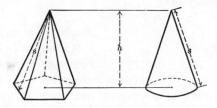

Fig. 1.23

Rectangular Parallelopiped (39) (Fig. 1.21)

$V = abc$; $d = \sqrt{a^2 + b^2 + c^2}$.
Total surface $= 2(ab + bc + ca)$.

Prism or Cylinder (40) (Fig. 1.22)

$V =$ (area of base) $\times$ (altitude).
Lateral area $=$ (perimeter of right section) $\times$ (lateral edge).

Pyramid or Cone (41) (Fig. 1.23)

$V = \frac{1}{3}$(area of base) $\times$ (altitude).
Lateral area of regular figure $= \frac{1}{2}$(perimeter of base) $\times$ (slant height).

Frustum of Pyramid or Cone (42) (Fig. 1.24)

$$V = \frac{1}{3}(A_1 + A_2 + \sqrt{A_1 \times A_2})h,$$

where A_1 and A_2 are areas of bases, and h is altitude.
Lateral area of regular figure $= \frac{1}{2}$(sum of perimeters of bases) $\times$ (slant height).

Prismatoid (Bases Are in Parallel Planes, Lateral Faces Are Triangles or Trapezoids) (43) (Fig. 1.25)

$$V = \frac{1}{6}(A_1 + A_2 + 4A_m)h,$$

where A_1, A_2 are areas of bases, A_m is area of midsection, and h is altitude.

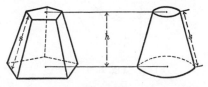

Fig. 1.24

Fig. 1.25

Sphere (44)

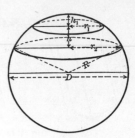

$$A(\text{sphere}) = 4\pi R^2 = \pi D^2.$$
$$A(\text{zone}) = 2\pi Rh = \pi Dh.$$
$$V(\text{sphere}) = \tfrac{4}{3}\pi R^3 = \tfrac{1}{6}\pi D^3.$$
$$V(\text{spherical sector}) = \tfrac{2}{3}\pi R^2 h = \tfrac{1}{6}\pi D^2 h.$$

$V(\text{spherical segment of one base})$
$$= \tfrac{1}{6}\pi h(3r_1^2 + 3r_2^2 + h^2).$$

$V(\text{spherical segment of two bases})$

Fig. 1.26

Ellipsoid (45)

$$V = \tfrac{4}{3}\pi abc.$$

Fig. 1.27

Paraboloidal Segment (46)

$$V(\text{segment of one base}) = \tfrac{1}{2}\pi r_1^2 h.$$
$$V(\text{segment of two bases}) = \tfrac{1}{2}\pi d(r_1^2 + r_2^2).$$

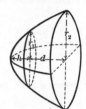

Torus (47) (Fig. 1.29)

Fig. 1.28

$$V = 2\pi^2 R r^2.$$
$$\text{Surface } (S) = 4\pi^2 R r.$$

Solid (V) or Surface (S) of Revolution (48) (Fig. 1.30)

Generated by revolving any plane area (A) or arc (s) about an axis in its plane, and not crossing the area or arc.

$$V = 2\pi RA; \ S = 2\pi Rs,$$

where R = distance of center of gravity (G) of area or arc from axis.

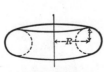

Fig. 1.29

Fig. 1.30

1.5   ANALYTIC GEOMETRY

1.5.1   Plane

Rectangular (or Cartesian) Coordinates (49)

Let two perpendicular lines, $X'X$ (x-axis) and $Y'Y$ (y-axis) meet in a point O (origin). The position of any point $P(x, y)$ is fixed by the distances x (abscissa) and y (ordinate) from $Y'Y$ and $X'X$, respectively, to P (Fig. 1.31).

NOTE. x is + to the right and − to the left of $Y'Y$, y is + above and − below $X'X$.

Polar Coordinates (50)

Let O (origin or pole) be a point in the plane and OX (initial line) be any line through O. The position of any point $P(r, \theta)$ is fixed by the distance r (radius vector) from O to the point and the angle θ (vectorial angle) measured from OX to OP (Fig. 1.31).

NOTE. r is + measured along terminal side of θ, r is − measured along the opposite side of θ; θ is + measured counterclockwise, θ is − measured clockwise.

Relations Connecting Rectangular and Polar Coordinates (51)

$x = r \cos \theta, y = r \sin \theta.$

$r = \sqrt{x^2 + y^2}, \theta = \tan^{-1} \dfrac{y}{x}, \sin \theta = \dfrac{y}{\sqrt{x^2 + y^2}}, \cos \theta = \dfrac{x}{\sqrt{x^2 + y^2}}, \tan \theta = \dfrac{y}{x}.$

Points and Slopes (52)

Let $P_1(x_1, y_1)$ and $P_2(x_2, y_2)$ be any two points, and let α be the angle from OX to P_1P_2, measured counterclockwise (Fig. 1.32).

$P_1P_2 = d = \sqrt{(x_2 - x_1)^2 + (y_2 - y_1)^2}.$

Midpoint of P_1P_2 is $\left(\dfrac{x_1 + x_2}{2}, \dfrac{y_1 + y_2}{2} \right).$

Point that divides P_1P_2 in the ratio $m_1 : m_2$ is $\left(\dfrac{m_1x_2 + m_2x_1}{m_1 + m_2}, \dfrac{m_1y_2 + m_2y_1}{m_1 + m_2} \right).$

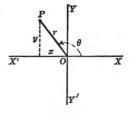

Fig.1.31

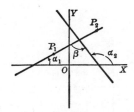

Fig. 1.32

Slope of $P_1P_2 = \tan \alpha = m = \dfrac{y_2 - y_1}{x_2 - x_1}.$

Angle between two lines of slopes m_1 and m_2 is $\beta = \tan^{-1} \dfrac{m_2 - m_1}{1 + m_1 m_2}.$

Two lines of slopes m_1 and m_2 are perpendicular if $m_2 = -\dfrac{1}{m_1}.$

Locus and Equation (53)

The collection of all points that satisfy a given condition is called the **locus** of that condition; the condition expressed by means of the variable coordinates of any point on the locus is called the **equation** of the locus.

The locus may be represented by equations of three kinds:

Rectangular equation involves the rectangular coordinates (x, y).

Polar equation involves the polar coordinates (r, θ).

Parametric equations express x and y or r and θ in terms of a third independent variable called a parameter.

The following equations are given in the system in which they are most simply expressed; sometimes several forms of the equation in one or more systems are given.

Straight Line (54)

Figure 1.33
$Ax + By + C = 0.$ $(-A \div B = \text{slope.})$
$y = mx + b.$ $(m = \text{slope}, b = \text{intercept on } OY.)$
$y - y_1 = m(x - x_1).$ $[m = \text{slope}, P_1(x_1, y_1) \text{ is a known point on line.}]$
$d = \dfrac{Ax_2 + By_2 + C}{\pm\sqrt{A^2 + B^2}}.$ $[d = \text{distance from a point } P_2(x_2, y_2) \text{ to the line } Ax +$
$By + C = 0.]$

Circle (55)

Locus of a point at a constant distance (radius) from a fixed point C (center). [For mensuration of circle, see (32).]

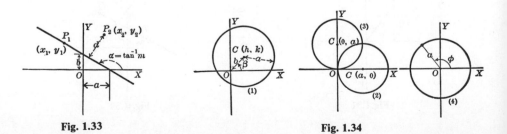

Fig. 1.33   Fig. 1.34

Circle (1) in Fig. 1.34
$(x - h)^2 + (y - k)^2 = a^2,$ $C(h, k),$ rad. $= a.$
$r^2 + b^2 - 2br \cos(\theta - \beta) = a^2,$ $C(b, \beta),$ rad. $= a.$
Circle (2) in Fig. 1.34
$x^2 + y^2 = 2ax,$ $C(a, 0),$ rad. $= a.$
$r = 2a \cos \theta,$ $C(a, 0),$ rad. $= a.$
Circle (3) in Fig. 1.34
$x^2 + y^2 = 2ay,$ $C(0, a),$ rad. $= a.$

$r = 2a \sin \theta,$ $C\left(a, \dfrac{\pi}{2}\right),$ rad. $= a.$

Circle (4) in Fig. 1.34
$x^2 + y^2 = a^2,$ $C(0, 0),$ rad. $= a.$
$r = a,$ $C(0, 0),$ rad. $= a.$
$x = a \cos \phi, y = a \sin \phi,$ $C(0, 0),$ rad. $= a, \phi =$ angle from OX to radius.

Conic (56)

Locus of a point whose distance from a fixed point (focus) is in a constant ratio, **e** (called **eccentricity**), to its distance from a fixed straight line (**directrix**) (Fig. 1.35).
$x^2 + y^2 = e^2(d + x)^2.$ $(d =$ distance from focus to directrix.)

$$r = \frac{de}{1 - e \cos \theta}.$$

The conic is called a **parabola** when $e = 1$, an **ellipse** when $e < 1$, a **hyperbola** when $e > 1$.

Parabola (57)

Conic where $e = 1$. [For mensuration of parabola, see (34).]
    Figure 1.36a
$(y - k)^2 = a(x - h).$ Vertex (h, k), axis $\parallel OX.$
$y^2 = ax.$ Vertex $(0, 0)$, axis along $OX.$
Figure 1.36b
$(x - h)^2 = a(y - k).$ Vertex (h, k), axis $\parallel OY.$
$x^2 = ay.$ Vertex $(0, 0)$, axis along $OY.$
Distance from vertex to focus $= VF = \frac{1}{4}a.$
Latus rectum $= LR = a.$

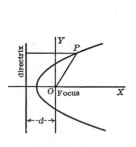

Fig. 1.35

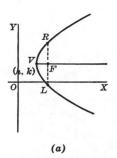

(a)   (b)

Fig. 1.36

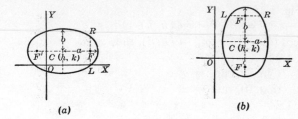

(a)   (b)

Fig. 1.37

Ellipse (58)

Conic where $e < 1$. [For mensuration of ellipse, see (33).]

$$\frac{(x - h)^2}{a^2} + \frac{(y + k)^2}{b^2} = 1, \quad \text{Center } (h, k), \text{ axes } \parallel OX, OY.$$

$$\frac{x^2}{a^2} + \frac{y^2}{b^2} = 1, \qquad\qquad \text{Center } (0, 0), \text{ axes along } OX, OY.$$

	$a > b$ Fig. 1.37a	$b > a$ Fig. 1.37b
Major axis	$2a$	$2b$
Minor axis	$2b$	$2a$
Distance from center to either focus	$\sqrt{a^2 - b^2}$	$\sqrt{b^2 - a^2}$
Latus rectum	$\dfrac{2b^2}{a}$	$\dfrac{2a^2}{b}$
Eccentricity, e	$\dfrac{\sqrt{a^2 - b^2}}{a}$	$\dfrac{\sqrt{b^2 - a^2}}{b}$
Sum of distances of any point from the foci, $PF' + PF$	$2a$	$2b$

Hyperbola (59)

Conic where $e > 1$.
  Figure 1.38a

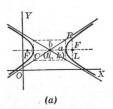   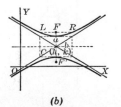   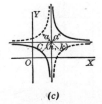

(a)   (b)   (c)

Fig. 1.38

$$\frac{(x-h)^2}{a^2} - \frac{(y-k)^2}{b^2} = 1, C(h, k), \text{ transverse axis } \| OX.$$

$$\frac{x^2}{a^2} - \frac{y^2}{b^2} = 1, C(0, 0), \text{ transverse axis along } OX.$$

Figure 1.38b

$$\frac{(y-k)^2}{a^2} - \frac{(x-h)^2}{b^2} = 1, C(h, k), \text{ transverse axis } \| OY.$$

$$\frac{y^2}{a^2} - \frac{x^2}{b^2} = 1, C(0, 0), \text{ transverse along } OY.$$

Transverse axis = $2a$; conjugate axis = $2b$.
Distance from center to either focus = $\sqrt{a^2 + b^2}$.

Latus rectum = $\dfrac{2b^2}{a}$.

Eccentricity, $e = \dfrac{\sqrt{a^2 + b^2}}{a}$.

Difference of distances of any point from the foci = $2a$.

Asymptotes are two lines through the center to which the branches of the hyperbola approach indefinitely near; their slopes are $\pm \dfrac{b}{a}$ (Fig. 38a) or $\pm \dfrac{a}{b}$ (Fig. 38b).

Rectangular (equilateral) hyperbola, $b = a$. The asymptotes are perpendicular.

Figure 1.38c

$$(x-h)(y-k) = \pm \frac{a^2}{2}, \text{ center } (h, k), \text{ asymptotes } \| OX, OY.$$

$$xy = \pm \frac{a^2}{2}, \text{ center } (0, 0), \text{ asymptotes along } OX, OY.$$

Where the $+$ sign gives the smooth curve in Fig. 1.38c.
Where the $-$ sign gives the dotted curve in Fig. 1.38c.

Sine Wave (60)

Figure 1.39
$y = a \sin(bx + c)$.

$y = a \cos(bx + c') = a \sin(bx + c)$, where $c = c' + \dfrac{\pi}{2}$.

$y = m \sin bx + n \cos bx = a \sin(bx + c)$, where $a = \sqrt{m^2 + n^2}$, $c = \tan^{-1} \dfrac{n}{m}$.

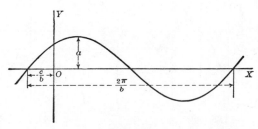

Fig. 1.39

The curve consists of a succession of waves, where
a = amplitude = maximum height of wave.

$\dfrac{2\pi}{b}$ = wave length = distance from any point on wave to the corresponding point

on the next wave.

$x = -\dfrac{c}{b}$ (called the phase) marks a point on OX from which the positive half

of the wave starts.

Tangent and Cotangent Curves (61)

Figure 1.40
1. $y = a \tan bx.$
2. $y = a \cot bx.$

Secant and Cosecant Curves (62)

Figure 1.41
1. $y = a \sec bx.$
2. $y = a \csc bx.$

Exponential or Logarithmic Curves (63)

Figure 1.42
1. $y = ab^x$ or $x = \log_b \dfrac{y}{a}.$

2. $y = ab^{-x}$ or $x = -\log_b \dfrac{y}{a}.$

3. $x = ab^y$ or $y = \log_b \dfrac{x}{a}.$

4. $x = ab^{-y}$ or $y = -\log_b \dfrac{x}{a}.$

The equations $y = ae^{\pm nx}$ and $x = ae^{\pm ny}$ are special cases of above.

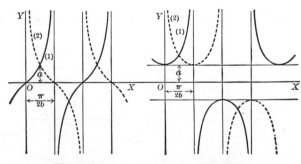

Fig. 1.40   Fig. 1.41

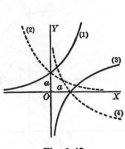

Fig. 1.42

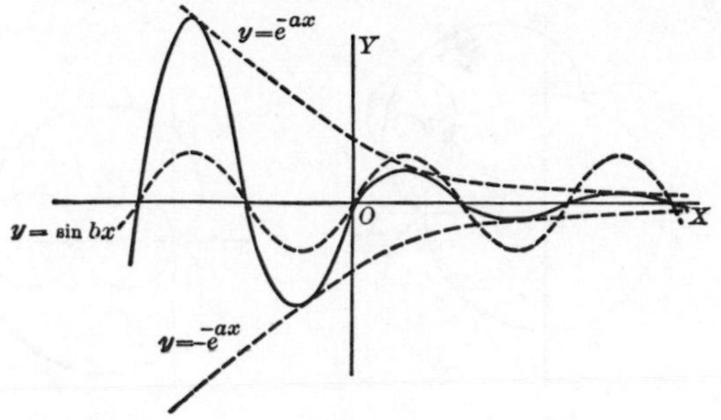

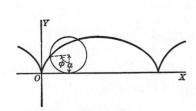

Fig. 1.43

Oscillatory Wave of Decreasing Amplitude (64)

Figure 1.43

$$y = e^{-ax} \sin bx.$$

NOTE.   The curve oscillates between $y = e^{-ax}$ and $y = -e^{-ax}$.

Catenary (65)

Curve made by a chain or cord of uniform weight suspended freely between two points at the same level (Fig. 1.44). [For mensuration of catenary, see (36).]

$$y = \frac{a}{2}(e^{x/a} + e^{-x/a}).$$

Cycloid (66)

Curve described by a point on a circle which rolls along a fixed straight line (Fig. 1.45).
$x = a(\phi - \sin \phi).$
$y = a(1 - \cos \phi).$

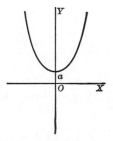

Fig. 1.44

Fig. 1.45

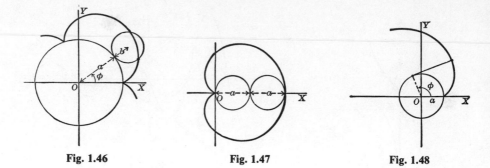

Fig. 1.46   Fig. 1.47   Fig. 1.48

Epicycloid (67)

Curve described by a point on a circle which rolls along the outside of a fixed circle (Fig. 1.46).

$$x = (a + b) \cos \phi - b \cos \left(\frac{a + b}{b} \phi \right).$$

$$y = (a + b) \sin \phi - b \sin \left(\frac{a + b}{b} \phi \right).$$

Hypocycloid (68)

Curve described by a point on a circle which rolls along the inside of a fixed circle.

$$x = (a - b) \cos \phi + b \cos \left(\frac{a - b}{b} \phi \right).$$

$$y = (a - b) \sin \phi - b \sin \left(\frac{a - b}{b} \phi \right).$$

(For a hypocycloid of n cusps, radius of fixed circle = nx radius of rolling circle; Fig. 1.47 = 4 cusps.)

Involute of the Circle (69)

Curve described by the end of a string which is kept taut while being unwound from a circle (Fig. 1.48).

$x = a \cos \phi + a \phi \sin \phi.$
$y = a \sin \phi - a \phi \cos \phi.$

1.5.2 Solid

Coordinates (70)

Let three mutually perpendicular planes, XOY, YOZ, ZOX (coordinate planes) meet in a point O (origin).

Rectangular System. The position of a point $P(x, y, z)$ in space is fixed by its three distances x, y, and z from the three coordinate planes.

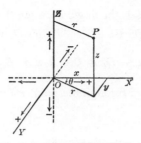

Fig. 1.49

Cylindrical System.   The position of any point $P(r, \theta, z)$ is fixed by z, its distance from the XOY plane, and by (r, θ), the polar coordinates of the projection of P in the XOY plane.

Relations connecting rectangular and cylindrical coordinates are the same as those given in (51).

Points, Lines, and Planes (71)

Distance (d) between two points $P_1(x_1, y_1, z_1)$ and $P_2(x_2, y_2, z_2)$,

$$d = \sqrt{(x_2 - x_1)^2 + (y_2 - y_1)^2 + (z_2 - z_1)^2}.$$

Direction cosines of a line (cosines of the angles α, β, γ which the line or any parallel line makes with the coordinate axes) are related by

$$\cos^2 \alpha + \cos^2 \beta + \cos^2 \gamma = 1.$$

If $\cos \alpha : \cos \beta : \cos \gamma = a : b : c$, then

$$\cos \alpha = \frac{a}{\sqrt{a^2 + b^2 + c^2}}, \quad \cos \beta = \frac{b}{\sqrt{a^2 + b^2 + c^2}}, \quad \cos \gamma = \frac{c}{\sqrt{a^2 + b^2 + c^2}}.$$

Direction cosines of the line joining $P_1(x_1, y_1, x_1)$ and $P_2(x_2, y_2, z_2)$,

$$\cos \alpha : \cos \beta : \cos \gamma = x_2 - x_1 : y_2 - y_1 : z_2 - z_1.$$

Angle (θ) between two lines, whose direction angles are $\alpha_1, \beta_1, \gamma_1$ and $\alpha_2, \beta_2, \gamma_2$,

$$\cos \theta = \cos \alpha_1 \cos \alpha_2 + \cos \beta_1 \cos \beta_2 + \cos \gamma_1 \cos \gamma_2.$$

Equation of a plane is of the first degree in x, y, and z,

$$Ax + By + Cz + D = 0,$$

where A, B, C are proportional to the direction cosines of a normal or perpendicular to the plane.

Angle between two planes is the angle between their normals.
Equations of a straight line are two equations of the first degree,

$$A_1x + B_1y + C_1z + D_1 = 0, \quad A_2x + B_2y + C_2z + D_2 = 0.$$

Equations of a straight line through the point $P_1(x_1, y_1, z_1)$ with direction cosines proportional to a, b, and c,

$$\frac{x - x_1}{a} = \frac{y - y_1}{b} = \frac{z - z_1}{c}.$$

Cylindrical Surfaces (72)

The locus in space of an equation containing only two of the coordinates x, y, z is a cylindrical surface with its elements perpendicular to the plane of the two coordinates. Considered as a plane geometry equation, the equation represents the curve of intersection of the cylinder with the plane of the two coordinates.

Circular cylinders. [For mensuration see (40).]
Figure 1.50a
$x^2 + y^2 = a^2$.
$r = a$.
Figure 1.50b
$x^2 + y^2 = 2ax$.
$r = 2a \cos \theta$.
Figure 1.50c
Parabolic cylinder $y^2 = ax$.

Surfaces of Revolution (73)

Equation of the surface of revolution obtained by revolving the plane curve $y = f(x)$ or $z = f(x)$ about OX,

$$y^2 + z^2 = [f(x)]^2.$$

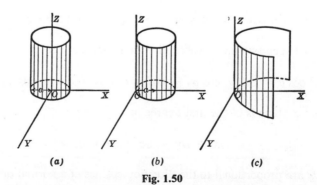

(a)   (b)   (c)

Fig. 1.50

Sphere (revolve circle $x^2 + y^2 = a^2$ about OX)

$$x^2 + y^2 + z^2 = a^2. \text{ [For mensuration of sphere, see (44).]}$$

Spheroid (revolve ellipse $\dfrac{x^2}{a^2} + \dfrac{y^2}{b^2} = 1$ about OX)

$$\frac{x^2}{a^2} + \frac{y^2 + z^2}{b^2} = 1 \text{ (prolate if } a > b, \text{ oblate if } b > a).$$

[For mensuration of ellipsoid, see (45).]

Cone (revolve line $y = mx$ about OX)

$$y^2 + z^2 = m^2x^2. \text{ [For mensuration of cone, see (41).]}$$

Paraboloid (revolve parabola $y^2 = ax$ about OX)

$$y^2 + z^2 = ax. \text{ [For mensuration of paraboloid, see (46).]}$$

Space Curves (74)

A curve in space may be represented by two equations connecting the coordinates x, y, z of any point on the curve, or by three equations expressing the coordinates x, y, z in terms of a fourth variable or parameter.

Helix.   Curve generated by a point moving on a cylinder so that the distance traversed parallel to the axis of the cylinder is proportional to the angle of rotation about the axis (Fig. 1.51).

$$x = a \cos \theta, \, y = a \sin \theta, \, z = k\theta,$$

where a = radius of cylinder, $2\pi k$ = pitch.

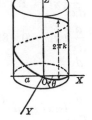

Fig. 1.51

1.6   DIFFERENTIAL CALCULUS

Definition of Function. Notation (75)

A variable y is said to be a function of another variable x, if, when x is given, y is determined.

The symbols $f(x)$, $F(x)$, $\phi(x)$, and so on, represent various functions of x.

The symbol $f(a)$ represents the value of $f(x)$ when $x = a$.

Definition of Derivative. Notation (76)

Let $y = f(x)$. If Δx is any increment (increase or decrease) given to x, and Δy is the corresponding increment in y, then the derivative of y with respect to x is the limit of the ratio of Δy to Δx as Δx approaches zero, that is:

$$\frac{dy}{dx} = \lim_{\Delta x \to 0} \frac{\Delta y}{\Delta x} = \lim_{\Delta x \to 0} \frac{f(x + \Delta x) - f(x)}{\Delta x} = f'(x).$$

$$\frac{d^2 y}{dx^2} = \frac{d}{dx}\left(\frac{dy}{dx}\right) = \frac{d}{dx} f'(x) = f''(x). \quad \text{(2nd derivative)}$$

$$\frac{d^n y}{dx^n} = \frac{d}{dx}\left(\frac{d^{n-1} y}{dx^{n-1}}\right) = \frac{d}{dx} f^{(n-1)}(x) = f^{(n)}(x). \quad \text{(nth derivative)}$$

The symbols $f'(a)$, $f''(a)$, $\cdots$, $f^{(n)}(a)$ represent the values of $f'(x)$, $f''(x)$ $\cdots$, $f^{(n)}(x)$, respectively, when $x = a$.

Some Relations among Derivatives (77)

If $x = f(y)$, then $\dfrac{dy}{dx} = 1 \div \dfrac{dx}{dy}$.

If $x = f(t)$, and $y = F(t)$, then $\dfrac{dy}{dx} = \dfrac{dy}{dt} \div \dfrac{dx}{dt}$.

If $y = f(u)$, and $u = F(x)$, then $\dfrac{dy}{dx} = \dfrac{dy}{du} \cdot \dfrac{du}{dx}$.

List of Derivatives (78)

Functions of x are represented by u and v, constants are represented by a, n, and e.

$$\frac{d}{dx}(x) = 1. \text{ (79)}$$

$$\frac{d}{dx}(a) = 0. \text{ (80)}$$

$$\frac{d}{dx}(u \pm v \pm \cdots) = \frac{du}{dx} \pm \frac{dv}{dx} \pm \cdots \cdots \text{ (81)}$$

$$\frac{d}{dx}(au) = a\frac{du}{dx}. \text{ (82)}$$

$$\frac{d}{dx}(uv) = u\frac{dv}{dx} + v\frac{du}{dx}. \text{ (83)}$$

$$\frac{d}{dx}\left(\frac{u}{v}\right) = \frac{v\dfrac{du}{dx} - u\dfrac{dv}{dx}}{v^2}. \text{ (84)}$$

$$\frac{d}{dx}(u^n) = nu^{n-1}\frac{du}{dx}. \text{ (85)}$$

$$\frac{d}{dx}\log_a u = \frac{\log_a e}{u}\frac{du}{dx}. \text{ (86)}$$

$$\frac{d}{dx}\ln u = \frac{1}{u}\frac{du}{dx}. \text{ (87)}$$

$$\frac{d}{dx}a^u = a^u \ln a \frac{du}{dx}. \text{ (88)}$$

$$\frac{d}{dx}e^u = e^u\frac{du}{dx}. \text{ (89)}$$

$$\frac{d}{dx}u^v = vu^{v-1}\frac{du}{dx} + u^v \ln u \frac{dv}{dx}. \text{ (90)}$$

$$\frac{d}{dx}\sin u = \cos u \frac{du}{dx}. \text{ (91)}$$

$$\frac{d}{dx}\cos u = -\sin u \frac{du}{dx}. \quad (92)$$

$$\frac{d}{dx}\tan u = \sec^2 u \frac{du}{dx}. \quad (93)$$

$$\frac{d}{dx}\cot u = -\csc^2 u \frac{du}{dx}. \quad (94)$$

$$\frac{d}{dx}\sin^{-1} u = \frac{1}{\sqrt{1-u^2}}\frac{du}{dx}\left(\text{where } \sin^{-1} u \text{ lies between } -\frac{\pi}{2} \text{ and } +\frac{\pi}{2}\right) (95)$$

$$\frac{d}{dx}\cos^{-1} u = -\frac{1}{\sqrt{1-u^2}}\frac{du}{dx} (\text{where } \cos^{-1} u \text{ lies between } 0 \text{ and } \pi). \quad (96)$$

$$\frac{d}{dx}\tan^{-1} u = \frac{1}{1+u^2}\frac{du}{dx}. \quad (97)$$

$$\frac{d}{dx}\cot^{-1} u = -\frac{1}{1+u^2}\frac{du}{dx}. \quad (98)$$

$$\frac{d}{dx}\sec^{-1} u = \frac{1}{u\sqrt{u^2-1}}\frac{du}{dx} (\text{where } \sec^{-1} u \text{ lies between } 0 \text{ and } \pi). \quad (99)$$

$$\frac{d}{dx}\csc^{-1} u = -\frac{1}{u\sqrt{u^2-1}}\frac{du}{dx}\left(\text{where } \csc^{-1} u \text{ lies between } -\frac{\pi}{2} \text{ and } +\frac{\pi}{2}\right). \quad (100)$$

$$\frac{d}{dx}\text{vers}^{-1} u = \frac{1}{\sqrt{2u-u^2}}\frac{du}{dx} (\text{where } \text{vers}^{-1} u \text{ lies between } 0 \text{ and } \pi). \quad (101)$$

The nth Derivative of Certain Functions (102)

$$\frac{d^n}{dx^n} e^{ax} = a^n e^{ax}. \quad (103)$$

$$\frac{d^n}{dx^n} a^x = (\ln a)^n a^x. \quad (104)$$

$$\frac{d^n}{dx^n} \ln x = \frac{(-1)^{n-1}\lfloor n-1}{x^n}, \quad \lfloor n-1 = 1 \cdot 2 \cdot 3 \cdots (n-1). \quad (105)$$

$$\frac{d^n}{dx^n} \sin ax = a^n \sin\left(ax + \frac{n\pi}{2}\right). \quad (106)$$

$$\frac{d^n}{dx^n} \cos ax = a^n \cos\left(ax + \frac{n\pi}{2}\right). \quad (107)$$

Slope of a Curve. Tangent and Normal (108)

The slope of the curve (slope of the tangent line to the curve) whose equation is $y = f(x)$ is

Slope $= m = \tan \phi = \dfrac{dy}{dx} = f'(x)$.

Slope at $x = x_1$ is $m_1 = f'(x_1)$.
    Equation of tangent line at $P_1(x_1, y_1)$ is

$$y - y_1 = m_1(x - x_1).$$

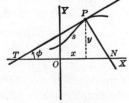

Fig. 1.52

Equation of normal at $P_1(x_1, y_1)$ is

$$y - y_1 = -\frac{1}{m_1}(x - x_1).$$

Angle (β) of intersection of two curves whose slopes at a common point are m_1 and m_2 is $\beta = \tan^{-1}\dfrac{m_2 - m_1}{1 + m_1 m_2}$.

Derivative of Length of Arc. Radius of Curvature (109)

If s is the length of arc measured along the curve $y = f(x)$ from some fixed point to any point $P(x, y)$, and ϕ is the inclination of the tangent line at P to OX, then (Fig. 1.52)

$$\frac{dx}{ds} = \cos\phi = \frac{1}{\sqrt{1 + \left(\dfrac{dy}{dx}\right)^2}}; \quad \frac{dy}{ds} = \sin\phi = \frac{1}{\sqrt{1 + \left(\dfrac{dx}{dy}\right)^2}}; \quad \left(\frac{dx}{ds}\right)^2 + \left(\frac{dy}{ds}\right)^2$$

Radius of curvature (ρ) at any point of the curve $y = f(x)$ or $r = f(\theta)$.

$$\rho = \frac{ds}{d\phi} = \frac{\left[1 + \left(\dfrac{dy}{dx}\right)^2\right]^{3/2}}{\left(\dfrac{d^2y}{dx^2}\right)} = \frac{\{1 + [f'(x)]^2\}^{3/2}}{f''(x)} = \frac{\left[1^2 + \left(\dfrac{dr}{d\theta}\right)^2\right]^{3/2}}{1^2 + 2\left(\dfrac{dr}{d\theta}\right)^2 - r\dfrac{d^2r}{d\theta^2}}.$$

ρ at $x = a$ is $\dfrac{\{1 + [f'(a)]^2\}^{3/2}}{f''(a)}$.

Curvature (k) at any point is $k = \dfrac{1}{\rho}$.

Maximum and Minimum Values of a Function (110)

The maximum (minimum) value of a function $f(x)$ in an interval $x = a$ to $x = b$, is the value of the function which is larger (smaller) than the values of the function in its immediate vicinity. Thus in Fig. 1.53, the value of the function at M_1 and M_2 is a maximum, its value at m_1 and m_2 is a minimum.

Test for a maximum at $x = x_1$: $f'(x_1) = 0$ or ∞, and $f''(x_1) < 0$.

Test for a minimum at $x = x_1$: $f'(x_1) = 0$ or ∞, and $f''(x_1) > 0$.

If $f'(x_1) = 0$ or ∞, then for a maximum, $f'''(x_1) = 0$ or ∞, and $f^{IV}(x_1) < 0$,

for a minimum, $f'''(x_1) = 0$ or ∞, and $f^{IV}(x_1) > 0$,

and similarly if $f^{IV}(x_1) = 0$ or ∞, and so on.

In a practical problem that suggests that the function $f(x)$ has a maximum or has a minimum in an interval $x = a$ to $x = b$, merely equate $f'(x)$ to zero and solve for the required value of x. To find the largest or smallest values of a function $f(x)$ in

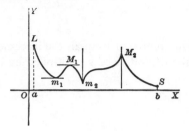

Fig. 1.53

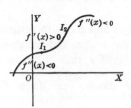

Fig. 1.54

an interval $x = a$ to $x = b$, find also the values $f(a)$ and $f(b)$, for (see Fig. 1.53 at L and S) these may be the largest and smallest values although they are not maximum or minimum values.

Points of Inflection of a Curve (111)

Wherever $f''(x) < 0$, the curve is concave down.
Wherever $f''(x) > 0$, the curve is concave up.
The curve is said to have a point of inflection at $x = x_1$ if $f''(x_1) = 0$ or ∞ and the curve is concave up on one side of $x = x_1$ and concave down on the other (see points I_1 and I_2 in Fig. 1.54).

Taylor's and Maclaurin's Theorems (112)

Any $f(x)$ may, in general, be expanded into a Taylor's Series.

$$f(x) = f(a) + f'(a)\frac{x - a}{1} + f''(a)\frac{(x - a)^2}{2!} + f'''(a)\frac{(x - a)^3}{3!} + \cdots$$
$$+ f^{(n-1)}(a)\frac{(x - a)^{n-1}}{(n - 1)!} + R_n,$$

where a is any quantity whatever so chosen that none of the expressions $f(a)$, $f'(a)$, $f''(a)$, ... become infinite. If the series is to be used for the purpose of computing the approximate value of $f(x)$ for a given value of x, a should be chosen so that $(x - a)$ is numerically very small, and thus only a few terms of the series need be used. $R_n = f^{(n)}(x_1)[(x - a)^n/n!]$, where x_1 lies between a and x, is the remainder after n terms, and gives the limits between which the error lies in using n terms of the series for the value of the function. $n! = 1 \cdot 2 \cdot 3 \cdots n$.
If $a = 0$, the foregoing series becomes Maclaurin's Series:

$$f(x) = f(0) + f'(0)\frac{x}{1} + f''(0)\frac{x^2}{2!} + f'''(0)\frac{x^3}{3!} + \cdots + f^{(n-1)}(0)\frac{x^{n-1}}{(n - 1)!} + R_n.$$

This series may be used for purposes of computation when x is numerically very small.

Differential of a Function (113)

If $y = f(x)$ and $\Delta x =$ increment in x, then the differential of x equals the increment of x, or $dx = \Delta x$; and the differential of y is the derivative of y multiplied by the differential of x, thus

$$dy = \frac{dy}{dx} dx = \frac{df(x)}{dx} dx = f'(x) dx,$$

and $\dfrac{dy}{dx} = dy \div dx.$

If $x = f_1(t)$ and $y = f_2(t)$, then $dx = f_1'(t) dt$, $dy = f_2'(t) dt$.

Every derivative formula has a corresponding differential formula; thus from the list of derivatives (78), we have, for example,

$$d(uv) = u\, dv + v\, du;\ d(\sin u) = \cos u\, du;\ d(\tan^{-1} u) = \frac{du}{1 + u^2},\ \text{and so on.}$$

Functions of Several Variables. Partial Derivatives Differentials (114)

Let z be a function of two variables, $z = f(x, y)$, then its partial derivatives are

$\dfrac{\partial z}{\partial x} = \dfrac{dz}{dx}$ when y is kept constant.

$\dfrac{\partial z}{\partial y} = \dfrac{dz}{dy}$ when x is kept constant.

$$\frac{\partial^2 z}{\partial x^2} = \frac{\partial}{\partial x}\left(\frac{\partial z}{\partial x}\right);\ \frac{\partial^2 z}{\partial y^2} = \frac{\partial}{\partial y}\left(\frac{\partial z}{\partial y}\right);\ \frac{\partial^2 z}{\partial x\, \partial y} = \frac{\partial}{\partial x}\left(\frac{\partial z}{\partial y}\right) = \frac{\partial}{\partial y}\left(\frac{\partial z}{\partial x}\right) = \frac{\partial^2 z}{\partial y\, \partial x}.$$

Similarly, if $z = f(x, y, u, \cdots)$, then, for example,

$\dfrac{\partial z}{\partial x} = \dfrac{dz}{dx}$ when $y, u, \ldots$ are kept constant.

If $z = f(x, y, \cdots)$ and $x, y, \ldots$ are functions of a single variable, t,

$$\frac{dz}{dt} = \frac{\partial z}{\partial x}\frac{dx}{dt} + \frac{\partial z}{\partial y}\frac{dy}{dt} + \cdots.$$

If $z = f(x, y, \cdots)$, then $dz = \dfrac{\partial z}{\partial x} dx + \dfrac{\partial z}{\partial y} dy + \cdots.$

If $F(x, y, z, \cdots) = 0$, then $\dfrac{\partial F}{\partial x} dx + \dfrac{\partial F}{\partial y} dy + \dfrac{\partial F}{\partial z} dz + \cdots = 0.$

If $f(x, y) = 0$, then $\dfrac{dy}{dx} = -\dfrac{\partial f}{\partial x} \div \dfrac{\partial f}{\partial y}.$

Maxima and Minima of Functions of Two Variables (115)

If $u = f(x, y)$, the values of x and y that make u a maximum or a minimum must satisfy the conditions

$$\frac{\partial u}{\partial x} = 0, \frac{\partial u}{\partial y} = 0, \left(\frac{\partial^2 u}{\partial x\, \partial y}\right)^2 < \left(\frac{\partial^2 u}{\partial x^2}\right)\left(\frac{\partial^2 u}{\partial y^2}\right).$$

A maximum (minimum) also requires both $\dfrac{\partial^2 u}{\partial x^2}$ and $\dfrac{\partial^2 u}{\partial y^2}$ to be negative (positive).

Space Curves. Surfaces (116) (See Analytic Geometry, Section 1.5)

Let $x = f_1(t)$, $y = f_2(t)$, $z = f_3(t)$ be the equations of any space curve. The direction cosines of the tangent line to the curve at any point are proportional to dx, dy, and dz, or to $\dfrac{dx}{dt}$, $\dfrac{dy}{dt}$, and $\dfrac{dz}{dt}$.

Equations of tangent line at a point (x_1, y_1, z_1) are

$$\frac{x - x_1}{(dx)_1} = \frac{y - y_1}{(dy)_1} = \frac{z - z_1}{(dz)_1}, \text{ where}$$

$(dx)_1$ = value of dx at (x_1, y_1, z_1), and so on.

Angle between two space curves is the angle between their tangent lines (see Analytic Geometry, Section 1.5).

Let $F(x, y, z) = 0$ be the equation of a surface.

Direction cosines of the normal to the surface at any point are proportional to $\dfrac{\partial F}{\partial x}$, $\dfrac{\partial F}{\partial y}$, $\dfrac{\partial F}{\partial z}$.

Equations of the normal at any point (x_1, y_1, z_1) are

$$\frac{x - x_1}{\left(\dfrac{\partial F}{\partial x}\right)_1} = \frac{y - y_1}{\left(\dfrac{\partial F}{\partial y}\right)_1} = \frac{z - z_1}{\left(\dfrac{\partial F}{\partial z}\right)_1}.$$

Equation of the tangent plane at any point (x_1, y_1, z_1) is

$$(x - x_1)\left(\frac{\partial F}{\partial x}\right)_1 + (y - y_1)\left(\frac{\partial F}{\partial y}\right)_1 + (z - z_1)\left(\frac{\partial F}{\partial z}\right)_1 = 0,$$

where $(\partial F/\partial x)_1$ is the value of $(\partial F/\partial x)$ at the point (x_1, y_1, z_1), and so on.

Angle between two surfaces is the angle between their normals.

1.7   INTEGRAL CALCULUS

Definition of Integral (117)

$F(x)$ is said to be the integral of $f(x)$ if the derivative of $F(x)$ is $f(x)$, or the differential of $F(x)$ is $f(x)\,dx$; in symbols:

$$F(x) = \int f(x)\,dx \text{ if } \frac{dF(x)}{dx} = f(x), \text{ or } dF(x) = f(x)\,dx.$$

In general: $\int f(x)\,dx = F(x) + C$, where C is an arbitrary constant.

Fundamental Theorems on Integrals (118)

$$\int df(x) = f(x) + C.$$

$$d \int f(x) \, dx = f(x) \, dx.$$

$$\int [f_1(x) \pm f_2(x) \pm \cdots] \, dx = \int f_1(x) \, dx \pm \int f_2(x) \, dx \pm \cdots .$$

$$\int af(x) \, dx = a \int f(x) \, dx, \text{ where } a \text{ is any constant.}$$

$$\int u^n \, du = \frac{u^{n+1}}{n+1} + C \, (n \neq -1); \, u \text{ is any function of } x.$$

$$\int \frac{du}{u} = \ln u + C; \, u \text{ is any function of } x.$$

$$\int u \, dv = uv - \int v \, du; \, u \text{ and } v \text{ are any functions of } x.$$

1.7.1 Selected Integrals

In the following list, the constant of integration (*c*) is omitted but should be added
to the result of every integration. The letter *x* represents any variable; the letter *u*
represents any function of *x;* all other letters represent constants which may have
any finite value unless otherwise indicated; ln = $\log_e$; all angles are in radians.

Functions Containing ax + b (119)

$$\int (ax + b)^n \, dx = \frac{1}{a(n+1)} (ax + b)^{n+1}. \, (n \neq -1) \, (120)$$

$$\int \frac{dx}{ax + b} = \frac{1}{a} \ln (ax + b). \, (121)$$

$$\int x(ax + b)^n \, dx = \frac{1}{a^2(n+2)} (ax + b)^{n+2} - \frac{b}{a^2(n+1)} (ax + b)^{n+1}. \, (122)$$
$$(n \neq -1, -2)$$

$$\int \frac{x \, dx}{ax + b} = \frac{x}{a} - \frac{b}{a^2} \ln (ax + b). \, (123)$$

$$\int \frac{x \, dx}{(ax + b)^2} = \frac{b}{a^2(ax + b)} + \frac{1}{a^2} \ln (ax + b). \, (124)$$

$$\int \frac{dx}{x(ax + b)} = \frac{1}{b} \ln \frac{x}{ax + b}. \, (125)$$

$$\int \frac{dx}{x^2(ax + b)} = -\frac{1}{bx} + \frac{a}{b^2} \ln \frac{ax + b}{x}. \, (126)$$

$$\int \frac{dx}{x(ax + b)^2} = \frac{1}{b(ax + b)} - \frac{1}{b^2} \ln \frac{ax + b}{x}. \, (127)$$

$$\int \frac{dx}{x^2(ax + b)^2} = -\frac{b + 2ax}{b^2 x(ax + b)} + \frac{2a}{b^3} \ln \frac{ax + b}{x}. \, (128)$$

$$\int \frac{dx}{x\sqrt{ax+b}} = \frac{1}{\sqrt{b}} \ln \frac{\sqrt{ax+b} - \sqrt{b}}{\sqrt{ax+b} + \sqrt{b}} \cdot (b \text{ pos.}) \ (129)$$

$$\int \frac{dx}{x\sqrt{ax+b}} = \frac{2}{\sqrt{-b}} \tan^{-1} \sqrt{\frac{ax+b}{-b}} \cdot (b \text{ neg.}) \ (130)$$

$$\int \frac{\sqrt{ax+b}}{x} dx = 2\sqrt{ax+b} + \sqrt{b} \ln \frac{\sqrt{ax+b} - \sqrt{b}}{\sqrt{ax+b} + \sqrt{b}} \cdot (b \text{ pos.}) \ (131)$$

$$\int \frac{\sqrt{ax+b}}{x} dx = 2\sqrt{ax+b} - 2\sqrt{-b} \tan^{-1} \sqrt{\frac{ax+b}{-b}} \cdot (b \text{ neg.}) \ (132)$$

$$\int \frac{px+q}{\sqrt{ax+b}} dx = \frac{2}{3a^2} (3aq - 2bp + apx)\sqrt{ax+b}. \ (133)$$

$$\int \frac{\sqrt{ax+b}}{px+q} dx = \frac{2\sqrt{ax+b}}{p} - \frac{2}{p} \sqrt{\frac{aq-bp}{p}} \tan^{-1} \sqrt{\frac{p(ax+b)}{aq-bp}}. \ (134)$$
$$(p \text{ pos.}, aq > bp)$$

$$\int \frac{\sqrt{ax+b}}{px+q} dx = \frac{2\sqrt{ax+b}}{p}$$

$$+ \frac{1}{p} \sqrt{\frac{bp-aq}{p}} \ln \frac{\sqrt{p(ax+b)} - \sqrt{bp-aq}}{\sqrt{p(ax+b)} + \sqrt{bp-aq}}. \ (135)$$
$$(p \text{ pos.}, bp > aq)$$

Functions Containing $ax^2 + b$ (136)

$$\int \frac{dx}{ax^2+b} = \frac{1}{\sqrt{ab}} \tan^{-1} \left(x \sqrt{\frac{a}{b}} \right). \ (a \text{ and } b \text{ pos.}) \ (137)$$

$$\int \frac{dx}{ax^2+b} = \frac{1}{2\sqrt{-ab}} \ln \frac{x\sqrt{a} - \sqrt{-b}}{x\sqrt{a} + \sqrt{-b}} \cdot (a \text{ pos.}, b \text{ neg.})$$

$$= \frac{1}{2\sqrt{-ab}} \ln \frac{\sqrt{b} + x\sqrt{-a}}{\sqrt{b} - x\sqrt{-a}} \cdot (a \text{ neg.}, b \text{ pos.}) \ (138)$$

$$\int \frac{dx}{(ax^2+b)^n} = \frac{1}{2(n-1)b} \frac{x}{(ax^2+b)^{n-1}} + \frac{2n-3}{2(n-1)b} \int \frac{dx}{(ax^2+b)^{n-1}} \ (139)$$
$$(n \text{ integ.} > 1)$$

$$\int (ax^2+b)^n x \, dx = \frac{1}{2a} \frac{(ax^2+b)^{n+1}}{n+1}. \ (n \neq -1) \ (140)$$

$$\int \frac{x \, dx}{ax^2+b} = \frac{1}{2a} \ln (ax^2+b). \ (141)$$

$$\int \frac{dx}{x(ax^2+b)} = \frac{1}{2b} \ln \frac{x^2}{ax^2+b}. \ (142)$$

$$\int \frac{x^2 \, dx}{ax^2+b} = \frac{x}{a} - \frac{b}{a} \int \frac{dx}{ax^2+b}. \ (143)$$

$$\int \frac{x^2\,dx}{(ax^2+b)^n} = -\frac{1}{2(n-1)a}\frac{x}{(ax^2+b)^{n-1}} + \frac{1}{2(n-1)a}\int \frac{dx}{(ax^2+b)^{n-1}}\cdot \quad (144)$$

$$(n \text{ integ.} > 1)$$

$$\int \frac{dx}{x^2(ax^2+b)^n} = \frac{1}{b}\int \frac{dx}{x^2(ax^2+b)^{n-1}} - \frac{a}{b}\int \frac{dx}{(ax^2+b)^n}\cdot \quad (n \text{ pos. integ.}) \;(145)$$

$$\int \sqrt{ax^2+b}\,dx = \frac{x}{2}\sqrt{ax^2+b} + \frac{b}{2\sqrt{a}}\ln(x\sqrt{a}+\sqrt{ax^2+b}).\,(a \text{ pos.}) \;(146)$$

$$\int \sqrt{ax^2+b}\,dx = \frac{x}{2}\sqrt{ax^2+b} + \frac{b}{2\sqrt{-a}}\sin^{-1}\left(x\,\sqrt{-\frac{a}{b}}\right).\,(a \text{ neg.}) \;(147)$$

$$\int \frac{dx}{\sqrt{ax^2+b}} = \frac{1}{\sqrt{a}}\ln(x\sqrt{a}+\sqrt{ax^2+b}).\,(a \text{ pos.}) \;(148)$$

$$\int \frac{dx}{\sqrt{ax^2+b}} = \frac{1}{\sqrt{-a}}\sin^{-1}\left(x\,\sqrt{-\frac{a}{b}}\right).\,(a \text{ neg.}) \;(149)$$

$$\int \sqrt{ax^2+b}\,x\,dx = \frac{1}{3a}(ax^2+b)^{3/2}.\;(150)$$

$$\int \frac{x\,dx}{\sqrt{ax^2+b}} = \frac{1}{a}\sqrt{ax^2+b}.\;(151)$$

$$\int \frac{\sqrt{ax^2+b}}{x}\,dx = \sqrt{ax^2+b} + \sqrt{b}\ln\frac{\sqrt{ax^2+b}-\sqrt{b}}{x}.\,(b \text{ pos.}) \;(152)$$

$$\int \frac{\sqrt{ax^2+b}}{x}\,dx = \sqrt{ax^2+b} - \sqrt{-b}\tan^{-1}\frac{\sqrt{ax^2+b}}{\sqrt{-b}}.\,(b \text{ neg.}) \;(153)$$

$$\int \frac{dx}{x\sqrt{ax^2+b}} = \frac{1}{\sqrt{b}}\ln\frac{\sqrt{ax^2+b}-\sqrt{b}}{x}.\,(b \text{ pos.}) \;(154)$$

$$\int \frac{dx}{x\sqrt{ax^2+b}} = \frac{1}{\sqrt{-b}}\sec^{-1}\left(x\,\sqrt{-\frac{a}{b}}\right).\,(b \text{ neg.}) \;(155)$$

$$\int \frac{\sqrt{ax^2+b}\,dx}{x^n} = -\frac{(ax^2+b)^{3/2}}{b(n-1)x^{n-1}} - \frac{(n-4)a}{(n-1)b}\int \frac{\sqrt{ax^2+b}}{x^{n-2}}\,dx.\;(n>1)\;(156)$$

$$\int \frac{dx}{x^n\sqrt{ax^2+b}} = -\frac{\sqrt{ax^2+b}}{b(n-1)x^{n-1}} - \frac{(n-2)a}{(n-1)b}\int \frac{dx}{x^{n-2}\sqrt{ax^2+b}}\cdot (n>1)\;(157)$$

$$\int \frac{dx}{x(ax^n+b)} = \frac{1}{bn}\ln\frac{x^n}{ax^n+b}.\;(158)$$

$$\int \frac{dx}{x\sqrt{ax^n+b}} = \frac{1}{n\sqrt{b}}\ln\frac{\sqrt{ax^n+b}-\sqrt{b}}{\sqrt{ax^n+b}+\sqrt{b}}.\,(b \text{ pos.}) \;(159)$$

$$\int \frac{dx}{x\sqrt{ax^n+b}} = \frac{2}{n\sqrt{-b}}\sec^{-1}\sqrt{\frac{-ax^n}{b}}.\,(b \text{ neg.}) \;(160)$$

Functions Containing $ax^2 + bx + c$ (161)

$$\int \frac{dx}{ax^2 + bx + c} = \frac{1}{\sqrt{b^2 - 4ac}} \ln \frac{2ax + b - \sqrt{b^2 - 4ac}}{2ax + b + \sqrt{b^2 - 4ac}} . \ (b^2 > 4ac) \ (162)$$

$$\int \frac{dx}{ax^2 + bx + c} = \frac{2}{\sqrt{4ac - b^2}} \tan^{-1} \frac{2ax + b}{\sqrt{4ac - b^2}} . \ (b^2 < 4ac) \ (163)$$

$$\int \frac{dx}{ax^2 + bx + c} = - \frac{2}{2ax + b} . \ (b^2 = 4ac) \ (164)$$

$$\int \frac{x \ dx}{ax^2 + bx + c} = \frac{1}{2a} \ln (ax^2 + bx + c) - \frac{b}{2a} \int \frac{dx}{ax^2 + bx + c} . \ (165)$$

$$\int \frac{x^2 \ dx}{ax^2 + bx + c} = \frac{x}{a} - \frac{b}{2a^2} \ln (ax^2 + bx + c)$$
$$+ \frac{b^2 - 2ac}{2a^2} \int \frac{dx}{ax^2 + bx + c} . \ (166)$$

$$\int \frac{dx}{\sqrt{ax^2 + bx + c}} = \frac{1}{\sqrt{a}} \ln (2ax + b + 2\sqrt{a}\sqrt{ax^2 + bx + c}). \ (a \text{ pos.}) \ (167)$$

$$\int \frac{dx}{\sqrt{ax^2 + bx + c}} = \frac{1}{\sqrt{-a}} \sin^{-1} \frac{-2ax - b}{\sqrt{b^2 - 4ac}} . \ (a \text{ neg.}) \ (168)$$

$$\int \sqrt{ax^2 + bx + c} \ dx = \frac{2ax + b}{4a} \sqrt{ax^2 + bx + c}$$
$$+ \frac{4ac - b^2}{8a} \int \frac{dx}{\sqrt{ax^2 + bx + c}} . \ (169)$$

$$\int \frac{x \ dx}{\sqrt{ax^2 + bx + c}} = \frac{\sqrt{ax^2 + bx + c}}{a} - \frac{b}{2a} \int \frac{dx}{\sqrt{ax^2 + bx + c}} . \ (170)$$

$$\int \sqrt{ax^2 + bx + c} \ x \ dx = \frac{(ax^2 + bx + c)^{3/2}}{3a}$$
$$- \frac{b}{2a} \int \sqrt{ax^2 + bx + c} \ dx. \ (171)$$

$$\int \frac{dx}{x\sqrt{ax^2 + bx + c}} = - \frac{1}{\sqrt{c}} \ln \left(\frac{\sqrt{ax^2 + bx + c} + \sqrt{c}}{x} + \frac{b}{2\sqrt{c}} \right) . \ (c \text{ pos.}) \ (172)$$

$$\int \frac{dx}{x\sqrt{ax^2 + bx + c}} = \frac{1}{\sqrt{-c}} \sin^{-1} \frac{bx + 2c}{x\sqrt{b^2 - 4ac}} . \ (c \text{ neg.}) \ (173)$$

$$\int \frac{dx}{x\sqrt{ax^2 + bx}} = - \frac{2}{bx} \sqrt{ax^2 + bx}. \ (174)$$

Functions Containing $\sin ax$ (175)

$$\int \sin u \ du = -\cos u. \ (u \text{ is any function of } x) \ (176)$$

$$\int \sin ax \ dx = - \frac{1}{a} \cos ax. \ (177)$$

$$\int \sin^2 ax \ dx = \frac{x}{2} - \frac{\sin 2ax}{4a} . \ (178)$$

$$\int \sin^n ax \, dx = -\frac{\sin^{n-1} ax \cos ax}{na} + \frac{n-1}{n} \int \sin^{n-2} ax \, dx. \text{ (} n \text{ pos. integ.) (179)}$$

$$\int \frac{dx}{\sin ax} = \frac{1}{a} \ln \tan \frac{ax}{2} = \frac{1}{a} \ln (\csc ax - \cot ax). \text{ (180)}$$

$$\int \frac{dx}{\sin^2 ax} = -\frac{1}{a} \cot ax. \text{ (181)}$$

$$\int \frac{dx}{\sin^n ax} = -\frac{1}{a(n-1)} \frac{\cos ax}{\sin^{n-1} ax} + \frac{n-2}{n-1} \int \frac{dx}{\sin^{n-2} ax} \cdot \text{ (} n \text{ integ.} > 1 \text{) (182)}$$

$$\int \frac{dx}{1 + \sin ax} = -\frac{1}{a} \tan \left(\frac{\pi}{4} - \frac{ax}{2} \right). \text{ (183)}$$

$$\int \frac{dx}{1 - \sin ax} = \frac{1}{a} \cot \left(\frac{\pi}{4} - \frac{ax}{2} \right). \text{ (184)}$$

Functions Containing cos ax (185)

$$\int \cos u \, du = \sin u. \text{ (} u \text{ is any function of } x \text{) (186)}$$

$$\int \cos ax \, dx = \frac{1}{a} \sin ax. \text{ (187)}$$

$$\int \cos^2 ax \, dx = \frac{x}{2} + \frac{\sin 2ax}{4a}. \text{ (188)}$$

$$\int \cos^n ax \, dx = \frac{\cos^{n-1} ax \sin ax}{na} + \frac{n-1}{n} \int \cos^{n-2} ax \, dx. \text{ (} n \text{ pos. integ.) (189)}$$

$$\int \frac{dx}{\cos ax} = \frac{1}{a} \ln \tan \left(\frac{ax}{2} + \frac{\pi}{4} \right) = \frac{1}{a} \ln (\tan ax + \sec ax). \text{ (190)}$$

$$\int \frac{dx}{\cos^2 ax} = \frac{1}{a} \tan ax. \text{ (191)}$$

$$\int \frac{dx}{\cos^n ax} = \frac{1}{a(n-1)} \frac{\sin ax}{\cos^{n-1} ax} + \frac{n-2}{n-1} \int \frac{dx}{\cos^{n-2} ax} \cdot \text{ (} n \text{ integ.} > 1 \text{) (192)}$$

$$\int \frac{dx}{1 + \cos ax} = \frac{1}{a} \tan \frac{ax}{2}. \text{ (193)}$$

$$\int \frac{dx}{1 - \cos ax} = -\frac{1}{a} \cot \frac{ax}{2}. \text{ (194)}$$

$$\int \sqrt{1 - \cos x} \, dx = \sqrt{2} \int \sin \frac{x}{2} \, dx. \text{ (195)}$$

$$\int \sqrt{1 + \cos x} \, dx = \sqrt{2} \int \cos \frac{x}{2} \, dx. \text{ (196)}$$

Functions Containing sin ax and cos ax (197)

$$\int \sin ax \cos bx \, dx = -\frac{1}{2} \left[\frac{\cos(a - b)x}{a - b} + \frac{\cos(a + b)x}{a + b} \right] . \text{ (} a^2 \neq b^2 \text{) (198)}$$

$$\int \sin^n ax \cos ax \, dx = \frac{1}{a(n + 1)} \sin^{n+1} ax. \text{ (} n \neq -1 \text{). (199)}$$

$$\int \frac{\cos ax}{\sin ax} dx = \frac{1}{a} \ln \sin ax. \quad (200)$$

$$\int (b + c \sin ax)^n \cos ax \, dx = \frac{1}{ac(n + 1)} (b + c \sin ax)^{n+1}. \, (n \neq -1) \quad (201)$$

$$\int \frac{\cos ax \, dx}{b + c \sin ax} = \frac{1}{ac} \ln (b + c \sin ax). \quad (202)$$

$$\int \cos^n ax \sin ax \, dx = - \frac{1}{a(n + 1)} \cos^{n+1} ax. \, (n \neq -1). \quad (203)$$

$$\int \frac{\sin ax}{\cos ax} dx = - \frac{1}{a} \ln \cos ax. \quad (204)$$

$$\int (b + c \cos ax)^n \sin ax \, dx = - \frac{1}{ac(n + 1)} (b + c \cos ax)^{n+1}. \, (n \neq -1) \quad (205)$$

$$\int \frac{\sin ax}{b + c \cos ax} dx = - \frac{1}{ac} \ln (b + c \cos ax). \quad (206)$$

$$\int \frac{dx}{b \sin ax + c \cos ax} = \frac{1}{a\sqrt{b^2 + c^2}} \ln \left[\tan \frac{1}{2} \left(ax + \tan^{-1} \frac{c}{b} \right) \right]. \quad (207)$$

$$\int \sin^2 ax \cos^2 ax \, dx = \frac{x}{8} - \frac{\sin 4ax}{32a}. \quad (208)$$

$$\int \frac{dx}{\sin ax \cos ax} = \frac{1}{a} \ln \tan ax. \quad (209)$$

$$\int \frac{dx}{\sin^2 ax \cos^2 ax} = \frac{1}{a} (\tan ax - \cot ax). \quad (210)$$

$$\int \frac{\sin^2 ax}{\cos ax} dx = \frac{1}{a} \left[-\sin ax + \ln \tan \left(\frac{ax}{2} + \frac{\pi}{4} \right) \right]. \quad (211)$$

$$\int \frac{\cos^2 ax}{\sin ax} dx = \frac{1}{a} \left[\cos ax + \ln \tan \frac{ax}{2} \right]. \quad (212)$$

Functions Containing $\tan ax \left(= \dfrac{1}{\cot ax} \right)$ **or** $\cot ax \left(= \dfrac{1}{\tan ax} \right)$ *(213)*

$$\int \tan u \, du = -\ln \cos u. \, (u \text{ is any function of } x) \quad (214)$$

$$\int \tan ax \, dx = - \frac{1}{a} \ln \cos ax. \quad (215)$$

$$\int \tan^2 ax \, dx = \frac{1}{a} \tan ax - x. \quad (216)$$

$$\int \cot u \, du = \ln \sin u. \, (u \text{ is any function of } x) \quad (217)$$

$$\int \cot ax \, dx = \int \frac{dx}{\tan ax} = \frac{1}{a} \ln \sin ax. \quad (218)$$

$$\int \cot^2 ax \, dx = \int \frac{dx}{\tan^2 ax} = - \frac{1}{a} \cot ax - x. \quad (219)$$

$$\int \frac{dx}{b + c \tan ax} = \int \frac{\cot ax \, dx}{b \cot ax + c}$$

$$= \frac{1}{b^2 + c^2} \left[bx + \frac{c}{a} \ln (b \cos ax + c \sin ax) \right]. \ (220)$$

$$\int \frac{dx}{b + c \cot ax} = \int \frac{\tan ax \, dx}{b \tan ax + c}$$

$$= \frac{1}{b^2 + c^2} \left[bx - \frac{c}{a} \ln (c \cos ax + b \sin ax) \right]. \ (221)$$

$$\int \frac{dx}{\sqrt{1 + \tan^2 ax}} = \frac{1}{a} \sin ax. \ (222)$$

$$\int \frac{dx}{\sqrt{b + c \tan^2 ax}} = \frac{1}{a\sqrt{b - c}} \sin^{-1} \left(\sqrt{\frac{b - c}{b}} \sin ax \right).$$
$$(b \text{ pos.}, \ b^2 > c^2) \ (223)$$

Functions Containing sec ax $\left(= \dfrac{1}{\cos ax} \right)$ **or csc ax** $\left(= \dfrac{1}{\sin ax} \right)$ **(224)**

$$\int \sec u \, du = \ln (\sec u + \tan u) = \ln \tan \left(\frac{u}{2} + \frac{\pi}{4} \right).$$
$$(u \text{ is any function of } x) \ (225)$$

$$\int \sec ax \, dx = \frac{1}{a} \ln \tan \left(\frac{ax}{2} + \frac{\pi}{4} \right). \ (226)$$

$$\int \sec^2 ax \, dx = \frac{1}{a} \tan ax. \ (227)$$

Functions Containing tan ax and sec ax or cot ax and csc ax (228)

$$\int \tan u \sec u \, du = \sec u. \ (u \text{ is any function of } x) \ (229)$$

$$\int \tan ax \sec ax \, dx = \frac{1}{a} \sec ax. \ (230)$$

$$\int \frac{\sec^2 ax \, dx}{\tan ax} = \frac{1}{a} \ln \tan ax. \ (231)$$

Inverse Trigonometric Functions (232)

$$\int \sin^{-1} ax \, dx = x \sin^{-1} ax + \frac{1}{a} \sqrt{1 - a^2x^2}. \ (233)$$

$$\int \cos^{-1} ax \, dx = x \cos^{-1} ax - \frac{1}{a} \sqrt{1 - a^2x^2}. \ (234)$$

$$\int \tan^{-1} ax \, dx = x \tan^{-1} ax - \frac{1}{2a} \ln (1 + a^2x^2). \ (235)$$

$$\int \cot^{-1} ax \, dx = x \cot^{-1} ax + \frac{1}{2a} \ln (1 + a^2x^2). \ (236)$$

Algebraic and Trigonometric Functions (237)

$$\int x \sin ax \, dx = \frac{1}{a^2} \sin ax - \frac{1}{a} x \cos ax. \ (238)$$

$$\int x^n \sin ax \, dx = -\frac{1}{a} x^n \cos ax + \frac{n}{a} \int x^{n-1} \cos ax \, dx. \ (n \text{ pos.}) \ (239)$$

$$\int \frac{\sin ax \, dx}{x} = ax - \frac{(ax)^3}{3 \cdot 3!} + \frac{(ax)^5}{5 \cdot 5!} - \cdots . \ (240)$$

$$\int x \cos ax \, dx = \frac{1}{a^2} \cos ax + \frac{1}{a} x \sin ax. \ (241)$$

$$\int \frac{\cos ax \, dx}{x} = \ln ax - \frac{(ax)^2}{2 \cdot 2!} + \frac{(ax)^4}{4 \cdot 4!} - \cdots . \ (242)$$

Exponential, Algebraic, Trigonometric, Logarithmic Functions (243)

$$\int b^u \, du = \frac{b^u}{\ln b} . \ (u \text{ is any function of } x) \ (244)$$

$$\int e^u \, du = e^u. \ (u \text{ is any function of } x) \ (245)$$

$$\int b^{ax} \, dx = \frac{b^{ax}}{a \ln b} . \ (246)$$

$$\int e^{ax} \, dx = \frac{1}{a} e^{ax}. \ (247)$$

$$\int \frac{dx}{b + ce^{ax}} = \frac{1}{ab} [ax - \ln (b + ce^{ax})]. \ (248)$$

$$\int \frac{e^{ax} \, dx}{b + ce^{ax}} = \frac{1}{ac} \ln (b + ce^{ax}). \ (249)$$

$$\int \frac{dx}{be^{ax} + ce^{-ax}} = \frac{1}{a\sqrt{bc}} \tan^{-1} \left(e^{ax} \sqrt{\frac{b}{c}} \right). \ (b \text{ and } c \text{ pos.}) \ (250)$$

$$\int x b^{ax} \, dx = \frac{x b^{ax}}{a \ln b} - \frac{b^{ax}}{a^2 (\ln b)^2} . \ (251)$$

$$\int x e^{ax} \, dx = \frac{e^{ax}}{a^2} (ax - 1). \ (252)$$

$$\int x^n b^{ax} \, dx = \frac{x^n b^{ax}}{a \ln b} - \frac{n}{a \ln b} \int x^{n-1} b^{ax} \, dx. \ (n \text{ pos.}) \ (253)$$

$$\int x^n e^{ax} \, dx = \frac{1}{a} x^n e^{ax} - \frac{n}{a} \int x^{n-1} e^{ax} \, dx. \ (n \text{ pos.}) \ (254)$$

$$\int \frac{e^{ax}}{x} \, dx = \ln x + ax + \frac{(ax)^2}{2 \cdot 2!} + \frac{(ax)^3}{3 \cdot 3!} + \cdots . \ (255)$$

$$\int \frac{e^{ax}}{x^n} \, dx = \frac{1}{n-1} \left(-\frac{e^{ax}}{x^{n-1}} + a \int \frac{e^{ax}}{x^{n-1}} \, dx \right). \ (n \text{ integ.} > 1) \ (256)$$

$$\int e^{ax} \ln x \, dx = \frac{1}{a} e^{ax} \ln x - \frac{1}{a} \int \frac{e^{ax}}{x} \, dx. \ (257)$$

$$\int e^{ax} \sin bx \, dx = \frac{e^{ax}}{a^2 + b^2} (a \sin bx - b \cos bx). \quad (258)$$

$$\int e^{ax} \cos bx \, dx = \frac{e^{ax}}{a^2 + b^2} (a \cos bx + b \sin bx). \quad (259)$$

$$\int \ln ax \, dx = x \ln ax - x. \quad (260)$$

$$\int (\ln ax)^n \, dx = x(\ln ax)^n - n \int (\ln ax)^{n-1} \, dx. \quad (n \text{ pos.}) \quad (261)$$

$$\int x^n \ln ax \, dx = x^{n+1} \left[\frac{\ln ax}{n + 1} - \frac{1}{(n + 1)^2} \right]. \quad (n \neq -1) \quad (262)$$

$$\int \frac{(\ln ax)^n}{x} \, dx = \frac{(\ln ax)^{n+1}}{n + 1}. \quad (n \neq -1) \quad (263)$$

$$\int \frac{dx}{x \ln ax} = \ln (\ln ax). \quad (264)$$

$$\int \frac{dx}{\ln ax} = \frac{1}{a} \left[\ln (\ln ax) + \ln ax + \frac{(\ln ax)^2}{2 \cdot 2!} + \frac{(\ln ax)^3}{3 \cdot 3!} + \cdots \right]. \quad (265)$$

$$\int \sin (\ln ax) \, dx = \frac{x}{2} [\sin (\ln ax) - \cos (\ln ax)]. \quad (266)$$

$$\int \cos (\ln ax) \, dx = \frac{x}{2} [\sin (\ln ax) + \cos (\ln ax)]. \quad (267)$$

Some Definite Integrals (268)

$$\int_0^a \sqrt{a^2 - x^2} \, dx = \frac{\pi a^2}{4}. \quad (269)$$

$$\int_0^a \sqrt{2ax - x^2} \, dx = \frac{\pi a^2}{4}. \quad (270)$$

$$\int_0^\infty \frac{dx}{ax^2 + b} = \frac{\pi}{2\sqrt{ab}}. \quad (a \text{ and } b \text{ pos.}) \quad (271)$$

$$\int_0^{\pi/2} \sin^n ax \, dx = \int_0^{\pi/2} \cos^n ax \, dx = \frac{1 \cdot 3 \cdot 5 \cdot \ldots (n - 1)}{2 \cdot 4 \cdot 6 \cdot \ldots n} \frac{\pi}{2a}. \quad (272)$$
$$(n \text{ pos. even integ.})$$

$$\int_0^{\pi/2} \sin^n ax \, dx = \int_0^{\pi/2} \cos^n ax \, dx = \frac{2 \cdot 4 \cdot 6 \cdot \ldots (n - 1)}{1 \cdot 3 \cdot 5 \cdot \ldots n} \frac{1}{a}. \quad (273)$$
$$(n \text{ pos. odd integ.})$$

$$\int_0^\pi \sin^2 ax \, dx = \int_0^\pi \cos^2 ax \, dx = \frac{\pi}{2}. \quad (274)$$

$$\int_0^\infty e^{-ax^2} \, dx = \frac{1}{2} \sqrt{\frac{\pi}{a}}. \quad (275)$$

$$\int_0^\infty x^n e^{-ax} \, dx = \frac{n!}{a^{n+1}}. \quad (n \text{ pos. integ.}) \quad (276)$$

1.7.2 Definite Integrals

Definition and Approximate Value of the Definite Integral (277)

If $f(x)$ is continuous from $x = a$ to $x = b$ inclusive, and this interval is divided into n equal parts by the points $a, x_1, x_2, \ldots x_{n-1}, b$ such that $\Delta x = (b - a) \div n$, then the definite integral of $f(x)\, dx$ between the limits $x = a$ to $x = b$ is

$$\int_a^b f(x)\, dx = \lim_{n \doteq \infty} [f(a)\, \Delta x + f(x_1)\, \Delta x + f(x_2)\, \Delta x + \cdots + f(x_{n-1})\, \Delta x]$$

$$= \left[\int f(x)\, dx \right]_a^b = [F(x)]_a^b = F(b) - F(a).$$

If $y_0, y_1, y_2, \ldots, y_{n-1}, y_n$ are the values of $f(x)$ when $x = a, x_1, x_2, \ldots, x_{n-1}, b$, respectively, and if $h = (b - a) \div n$, then approximate values of this definite integral are given by the Trapezoidal, Durand's, and Simpson's rules (37).

Some Fundamental Theorems on Definite Integrals (278)

$$\int_a^b [f_1(x) + f_2(x) + \cdots]\, dx = \int_a^b f_1(x)\, dx + \int_a^b f_2(x)\, dx + \cdots . \quad (279)$$

$$\int_a^b kf(x)\, dx = k \int_a^b f(x)\, dx. \ (k \text{ is any constant}) \ (280)$$

$$\int_a^b f(x)\, dx = - \int_b^a f(x)\, dx. \ (281)$$

$$\int_a^b f(x)\, dx = \int_a^c f(x)\, dx + \int_c^b f(x)\, dx. \ (282)$$

$$\int_a^b f(x)\, dx = (b - a)f(x_1), \text{ where } x_1 \text{ lies between } a \text{ and } b. \ (283)$$

$$\int_a^\infty f(x)\, dx = \lim_{b \doteq \infty} \int_a^b f(x)\, dx. \ (284)$$

Some Applications of the Definite Integral

Plane Area (285). Area (A) bounded by the curve $y = f(x)$, the axis OX, and the ordinates $x = a, x = b$ (Fig. 1.55a).

$$dA = y\, dx, \qquad A = \int_a^b f(x)\, dx.$$

Area (A) bounded by the curve $x = f(y)$, the axis OY, and the abscissas $y = c$, $y = d$ (Fig. 1.55b).

$$dA = x\, dy, \qquad A = \int_c^d f(y)\, dy.$$

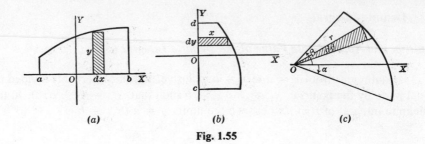

Fig. 1.55

Area (A) bounded by the curve $x = f_1(t)$, $y = f_2(t)$, the axis OX, and $t = a$, $t = b$.

$$dA = y\,dx, \qquad A = \int_a^b f_2(t)f_1(t)\,dt.$$

Area (A) bounded by the curve $r = f(\theta)$ and two radii $\theta = \alpha$, $\theta = \beta$ (Fig. 1.55c).

$$dA = \tfrac{1}{2}r^2\,d\theta, \qquad A = \tfrac{1}{2}\int_\alpha^\beta [f(\theta)]^2\,d\theta.$$

Length of Arc (286). Length (s) of arc of curve $f(x, y) = 0$ from the point (a, c) to the point (b, d) (Fig. 1.56a).

$$ds = \sqrt{(dx)^2 + (dy)^2}, \qquad s = \int_a^b \sqrt{1 + \left(\frac{dy}{dx}\right)^2}\,dx$$

$$= \int_c^d \sqrt{1 + \left(\frac{dx}{dy}\right)^2}\,dy.$$

Length (s) of arc of curve $x = f_1(t)$, $y = f_2(t)$ from $t = a$ to $t = b$.

$$ds = \sqrt{(dx)^2 + (dy)^2}, \qquad s = \int_a^b \sqrt{\left(\frac{dx}{dt}\right)^2 + \left(\frac{dy}{dt}\right)^2}\,dt.$$

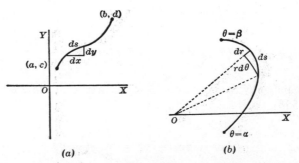

Fig. 1.56

Length (s) of arc of curve $r = f(\theta)$ from $\theta = \alpha$ to $\theta = \beta$ (Fig. 1.56b).

$$ds = \sqrt{(dr)^2 + (r\,d\theta)^2}, \qquad s = \int_\alpha^\beta \sqrt{r^2 + \left(\frac{dr}{d\theta}\right)^2}\,d\theta.$$

Length (s) of arc of space curve $x = f_1(t)$, $y = f_2(t)$, $z = f_3(t)$ from $t = a$ to $t = b$.

$$ds = \sqrt{(dx)^2 + (dy)^2 + (dz)^2}, \qquad s = \int_a^b \sqrt{\left(\frac{dx}{dt}\right)^2 + \left(\frac{dy}{dt}\right)^2 + \left(\frac{dz}{dt}\right)^2}\,dt.$$

Volume of Revolution (287). Volume (V) of revolution generated by revolving about the line $y = k$ the area enclosed by the curve $y = f(x)$, the ordinates $x = a$, $x = b$, and the line $y = k$ (Fig. 1.57).

$$dV = \pi R^2\,dx = \pi(y - k)^2\,dx,$$
$$V = \pi \int_a^b [f(x) - k]^2\,dx.$$

Volume (V) of revolution generated by revolving about the line $x = k$ the area enclosed by the curve $x = f(y)$, the abscissas $y = c$, $y = d$, and the line $x = k$.

$$dV = \pi R^2\,dy = \pi(x - k)^2\,dy, \qquad V = \pi \int_c^d [f(y) - k]^2\,dy.$$

Area of Surface of Revolution (288). Area (S) of surface of revolution generated by revolving the arc of the curve $f(x, y) = 0$ from the point (a, c) to the point (b, d) (Fig. 1.58).

About $y = k$: $dS = 2\pi R\,ds$,
$$S = 2\pi \int_a^b (y - k) \sqrt{1 + \left(\frac{dy}{dx}\right)^2}\,dx.$$
About $x = k$: $dS = 2\pi R\,ds$,
$$S = 2\pi \int_c^d (x - k) \sqrt{1 + \left(\frac{dx}{dy}\right)^2}\,dy.$$

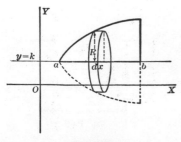

Fig. 1.57

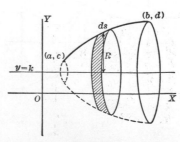

Fig. 1.58

Area (S) of surface of revolution generated by revolving the arc of the curve $r = f(\theta)$ from $\theta = \alpha$ to $\theta = \beta$.

About OX: $dS = 2\pi R \, ds$, $S = 2\pi \displaystyle\int_{\alpha}^{\beta} r \sin\theta \sqrt{r^2 + \left(\dfrac{dr}{d\theta}\right)^2} \, d\theta.$

About OY: $dS = 2\pi R \, ds$, $S = 2\pi \displaystyle\int_{\alpha}^{\beta} r \cos\theta \sqrt{r^2 + \left(\dfrac{dr}{d\theta}\right)^2} \, d\theta.$

Volume by Parallel Sections (289).   Volume (V) of a solid generated by moving a plane section of area A_x perpendicular to OX from $x = a$ to $x = b$ (Fig. 1.59).

$$dV = A_x \, dx, \qquad V = \int_{a}^{b} A_x \, dx,$$

where A_x must be expressed as a function of x.

Mass (290).   Mass (m) constant or variable density (δ).

$$dm = \delta \, dA \quad \text{or} \quad \delta \, ds \quad \text{or} \quad \delta \, dV \quad \text{or} \quad \delta \, dS, \qquad m = \int dm,$$

where dA, ds, dV, dS are the elements of area, length, volume, surface in (281) to (285), and $\delta = $ mass per unit element.

Moment (291).   Moment (M) of a mass (m).

About OX: $M_x = \displaystyle\int y \, dm = \int r \sin\theta \, dm.$

About OY: $M_y = \displaystyle\int x \, dm = \int r \cos\theta \, dm.$

About O: $M_0 = \displaystyle\int \sqrt{x^2 + y^2} \, dm = \int r \, dm.$

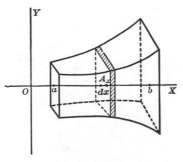

Fig. 1.59

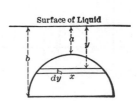

Fig. 1.60

Moment of Inertia (292). Moment of inertia (J) of a mass (m).

$$\text{About } OX: J_x = \int y^2 \, dm = \int r^2 \sin^2 \theta \, dm.$$

$$\text{About } OY: J_y = \int x^2 \, dm = \int r^2 \cos^2 \theta \, dm.$$

$$\text{About } O: \quad J_0 = \int (x^2 + y^2) \, dm = \int r^2 \, dm.$$

Center of Gravity (293). Coordinates ($\bar{x}, \bar{y}$) of the center of gravity of a mass (m).

$$\bar{x} = \frac{\int x \, dm}{\int dm}, \qquad \bar{y} = \frac{\int y \, dm}{\int dm}.$$

NOTE. The center of gravity of the element of area may be taken at its midpoint. In the above equations x and y are the coordinates of the center of gravity of the element.

Work (294). Work (W) done in moving a particle from $s = a$ to $s = b$ against a force whose component in the direction of motion is F_s.

$$dW = F_s \, ds, \qquad W = \int_a^b F_s \, ds,$$

where F_s must be expressed as a function of s.

Pressure (295). Pressure (p) against an area vertical to the surface of the liquid and between depths a and b (Fig. 1.60).

$$dp = wyx \, dy, \qquad p = \int_a^b wyx \, dy,$$

where w = weight of liquid per unit volume, y = depth beneath surface of liquid of a horizontal element of area, and x = length of horizontal element of area; x must be expressed in terms of y.

Center of Pressure (296). The depth ($\bar{y}$) of the center of pressure against an area vertical to the surface of the liquid and between depths a and b.

$$\bar{y} = \frac{\int_a^b y \, dp}{\int_a^b dp} . \quad \text{[for } dp \text{ see (295).]}$$

1.8   DIFFERENTIAL EQUATIONS

Definitions and Notation

A **differential equation** is an equation involving differentials or derivatives.

The **order** of a differential equation is the same as that of the derivative of highest order which it contains.

The **degree** of a differential equation is the same as the power to which the derivative of highest order in the equation is raised, that derivative entering the equation free from radicals.

The **solution** of a differential equation is the relation involving only the variables (but not their derivatives) and arbitrary constants, consistent with the given differential equation.

The most **general solution** of a differential equation of the *n*th order contains *n* arbitrary constants. If particular values are assigned to these arbitrary constants, the solution is called a particular solution.

Notation: M and N denote functions of x and y; X denotes a function of x alone or a constant, Y denotes a function of y alone or a constant; $C, C_1, C_2, \ldots, C_n$ denote arbitrary constants of integration, $a, b, k, l, m, n, \ldots$ denote given constants.

Equations of First Order and First Degree. $M\ dx + N\ dy = 0$ (297)

Variables Separable: $X_1 Y_1\ dx + X_2 Y_2\ dy = 0.$ (298)

Solution:

$$\int \frac{X_1}{X_2}\ dx + \int \frac{Y_2}{Y_1}\ dy = C.$$

Homogeneous Equation.

$$dy - f\left(\frac{y}{x}\right) dx = 0.\ (299)$$

Solution:

$$x = Ce^{\int [dv/f(v)-v]} \text{ and } v = \frac{y}{x}.$$

NOTE.   Here, $M \div N$ can be written in a form such that x and y occur only in the combination $y \div x$; this can always be done if every term in M and N is of the same degree in x and y.

Linear Equation.

$$dy + (X_1 y - X_2)\ dx = 0.\ (300)$$

Solution:

$$y = e^{-\int X_1 dx} \left(\int X_2 e^{\int X_1 dx} \, dx + C \right).$$

NOTE. A similar solution exists for $dx + (Y_1 x - Y_2) \, dy = 0$.

Exact Equation.

$$M \, dx + N \, dy = 0, \text{ where } \frac{\partial M}{\partial y} = \frac{\partial N}{\partial x}. \ (301)$$

Solution:

$$\int M \, dx + \int \left[N - \frac{\partial}{\partial y} \int M \, dx \right] dy = C,$$

where y is constant when integrating with respect to x.

Nonexact Equation.

$$M \, dx + N \, dy = 0, \text{ where } \frac{\partial M}{\partial y} \neq \frac{\partial N}{\partial x}. \ (302)$$

Solution:

The equation may be made exact by multiplying by an integrating factor $\mu(x, y)$. The form of this factor is readily recognized in a large number of cases. Then solve by (297).

Certain Special Equations of the Second Order.

$$\frac{d^2 y}{dx^2} = f\left(x, y, \frac{dy}{dx} \right). \ (303)$$

Equation.

$$\frac{d^2 y}{dx^2} = X. \ (304)$$

Solution:

$$y = x \int X \, dx - \int x X \, dx + C_1 x + C_2.$$

Equation.

$$\frac{d^2 y}{dx^2} = Y. \ (305)$$

Solution:

$$x = \int \frac{dy}{\sqrt{2 \int Y\, dy + C_1}} + C_2.$$

Equation.

$$\frac{d^2y}{dx^2} = f\left(\frac{dy}{dx}\right). \quad (306)$$

Solution:

$$x = \int \frac{dp}{f(p)} + C_1 \text{ and } y = \int \frac{p\, dp}{f(p)} + C_2.$$

From these two equations eliminate $p = \dfrac{dy}{dx}$ if necessary.

Equation.

$$\frac{d^2y}{dx^2} = f\left(x, \frac{dy}{dx}\right). \quad (307)$$

Solution:

Place $\dfrac{dy}{dx} = p$ and $\dfrac{d^2y}{dx^2} = \dfrac{dp}{dx}$, thus bringing the equation into the form $\dfrac{dp}{dx} = f(x,\,p)$. This is of the first order and may be solved for p by (297) to (301).

Then replace p by $\dfrac{dy}{dx}$ and integrate for y.

Equation.

$$\frac{d^2y}{dx^2} = f\left(y, \frac{dy}{dx}\right). \quad (308)$$

Solution:

Place $\dfrac{dy}{dx} = p$ and $\dfrac{d^2y}{dx^2} = p\dfrac{dp}{dy}$, thus bringing the equation into the form $p\dfrac{dp}{dy} = f(y,\,p)$. This is of the first order and may be solved for p by (297) to (301).

Then replace p by $\dfrac{dy}{dx}$ and integrate for y.

Linear Equations of Physics. Second Order with Constant Coefficients.

$$\frac{d^2x}{dt^2} + 2l\frac{dx}{dt} \pm k^2x = f(t). \text{ (309)}$$

Equation.

$$\frac{d^2x}{dt^2} - k^2x = 0. \text{ (309a)}$$

Solution:

$$x = C_1e^{kt} + C_2e^{-kt}.$$

Equation of Simple Harmonic Motion.

$$\frac{d^2x}{dt^2} + k^2x = 0. \text{ (310)}$$

Solution:

This may be written in the following forms:

1. $x = C_1e^{kt\sqrt{-1}} + C_2e^{-kt\sqrt{-1}}.$
2. $x = C_1 \cos kt + C_2 \sin kt,$
3. $x = C_1 \sin (kt + C_2).$
4. $x = C_1 \cos (kt + C_2).$

Equation of Harmonic Motion with Constant Disturbing Force.

$$\frac{d^2x}{dt^2} + k^2x = a. \text{ (311)}$$

Solution:

$$x = C_1 \cos kt + C_2 \sin kt + \frac{a}{k^2},$$

or

$$x = C_1 \sin (kt + C_2) + \frac{a}{k^2}.$$

Equation of Forced Vibration (312).

1. $\dfrac{d^2x}{dt^2} + k^2x = a \cos nt + b \sin nt,$ where $n \neq k.$

Solution:

$$x = C_1 \cos kt + C_2 \sin kt + \frac{1}{k^2 - n^2} (a \cos nt + b \sin nt).$$

2. $\dfrac{d^2x}{dt^2} + k^2x = a \cos kt + b \sin kt.$

Solution:

$$x = C_1 \cos kt + C_2 \sin kt + \frac{t}{2k} (a \sin kt - b \cos kt).$$

Equation of Damped Vibration.

$$\frac{d^2x}{dt^2} + 2l\frac{dx}{dt} + k^2x = 0. \text{ (313)}$$

Solution:

If $l^2 = k^2,$ $x = e^{-lt}(C_1 + C_2t).$
If $l^2 > k^2,$ $x = e^{-lt}(C_1 e^{\sqrt{l^2-k^2}\,t} + C_2 e^{-\sqrt{l^2-k^2}\,t}).$
If $l^2 < k^2,$ $x = e^{-lt}(C_1 \cos \sqrt{k^2 - l^2}\,t + C_2 \sin \sqrt{k^2 - l^2}\,t)$
    or $x = C_1 e^{-lt} \sin (\sqrt{k^2 - l^2}\,t + C_2).$

Equation of Damped Vibration with Constant Disturbing Force.

$$\frac{d^2x}{dt^2} + 2l\frac{dx}{dt} + k^2x = a. \text{ (314)}$$

Solution:

$$x = x_1 + \frac{a}{k^2},$$

where x_1 is the solution of (309).

General Equation.

$$\frac{d^2x}{dt^2} + 2l\frac{dx}{dt} + k^2x = f(t) = T. \text{ (315)}$$

Solution:

$$x = x_1 + I,$$

where x_1 is the solution of (313) and I is given by

CASE I. $l^2 = k^2$,

$$I = e^{-lt} \left[t \int e^{lt} T \, dt - \int e^{lt} Tt \, dt \right].$$

CASE II. $l^2 > k^2$,

$$I = \frac{1}{\alpha - \beta} \left[e^{\alpha t} \int e^{-\alpha t} T \, dt - e^{\beta t} \int e^{-\beta t} T \, dt \right],$$

where $\alpha = -1 + \sqrt{l^2 - k^2}$, $\beta = -1 - \sqrt{l^2 - k^2}$.

CASE III. $l^2 < k^2$,

$$I = \frac{e^{\alpha t}}{\beta} \left[\sin \beta t \int e^{-\alpha t} \cos \beta t \, T \, dt - \cos \beta t \int e^{-\alpha t} \sin \beta t \, T \, dt \right],$$

where $\alpha = -l$, $\beta = \sqrt{k^2 - l^2}$.

NOTE. *I* may also be found by the method indicated in (317).

Linear Equations with Constant Coefficients: *n*th Order Equation (316).

$$a_n \frac{d^n x}{dt^n} + a_{n-1} \frac{d^{n-1} x}{dt^{n-1}} + a_{n-2} \frac{d^{n-2} x}{dt^{n-2}} + \cdots + a_1 \frac{dx}{dt} + a_0 x = 0.$$

Solution:

Let $D = \alpha_1, \alpha_2, \alpha_3, \ldots, \alpha_n$ be the *n* roots of the auxiliary algebraic equation $a_n D^n + a_{n-1} D^{n-1} + a_{n-2} D^{n-2} + \cdots + a_1 D + a_0 = 0$.
If all roots are real and distinct,

$$x = C_1 e^{\alpha_1 t} + C_2 e^{\alpha_2 t} + \cdots + C_n e^{\alpha_n t}.$$

If two roots are equal: $\alpha_1 = \alpha_2$, the rest real and distinct,

$$x = e^{\alpha_1 t}(C_1 + C_2 t) + C_3 e^{\alpha_3 t} + \cdots + C_n e^{\alpha_n t}.$$

If *p* roots are equal: $\alpha_1 = \alpha_2 = \cdots = \alpha_p$, the rest real and distinct,

$$x = e^{\alpha_1 t}(C_1 + C_2 t + C_3 t^2 + \cdots + C_p t^{p-1}) + \cdots + C_n e^{\alpha_n t}.$$

If two roots are conjugate imaginary: $\alpha_1 = \beta + \gamma \sqrt{-1}, \alpha_2 = \beta - \gamma \sqrt{-1}$,

$$x = e^{\beta t}(C_1 \cos \gamma t + C_2 \sin \gamma t) + C_3 e^{\alpha_3 t} + \cdots + C_n e^{\alpha_n t}.$$

If there is a pair of conjugate imaginary double roots:

$$\alpha_1 = \beta + \gamma\sqrt{-1} = \alpha_2, \qquad \alpha_3 = \beta - \gamma\sqrt{-1} = \alpha_4,$$

$$x = e^{\beta t}[(C_1 + C_2 t)\cos \gamma t + (C_3 + C_4 t)\sin \gamma t] + \cdots + C_n e^{\alpha_n t}.$$

Equation (317).

$$a_n \frac{d^n x}{dt^n} + a_{n-1}\frac{d^{n-1}x}{dt^{n-1}} + \cdots + a_1 \frac{dx}{dt} + a_0 x = f(t). \quad (317)$$

Solution:

$$x = x_1 + I,$$

where x_1 is the solution of Equation (316) and where I may be found by the following method.

Let $f(t) = T_1 + T_2 + T_3 + \cdots$. Find the 1st, 2nd, 3rd, $\cdots$ derivatives of these terms. If $\tau_1, \tau_2, \tau_3, \cdots \tau_n$ are the resulting expressions which have different functional form (disregarding constant coefficients), assume

$$I = A\tau_1 + B\tau_2 + C\tau_3 + \cdots + K\tau_k + \cdots + N\tau_n.$$

NOTE. Thus, if $T = a \sin nt + bt^2 e^{kt}$, all possible successive derivatives of $\sin nt$ and $t^2 e^{kt}$ give terms of the form: $\sin nt$, $\cos nt$, e^{kt}, te^{kt}, $t^2 e^{kt}$; hence assume $I = A \sin nt + B \cos nt + Ce^{kt} + Dte^{kt} + Et^2 e^{kt}$.

Substitute this value of I for x in the given equation, expand, equate coefficients of like terms in the left and right members of the equation, and solve for $A, B, C, \ldots N$.

NOTE. If a root, α_k, occurring m times, of the algebraic equation in D [see (316)] gives rise to a term of the form τ_k in x_1, then the corresponding term in the assumed value of I is $Kt^m \tau_k$.

Simultaneous Equations (318).

$$a_n \frac{d^n x}{dt^n} + b_m \frac{d^m y}{dt^m} + \cdots + a_1 \frac{dx}{dt} + b_1 \frac{dy}{dt} + a_0 x + b_0 y = f_1(t).$$

$$c_k \frac{d^k x}{dt^k} + g_l \frac{d^l y}{dt^l} + \cdots + c_1 \frac{dx}{dt} + g_1 \frac{dy}{dt} + c_0 x + g_0 y = f_2(t).$$

Solution:

Write the equations in the form:

$$(a_n D^n + \cdots + a_1 D + a_0)x + (b_m D^m + \cdots + b_1 D + b_0)y = f_1(t),$$
$$(c_k D^k + \cdots + c_1 D + c_0)x + (g_l D^l + \cdots + g_1 D + g_0)y = f_2(t),$$

where
$$D = \frac{d}{dt}, \ldots, D^i = \frac{d^i}{dt^i}, \ldots.$$

Regarding this set of equations as a pair of simultaneous algebraic equations in x and y, eliminate y and x in turn, getting two linear differential equations of the form of (317) whose solutions are

$$x = x_1 + I_1, \qquad y = y_1 + I_2.$$

Substitute these values of x and y in the original equations, equate coefficients of like terms, and thus express the arbitrary constants in y_1, say, in terms of those in x_1.

Partial Differential Equations (319)

Equation of Oscillation (320).

$$\frac{\partial^2 y}{\partial t^2} = a^2 \frac{\partial^2 y}{\partial x^2}.$$

Solution:

$$y = \sum_{i=1}^{i=\infty} C_i e^{(x+at)\alpha_i} + \sum_{i=1}^{i=\infty} C_i' e^{(x-at)\alpha_i},$$

where C_i, C_i', α_i are arbitrary constants.

Equation of Thermodynamics (321).

$$\frac{\partial u}{\partial t} = a^2 \frac{\partial^2 u}{\partial x^2}.$$

Solution:

$$u = \sum_{i=1}^{i=\infty} C_i e^{\alpha_i x} e^{a^2 \alpha_i^2 t},$$

where C_i and α_i are arbitrary constants.

Equation of Laplace or Condition of Continuity of Incompressible Liquids (322).

$$\frac{\partial^2 u}{\partial x^2} + \frac{\partial^2 u}{\partial y^2} = 0.$$

Solution:

$$u = \sum_{i=1}^{i=\infty} C_i e^{(x+y\sqrt{-1})\alpha_i} + \sum_{i=1}^{i=\infty} C_i' e^{(x-y\sqrt{-1})\alpha_i},$$

where C_i, C_i', α_i are arbitrary constants.

1.9 COMPLEX QUANTITIES

Definition and Representation of a Complex Quantity (323)

If $z = x + jy$, where $j = \sqrt{-1}$ and x and y are real, z is called a complex quantity. z is completely determined by x and y.

If $P(x, y)$ is a point in the plane (Fig. 1.61), then the segment OP in magnitude and direction is said to represent the complex quantity $z = x + jy$.

If θ is the angle from OX to OP and r is the length of OP, then

$$z = x + jy = r(\cos \theta + j \sin \theta) = re^{j\theta},$$

where $\theta = \tan^{-1} \dfrac{y}{x}$, $r = + \sqrt{x^2 + y^2}$, and e is the base of natural logarithms. $x + jy$ and $x - jy$ are called conjugate complex quantities.

Properties of Complex Quantities (324)

Let z, z_1, z_2 represent complex quantities, then:

Sum or Difference: $z_1 \pm z_2 = (x_1 \pm x_2) + j(y_1 \pm y_2)$.

Product: $\begin{aligned}z_1 \cdot z_2 &= r_1 r_2 [\cos (\theta_1 + \theta_2) + j \sin (\theta_1 + \theta_2)] \\ &= r_1 r_2 e^{j(\theta_1 + \theta_2)} = (x_1 x_2 - y_1 y_2) + j(x_1 y_2 + x_2 y_1).\end{aligned}$

Quotient: $\begin{aligned}\frac{z_1}{z_2} &= \frac{r_1}{r_2} [\cos (\theta_1 - \theta_2) + j \sin (\theta_1 - \theta_2)] \\ &= \frac{r_1}{r_2} e^{j(\theta_1 - \theta_2)} = \frac{x_1 x_2 + y_1 y_2}{x_2^2 + y_2^2} + j \frac{x_2 y_1 - x_1 y_2}{x_2^2 + y_2^2}.\end{aligned}$

Power: $\quad z^n = r^n [\cos n\theta + j \sin n\theta] = r^n e^{jn\theta}$.

Root: $\quad \sqrt[n]{z} = \sqrt[n]{r} \left[\cos \dfrac{\theta + 2k\pi}{n} + j \sin \dfrac{\theta + 2k\pi}{n} \right] = \sqrt[n]{r} e^{j(\theta + 2k\pi/n)}$,

where k takes in succession the values $0, 1, 2, 3, \ldots, n - 1$.

Equation: If $z_1 = z_2$, then $x_1 = x_2$ and $y_1 = y_2$.

Periodicity: $\quad z = r(\cos \theta + j \sin \theta) = r[\cos (\theta + 2k\pi) + j \sin (\theta + 2k\pi)]$,

or $\quad\quad\quad z = re^{j\theta} = re^{j(\theta + 2k\pi)}$ and $e^{j2k\pi} = 1$, where k is any integer.

Exponential–Trigonometric Relations:

$$e^{jz} = \cos z + j \sin z, \quad e^{-jz} = \cos z - j \sin z,$$

$$\cos z = \frac{1}{2} (e^{jz} + e^{-jz}), \quad \sin z = \frac{1}{2j} (e^{jz} - e^{-jz}).$$

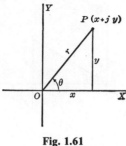

Fig. 1.61   Fig. 1.62

1.10   VECTORS

Definition and Graphical Representation of a Vector (325) (Fig. 1.62)

A vector (**V**) is a quantity which is completely specified by a magnitude and a direction. A scalar (*s*) is a quantity which is completely specified by a magnitude.

The vector (**V**) may be represented geometrically by the segment $\overrightarrow{OA}$, the length of *OA* signifying the magnitude of **V** and the arrow carried by *OA* signifying the direction of **V**.

The segment $\overrightarrow{AO}$ represents the vector $-\mathbf{V}$.

Graphical Summation of Vectors (326)

If $\mathbf{V}_1$, $\mathbf{V}_2$ are two vectors, their graphical sum, $\mathbf{V} = \mathbf{V}_1 + \mathbf{V}_2$, is formed by drawing the vector $\mathbf{V}_1 = \overrightarrow{OA}$ from any point *O*, and the vector $\mathbf{V}_2 = \overrightarrow{AB}$ from the end of $\mathbf{V}_1$, and joining *O* and *B*; then $\mathbf{V} = \overrightarrow{OB}$. Also $\mathbf{V}_1 + \mathbf{V}_2 = \mathbf{V}_2 + \mathbf{V}_1$ and $\mathbf{V}_1 + \mathbf{V}_2 - \mathbf{V} = 0$ (Fig. 1.63*a*).

Similarly, if $\mathbf{V}_1$, $\mathbf{V}_2$, $\mathbf{V}_3$, ... $\mathbf{V}_n$ are any number of vectors drawn so that the initial point of one is the end point of the preceding one, then their graphical sum, $\mathbf{V} = \mathbf{V}_1 + \mathbf{V}_2 + \cdots + \mathbf{V}_n$, is the vector joining the initial point of $\mathbf{V}_1$ with the end point of $\mathbf{V}_n$ (Fig. 1.63*b*).

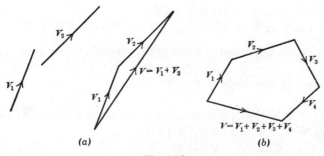

(a)   (b)

Fig. 1.63

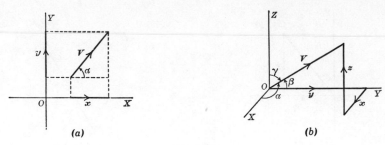

Fig. 1.64

Components of a Vector. Analytic Representation (327)

A vector (**V**) considered as lying in the xy coordinate plane is completely determined by its horizontal and vertical components x and y. If **i** and **j** represent vectors of unit magnitude along OX and OY, respectively, and a and b are the magnitudes of the components x and y, then **V** may be represented by $\mathbf{V} = a\mathbf{i} + b\mathbf{j}$, its magnitude by $|\mathbf{V}| = +\sqrt{a^2 + b^2}$, and its direction by $\alpha = \tan^{-1}(b/a)$.

A vector (**V**) considered as lying in space is completely determined by its components x, y, and z along three mutually perpendicular lines OX, OY, and OZ, directed as in Fig. 1.64b. If **i, j, k** represent vectors of unit magnitude along OX, OY, OZ, respectively, and a, b, c are the magnitudes of the components x, y, z, respectively, then **V** may be represented by $\mathbf{V} = a\mathbf{i} + b\mathbf{j} + c\mathbf{k}$, its magnitude by $|\mathbf{V}| = +\sqrt{a^2 + b^2 + c^2}$, and its direction by $\cos \alpha : \cos \beta : \cos \gamma = a : b : c$.

Properties of Vectors (328)

$$\mathbf{V} = a\mathbf{i} + b\mathbf{j} \quad \text{or} \quad \mathbf{V} = a\mathbf{i} + b\mathbf{j} + c\mathbf{k}.$$

Vector Sum (V) of Any Number of Vectors, $V_1, V_2, V_3, \cdots$. (329)

$$\mathbf{V} = \mathbf{V}_1 + \mathbf{V}_2 + \mathbf{V}_3 + \cdots = (a_1 + a_2 + a_3 + \cdots)\mathbf{i}$$
$$+ (b_1 + b_2 + b_3 + \cdots)\mathbf{j} + (c_1 + c_2 + c_3 + \cdots)\mathbf{k}.$$

Product of a Vector (V) by a Scalar (s) (330)

$$s\mathbf{V} = (sa)\mathbf{i} + (sb)\mathbf{j} + (sc)\mathbf{k}.$$
$$(s_1 + s_2)\mathbf{V} = s_1\mathbf{V} + s_2\mathbf{V}; \quad (\mathbf{V}_1 + \mathbf{V}_2)s = \mathbf{V}_1 s + \mathbf{V}_2 s.$$

NOTE. $s\mathbf{V}$ has the same direction as **V** and its magnitude is s times the magnitude of **V**.

Scalar Product of Two Vectors: $V_1 \cdot V_2$. (331)

$\mathbf{V}_1 \cdot \mathbf{V}_2 = |\mathbf{V}_1||\mathbf{V}_2| \cos \phi$, where ϕ is the angle between $\mathbf{V}_1$ and $\mathbf{V}_2$.
$\mathbf{V}_1 \cdot \mathbf{V}_2 = \mathbf{V}_2 \cdot \mathbf{V}_1; \quad \mathbf{V}_1 \cdot \mathbf{V}_1 = |\mathbf{V}_1|^2.$
$(\mathbf{V}_1 + \mathbf{V}_2) \cdot \mathbf{V}_3 = \mathbf{V}_1 \cdot \mathbf{V}_3 + \mathbf{V}_2 \cdot \mathbf{V}_3;$

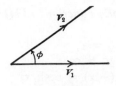

Fig. 1.65

$$(\mathbf{V}_1 + \mathbf{V}_2) \cdot (\mathbf{V}_3 + \mathbf{V}_4) = \mathbf{V}_1 \cdot \mathbf{V}_3 + \mathbf{V}_1 \cdot \mathbf{V}_4 + \mathbf{V}_2 \cdot \mathbf{V}_3 + \mathbf{V}_2 \cdot \mathbf{V}_4.$$
$$\mathbf{i} \cdot \mathbf{i} = \mathbf{j} \cdot \mathbf{j} = \mathbf{k} \cdot \mathbf{k} = 1; \quad \mathbf{i} \cdot \mathbf{j} = \mathbf{j} \cdot \mathbf{k} = \mathbf{k} \cdot \mathbf{i} = 0.$$

In plane: $\mathbf{V}_1 \cdot \mathbf{V}_2 = a_1 a_2 + b_1 b_2$; in space: $\mathbf{V}_1 \cdot \mathbf{V}_2 = a_1 a_2 + b_1 b_2 + c_1 c_2$.

NOTE. The scalar product of two vectors $\mathbf{V}_1 \cdot \mathbf{V}_2$ is a scalar quantity and may physically be represented by the work done by a constant force of magnitude $|\mathbf{V}_1|$ on a unit particle moving through a distance $|\mathbf{V}_2|$, where ϕ is the angle between the line of force and the direction of motion.

1.11 HYPERBOLIC FUNCTIONS

Definitions of Hyperbolic Functions [See Table 10.3(b)] (332)

Hyperbolic sine (sinh) $x = \dfrac{1}{2}(e^x - e^{-x})$; $\operatorname{csch} x = \dfrac{1}{\sinh x}$

Hyperbolic cosine (cosh) $x = \dfrac{1}{2}(e^x + e^{-x})$; $\operatorname{sech} x = \dfrac{1}{\cosh x}$

Hyperbolic tangent (tanh) $x = \dfrac{e^x - e^{-x}}{e^x + e^{-x}}$; $\coth x = \dfrac{1}{\tanh x}$

where e = base of natural logarithms.

NOTE. The circular or ordinary trigonometric functions were defined with reference to a circle; in a similar manner, the hyperbolic functions may be defined with reference to a hyperbola. In the above definitions the hyperbolic functions are abbreviations for certain exponential functions.

Graphs of Hyperbolic Functions (333)

Figure 1.66

(a) $y = \sinh x$;
(b) $y = \cosh x$;
(c) $y = \tanh x$.

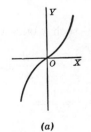

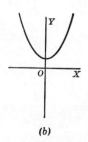

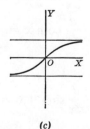

(a)   (b)   (c)

Fig. 1.66

Some Relations among Hyperbolic Functions (334)

$\sinh 0 = 0$, $\cosh 0 = 1$, $\tanh 0 = 0$.
$\sinh \infty = \infty$, $\cosh \infty = \infty$, $\tanh \infty = 1$.
$\sinh (-x) = -\sinh x$, $\cosh (-x) = \cosh x$, $\tanh (-x) = -\tanh x$.
$\cosh^2 x - \sinh^2 x = 1$, $\text{sech}^2 x + \tanh^2 x = 1$, $\text{csch}^2 x - \coth^2 x = -1$.
$\sinh 2x = 2 \sinh x \cosh x$, $\cosh 2x = \cosh^2 x + \sinh^2 x$.

$$2 \sinh^2 \frac{x}{2} = \cosh x - 1, \qquad 2 \cosh^2 \frac{x}{2} = \cosh x + 1.$$

$\sinh (x \pm y) = \sinh x \cosh y \pm \cosh x \sinh y$.
$\cosh (x \pm y) = \cosh x \cosh y \pm \sinh x \sinh y$.

$$\tanh (x \pm y) = \frac{\tanh x \pm \tanh y}{1 \pm \tanh x \tanh y}.$$

Hyperbolic Functions of Pure Imaginary and Complex Quantities (335)

$\sinh jy = j \sin y$; $\cosh jy = \cos y$; $\tanh jy = j \tan y$.
$\sinh (x + jy) = \sinh x \cos y + j \cosh x \sin y$.
$\cosh (x + jy) = \cosh x \cos y + j \sinh x \sin y$.
$\sinh (x + 2j\pi) = \sinh x$; $\cosh (x + 2j\pi) = \cosh x$.
$\sinh (x + j\pi) = -\sinh x$; $\cosh (x + j\pi) = -\cosh x$.
$\sinh (x + \frac{1}{2}j\pi) = j \cosh x$; $\cosh (x + \frac{1}{2}j\pi) = j \sinh x$.

Inverse or Anti-hyperbolic Functions (336)

If $x = \sinh y$, then y is the anti-hyperbolic sine of x or $y = \sinh^{-1} x$.

$$\sinh^{-1} x = \ln (x + \sqrt{x^2 + 1}); \qquad \text{csch}^{-1} x = \sinh^{-1} \frac{1}{x}.$$

$$\cosh^{-1} x = \ln (x + \sqrt{x^2 - 1}); \qquad \text{sech}^{-1} x = \cosh^{-1} \frac{1}{x}.$$

$$\tanh^{-1} x = \frac{1}{2} \ln \frac{1 + x}{1 - x}; \qquad \coth^{-1} x = \tanh^{-1} \frac{1}{x}.$$

Derivatives of Hyperbolic Functions (337)

$$\frac{d}{dx} \sinh x = \cosh x; \qquad \frac{d}{dx} \cosh x = \sinh x; \qquad \frac{d}{dx} \tanh x = \text{sech}^2 x.$$

$$\frac{d}{dx} \coth x = -\text{csch}^2 x; \qquad \frac{d}{dx} \text{sech} x = -\text{sech} x \tanh x;$$

$$\frac{d}{dx} \text{csch} x = -\text{csch} x \coth x.$$

$$\frac{d}{dx} \sinh^{-1} x = \frac{1}{\sqrt{x^2 + 1}}; \quad \frac{d}{dx} \cosh^{-1} x = \frac{1}{\sqrt{x^2 - 1}};$$

$$\frac{d}{dx} \tanh^{-1} x = \frac{1}{1 - x^2}.$$

$$\frac{d}{dx}\coth^{-1}x = -\frac{1}{x^2-1}; \frac{d}{dx}\operatorname{sech}^{-1}x = -\frac{1}{x\sqrt{1-x^2}};$$

$$\frac{d}{dx}\operatorname{csch}^{-1}x = -\frac{1}{x\sqrt{x^2+1}}.$$

Some Integrals Leading to Hyperbolic Functions (338)

$$\int \sinh x\,dx = \cosh x; \quad \int \cosh x\,dx = \sinh x; \quad \int \tanh x\,dx = \ln\cosh x.$$

$$\int \coth x\,dx = \ln\sinh x; \quad \int \operatorname{sech} x\,dx = \sin^{-1}(\tanh x);$$

$$\int \operatorname{csch} x\,dx = \ln\tanh\frac{x}{2}.$$

$$\int \frac{dx}{\sqrt{x^2+a^2}} = \sinh^{-1}\frac{x}{a}; \quad \int \frac{dx}{\sqrt{x^2-a^2}} = \cosh^{-1}\frac{x}{a};$$

$$\int \frac{dx}{a^2-x^2} = \frac{1}{a}\tanh^{-1}\frac{x}{a}.\,(x<a)$$

$$\int \frac{dx}{x\sqrt{a^2+x^2}} = -\frac{1}{a}\sinh^{-1}\frac{a}{x}; \quad \int \frac{dx}{x\sqrt{a^2-x^2}} = -\frac{1}{a}\cosh^{-1}\frac{a}{x};$$

$$\int \frac{dx}{x^2-a^2} = -\frac{1}{a}\tanh^{-1}\frac{a}{x}.\,(x>a)$$

$$\int \sqrt{x^2-a^2}\,dx = \frac{x}{2}\sqrt{x^2-a^2} - \frac{a^2}{2}\cosh^{-1}\frac{x}{a}.$$

$$\int \sqrt{x^2+a^2}\,dx = \frac{x}{2}\sqrt{x^2+a^2} + \frac{a^2}{2}\sinh^{-1}\frac{x}{a}.$$

Expansions of Hyperbolic Functions into Series (339)

$$\sinh x = x + \frac{x^3}{3!} + \frac{x^5}{5!} + \cdots.$$

$$\cosh x = 1 + \frac{x^2}{2!} + \frac{x^4}{4!} + \cdots.$$

$$\tanh x = x - \frac{x^3}{3} + \frac{2x^5}{15} + \frac{17x^7}{315} + \cdots.$$

$$\sinh^{-1} x = x - \frac{1}{2}\frac{x^3}{3} + \frac{1\cdot3}{2\cdot4}\frac{x^5}{5} - \frac{1\cdot3\cdot5}{2\cdot4\cdot6}\frac{x^7}{7} + \cdots.\,(x<1)$$

$$\sinh^{-1} x = \ln 2x + \frac{1}{2}\frac{1}{2x^2} - \frac{1\cdot3}{2\cdot4}\frac{1}{4x^4} + \frac{1\cdot3\cdot5}{2\cdot4\cdot6}\frac{1}{6x^6} - \cdots.\,(x>1)$$

$$\cosh^{-1} x = \ln 2x - \frac{1}{2}\frac{1}{2x^2} - \frac{1\cdot3}{2\cdot4}\frac{1}{4x^4} - \frac{1\cdot3\cdot5}{2\cdot4\cdot6}\frac{1}{6x^6} - \cdots.$$

$$\tanh^{-1} x = x + \frac{x^3}{3} + \frac{x^5}{5} + \frac{x^7}{7} + \cdots.$$

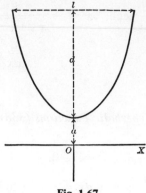

Fig. 1.67

The Catenary (340) (Fig. 1.67)

For definition, see (36).

Equation.

$$y = \frac{a}{2}(e^{x/a} + e^{-x/a}) = a \cosh \frac{x}{a}.$$

If the width of the span is l and the sag is d, then the length of the arc (s) is found by means of the equations:

$$\cosh z = \frac{2d}{l}z + 1, \qquad s = \frac{l}{z}\sinh z,$$

where z is to be found approximately by means of Table 10.3(b), from the first of these equations and this value substituted in the second.

If s and l are known, d may be found similarly by means of

$$\sinh z = \frac{s}{l}z, \qquad d = \frac{l}{2z}(\cosh z - 1).$$

1.12   PROGRESSIONS

An **arithmetic progression** is a sequence of terms each of which differs from the preceding by the same number d, called the common difference. If n = number of terms, a = first term, l = last term, s = sum of n terms, then $l = a + (n - 1)d$, and $s = \frac{n}{2}(a + 1)$. The arithmetic mean of two numbers is the number which placed between them would make them an arithmetic progression. Thus, the arithmetic mean of m, n is $(m + n)/2$.

$a, a + d, a + 2d, a + 3d, \ldots$, where d = common difference. (341)
The nth term, $t_n = a + (n - 1)d$.

The sum of n terms, $S_n = \dfrac{n}{2}[2a + (n-1)d] = \dfrac{n}{2}(a + t_n)$.

The arithmetic mean of a and $b = \dfrac{a+b}{2}$.

A **geometric progression** is a sequence of terms each of which is obtained from the preceding by multiplying it by a fixed number r, called the ratio. If n = number of terms, a = first term, l = last term, s = sum of n terms, then $l = ar^{n-1}$, $s = (rl - a)/(r - 1) = a(1 - r^n)/(1 - r)$. The geometric mean of two numbers is the number which, placed between them, <u>would</u> make them a geometric progression. Thus, the geometric mean of m, n is $\sqrt{m \cdot n}$.

$a, ar, ar^2, ar^3, \ldots$, where r = common ratio. (342)
The nth term, $t_n = ar^{n-1}$.

The sum of n terms, $S_n = a\left(\dfrac{r^n - 1}{r - 1}\right) = \dfrac{rt_n - a}{r - 1}$.

If $r^2 < 1$, S_n approaches a definite limit as n increases indefinitely, and

$$S_\infty = \frac{a}{1 - r}.$$

The geometric mean of a and $b = \sqrt{ab}$.

A **harmonic progression** is a sequence of terms whose reciprocals form an arithmetic progression. The harmonic mean of two numbers is the number which, placed between them, would make with them a harmonic progression. Thus, the harmonic mean of m, n is $2mn/(m + n)$. (343)

The relation between **arithmetic, geometric,** and **harmonic means** of two numbers is expressed by the equality $G^2 = AH$, where G = geometric mean, A = arithmetic mean, and H = harmonic mean. (344)

1.13 SOME STANDARD SERIES

The following series are obtained through expansions of the functions by Taylor's or Maclaurin's theorems. The expression in brackets following each series gives the region of convergence of the series, that is, the values of x for which the remainder, R_n, approaches zero as n increases, so that a number of terms of the series may be used for an approximation of the function. If the region of convergence is not indicated, it is to be understood that the series converges for all finite values of x. ($n! = 1 \cdot 2 \cdot 3 \cdots n$.)

Binomial Series (345)

$$(a + x)^n = a^n + na^{n-1}x + \frac{n(n-1)}{2!}a^{n-2}x^2 + \frac{n(n-1)(n-2)}{3!}a^{n-3}x^3 + \cdots .$$

$$(x^2 < a^2)$$

NOTE.   The series consists of $(n + 1)$ terms when n is a positive integer; the number of terms is infinite when n is a negative or fractional number.

$$(a - bx)^{-1} = \frac{1}{a}\left(1 + \frac{bx}{a} + \frac{b^2x^2}{a^2} + \frac{b^3x^3}{a^3} + \cdots\right).$$

$$(b^2x^2 < a^2)$$

Exponential Series (346)

$$a^x = 1 + x \ln a + \frac{(x \ln a)^2}{2!} + \frac{(x \ln a)^3}{3!} + \cdots . \quad (347)$$

$$e^x = 1 + x + \frac{x^2}{2!} + \frac{x^3}{3!} + \cdots . \quad (348)$$

$$\frac{1}{2}(e^x + e^{-x}) = 1 + \frac{x^2}{2!} + \frac{x^4}{4!} + \cdots . \quad (349)$$

$$\frac{1}{2}(e^x - e^{-x}) = x + \frac{x^3}{3!} + \frac{x^5}{5!} + \cdots . \quad (350)$$

$$e^{-x^2} = 1 - x^2 + \frac{x^4}{2!} - \frac{x^6}{3!} + \frac{x^8}{4!} - \cdots . \quad (351)$$

Logarithmic Series (352)

$$\ln x = (x - 1) - \tfrac{1}{2}(x - 1)^2 + \tfrac{1}{3}(x - 1)^3 - \cdots . \quad (0 < x < 2) \ (353)$$

$$\ln x = \frac{x - 1}{x} + \frac{1}{2}\left(\frac{x - 1}{x}\right)^2 + \frac{1}{3}\left(\frac{x - 1}{x}\right)^3 + \cdots . \quad (x > \tfrac{1}{2}) \ (354)$$

$$\ln x = 2\left[\frac{x - 1}{x + 1} + \frac{1}{3}\left(\frac{x - 1}{x + 1}\right)^3 + \frac{1}{5}\left(\frac{x - 1}{x + 1}\right)^5 + \cdots \right]. \quad (x \text{ positive}) \ (355)$$

$$\ln (1 + x) = x - \frac{x^2}{2} + \frac{x^3}{3} - \frac{x^4}{4} + \cdots . \quad (356)$$

$$\ln (a + x) = \ln a + 2\left[\frac{x}{2a + x} + \frac{1}{3}\left(\frac{x}{2a + x}\right)^3 + \frac{1}{5}\left(\frac{x}{2a + x}\right)^5 + \cdots \right].$$

$$\binom{a \text{ positive}}{x \text{ between } -a \text{ and } +\infty} \ (357)$$

$$\ln \left(\frac{1 + x}{1 - x}\right) = 2\left(x + \frac{x^3}{3} + \frac{x^5}{5} + \frac{x^7}{7} + \cdots \right). \quad (x^2 < 1) \ (358)$$

$$\ln \left(\frac{x + 1}{x - 1}\right) = 2\left[\frac{1}{x} + \frac{1}{3}\left(\frac{1}{x^3}\right) + \frac{1}{5}\left(\frac{1}{x}\right)^5 + \frac{1}{7}\left(\frac{1}{x}\right)^7 + \cdots \right]. \quad (x^2 > 1) \ (359)$$

$$\ln \left(\frac{x + 1}{x}\right) = 2\left[\frac{1}{2x + 1} + \frac{1}{3(2x + 1)^3} + \frac{1}{5(2x + 1)^5} + \cdots \right].$$

$$(x \text{ positive}) \ (360)$$

$$\ln (x + \sqrt{1 + x^2}) = x - \frac{1}{2}\frac{x^3}{3} + \frac{1 \cdot 3}{2 \cdot 4}\frac{x^5}{5} - \frac{1 \cdot 3 \cdot 5}{2 \cdot 4 \cdot 6}\frac{x^7}{7} + \cdots .$$

$$(x^2 < 1) \ (361)$$

Trigonometric Series (362)

$$\sin x = x - \frac{x^3}{3!} + \frac{x^5}{5!} - \frac{x^7}{7!} + \cdots . \quad (363)$$

$$\cos x = 1 - \frac{x^2}{2!} + \frac{x^4}{4!} - \frac{x^6}{6!} + \cdots . \quad (364)$$

$$\tan x = x + \frac{x^3}{3} + \frac{2x^5}{15} + \frac{17x^7}{315} + \frac{62x^9}{2835} + \cdots . \left(x^2 < \frac{\pi^2}{4} \right) (365)$$

$$\sin^{-1} x = x + \frac{1}{2} \frac{x^3}{3} + \frac{1 \cdot 3}{2 \cdot 4} \frac{x^5}{5} + \frac{1 \cdot 3 \cdot 5}{2 \cdot 4 \cdot 6} \frac{x^7}{7} + \cdots . \quad (x^2 < 1) \ (366)$$

$$\tan^{-1} x = x - \frac{1}{3} x^3 + \frac{1}{5} x^5 - \frac{1}{7} x^7 + \cdots . \ (x^2 \leqq 1) \ (367)$$

Other Series

1. $1 + 2 + 3 + 4 + \cdots + (n - 1) + n = n(n + 1)/2.$ (368)
2. $p + (p + 1) + (p + 2) + \cdots + (q - 1) + q = (q + p)(q - p + 1)/2.$
   (369)
3. $2 + 4 + 6 + 8 + \cdots + (2n - 2) + 2n = n(n + 1).$ (370)
4. $1 + 3 + 5 + 7 + \cdots + (2n - 3) + (2n - 1) = n^2.$ (371)
5. $1^2 + 2^2 + 3^2 + 4^2 + \cdots + (n - 1)^2 + n^2 = n(n + 1)(2n + 1)/6.$ (372)
6. $1^3 + 2^3 + 3^3 + 4^3 + \cdots + (n - 1)^3 + n^3 = n^2(n + 1)^2/4.$ (373)
7. $\dfrac{1 + 2 + 3 + 4 + 5 \cdots + n}{n^2} \rightarrow \dfrac{1}{2}$
8. $\dfrac{1 + 2^2 + 3^2 + 4^2 + \cdots + n^2}{n^3} \rightarrow \dfrac{1}{3}$ as $n \rightarrow \infty.$ (374)
9. $\dfrac{1 + 2^3 + 3^3 + 4^3 + \cdots + n^3}{n^4} \rightarrow \dfrac{1}{4}$

1.14 APPROXIMATIONS OF EXPRESSIONS CONTAINING SMALL TERMS

These may be derived from various infinite series given in Section 1.13. Some first approximations derived by neglecting all powers but the first of the small positive or negative quantity $x = s$ are given below. The expression in brackets gives the next term beyond that which is used and by means of it the accuracy of the approximation may be estimated.

$$\frac{1}{1 + s} = 1 - s. \qquad\qquad [+s^2] \ (375)$$

$$(1 + s)^n = 1 + ns. \qquad\qquad \left[+ \frac{n(n - 1)}{2} s^2 \right] (376)$$

$$e^s = 1 + s. \qquad\qquad \left[+ \frac{s^2}{2} \right] (377)$$

$$\ln (1 + s) = s. \qquad \left[-\frac{s^2}{2} \right] \text{(378)}$$

$$\sin s = s. \qquad \left[-\frac{s^3}{6} \right] \text{(379)}$$

$$\cos s = 1. \qquad \left[-\frac{s^2}{2} \right] \text{(380)}$$

$$(1 + s_1)(1 + s_2) = (1 + s_1 + s_2). \quad [+s_1 s_2] \text{ (381)}$$

The following expressions are some that may be approximated by $1 + s$, where s is a small positive or negative quantity and n is any number.

$$\left(1 + \frac{s}{n} \right)^n . \text{(382)} \quad e^s. \text{(385)} \quad 1 + \ln \sqrt{\frac{1 + s}{1 - s}}. \text{(388)}$$

$$\sqrt[n]{1 + ns}. \text{(383)} \quad 2 - e^{-s}. \text{(386)} \quad 1 + n \sin \frac{s}{n}. \text{(389)}$$

$$\sqrt[n]{\frac{1 + \dfrac{ns}{2}}{1 - \dfrac{ns}{2}}}. \text{(384)} \quad 1 + n \ln \left(1 + \frac{s}{n} \right). \text{(387)} \quad \cos \sqrt{-2s}. \text{(390)}$$

1.15   INDETERMINATE FORMS (SEE SECTION 1.2)

Let $f(x)$ and $F(x)$ be two functions of x, and let a be a value of x.

If $\dfrac{f(a)}{F(a)} = \dfrac{0}{0}$ or $\dfrac{\infty}{\infty}$, use $\dfrac{f'(a)}{F'(a)}$ for the value of this fraction. (391)

If $\dfrac{f'(a)}{F'(a)} = \dfrac{0}{0}$ or $\dfrac{\infty}{\infty}$, use $\dfrac{f''(a)}{F''(a)}$ for the value of this fraction, and so on. (392)

If $f(a) \cdot F(a) = 0 \cdot \infty$ or if $f(a) - F(a) = \infty - \infty$, evaluate by changing the product or difference to the form $0/0$ or ∞/∞ and use (391) or (392). (393)

If $f(a)^{F(a)} = 0^0$ or ∞^0 or 1^∞, then $f(a)^{F(a)} = e^{F(a) \cdot \ln f(a)}$, and the exponent, being of the form $0 \cdot \infty$, may be evaluated by (2). (394)

SECTION 2
MECHANICS/MATERIALS

2.1 KINEMATICS

2.1.1 Linear Motion

Velocity (v) of a particle which moves uniformly s feet in t seconds.

$$v = \frac{s}{t} \text{ feet/second.} \tag{2.1}$$

NOTE.   The velocity v of a moving particle at any instant equals ds/dt. The speed of a moving particle equals the magnitude of its velocity but has no direction.

Acceleration (a) of a particle whose velocity increases uniformly v feet per second in t seconds.

$$a = \frac{v}{t} \text{ feet/second}^2 \tag{2.2}$$

NOTE.   The acceleration a of a moving particle at any instant equals dv/dt or d^2s/dt^2. The acceleration g of a falling body *in vacuo* at sea level and latitude 45° equals 32.17 feet (9.805 m)/second^2.

Velocity (v_t) at the end of t seconds acquired by a particle having an initial velocity of v_0 feet per second and a uniform acceleration of a feet/second^2.

$$v_t = v_0 + at \text{ feet/second.} \tag{2.3}$$

NOTE. a is negative if the initial velocity and the acceleration act in opposite directions.

Distance (s) traversed in t seconds by a particle having an initial velocity of v_0 feet/second and a uniform acceleration of a feet/second^2.

$$s = v_0 t + \tfrac{1}{2}at^2 \text{ feet.} \tag{2.4}$$

Distance (s) required for a particle with an initial velocity of v_0 feet/second and a uniform acceleration of a feet/second^2 to reach a velocity of v_t feet/second.

$$s = \frac{v_t^2 - v_0^2}{2a} \text{ feet.} \tag{2.5}$$

Velocity (v_t) acquired, in traveling s feet, by a particle having an initial velocity of v_0 feet/second and a uniform acceleration of a feet/second^2.

$$v_t = \sqrt{v_0^2 + 2as} \text{ feet/second.} \tag{2.6}$$

Time (t) required for a particle having an initial velocity of v_0 feet/second and a uniform acceleration of a feet/second^2 to travel s feet.

$$t = \frac{-v_0 + \sqrt{v_0^2 + 2as}}{a} \text{ seconds.} \tag{2.7}$$

Uniform acceleration (a) required to move a particle, with an initial velocity of v_0 feet/second, s feet in t seconds.

$$a = \frac{2(s - v_0 t)}{t^2} \text{ feet/second}^2. \tag{2.8}$$

2.1.2   Circular Motion

Angular velocity (ω) of a particle moving uniformly through θ radians in t seconds.

$$\omega = \frac{\theta}{t} \text{ radians/second.} \tag{2.9}$$

NOTE.   The angular velocity (ω) of a moving particle at any instant equals $d\theta/dt^2$.

Normal acceleration (a) toward the center of its path of a particle moving uniformly with v feet/second tangential velocity and r feet radius of curvature of path.

$$a = \frac{v^2}{r} \text{ feet/second}^2. \tag{2.10}$$

NOTE.   The tangential acceleration of a particle moving with constant speed in a circular path is zero.

Angular acceleration (α) of a particle whose angular velocity increases uniformly ω radians/second in t seconds.

$$\alpha = \frac{\omega}{t} \text{ radians/second}^2. \tag{2.11}$$

NOTE.   The angular acceleration α of a moving particle at any instant equals $d\omega/dt$ or $d^2\theta/dt^2$.

Angular velocity (ω_t) at the end of t seconds acquired by a particle having an initial angular velocity of ω_0 radians/second and a uniform angular acceleration of α radians/second^2.

$$\omega_t = \omega_0 + \alpha t \text{ radians/second.} \tag{2.12}$$

Angle (θ) subtended in t seconds by a particle having an initial angular velocity of ω_0 radians/second and a uniform angular acceleration of α radians/second^2.

$$\theta = \omega_0 t + \tfrac{1}{2}\alpha t^2 \text{ radians.} \tag{2.13}$$

Angle (θ) subtended by a particle with an initial angular velocity of ω_0 radians/second and a uniform angular acceleration of α radians/second^2 in acquiring an angular velocity of ω_t radians/second.

$$\theta = \frac{\omega_t^2 - \omega_0^2}{2\alpha} \text{ radians.} \tag{2.14}$$

Angular velocity (ω_t) acquired in subtending θ radians by a particle having an initial angular velocity of ω_0 radians/second and a uniform angular acceleration of α radians/second^2.

$$\omega_t = \sqrt{\omega_0^2 + 2\alpha\theta} \text{ radians/second.} \tag{2.15}$$

Time (t) required for a particle having an initial angular velocity of ω_0 radians/second and a uniform angular acceleration of α radians/second^2 to subtend θ radians.

$$t = \frac{-\omega_0 + \sqrt{\omega_0^2 + 2\alpha\theta}}{\alpha} \text{ seconds.} \tag{2.16}$$

Uniform angular acceleration (α) required for a particle with an initial angular velocity of ω_0 radians/second to subtend θ radians in t seconds.

$$\alpha = \frac{2(\theta - \omega_0 t)}{t^2} \text{ radians/second}^2. \tag{2.17}$$

Velocity (v) of a particle r feet from the axis of rotation in a body making n revolutions/second.

$$v = 2\pi r n \text{ feet/second.} \tag{2.18}$$

Velocity (v) of a particle r feet from the axis of rotation in a body rotating with an angular velocity of ω radians/second.

$$v = \omega r \text{ feet/second.} \tag{2.19}$$

Angular velocity (ω) of a body making n revolutions/second.

$$\omega = 2\pi n \text{ radians/second.} \tag{2.20}$$

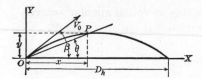

Fig. 2.1 Path of a projectile.

2.1.3 Path of a Projectile*

Horizontal component of velocity (v_x) of a particle having an initial velocity of v_0 feet/second in a direction making an angle of β degrees with the horizontal.*

$$v_x = v_0 \cos \beta \text{ feet/second.} \tag{2.21}$$

Horizontal distance (x) traveled in t seconds by a particle having an initial velocity of v_0 feet/second at β degrees with the horizontal and a uniform downward acceleration of a feet/second^2.

$$x = v_0 t \cos \beta \text{ feet.} \tag{2.22}$$

Vertical component of velocity (v_y) at the end of t seconds of a particle having an initial velocity of v_0 feet/second at β degrees with the horizontal and a uniform downward acceleration of a feet/second^2.

$$v_y = v_0 \sin \beta - at \text{ feet/second.} \tag{2.23}$$

Vertical distance (y) traveled in t seconds by a particle having an initial velocity of v_0 feet/second at β degrees with the horizontal and a uniform downward acceleration of a feet/second^2.

$$y = v_0 t \sin \beta - \tfrac{1}{2} a t^2 \text{ feet.} \tag{2.24}$$

Time (t_v) to reach the highest point of the path of a particle having an initial velocity of v_0 feet/second at β degrees with the horizontal and a uniform downward acceleration of a feet/second^2.

$$t_v = \frac{v_0 \sin \beta}{a} \text{ seconds.} \tag{2.25}$$

Vertical distance (d_v) from the horizontal to the highest point of the path of a particle having an initial velocity of v_0 feet/second at β degrees with the horizontal and a uniform downward acceleration of a feet/second^2.

$$d_v = \frac{v_0^2 \sin^2 \beta}{2a} \text{ feet.} \tag{2.26}$$

*Friction of the air is neglected throughout.

Velocity (v) at the end of t seconds of a particle having an initial velocity of v_0 feet/second at β degrees with the horizontal and a uniform downward acceleration of a feet/second^2.

$$v = \sqrt{v_x^2 + v_y^2} = \sqrt{v_0^2 - 2v_0\, at \sin \beta + a^2 t^2} \text{ feet/second.} \qquad (2.27)$$

Time (t_h) to reach the same horizontal as at start for a particle having an initial velocity of v_0 feet/second at β degrees with the horizontal and a uniform downward acceleration of a feet/second^2.

$$t_h = \frac{2v_0 \sin \beta}{a} \text{ seconds.} \qquad (2.28)$$

Horizontal distance (d_h) traveled by a particle having an initial velocity of v_0 feet/second at β degrees with the horizontal and a uniform downward acceleration of a feet/second^2 in returning to the same horizontal as at start.

$$d_h = \frac{v_0^2 \sin 2\beta}{a} \text{ feet.} \qquad (2.29)$$

Time (t) to reach any point P for a particle having an initial velocity of v_0 feet/second at β degrees with the horizontal and a uniform downward acceleration of a feet/second^2, if a line through P and the point of starting makes θ degrees with the horizontal.

$$t = \frac{2v_0 \sin (\beta - \theta)}{a \cos \theta} \text{ seconds.} \qquad (2.30)$$

2.1.4 Harmonic Motion

Simple harmonic motion is the motion of the projection, on the diameter of a circle, of a particle moving with constant speed around the circumference of the circle. **Amplitude** is one-half the projection of the path of the particle or equal to the radius of the circle. **Frequency** is the number of complete oscillations per unit time.

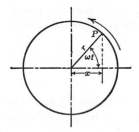

Fig. 2.2 Harmonic motion.

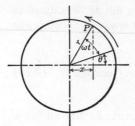

Fig. 2.3 Acceleration.

Displacement (x) from the center t seconds after starting, of the projection on the diameter, of a particle moving with a uniform angular velocity of ω radians/second about a circle r feet in radius.

$$x = r \cos \omega t \text{ feet.} \tag{2.31}$$

Velocity (v), t seconds after starting, of the projection on the diameter, of a particle moving with a uniform angular velocity of ω radians/second about a circle r feet in radius.

$$v = -\omega r \sin \omega t \text{ feet/second.} \tag{2.32}$$

Acceleration (a), t seconds after starting, of the projection on the diameter, of a particle moving with a uniform angular velocity of ω radians/second about a circle r feet in radius.

$$a = -\omega^2 r \cos \omega t = -\omega^2 x \text{ feet/second}^2. \tag{2.33}$$

NOTE.   If the time t is reckoned from a position displaced by θ radians from the horizontal (called lead if positive and lag if negative) the formulas become: $x = r \cos (\omega t + \theta)$ feet, $v = -\omega r \sin (\omega t + \theta)$ feet/second, and $a = -\omega^2 r \cos (\omega t + \theta)$ feet/second^2.

2.2   RELATIONS OF MASS AND SPACE

2.2.1   Mass

Mass (m) of a body weighing w pounds.

$$m = \frac{w}{g} \text{ pounds (mass).} \tag{2.34}$$

NOTE.   The mass m of a body may be measured by its weight w, designated "pounds (abs.)," or by its weight w divided by the acceleration due to gravity g, designated "pounds (mass)." In this text the latter unit is used throughout.

2.2.2   Center of Gravity

Center of gravity of a body or system of bodies is that point through which the resultant of the weights of the component particles passes, whatever position be given the body or system.

NOTE. The center of mass of a body is the same as the center of gravity. The center of gravity of a line, surface, or volume is obtained by considering it to be the center of gravity of a slender rod, thin plate, or homogeneous body and is often called the centroid.

Moment (M) of a body of weight w, or of mass m, about a plane if x is the horizontal distance from the center of gravity of the body to the plane.

$$M = wx \text{ or } M = mx. \tag{2.35}$$

Moment (S) of an area A, about an axis X if x is the horizontal distance from the center of gravity of the area to the axis.

$$S = Ax. \tag{2.36}$$

NOTE. The moment of an area about an axis through its center of gravity is zero.

Distances (x_0, y_0, z_0) from each of three coordinate planes (X, Y, Z) to the center of gravity or mass of a system of bodies, if Σw is the sum of their weights or Σm is the sum of their masses and Σwx, Σwy, Σwz, or Σmx, Σmy, Σmz are the algebraic sums of moments of the separate bodies about the X, Y, and Z planes.

$$x_0 = \frac{\Sigma wx}{\Sigma w} = \frac{\Sigma mx}{\Sigma m}.$$

$$y_0 = \frac{\Sigma wy}{\Sigma w} = \frac{\Sigma my}{\Sigma m}. \tag{2.37}$$

$$z_0 = \frac{\Sigma wz}{\Sigma w} = \frac{\Sigma mz}{\Sigma m}.$$

Distances (x_0, y_0, z_0) from each of three coordinate planes to the center of gravity of a volume, if Σv is the sum of the component volumes and Σvx, Σvy, and Σvz are the algebraic sums of the moments of these component volumes about the X, Y, and Z planes.

$$x_0 = \frac{\Sigma vx}{\Sigma v}. \qquad y_0 = \frac{\Sigma vy}{\Sigma v}. \qquad z_0 = \frac{\Sigma vz}{\Sigma v}. \tag{2.38}$$

Distances (x_0, y_0) from each of two coordinate axes to the center of gravity of an area, if ΣA is the sum of the component areas and ΣAx and ΣAy are the algebraic sums of the moments of these component areas about the X and Y axes.

$$x_0 = \frac{\Sigma Ax}{\Sigma A}. \qquad y_0 = \frac{\Sigma Ay}{\Sigma A}. \tag{2.39}$$

NOTE. The general method of finding the center of gravity of an irregular area is to divide it into component areas, the centers of gravity of which may be calculated or determined from Table 2.1, then find the sum of statical moments of the component areas about some convenient axis, and divide by the total area to obtain the distance from that axis to the center of gravity of the whole area. In numerical problems it is often convenient to take the axis of reference through the center of gravity of one of the component areas, thereby eliminating the moment of that area and simplifying the numerical work.

2.2.3 Moment of Inertia

Moment of inertia (J) of an area about an axis is the sum of the products of the component areas into the square of their distances from the axis (ΣAx^2).

$$J = \Sigma Ax^2. \tag{2.40}$$

NOTE.   In general an expression for moment of inertia involves the use of calculus, the area being considered as divided into differential areas dA. $J_x = \int x^2 \, dA$ and $J_y = \int y^2 \, dA$. The unit of moment of inertia of an area is inches, feet, and so on, to the fourth power.

Moment of inertia (J_x) of an area A about any axis in terms of the moment of inertia J_0 about a parallel axis through the center of gravity of the area, if x_0 is the distance between the two axes.

$$J_x = J_0 + Ax_0^2. \tag{2.41}$$

Radius of gyration (K) of an area A from an axis about which the moment of inertia is J.

$$K = \sqrt{\frac{J}{A}}. \tag{2.42}$$

Radius of gyration (K_x) of an area A about any axis in terms of the radius of gyration K_0 about a parallel axis through the center of gravity of the area, if x_0 is the distance between the two axes.

$$K_x^2 = K_0^2 + x_0^2. \tag{2.43}$$

Product of inertia (U) of an area with respect to two rectangular coordinate axes is the sum of the products of the component areas into the product of their distances from the two axes (ΣAxy).

$$U = \Sigma Axy. \tag{2.44}$$

NOTE.   Product of inertia, like moment of inertia, is generally expressed by use of calculus:

$$U = \int xy \, dA. \tag{2.45}$$

In case one of the areas is an axis of symmetry the product of inertia is zero.

Product of inertia (U_{xy}) of an area A about any two rectangular axes in terms of the product of inertia U_0 about two parallel rectangular axes through the center of gravity of the area, if x_0 and y_0 are the distances between these two sets of axes.

$$U_{xy} = U_0 + Ax_0y_0. \tag{2.46}$$

Moment of inertia ($J_{x'}$ and $J_{y'}$) and **product of inertia** ($U_{x'y'}$) of an area A about each of two rectangular coordinate axes (X' and Y') in terms of the moments and product of inertia (J_x, J_y, U_{xy}) about two other rectangular coordinate axes making an angle α with X' and Y'.

$$J_{x'} = J_y \sin^2 \alpha + J_x \cos^2 \alpha - 2U_{xy} \cos \alpha \sin \alpha. \tag{2.47}$$
$$J_{y'} = J_y \cos^2 \alpha + J_x \sin^2 \alpha + 2U_{xy} \cos \alpha \sin \alpha. \tag{2.48}$$
$$U_{x'y'} = (J_x - J_y) \cos \alpha \sin \alpha + U_{xy}(\cos^2 \alpha - \sin^2 \alpha). \tag{2.49}$$

Principal axes of an area are those axes, through any point, about one of which the moment of inertia is a maximum, the moment of inertia about the other being a minimum. The axes are at right angles to each other.

Angle (α) between the rectangular coordinate axes X and Y, about which the moments and products of inertia are J_x, J_y, and U_{xy}, and the principal axes through the point of intersection of X and Y.

$$\tan 2\alpha = \frac{2U_{xy}}{J_y - J_x}. \tag{2.50}$$

NOTE. An axis of symmetry is a principal axis. The product of inertia about principal axes is zero. If J_y and J_x are moments of inertia about principal axes the equations for the moments of inertia about rectangular axes making an angle α with these principal axes are: $J_{x'} = J_y \sin^2 \alpha + J_x \cos^2 \alpha$ and $J_{y'} = J_y \cos^2 \alpha + J_x \sin^2 \alpha$. The sum of the moments of inertia about rectangular coordinate axes is a constant for all pairs of axes intersecting at the same point, that is, $J_x + J_y = J_{x'} + J_{y'}$.

Polar moment of inertia (J_p) of an area is the moment of inertia about an axis perpendicular to the plane of the area and is equal to the sum of the products of the component areas into the squares of their distances from the axis (ΣAr^2).

$$J_p = \Sigma Ar^2. \tag{2.51}$$

NOTE. Polar moment of inertia is generally expressed by use of calculus: $J_p = \int r^2 \, dA$.

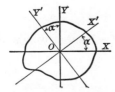

Fig. 2.4 Moment of inertia and product of inertia.

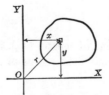

Fig. 2.5 Polar moment of inertia.

Polar moment of inertia (J_p) of an area A in terms of the moments of inertia J_x and J_y about two rectangular coordinate axes intersecting on the polar axis.

$$J_p = J_x + J_y. \tag{2.52}$$

Moment of inertia (J_m) of a body about an axis, in terms of the mass, is the sum of the products of the component masses and the squares of their distances from the axis (Σmr^2).

$$J_m = \Sigma mr^2. \tag{2.53}$$

Moment of inertia (J) of a body about an axis, in terms of the weight, is the sum of the products of the component weights and the squares of their distances from the axis (Σwr^2).

$$J = \Sigma wr^2. \tag{2.54}$$

Moment of inertia (J_m) in terms of the mass for a case where the moment of inertia in terms of the weight is J.

$$J_m = \frac{J}{g}. \tag{2.55}$$

NOTE.   The moment of inertia of a body is generally expressed by calculus. $J_m = \int r^2 \, dm$. $J = \int r^2 \, dw$. The unit of moment of inertia of solid is pound-feet^2.

Moment of inertia (J_x) of a body of weight W about any axis in terms of the moment of inertia (J_0) about a parallel axis through the center of gravity of the body, if x_0 is the distance between the axes.

$$J_x = J_0 + Wx_0^2. \tag{2.56}$$

Radius of gyration (K) of a body of weight W from an axis about which the moment of inertia is J.

$$K = \sqrt{\frac{J}{W}}. \tag{2.57}$$

Moment of inertia (J_m), in terms of the mass, of a body of weight W about an axis for which the radius of gyration is K.

$$J_m = \frac{W}{g} K^2. \tag{2.58}$$

Table 2.1 Properties of Various Plane Sections

Section	Distance to center of gravity, x	Moment of inertia, J*	Radius of gyration, K
Square	$x_a = x_b = \dfrac{b}{2}.$ $x_d = \dfrac{b}{\sqrt{2}}.$	$J_{AA} = J_{BB} = J_{DD} = \dfrac{b^4}{12}.$ $J_{CC} = \dfrac{b^4}{3}.$ $J_p = \dfrac{b^4}{6}.$	$K_{AA} = K_{BB} = K_{DD} = \dfrac{b}{\sqrt{12}}$ $= 0.289\,b.$ $K_{CC} = \dfrac{b}{\sqrt{3}} = 0.577\,b.$
Hollow Square	$x_a = x_b = \dfrac{b}{2}.$ $x_d = \dfrac{b}{\sqrt{2}}.$	$J_{AA} = J_{BB} = J_{DD} = \dfrac{b^4 - b_1^4}{12}.$ $J_{CC} = \dfrac{b^4}{3} - \dfrac{b_1^2\,(3\,b^2 + b_1^2)}{12}.$ $J_p = \dfrac{b^4 - b_1^4}{6}.$	$K_{AA} = K_{BB} = K_{DD} = \sqrt{\dfrac{b^2 + b_1^2}{12}}$ $= 0.289\,\sqrt{b^2 + b_1^2}.$

* J_p, polar moment of inertia, refers to an axis through the center of gravity.

Table 2.1 (*Continued*)

Section	Distance to center of gravity, x	Moment of inertia, J*	Radius of gyration, K
Rectangle	$x_a = \dfrac{h}{2}$ $x_b = \dfrac{b}{2}$ $x_d = \dfrac{bh}{\sqrt{b^2+h^2}}$	$J_{AA} = \dfrac{bh^3}{12}$ $J_{BB} = \dfrac{hb^3}{12}$ $J_{CC} = \dfrac{bh^3}{3}$ $J_{DD} = \dfrac{b^3h^3}{6(b^2+h^2)}$ $J_p = \dfrac{bh^3 + hb^3}{12}$	$K_{AA} = \dfrac{h}{\sqrt{12}} = 0.289\,h$ $K_{BB} = \dfrac{b}{\sqrt{12}} = 0.289\,b$ $K_{CC} = \dfrac{h}{\sqrt{3}} = 0.577\,h$ $K_{DD} = \dfrac{bh}{\sqrt{6(b^2+h^2)}}$
Rectangle	$x = \dfrac{b\sin\alpha + h\cos\alpha}{2}$	$J_{AA} = \dfrac{bh(b^2\sin^2\alpha + h^2\cos^2\alpha)}{12}$	$K_{AA} = \sqrt{\dfrac{b^2\sin^2\alpha + h^2\cos^2\alpha}{12}}$

* J_p polar moment of inertia, refers to an axis through the center of gravity.

Table 2.1 (*Continued*).

Section	Distance to center of gravity, x	Moment of inertia, J*	Radius of gyration, K
Hollow Rectangle	$x_a = \frac{h}{2}$ $x_b = \frac{b}{2}$	$J_{AA} = \frac{bh^3 - b_1 h_1^3}{12}$ $J_{BB} = \frac{hb^3 - h_1 b_1^3}{12}$ $J_{CC} = \frac{bh^3}{3} - \frac{b_1 h_1}{12}(3h^2 + h_1^2)$	$K_{AA} = \sqrt{\frac{bh^3 - b_1 h_1^3}{12(bh - b_1 h_1)}}$ $K_{BB} = \sqrt{\frac{hb^3 - h_1 b_1^3}{12(hb - h_1 b_1)}}$
Triangle	$x_a = \frac{2}{3}h$	$J_{AA} = \frac{bh^3}{36}$ $J_{BB} = \frac{bh^3}{12}$	$K_{AA} = \frac{h}{\sqrt{18}} = 0.236\,h$ $K_{BB} = \frac{h}{\sqrt{6}} = 0.408\,h$
Trapezoid	$x_a = \frac{h(b_1 + 2b)}{3(b + b_1)}$ $x_b = \frac{h(b + 2b_1)}{3(b + b_1)}$	$J_{AA} = \frac{h^3(b^2 + 4bb_1 + b_1^2)}{36(b + b_1)}$ $J_{BB} = \frac{h^3(b + 3b_1)}{12}$	$K_{AA} = \frac{h\sqrt{2(b^2 + 4bb_1 + b_1^2)}}{6(b + b_1)}$ $K_{BB} = \frac{h\sqrt{b + 3b_1}}{\sqrt{6(b + b_1)}}$

* J_p, polar moment of inertia, refers to an axis through the center of gravity.

81

Table 2.1 *(Continued)*

Section	Distance to center of gravity, x	Moment of inertia, J*	Radius of gyration, K
Circle	$x_a = x_b = \dfrac{d}{2} = r.$	$J_{AA} = \dfrac{\pi d^4}{64} = 0.0491\,d^4$ $= \dfrac{\pi r^4}{4} = 0.7854\,r^4.$ $J_p = \dfrac{\pi r^4}{2}.$	$K_{AA} = \dfrac{d}{4} = \dfrac{r}{2}.$
Hollow circle	$x_a = x_b = \dfrac{d}{2} = r.$	$J_{AA} = \dfrac{\pi\,(d^4 - d_i^4)}{64}$ $= 0.0491\,(d^4 - d_i^4)$ $= \dfrac{\pi\,(r^4 - r_i^4)}{4}$ $= 0.7854\,(r^4 - r_i^4).$ $J_p = \dfrac{\pi\,(r^4 - r_i^4)}{2}.$	$K_{AA} = \dfrac{\sqrt{d^2 + d_i^2}}{4}$ $= \dfrac{\sqrt{r^2 + r_i^2}}{2}.$
Semi-circle	$x_a = \dfrac{d\,(3\pi - 4)}{6\pi} = 0.288\,d$ $= 0.576\,r.$ $x_b = \dfrac{2\,d}{3\pi} = 0.212\,d = \dfrac{4\,r}{3\pi} = 0.424\,r.$	$J_{AA} = \dfrac{d^4\,(9\pi^2 - 64)}{1152\,\pi} = 0.00686\,d^4$ $= 0.1098\,r^4.$	$K_{AA} = \dfrac{d}{12\,\pi}\,\sqrt{(9\,\pi^2 - 64)}$ $= 0.132\,d.$

* J_p, polar moment of inertia, refers to an axis through the center of gravity.

Table 2.1 (*Continued*)

Section	Distance to center of gravity, x	Moment of inertia, J*	Radius of gyration, K
Hollow Half Circle	$x_b = \dfrac{2(d^3 - d_i^3)}{3\pi(d^2 - d_i^2)}.$	$J_{AA} = \dfrac{\pi(d^4 - d_i^4)}{128} - \dfrac{4(d^3 - d_i^3)^2}{72\pi(d^2 - d_i^2)}.$	$K_{AA} = \sqrt{\dfrac{(d^4 - d_i^4)}{16(d^2 - d_i^2)} - \dfrac{4(d^3 - d_i^3)^2}{9\pi^2(d^2 - d_i^2)^2}}$
Circular Segment	$x = \dfrac{2}{3}\dfrac{r^3 \sin^3 \alpha}{A}.$ $[A = \tfrac{1}{2}r^2(2\alpha - \sin 2\alpha)$ where first α is in radians$]$	$J_{AA} = \dfrac{1}{4}Ar^2\left[1 - \dfrac{2}{3}\dfrac{\sin^3 \alpha \cos \alpha}{\alpha - \sin \alpha \cos \alpha}\right].$ $J_{BB} = \dfrac{1}{4}Ar^2\left[1 + \dfrac{2\sin^3 \alpha \cos \alpha}{\alpha - \sin \alpha \cos \alpha}\right].$	$K = \sqrt{\dfrac{J}{A}}.$
Circular Sector	$x = \dfrac{2}{3}\dfrac{r \sin \alpha}{\alpha}.$	$J_{AA} = \dfrac{1}{4}Ar^2\left(1 - \dfrac{\sin \alpha \cos \alpha}{\alpha}\right).$ $J_{BB} = \dfrac{1}{4}Ar^2\left(1 + \dfrac{\sin \alpha \cos \alpha}{\alpha}\right).$	$K = \sqrt{\dfrac{J}{A}}.$

* J_p, polar moment of inertia, refers to an axis through the center of gravity.

Table 2.1 (Continued)

Section	Distance to center of gravity, x	Moment of inertia, J*	Radius of gyration, K
Parabolic Segment	$x = \tfrac{3}{5}\,a.$ (for half segment, $y = \tfrac{3}{8}\,b$)	$J_{AA} = \tfrac{4}{15}\,ab^3.$ $J_{BB} = \tfrac{4}{7}\,ba^3.$	$K_{AA} = \dfrac{b}{\sqrt{5}} = 0.447\,b.$ $K_{BB} = a\sqrt{\tfrac{3}{7}} = 0.654\,a.$
Ellipse	$x_a = a.$ $x_b = b.$	$J_{AA} = \dfrac{\pi a^3 b}{4} = 0.7854\,a^3 b.$ $J_{BB} = \dfrac{\pi b^3 a}{4} = 0.7854\,ab^3.$ $J_p = \dfrac{\pi ab\,(a^2 + b^2)}{4}.$	$K_{AA} = \dfrac{a}{2}.$ $K_{BB} = \dfrac{b}{2}.$
Elliptical Ring	$x_a = a.$ $x_b = b.$	$J_{AA} = \dfrac{\pi}{4}\,(a^3 b - a_1^3 b_1)$ $= 0.7854\,(a^3 b - a_1^3 b_1).$ $J_{BB} = \dfrac{\pi}{4}\,(b^3 a - b_1^3 a_1)$ $= 0.7854\,(b^3 a - b_1^3 a_1).$	$K_{AA} = \dfrac{1}{2}\sqrt{\dfrac{a^3 b - a_1^3 b_1}{ab - a_1 b_1}}.$ $K_{BB} = \dfrac{1}{2}\sqrt{\dfrac{b^3 a - b_1^3 a_1}{ba - b_1 a_1}}.$

* J_P, polar moment of inertia, refers to an axis through the center of gravity.

Table 2.1 (*Continued*)

Section	Distance to center of gravity, x	Moment of inertia, J*	Radius of gyration, K
Equal Angle 	$x_a = x_b = \dfrac{a^2 + (a-t)\,t}{2\,(2\,a-t)}$ $[\alpha = 45°]$	$J_{AA} = \dfrac{t\,(a-x)^3 + ax^3 - a\,(x-t)^3}{3}$ $J_{BB} = J_{AA}.$ $J_{CC} = \dfrac{bt^3 + b^3t + 3\,a^2bt + t^4}{12}.$ $J_{DD} = \dfrac{bt^3 + b^3t + 3bt(a-4x+2t)^2 + t^4 + 6t^2\,(2x-t)^2}{12}.$	$K = \sqrt{\dfrac{J}{A}}.$
Unequal Angle 	$x_a = \dfrac{t\,(b+2\,c) + c^2}{2\,(b+c)}.$ $x_b = \dfrac{t\,(2\,d+a) + d^2}{2\,(a+d)}.$ $\tan 2\,d = \dfrac{t(2x_b-t)a(a-2x_a) + d(2x_a-t)(b+t-2x_b)}{2\,(J_{AA} - J_{BB})}$	$J_{AA} = \dfrac{t\,(a-x_a)^3 + bx_a^3 - d\,(x_a - t)^3}{3}.$ $J_{BB} = \dfrac{t\,(b-x_b)^3 + ax_b^3 - c\,(x_b - t)^3}{3}.$ $J_{CC} = \dfrac{J_{AA}\cos^2\alpha - J_{BB}\sin^2\alpha}{\cos 2\,\alpha}$ $J_{DD} = \dfrac{J_{BB}\cos^2\alpha - J_{AA}\sin^2\alpha}{\cos 2\,\alpha}.$	$K = \sqrt{\dfrac{J}{A}}.$
I-Beam 	$x_a = \dfrac{d}{2}.$ $x_b = \dfrac{b}{2}.$	$J_{AA} = \dfrac{bd^3 - c^3\,(b-t)}{12}.$ $J_{BB} = \dfrac{2\,mb^3 + ct^3}{12}.$	$K_{AA} = \sqrt{\dfrac{bd^3 - c^3\,(b-t)}{12\,[bd - c\,(b-t)]}}.$ $K_{BB} = \sqrt{\dfrac{2\,mb^3 + ct^3}{12\,[bd - c\,(b-t)]}}.$

* J_p, polar moment of inertia, refers to an axis through the center of gravity.

Table 2.1 (*Continued*)

Section	Distance to center of gravity, x	Moment of inertia, J*	Radius of gyration, K
Channel	$x_a = \dfrac{d}{2}.$ $x_b = \dfrac{\dfrac{dt^2}{2} + 2\,am\left(t + \dfrac{a}{2}\right)}{dt + 2\,am}.$	$J_{AA} = \dfrac{bd^3 - ac^3}{12}.$ $J_{BB} =$ $\dfrac{dx_b^3 - d(x_b - t)^3 + 2\,m(b - x_b)^3}{3}.$	$K_{AA} = \sqrt{\dfrac{bd^3 - ac^3}{12\,(bd - ac)}}.$ $K_{BB} =$ $\sqrt{\dfrac{dx_b^3 - d(x_b - t)^3 + 2\,m(b - x_b)^3}{3\,(bd - ac)}}.$

* J_P, polar moment of inertia, refers to an axis through the center of gravity.

Table 2.1 (*Continued*)

Tee

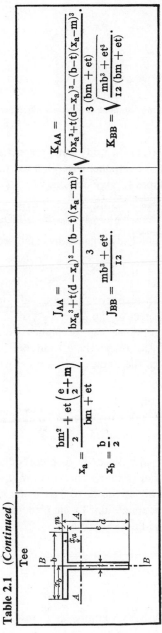

$$x_a = \frac{\dfrac{bm^2}{2} + et\left(\dfrac{e}{2} + m\right)}{bm + et}.$$

$$x_b = \frac{b}{2}.$$

$$J_{AA} = \frac{bx_a{}^3 + t(d - x_a)^3 - (b-t)(x_a - m)^3}{3}.$$

$$J_{BB} = \frac{mb^3 + et^3}{12}.$$

$$K_{AA} = \sqrt{\frac{bx_a{}^2 + t(d - x_a)^3 - (b-t)(x_a - m)^3}{3\ (bm + et)}}.$$

$$K_{BB} = \sqrt{\frac{mb^3 + et^3}{12\ (bm + et)}}.$$

* J_p, polar moment of inertia, refers to an axis through the center of gravity.

87

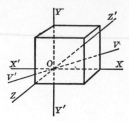

Fig. 2.6   Moment of inertia with respect to the axis $V'V$.

Product of inertia (U or U_m) of a body with respect to two coordinate planes is the sum of the products of the component weights (or masses) and the products of their distances from these planes (Σwxy or Σmxy).

$$U = \Sigma wxy, \qquad U_m = \Sigma mxy. \tag{2.59}$$

NOTE.   The product of inertia of a body is generally expressed by calculus. $U = \int xy \, dw$. $U_m = \int xy \, dm$.

Moment of inertia (J) with respect to the axis $V'V$ in terms of the moments of inertia J_x, J_y, and J_z with respect to the axes $X'X$, $Y'Y$, and $Z'Z$ and the products of inertia U_{xy}, U_{xz}, and U_{yz} with respect to the planes Y_{0y} and X_{0x}, the planes Y_{0z} and X_{0x}, and the planes X_{0z} and X_{0y}, respectively, where $V'V$ passes through the origin of these three axes and makes the angles α, β, and γ with the axes $X'X$, $Y'Y$, and $Z'Z$, respectively.

$$J = J_x \cos^2 \alpha + J_y \cos^2 \beta + J_z \cos^2 \gamma - 2U_{xy} \cos \alpha \cos \beta$$
$$- 2U_{xz} \cos \alpha \cos \gamma - 2U_{yz} \cos \beta \cos \gamma. \tag{2.60}$$

Principal axes of a body are those three rectangular axes through any point, about one of which the moment of inertia is a maximum and about another a minimum, the moment of inertia about the third axis being intermediate in value. **Principal planes** are the planes perpendicular to the principal axes. The products of inertia with respect to the principal planes are zero.

2.3   KINETICS

2.3.1   Translation

Three laws of motion: (1) A body remains in a state of rest or of uniform motion except under the action of some unbalanced force. (2) A single force acting on a body causes it to move with accelerated motion in the direction of the force. The acceleration is directly proportional to the force and inversely proportional to the mass of the body. (3) To every action there is an equal and opposite reaction.

Force (F) imparting an acceleration of a feet/second^2 to a weight of m pounds (mass).

$$F = ma \text{ pounds.} \tag{2.61}$$

NOTE.   In terms of the weight w, $F = (w/g)a$.

Table 2.2 Properties of Various Solids*

Solids	Moment of inertia, J	Radius of gyration, K
Straight Rod 	$J_{AA} = \tfrac{1}{12} Wl^2.$ $J_{BB} = \tfrac{1}{3} Wl^2.$ $J_{CC} = \tfrac{1}{3} Wl^2 \sin^2 \alpha.$	$K_{AA} = \dfrac{1}{\sqrt{12}}.$ $K_{BB} = \dfrac{1}{\sqrt{3}}.$ $K_{CC} = 1\sqrt{\dfrac{\sin \alpha}{3}}.$
Rod bent into a Circular Arc 	$J_{AA} =$ $\dfrac{1}{2} Wr^2 \left[1 - \dfrac{\sin \alpha \cos \alpha}{\alpha} \right].$ $J_{BB} =$ $\dfrac{1}{2} Wr^2 \left[1 + \dfrac{\sin \alpha \cos \alpha}{\alpha} \right].$	$K_{AA} =$ $r\sqrt{\dfrac{1}{2}\left(1 - \dfrac{\sin \alpha \cos \alpha}{\alpha} \right)}.$ $K_{BB} =$ $r\sqrt{\dfrac{1}{2}\left(1 + \dfrac{\sin \alpha \cos \alpha}{\alpha} \right)}.$
Cube 	$J_{AA} = J_{BB} = \tfrac{1}{6} Wa^2.$	$K_{AA} = K_{BB} = \dfrac{a}{\sqrt{6}}.$
Rectangular Prism 	$J_{AA} = \tfrac{1}{12} W (a^2 + b^2).$ $J_{BB} = \tfrac{1}{12} W (b^2 + c^2).$	$K_{AA} = \sqrt{\dfrac{a^2 + b^2}{12}}.$ $K_{BB} = \sqrt{\dfrac{b^2 + c^2}{12}}.$

* All axes pass through the center of gravity unless otherwise noted. $J_m = \dfrac{J}{g}.$ W = total weight of the body.

Table 2.2 (*Continued*)

Solids	Moments of inertia, J	Radius of gyration, K
Right Circular Cylinder	$J_{AA} = \frac{1}{2} W r^2.$ $J_{BB} = \frac{1}{12} W (3 r^2 + h^2).$	$K_{AA} = \dfrac{r}{\sqrt{2}}.$ $K_{BB} = \sqrt{\dfrac{3 r^2 + h^2}{12}}.$
Hollow Right Circular Cylinder	$J_{AA} = \frac{1}{2} W (R^2 + r^2).$ $J_{BB} = \frac{1}{4} W \left(R^2 + r^2 + \dfrac{h^2}{3} \right).$	$K_{AA} = \sqrt{\dfrac{R^2 + r^2}{2}}.$ $K_{BB} = \sqrt{\dfrac{3 R^2 + 3 r^2 + h^2}{12}}.$
Thin Hollow Cylinder	$J_{AA} = W r^2.$ $J_{BB} = \dfrac{W}{2} \left(r^2 + \dfrac{h^2}{6} \right).$	$K_{AA} = r.$ $K_{BB} = \sqrt{\dfrac{6 r^2 + h^2}{12}}.$

* All axes pass through the center of gravity unless otherwise noted. $J_m = \dfrac{J}{g}.$ W = total weight of the body.

Table 2.2 *(Continued)*

Solids	Moments of inertia, J	Radius of gyration, K
Elliptical Cylinder	$J_{AA} = \frac{1}{4} W (a^2 + b^2).$ $J_{BB} = \frac{1}{12} W (3 b^2 + h^2).$ $J_{CC} = \frac{1}{12} W (3 a^2 + h^2).$	$K_{AA} = \sqrt{\dfrac{a^2 + b^2}{2}}.$ $K_{BB} = \sqrt{\dfrac{3 b^2 + h^2}{12}}.$ $K_{CC} = \sqrt{\dfrac{3 a^2 + h^2}{12}}.$
Sphere	$J_{AA} = \frac{2}{5} W r^2.$	$K_{AA} = \dfrac{2 r}{\sqrt{10}}.$
Hollow Sphere	$J_{AA} = \dfrac{2}{5} W \dfrac{R^5 - r^5}{R^3 - r^3}.$	$K_{AA} = \sqrt{\dfrac{2}{5}\left(\dfrac{R^5 - r^5}{R^3 - r^3}\right)}.$

* All axes pass through the center of gravity unless otherwise noted. $J_m = \dfrac{J}{g}.$ W = total
ht of the body.

Table 2.2 *(Continued)*

Solids	Moment of inertia, J	Radius of gyration, K
Thin Hollow Sphere 	$J_{AA} = \frac{2}{3} W r^2.$	$K_{AA} = \frac{2\,r}{\sqrt{6}}.$
Ellipsoid 	$J_{AA} = \frac{1}{5} W (b^2 + c^2).$ $J_{BB} = \frac{1}{5} W (a^2 + c^2).$ $J_{CC} = \frac{1}{5} W (a^2 + b^2).$	$K_{AA} = \sqrt{\dfrac{b^2 + c^2}{5}}.$ $K_{BB} = \sqrt{\dfrac{a^2 + c^2}{5}}.$ $K_{CC} = \sqrt{\dfrac{a^2 + b^2}{5}}.$
Torus 	$J_{AA} = W (R^2 + \frac{3}{4} r^2).$ $J_{BB} = W \left(\dfrac{R^2}{2} + \dfrac{5}{8} r^2 \right).$	$K_{AA} = \frac{1}{2} \sqrt{4 R^2 + 3 r^2}.$ $K_{BB} = \sqrt{\dfrac{4 R^2 + 5 r^2}{8}}.$

* All axes pass through the center of gravity unless otherwise noted. $J_m = \dfrac{J}{g}.$ W = total weight of the body.

Table 2.2 *(Continued)*

Solids	Distance to center of gravity, x	Moment of inertia, J	Radius of gyration, K
Right Rectangular Pyramid	$x = \dfrac{h}{4}.$	$J_{AA} = \frac{1}{20} W (a^2 + b^2).$ $J_{BB} =$ $\frac{1}{20} W \left(b^2 + \dfrac{3h^2}{4} \right).$	$K_{AA} =$ $\sqrt{\dfrac{a^2 + b^2}{20}}.$ $K_{BB} =$ $\sqrt{\frac{1}{80}(4b^2 + 3h^2)}.$
Right Circular Cone	$x = \dfrac{h}{4}.$	$J_{AA} = \frac{3}{10} Wr^2.$ $J_{BB} =$ $\frac{3}{20} W \left(r^2 + \dfrac{h^2}{4} \right).$	$K_{AA} = \dfrac{3\,r}{\sqrt{30}}.$ $K_{\overline{BB}} =$ $\sqrt{\frac{3}{80}(4r^2 + h^2)}.$
Frustum of a Cone	$x =$ $\dfrac{h(R^2 + 2Rr + 3r^2)}{4(R^2 + Rr + r^2)}$	$J_{AA} =$ $\frac{3}{10} W \dfrac{(R^5 - r^5)}{(R^3 - r^3)}.$	$K_{AA} =$ $\sqrt{\dfrac{3}{10} \dfrac{(R^5 - r^5)}{(R^3 - r^3)}}.$
Paraboloid	$x = \frac{1}{3} h.$	$J_{AA} = \frac{1}{3} Wr^2.$ $J_{BB} =$ $\frac{1}{18} W (3 r^2 + h^2).$	$K_{AA} = \dfrac{r}{\sqrt{3}}.$ $K_{BB} =$ $\sqrt{\frac{1}{18}(3r^2 + h^2)}.$

* All axes pass through the center of gravity unless otherwise noted. $J_m = \dfrac{J}{g}.$ W = total weight of the body.

Table 2.2 (*Continued*)

Solids	Distance to center of gravity, x	Moment of inertia, J	Radius of gyration, K
Spherical Sector 	$x = \frac{3}{8}(2r - h)$.	$J_{AA} =$ $\frac{1}{5}W(3rh - h^2)$.	$K_{AA} =$ $\sqrt{\dfrac{3rh - h^2}{5}}$.
Spherical Segment 	$x = \dfrac{3}{4}\dfrac{(2r - h)^2}{(3r - h)}$. For half sphere $x = \frac{3}{8}r$.	$J_{AA} = W\left(r^2 - \dfrac{3rh}{4}\right.$ $\left. + \dfrac{3h^2}{20}\right)\dfrac{2h}{3r - h}$.	$K_{AA} = \sqrt{\dfrac{J}{W}}$.

* All axes pass through the center of gravity unless otherwise noted. $J_m = \frac{J}{g}$. $W = $ total weight of the body.

Impulse (I) of a force of F pounds acting for t seconds.

$$I = Ft \text{ pound-seconds.} \tag{2.62}$$

Momentum ($\mathfrak{M}$) of a body of m pounds (mass) moving with a velocity of v feet/second.

$$\mathfrak{M} = mv \text{ pounds (mass)-feet/second.} \tag{2.63}$$

Force (F) required to change the velocity of m pounds (mass) from v_1 feet/second to v_2 feet/second in t seconds.

$$F = \frac{m(v_1 - v_2)}{t} \text{ pounds.} \tag{2.64}$$

NOTE.   The change in momentum of a body during any time interval equals the impulse of the force acting on the body for that time.

Work (W) done by a force of F pounds acting through a distance of s feet.

$$W = Fs \text{ foot-pounds.} \tag{2.65}$$

NOTE.   If the force is variable, $W = \displaystyle\int_0^s F\,ds$.

Power (P) required to do W foot-pounds of work at a constant rate in t seconds.

$$P = \frac{W}{t} \text{ foot-pounds/second.} \qquad (2.66)$$

Potential energy (W), referred to a certain datum, of a body of w pounds weight and at an elevation of h feet above the datum.

$$W = wh \text{ foot-pounds.} \qquad (2.67)$$

Kinetic energy (W) of a body of m pounds (mass) mass having a velocity of translation of v feet/second.

$$W = \frac{mv^2}{2} \text{ foot-pounds.} \qquad (2.68)$$

Force (F) required to change the velocity of a mass of m pounds (mass) from v_1 feet/second to v_2 feet/second in s feet.

$$F = \frac{m(v_1^2 - v_2^2)}{2s} \text{ pounds.} \qquad (2.69)$$

NOTE. The change in kinetic energy of the body equals the work done on the body.

Force (F) required to move m pounds (mass) in a circular path of r feet radius with a constant speed of v feet/second.

$$F = \frac{mv^2}{r} \text{ pounds.} \qquad (2.70)$$

NOTE. This force acts along the normal to the path of the body toward the center of curvature, and is called the centripetal or deviating force. The reaction to this force along the normal to the path of the body away from the center of curvature is called the centrifugal force.

2.3.2 Rotation

Torque or moment (T) about the axis of rotation imparting an angular acceleration of α radians/second^2 to a body with a mass moment of inertia of J_m pound(mass)-feet squared about the axis of rotation.

$$T = J_m \alpha \text{ pound-feet.} \qquad (2.71)$$

NOTE. In terms of the weight, w pounds, of the body and its radius of gyration, K feet, about the axis of rotation, $T = (w/g)K^2\alpha$ pound-feet.

Angular impulse (I_a) of a torque of T pound-feet acting for t seconds.

$$I_a = Tt \text{ pound-feet-seconds.} \qquad (2.72)$$

Angular momentum ($\mathfrak{M}_a$) of a body with a mass moment of inertia of J_m pound (mass)-feet^2 about the axis of rotation and an angular velocity of ω radians/second.

$$\mathfrak{M}_a = J_m\omega \text{ pound(mass)-feet}^2/\text{second.} \qquad (2.73)$$

NOTE. The angular momentum of a body is sometimes called its moment of momentum. The angular momentum of a body moving in a plane perpendicular to the axis of rotation is given by $\mathfrak{M}_a = \mathfrak{M}r$ pound(mass)-feet^2/second where $\mathfrak{M}$ equals the momentum of the body in pound(mass)-feet/second, and r equals the perpendicular distance in feet from the line of direction of the momentum to the axis of rotation.

Torque (T) required to change the angular velocity of a body of mass moment of inertia of J_m pound(mass)-feet^2 about the axis of rotation from ω_1 radians/second to ω_2 radians/second in t seconds.

$$T = \frac{J_m(\omega_1 - \omega_2)}{t} \text{ pound-feet.} \qquad (2.74)$$

NOTE. The change in angular momentum of a body is equal to the angular impulse.

Work (W) done by a torque of T pound-feet acting through an angle of θ radians.

$$W = T\theta \text{ foot-pounds.} \qquad (2.75)$$

NOTE. If the torque is variable, $W = \int_0^\theta T d\theta$. The work done by a torque of T pound-feet in N revolutions is given by $W = T\,2\pi N$ foot-pounds.

Kinetic energy (W) of a body which has an angular velocity of ω radians per second and a mass moment of inertia of J_m pound(grav.)-feet mass squared about the axis of rotation.

$$W = \frac{J_m\omega^2}{2} \text{ foot-pounds.} \qquad (2.76)$$

NOTE. In terms of the weight, w pounds, of the body and its radius of gyration, K feet, about the axis of rotation, $W = (wK^2\omega^2)/2g$ foot-pounds.

Torque (T) required to change the angular velocity of a body of mass moment of inertia of J_m pound(mass)-feet^2 about the axis of rotation from ω_1 radians/second to ω_2 radians/second, the torque acting through an angle of θ radians.

$$T = \frac{J_m(\omega_1^2 - \omega_2^2)}{2\theta} \text{ pound-feet.} \qquad (2.77)$$

NOTE. The change in kinetic energy of a body equals the work done on the body.

Center of percussion with respect to the axis of rotation is the point through which the line of action of the resultant of all the external forces acting on the rotating body passes.

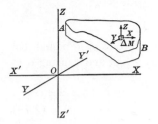

Fig. 2.7 Rotation about a fixed axis.

Distance (l) from the axis of rotation to the center of percussion of a body with a mass moment of inertia J_m pound(mass)-feet² about the axis of rotation, m pounds (mass), and x_0 feet between the axis and the center of gravity.

$$l = \frac{J_m}{x_0 m} \text{ feet.} \tag{2.78}$$

NOTE. In terms of the radius of gyration K, $l = K^2/x_0$.

General Formulas for Rotation about a Fixed Axis

Assume a body AB rotating about the axis $Z'Z$. Let m = mass of the body; α = angular acceleration at any instant; ω = angular velocity at any instant, and x_0, y_0, z_0 = the coordinates of the center of gravity of the body.

Considering the forces and motions of the small particles (as Δm) of which it may be composed, if ΣX, ΣY, ΣZ = the sums of the components of the forces parallel to the axes $X'X$, $Y'Y$, $Z'Z$, respectively; ΣT_x, ΣT_y, ΣT_z = the sums of the torques about the axes $X'X$, $Y'Y$, $Z'Z$, respectively; ΣJ_m = the moment of inertia of the mass about the axis $Z'Z$; ΣU_{xzm}, ΣU_{yzm}, = the products of inertia of mass with respect to the planes YOZ and XOY and the planes XOZ and XOY, respectively.

$$\Sigma X = -\alpha y_0 m - \omega^2 x_0 m. \qquad \Sigma T_x = -\alpha U_{xzm} + \omega^2 U_{yzm}.$$
$$\Sigma Y = +\alpha x_0 m - \omega^2 y_0 m. \qquad \Sigma T_y = -\alpha U_{yzm} - \omega^2 U_{xzm}.$$
$$\Sigma Z = 0. \qquad\qquad\qquad \Sigma T_z = \alpha J_m.$$

Table 2.3 Formulas for Translation and Rotation

Translation		Rotation	
Force.........	$F = ma$	Torque.........	$T = J_m a$
Impulse.......	$I = Ft$	Angular impulse..	$Ia = Tt$
Momentum....	$\mathfrak{M} = mv$	Angular momentum	$\mathfrak{M}a = J_m \omega$
Change of momentum.....	$m(v_1 - v_0) = Ft$	Change of angular momentum.....	$J_m(\omega_1 - \omega_0) = Tt$
Work.........	$W = Fs$	Work..........	$W = T\theta$
Kinetic energy.	$W = \frac{1}{2}mv^2$	Kinetic energy...	$W = \frac{1}{2}J_m\omega^2$
Change of kinetic energy.....	$\frac{1}{2}m(v_1^2 - v_0^2) =$ $F(s_1 - s_0)$	Change of kinetic energy........	$\frac{1}{2}J_m(\omega_1^2 - \omega_0^2) =$ $T(\theta_1 - \theta_0)$

2.3.3   Translation and Rotation

Work (W) done on a body by a force of F pounds having a torque of T pound-feet about the center of gravity of the body in moving the body s feet and causing it to rotate through an angle of θ radians.

$$W = Fs + T\theta. \tag{2.79}$$

Kinetic energy (W) of a body of m pounds (mass) with a mass moment of inertia of J_m pound(mass)-feet^2 about its center of gravity and having a velocity of translation of v feet/second and an angular velocity of ω radians/second.

$$W = \tfrac{1}{2}mv^2 + \tfrac{1}{2}\omega^2 J_m \text{ foot-pounds.} \tag{2.80}$$

NOTE.   If the body weighs w pounds and has K feet radius of gyration about the center of gravity, $W = [\tfrac{1}{2}(w/g)v^2] + [\tfrac{1}{2}(w/g)K^2\omega^2]$.

Kinetic energy developed in a body during any displacement is equal to the external work done upon it.

$$Fs + T\theta = \tfrac{1}{2}mv^2 + \tfrac{1}{2}\omega^2 J_m \text{ foot-pounds.} \tag{2.81}$$

Instantaneous axis. Any plane motion may be considered as a rotation about an axis which may be constantly changing to successive parallel positions. This axis at any instant is called the instantaneous axis.

NOTE.   If the velocities, at any instant, of two points in a body are known, the instantaneous axis passes through the intersection of the perpendiculars to the lines of motion of these two points.

Distance (l) from the instantaneous axis to the center of percussion of a body of m pounds (mass) and mass moment of inertia of J_m pound(mass)-feet^2 about its center of gravity, for a position of the instantaneous axis of x_0 feet distance from the center of gravity of the body.

$$l = \frac{J_m}{x_0 m} + x_0. \tag{2.82}$$

Velocity of translation (v_c) of the center of gravity of a body having an angular velocity of ω radians/second about the instantaneous axis which is x_0 feet from the center of gravity.

$$v_c = \omega x_0 \text{ feet/second.} \tag{2.83}$$

Kinetic energy (W) of a body with a mass moment of inertia of J_m' pound(mass)-feet^2 about the instantaneous axis and an angular velocity of ω radians/second about the instantaneous axis.

$$W = \tfrac{1}{2}\omega^2 J_m' \text{ foot-pounds.} \tag{2.84}$$

2.3.4 Pendulum

The imaginary pendulum conceived as a material point suspended by a weightless cord is called a **simple pendulum.** A real pendulum is called a **compound pendulum.**

Period (p) of oscillation (from a maximum deflection to the right to a maximum deflection to the left) of a simple pendulum l feet in length.

$$p = \pi \sqrt{\frac{l}{g}} \text{ seconds (for small oscillations).} \qquad (2.85)$$

NOTE. An approximate expression for all arcs is $t = \pi(l/g)^{1/2}[1 + (h/81)]$, where h is the vertical distance between the highest and lowest points of the path.

Length (l) of a simple seconds pendulum (one whose period of oscillation is one second).

$$l = \frac{g}{\pi^2} \text{ feet.} \qquad (2.86)$$

Period (p) of oscillation of a compound pendulum of K feet radius of gyration with respect to the axis of suspension and l feet length from the axis of suspension to the center of gravity of the pendulum.

$$p = \pi \sqrt{\frac{K^2}{lg}} \text{ (for small oscillations).} \qquad (2.87)$$

Distance (d) from the center of suspension to the center of oscillation, of a compound pendulum, of K feet radius of gyration about the center of suspension, the distance from the center of suspension to the center of gravity being l feet.

$$d = \frac{K^2}{l} \text{ feet.} \qquad (2.88)$$

NOTE. The period of oscillation, for a small oscillation, about an axis through the center of suspension is the same as that of a small oscillation about a parallel axis through the center of oscillation.

Tension (T) in the cord of a conical pendulum with a weight of W pounds and l feet length of cord, rotating with n revolutions/second.

$$T = \frac{Wl4\pi^2 n^2}{g} \text{ pounds.} \qquad (2.89)$$

NOTE. In terms of the angular velocity ω radians/second; $T = (Wl\omega^2)/g$ pounds.

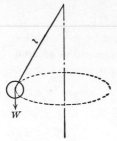

Fig. 2.8 Oscillation.

Period (p) of oscillation of a simple cycloidal pendulum swinging on the arc of a cycloid described by a circle of r feet radius.

$$p = 2\pi \sqrt{\frac{r}{g}} \text{ seconds.} \tag{2.90}$$

2.3.5   Prony Brake

Power (P) indicated by a Prony brake when the perpendicular distance from the center of the pulley to the direction of a force of F pounds applied at the end of the brake arm is l feet and the pulley revolves at a speed of S revolutions/minute.

$$P = 1.903 \ lSF \times 10^{-4} \text{ horsepower.} \tag{2.91}$$

NOTE.   The torque of the pulley equals lF pound-feet. If l is made 5 feet 3 inches, $P = SF/1000$ horsepower.

2.3.6   Friction

Static friction is the force, in addition to that overcoming inertia, required to set in motion one body in contact with another.

Coefficient of static friction (f) between two bodies, when N is the normal pressure between them and F is the corresponding static friction (N and F in the same units).

$$f = \frac{F}{N}. \tag{2.92}$$

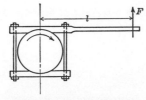

Fig. 2.9   Prony brake.

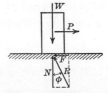

Fig. 2.10   Friction.

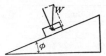

Fig. 2.11 Angle of static friction.

Resultant force (R) between two bodies starting from relative rest with a normal pressure of N pounds and a static friction of F pounds between them.

$$R = \sqrt{F^2 + N^2} \text{ pounds.} \qquad (2.93)$$

Angle of static friction (ϕ) for two surfaces with a normal pressure N and a static friction F between them (N and F in the same units).

$$\tan \phi = \frac{F}{N} = f. \qquad (2.94)$$

NOTE. The angle of repose is the angle of inclination of the surface of one body at which the other body will begin to slide along it, under the action of its own weight. The angle of repose (ϕ) is equal to the angle of static friction.

Sliding friction is the force, in addition to that overcoming inertia, required to maintain relative motion between two bodies.

NOTE. (1) For moderate pressures the friction is proportional to the normal pressure between the surfaces. (2) For moderate pressures the friction is independent of the extent of the surface in contact. (3) At low velocities the friction is independent of the velocity of rubbing. The friction decreases as the velocity increases. (4) Sliding friction is usually less than static friction.

Coefficient of sliding friction (f) between two bodies when N is the normal pressure between them and F is the corresponding sliding friction (N and F in the same units).

$$f = \frac{F}{N}. \qquad (2.95)$$

Angle of sliding friction (ϕ) for two surfaces with a normal pressure N and a sliding friction F between them (N and F in the same units).

NOTE. See Equation (2.94). The angle of sliding friction is the angle of inclination of the surface of one body, at which the motion of another body sliding upon it will be maintained. The angle of sliding friction is in general less than the angle of static friction.

Applications of Principles of Friction

Inclined plane. Let W = weight in pounds of a body sliding on the plane, α = angle of inclination of plane, β = angle between force F and plane, ϕ = angle of repose, f = coefficient of friction ($\tan \phi = f$), and F = force applied to the body along the line of action indicated.

Table 2.4 Coefficients of Static and Sliding Friction[4]

(Reference letters indicate the lubricant used; numbers in parentheses give the sources. See footnote)

Materials	Static Dry	Static Greasy	Sliding Dry	Sliding Greasy
Hard steel on hard steel	0.78 (1)	0.11 (1, a)	0.42 (2)	0.029 (5, h)
		0.23 (1, b)		0.081 (5, c)
		0.15 (1, c)		0.080 (5, i)
		0.11 (1, d)		0.058 (5, j)
		0.0075 (18, p)		0.084 (5, d)
		0.0052 (18, h)		0.105 (5, k)
				0.096 (5, l)
				0.108 (5, m)
				0.12 (5, a)
Mild steel on mild steel	0.74 (19)		0.57 (3)	0.09 (3, a)
				0.19 (3, u)
Hard steel on graphite	0.21 (1)	0.09 (1, a)		
Hard steel on babbitt (ASTM No. 1)	0.70 (11)	0.23 (1, b)	0.33 (6)	0.16 (1, b)
		0.15 (1, c)		0.06 (1, c)
		0.08 (1, d)		0.11 (1, d)
		0.085 (1, e)		
Hard steel on babbitt (ASTM No. 8)	0.42 (11)	0.17 (1, b)	0.35 (11)	0.14 (1, b)
		0.11 (1, c)		0.065 (1, c)
		0.09 (1, d)		0.07 (1, d)
		0.08 (1, e)		0.08 (11, h)
Hard steel on babbitt (ASTM No. 10)		0.25 (1, b)		0.13 (1, b)
		0.12 (1, c)		0.06 (1, c)
		0.10 (1, d)		0.055 (1, d)
		0.11 (1, e)		
Mild steel on cadmium silver				0.097 (2, f)
Mild steel on phosphor bronze			0.34 (3)	0.173 (2, f)
Mild steel on copper lead				0.145 (2, f)
Mild steel on cast iron		0.183 (15, c)	0.23 (6)	0.133 (2, f)
Mild steel on lead	0.95 (11)	0.5 (1, f)	0.95 (11)	0.3 (11, f)
Nickel on mild steel			0.64 (3)	0.178 (3, x)
Aluminum on mild steel	0.61 (8)		0.47 (3)	
Magnesium on mild steel			0.42 (3)	
Magnesium on magnesium	0.6 (22)	0.08 (22, y)		
Teflon on Teflon	0.04 (22)			0.04 (22, f)
Teflon on steel	0.04 (22)			0.04 (22, f)
Tungsten carbide on tungsten carbide	0.2 (22)	0.12 (22, a)		
Tungsten carbide on steel	0.5 (22)	0.08 (22, a)		
Tungsten carbide on copper	0.35 (23)			
Tungsten carbide on iron	0.8 (23)			
Bonded carbide on copper	0.35 (23)			
Bonded carbide on iron	0.8 (23)			
Cadmium on mild steel			0.46 (3)	
Copper on mild steel	0.53 (8)		0.36 (3)	0.18 (17, a)
Nickel on nickel	1.10 (16)		0.53 (3)	0.12 (3, w)
Brass on mild steel	0.51 (8)		0.44 (6)	
Brass on cast iron			0.30 (6)	
Zinc on cast iron	0.85 (16)		0.21 (7)	
Magnesium on cast iron			0.25 (7)	
Copper on cast iron	1.05 (16)		0.29 (7)	
Tin on cast iron			0.32 (7)	
Lead on cast iron			0.43 (7)	
Aluminum on aluminum	1.05 (16)		1.4 (3)	
Glass on glass	0.94 (8)	0.01 (10, p)	0.40 (3)	0.09 (3, a)
		0.005 (10, q)		0.116 (3, v)
Carbon on glass			0.18 (3)	
Garnet on mild steel			0.39 (3)	
Glass on nickel	0.78 (8)		0.56 (3)	
Copper on glass	0.68 (8)		0.53 (3)	
Cast iron on cast iron	1.10 (16)		0.15 (9)	0.070 (9, d)
				0.064 (9, n)
Bronze on cast iron			0.22 (9)	0.077 (9, n)
Oak on oak (parallel to grain)	0.62 (9)		0.48 (9)	0.164 (9, r)
				0.067 (9, s)
Oak on oak (perpendicular)	0.54 (9)		0.32 (9)	0.072 (9, s)
Leather on oak (parallel)	0.61 (9)		0.52 (9)	
Cast iron on oak			0.49 (9)	0.075 (9, n)
Leather on cast iron			0.56 (9)	0.36 (9, t)
				0.13 (9, n)
Laminated plastic on steel			0.35 (12)	0.05 (12, t)
Fluted rubber bearing on steel				0.05 (13, t)

(1) Campbell, *Trans. ASME*, 1939; (2) Clarke, Lincoln, and Sterrett, *Proc. API*, 1935; (3) Beare and Bowden, *Phil. Trans. Roy. Soc.*, 1935; (4) Dokos, *Trans. ASME*, 1946; (5) Boyd and Robertson, *Trans. ASME*, 1945; (6) Sachs, *Zeit. f. angew. Math. und Mech.*, 1924; (7) Honda and Yama la, *Jour. I of M*, 1925; (8) Tomlinson, *Phil. Mag.*, 1929; (9) Morin, *Acad. Roy. des Sciences*, 1838; (10) Claypoole, *Trans. ASME*, 1943; (11) Tabor, *Jour. Applied Phys.*, 1945; (12) Eyssen, General Discussion on Lubrication, *ASME*, 1937; (13) Brazier and Holland-Bowyer, General Discussion on Lubrication, *ASME*, 1937; (14) Burwell, *Jour. SAE*, 1942; (15) Stanton, "Friction," Longmans; (16) Ernst and Merchant, Conference on Friction and Surface Finish, M.I.T., 1940; (17) Gongwer, Conference on Friction and Surface Finish, M.I.T., 1940; (18) Hardy and Bircumshaw, *Proc. Roy. Soc.*, 1925; (19) Hardy and Hardy, *Phil. Mag.*, 1919; (20) Bowden and Young, *Proc. Roy. Soc.*, 1951; (21) Hardy and Doubleday, *Proc. Roy. Soc.*, 1923; (22) Bowden and Tabor, "The Friction and Lubrication of Solids," Oxford; (23) Shooter, *Research*, 4, 1951.

(a) Oleic acid; (b) Atlantic spindle oil (light mineral); (c) castor oil; (d) lard oil; (e) Atlantic spindle oil plus 2 percent oleic acid; (f) medium mineral oil; (g) medium mineral oil plus ½ percent oleic acid; (h) stearic acid; (i) grease (zinc oxide base); (j) graphite; (k) turbine oil plus 1 percent graphite; (l) turbine oil plus 1 percent stearic acid; (m) turbine oil (medium mineral); (n) olive oil; (p) palmitic acid; (q) ricinoleic acid; (r) dry soap; (s) lard; (t) water; (u) rape oil; (v) 3-in-1 oil; (w) octyl alcohol; (x) triolein; (y) 1 percent lauric acid in paraffin oil.

Source: Mark's Standard Handbook for Mechanical Engineers, 8th Edition, 1978. Used with permission from McGraw-Hill Book Co.

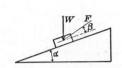

Fig. 2.12 Inclined plane.

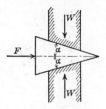

Fig. 2.13 Wedge.

1. Force (F) to prevent slipping. $(\alpha > \phi)$

$$F = W \frac{\sin (\alpha - \phi)}{\cos (\beta + \phi)} \text{ pounds.} \qquad (2.96)$$

2. Force (F) to start the body up the plane. $(\alpha > \phi)$

$$F = W \frac{\sin (\alpha + \phi)}{\cos (\beta - \phi)} \text{ pounds.} \qquad (2.97)$$

3. Force (F) to start the body down the plane. $(\alpha < \phi)$

$$F = W \frac{\sin (\phi - \alpha)}{\cos (\beta + \phi)} \text{ pounds.} \qquad (2.98)$$

Wedge. Let W = force in pounds opposing motion, α = angle of inclination of sides of wedge, ϕ = angle of friction, and F = force applied to wedge.
1. Force (F) to push wedge.

$$F = 2W \tan (\alpha + \phi) \text{ pounds.} \qquad (2.99)$$

2. Force (F) to draw wedge $(\alpha > \phi)$.

$$F = 2W \tan (\phi - \alpha) \text{ pounds.} \qquad (2.100)$$

Square-threaded Screw. Let r = mean radius of screw, p = pitch of screw, α = angle of pitch [$\tan \alpha = p/(2\pi r)$], F = force applied to screw at end of arm a, W = total weight in pounds to be moved, and ϕ = angle of friction (r and a in same units).
1. Force (F) to lower screw.

$$F = \frac{Wr(\tan \phi - \tan \alpha)}{a} \text{ pounds (approx.).} \qquad (2.101)$$

2. Force (F) to raise screw.

$$F = \frac{Wr(\tan \phi + \tan \alpha)}{a} \text{ pounds (approx.).} \qquad (2.102)$$

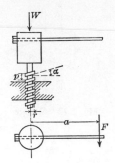
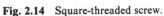

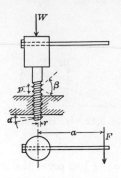

Fig. 2.14 Square-threaded screw.   Fig. 2.15 Sharp-threaded screw.

Sharp-threaded Screw. Let r = mean radius of screw, α = angle of pitch, β = angle between faces of the screw, F = force in pounds applied to screw at end of arm a, W = total weight in pounds to move, and ϕ = angle of friction (r and a in same units).

1. Force (F) to lower screw.

$$F = \frac{Wr}{a}\left(\frac{\tan\phi\cos\alpha}{\cos\dfrac{\beta}{2}} - \tan\alpha\right) \text{ pounds (approx.).} \qquad (2.103)$$

2. Force (F) to raise screw.

$$F = \frac{Wr}{a}\left(\frac{\tan\phi\cos\alpha}{\cos\dfrac{\beta}{2}} + \tan\alpha\right) \text{ pounds (approx.).} \qquad (2.104)$$

Coefficient of rolling friction (c) of a wheel with a load of W pounds and with r inches radius, moved at a uniform speed by a force of F pounds applied at its center.

$$c = \frac{Fr}{W} \text{ inches.} \qquad (2.105)$$

NOTE. Coefficients of rolling friction.

Lignum vitæ roller on oak track	$c = 0.019$ inches
Elm roller on oak track	$c = 0.032$ inches
Iron on iron (and steel on steel)	$c = 0.020$ inches

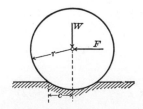

Fig. 2.16 Coefficient of rolling friction.

Table 2.5   Pivot Friction

f = coefficient of friction. W = load in pounds.
T = torque of friction about the axis of the shaft.
r = radius in inches. n = revolutions per second.

Type of Pivot	Torque T in pound-inches	Power P lost by friction in ft.-lbs. per second
Shafts and Journals ($180°$ bearing)	$T = fWr.$	$P = \dfrac{2\,\pi n}{12}\, fWr.$
Flat Pivot 	$T = \tfrac{2}{3} fWr.$	$P = \dfrac{4\,\pi n}{3 \times 12}\, fWr.$
Collar-bearing 	$T = \dfrac{2}{3}\, fW\, \dfrac{R^3 - r^3}{R^2 - r^2}.$	$P = \dfrac{4\,\pi n}{3 \times 12}\, fw\, \dfrac{R^3 - r^3}{R^2 - r^2}.$
Conical Pivot 	$T = \dfrac{2}{3}\, fW\, \dfrac{r}{\sin \alpha}.$	$P = \dfrac{4\,\pi n fWr}{3 \times 12 \sin \alpha}.$
Truncated-cone Pivot 	$T = \dfrac{2}{3} fW\, \dfrac{(R^3 - r^3)}{(R^2 - r^2)\sin \alpha}.$	$P = \dfrac{4\,\pi n fW\,(R^3 - r^3)}{3 \times 12\,(R^2 - r^2)\sin \alpha}.$

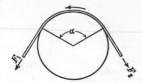

Fig. 2.17 Belt friction.

Belt Friction

Ratio (F_1/F_2) of the pull F_1 on the driving side of a belt to the pull F_2 on the driven side of the belt, when slipping is impending, in terms of the coefficient of friction f and the angle of contact α, in radians ($\epsilon = 2.718$).

$$\frac{F_1}{F_2} = \epsilon^{f\alpha}. \tag{2.106}$$

NOTE.   Mean values of f are as follows:

Leather on wood (somewhat oily)	0.47
Leather on cast iron (somewhat oily)	0.28
Leather on cast iron (moist)	0.38
Hemp-rope on iron drum	0.25
Hemp-rope on wooden drum	0.40
Hemp-rope on polished drum	0.33
Hemp-rope on rough wood	0.50

Values of $\dfrac{F_1}{F_2}$ (Slipping impending)

$\dfrac{\alpha}{2\pi}$	$f = 0.25$	$f = 0.33$	$f = 0.40$	$f = 0.50$	$\dfrac{\alpha}{2\pi}$	$f = 0.25$	$f = 0.33$	$f = 0.40$	$f = 0.50$
0.1	1.17	1.23	1.29	1.37	0.6	2.57	3.47	4.52	6.59
0.2	1.37	1.51	1.65	1.87	0.7	3.00	4.27	5.81	9.00
0.3	1.60	1.86	2.13	2.57	0.8	3.51	5.25	7.47	12.34
0.4	1.87	2.29	2.73	3.51	0.9	4.11	6.46	9.60	16.90
0.425	1.95	2.41	2.91	3.80	1.0	4.81	7.95	12.35	23.14
0.45	2.03	2.54	3.10	4.11	1.5	10.55	22.42	43.38	111.2
0.475	2.11	2.68	3.30	4.45	2.0	23.14	63.23	152.4	535.5
0.5	2.19	2.82	3.51	4.81	2.5	50.75	178.5	535.5	2,576
0.525	2.28	2.97	3.74	5.20	3.0	111.3	502.9	1881	12,392
0.55	2.37	3.31	3.98	5.63	3.5	244.2	1418	6611	59,610

2.3.7 Impact*

Common velocity (v'), after direct central impact, of two inelastic bodies of mass m_1 and m_2 and initial velocities v_1 and v_2, respectively.

$$v' = \frac{m_1 v_1 + m_2 v_2}{m_1 + m_2} \qquad (2.107)$$

Final velocities (v_1' and v_2'), after direct central impact, of two perfectly elastic bodies of mass m_1 and m_2 and initial velocities v_1 and v_2, respectively.

$$v_1' = \frac{m_1 v_1 - m_2 v_1 + 2m_2 v_2}{m_1 + m_2},$$

$$\qquad (2.108)$$

$$v_2' = \frac{m_2 v_2 - m_1 v_2 + 2m_1 v_1}{m_1 + m_2}.$$

Final velocities (v_1' and v_2'), after direct central impact, of two partially but equally inelastic bodies of mass m_1 and m_2 and initial velocities v_1 and v_2, respectively, and constant e depending on the elasticity of bodies.

$$v_1' = \frac{m_1 v_1 + m_2 v_2 - em_2(v_1 - v_2)}{m_1 + m_2},$$

$$\qquad (2.109)$$

$$v_2' = \frac{m_1 v_1 + m_2 v_2 - em_1(v_2 - v_1)}{m_1 + m_2}.$$

NOTE. $e = (H/h)^{1/2}$ where H is the height of rebound of a sphere dropped from a height h onto a horizontal surface of a rigid mass. If the bodies are inelastic, $e = 0$, and if bodies are perfectly elastic, $e = 1$.

2.4 STATICS

2.4.1 Forces and Resultants

Components of a force F (F_x and F_y) parallel to two rectangular axes $X'X$ and $Y'Y$, the force F making an angle α with the axis $X'X$.

$$F_x = F \cos \alpha, \qquad F_y = F \sin \alpha. \qquad (2.110)$$

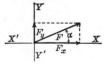

Fig. 2.18 Components of force F.

*m_1 and m_2, v_1 and v_2 in the same units.

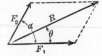

Fig. 2.19   Resultant force R.

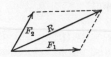

Fig. 2.20   Parallelogram of forces.

Moment or torque (M) of a force of F pounds about a given point, the perpendicular distance from the point to the direction of the force being d feet.

$$M = Fd \text{ pound-feet.} \tag{2.111}$$

NOTE.   A couple is formed by two equal, opposite, parallel forces acting in the same plane but not in the same straight line. The moment (M) of a couple of two forces, each of F pounds, with a perpendicular distance of d feet between them is Fd pound-feet. The moment, about any point, of the resultant of several forces, lying in the same plane, is the algebraic sum of the moments of the separate forces about that point.

Resultant force (R) of two forces, F_1 and F_2, which make an angle α with each other, the angle between the resultant force R and the force F_1 being θ.

$$R = \sqrt{F_1^2 + F_2^2 + 2F_1F_2 \cos \alpha}. \tag{2.112}$$

$$\tan \theta = \frac{F_2 \sin \alpha}{F_1 + F_2 \cos \alpha}, \quad \text{or} \quad \sin \theta = \frac{F_1 \sin \alpha}{R}. \tag{2.113}$$

Parallelogram of Forces.   The resultant force (R) of two forces F_1 and F_2 is represented in magnitude and direction by the diagonal lying between those two sides of a parallelogram which represent F_1 and F_2 in magnitude and direction.

Triangle of Forces.   The resultant force (R) of two forces F_1 and F_2 is represented in magnitude and direction by the third side of a triangle in which the other two sides represent F_1 and F_2 in magnitude and direction.

Resultant force (R) of three forces F_1, F_2, and F_3 mutually at right angles to each other and not lying in the same plane, the angles between the resultant force R and the forces F_1, F_2, and F_3 being α, β, and γ, respectively.

$$R = \sqrt{F_1^2 + F_2^2 + F_3^2}. \tag{2.114}$$

$$\cos \alpha = \frac{F_1}{R}, \quad \cos \beta = \frac{F_2}{R}, \quad \cos \gamma = \frac{F_3}{R}. \tag{2.115}$$

NOTE.   If three forces not in the same plane are not mutually at right angles to each other, the resultant force may be found by Equation (2.121).

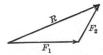

Fig. 2.21   Triangle of forces.

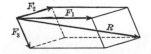

Fig. 2.22   Parallelopiped of forces.

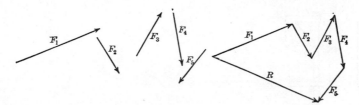

Fig. 2.23 Force polygon. The arrows indicate the directions of the forces, and for the given forces, they must point in the same direction around the polygon; but for the resultant force, they must point in the opposite direction or leading from the starting point of the first force to the end point of the last force.

Parallelopiped of Forces. The resultant force (R) of three forces F_1, F_2, and F_3, not lying in the same plane, is represented in magnitude and direction by the diagonal lying between those three sides of a parallelopiped which represent F_1, F_2, and F_3 in magnitude and direction.

Resultant force (R) of several forces lying in the same plane, if ΣF_x and ΣF are the algebraic sums of the components of the forces parallel to two rectangular axes $X'X$ and $Y'Y$, the angle between the resultant force and the axis $X'X$ being α.

$$R = \sqrt{(\Sigma F_x)^2 + (\Sigma F_y)^2} . \tag{2.116}$$

$$\tan \alpha = \frac{\Sigma F_y}{\Sigma F_x}, \quad \sin \alpha = \frac{\Sigma F_y}{R}, \quad \cos \alpha = \frac{\Sigma F_x}{R}. \tag{2.117}$$

Perpendicular distance (d) from a given point to the resultant force (R) of several forces lying in the same plane, if ΣM is the algebraic sum of the moments, about that point, of the separate forces.

$$d = \frac{\Sigma M}{R}. \tag{2.118}$$

NOTE. The resultant of several parallel forces is the algebraic sum of the forces (ΣF). If $\Sigma F = 0$ the resultant is a couple whose moment is ΣM.

Force Polygon. The resultant force (R) of several forces F_1, F_2, ..., F_n, lying in the same plane, is represented in magnitude and direction by the closing side of a polygon in which the remaining sides represent the forces F_1, F_2, ..., F_n in magnitude and direction (Fig. 2.23).

Moment (M) of a force F, about a line, is the product of the rectangular component of the force perpendicular to the line (the other component being parallel to the line) into the perpendicular distance between the line and this rectangular component; or the force F may be resolved into three rectangular components, one parallel and the other two perpendicular to the line, as in Fig. 2.24. The moment of the force about each axis is then obtained as follows:

$$M_x = yF \cos \gamma - zF \cos \beta.$$
$$M_y = zF \cos \alpha - xF \cos \gamma. \tag{2.119}$$
$$M_z = xF \cos \beta - yF \cos \alpha.$$

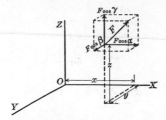

Fig. 2.24 Moment M of a force F.

Resultant force (R) of several parallel forces, not lying in the same plane, is the algebraic sum (ΣF) of the forces.

NOTE.   If $\Sigma F = 0$, the resultant is a couple whose moments are ΣM_x, ΣM_y, and so on.

Perpendicular distances (d_x) and (d_y) from each of two axes $X'X$ and $Y'Y$ to the resultant force (R) of several parallel forces, not lying in the same plane, if ΣM_x and ΣM_y are the algebraic sums of the moments of the separate forces about the axes $X'X$ and $Y'Y$, respectively.

$$d_x = \frac{\Sigma M_x}{R}, \qquad d_y = \frac{\Sigma M_y}{R}. \tag{2.120}$$

Resultant force (R) and direction (α, β, γ) of the resultant force of several forces, not lying in the same plane, if ΣF_x, ΣF_y, and ΣF_z are the algebraic sums of the components parallel to three rectangular axes $X'X$, $Y'Y$, and $Z'Z$, and α, β, and γ are the angles which the resultant force makes with the axes $X'X$, $Y'Y$, and $Z'Z$, respectively.

$$R = \sqrt{(\Sigma F_x)^2 + (\Sigma F_y)^2 + (\Sigma F_z)^2}. \tag{2.121}$$

$$\cos \alpha = \frac{\Sigma F_x}{R}, \quad \cos \beta = \frac{\Sigma F_y}{R}, \quad \cos \gamma = \frac{\Sigma F_z}{R}. \tag{2.122}$$

Resultant couple (M) and direction (α_m, β_m, γ_m) of the axis of the resultant couple of several forces, not acting in the same plane, if ΣM_x, ΣM_y, and ΣM_z are the algebraic sums of the moments about three rectangular axes $X'X$, $Y'Y$, and $Z'Z$, and α_m, β_m, and γ_m are the angles which the moment axis of the resultant couple makes with the axes $X'X$, $Y'Y$, and $Z'Z$, respectively.

$$M = \sqrt{(\Sigma M_x)^2 + (\Sigma M_y)^2 + (\Sigma M_z)^2}. \tag{2.123}$$

$$\cos \alpha_m = \frac{\Sigma M_x}{\Sigma M}, \quad \cos \beta_m = \frac{\Sigma M_y}{\Sigma M}, \quad \cos \gamma_m = \frac{\Sigma M_z}{\Sigma M}. \tag{2.124}$$

NOTE.   In general the resultant of several nonparallel forces, not in the same plane, is not a single force, but by the use of the above principles the system may be reduced to a single force and a couple.

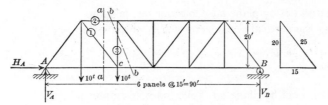

Fig. 2.25   Pratt truss.

Conditions of equilibrium of several forces, lying in the same plane, if ΣF_x and ΣF_y are the algebraic sums of the components parallel to two axes $X'X$ and $Y'Y$, and ΣM is the algebraic sum of the moments of the forces about any point.

$$\Sigma F_x = 0, \qquad \Sigma F_y = 0, \qquad \Sigma M = 0. \tag{2.125}$$

Conditions of equilibrium of several forces, not lying in the same plane, if ΣF_x, ΣF_y, and ΣF_z are the algebraic sums of the components parallel to three axes $X'X$, $Y'Y$, and $Z'Z$ which intersect at a common point but do not lie in the same plane, and ΣM_x, ΣM_y, and ΣM_z are the algebraic sums of the moments of the forces about these three axes.

$$\Sigma F_x = 0, \qquad \Sigma F_y = 0, \qquad \Sigma F_z = 0. \tag{2.126}$$

$$\Sigma M_x = 0, \qquad \Sigma M_y = 0, \qquad \Sigma M_z = 0. \tag{2.127}$$

2.4.2   Stresses in Framed Structures*

Pratt Truss.   Two live loads of 10 tons each as shown in Fig. 2.25.

(a) Reactions [use conditions of equilibrium, Equation (2.126)].

By $\Sigma M = 0$, $\Sigma M_A = 0 = 10 \times 15 + 10 \times 30 - V_B \times 90$, $V_B = 5$ tons.

By $\Sigma F_y = 0$, $20 - V_B = V_A$, $V_A = 15$ tons.

By $\Sigma F_x = 0$, $H_A = 0$. (Note that a roller is used at B, fixing the reaction there in a vertical direction.)

(b) Stresses in bars.

To find the stress in a bar consider a plane cutting the bar to divide the truss into two parts; remove one part and replace that portion of the bars that are removed by their stresses which may now be treated as outer forces. These stresses are found by applying the equations of equilibrium. It is essential that only three of the bars that are cut shall have unknown stresses.

NOTE.   If tension is called positive and all unknown stresses are assumed to be tension stresses, a positive sign for the result indicates tension and a negative sign compression.

*Due to live loads only. Weight of structure is neglected.

Bar ①. Truss cut by plane aa. Consider left portion.

Let $V_①$ = the vertical component of $S_①$, the stress in bar ①.

By $\Sigma F_y = 0, -V_A + 10 + V_① = 0, V_① = 5, S_① = \dfrac{25}{20} \times 5 = 6.25$ tons tension.

Bar ②. Truss cut by plane aa. Take moments about joint c.

By $\Sigma M = 0, \Sigma M_c = 0 = V_A \times 30 - 10 \times 15 + S_② \times 20$.

$$S_② = \frac{-450 + 150}{20} = -15 = 15 \text{ tons compression.}$$

Bar ③. Truss cut by plane bb.

By $\Sigma F_y = 0, -V_A + 20 + S_③ = 0, S_③ = -5 = 5$ tons compression.

Roof Truss. Two live loads of 3 tons each as shown in Fig. 2.26.

(a) Reactions [use conditions of equilibrium, Equation (2.126)].

By $\Sigma M = 0, \Sigma M_A = 0 = 3 \times 13.4 + 3 \times 26.8 - V_B \times 72, V_B = 1.67$ tons.

By $\Sigma F_y = 0, 2 \times 2.69 - V_B - V_A = 0,$ $\qquad\qquad V_A = 3.71$ tons.

By $\Sigma F_x = 0, 2.70 + H_A = 0, H_A = -2.70$ tons (i.e., acting to the left). (Note that a roller is used at B, fixing the reaction n a vertical direction.)

(b) Stresses in bars. (See b under Pratt Truss.)

Bar ①. Truss cut by plane aa. Consider left portion. Take moments about joint c. Let $H_①$ = horizontal component of $S_①$ (Fig. 2.27).

By $\Sigma M = 0, \Sigma M_c = 0 = V_A \times 24 - 2.69 \times 12 + 1.35 \times 6 + H_① \times 12$.

$$H_① = -5.34, S_① = \frac{\sqrt{5}}{2} \times 5.34 = 5.96 \text{ tons compression.}$$

Bar ②. Truss cut by plane aa. Take moments about A.

Let $V_②$ = vertical component of $S_②$.

By $\Sigma M = 0, \Sigma M_A = 0 = 3 \times 13.4 + V_② \times 24, V_② = -1.67$ tons.

$S_② = \sqrt{5} \times 1.67 = 3.73$ tons compression.

Bar ③. Truss cut by plane bb. Consider right portion, as fewer loads lie to the right of cutting plane.

Take moments about joint d.

By $\Sigma M = 0, \Sigma M_d = 0 = -V_B \times 48 + S_③ \times 12, S_③ = 6.68$ tons tension.

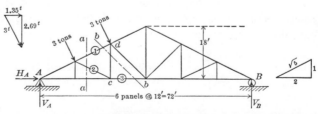

Fig. 2.26 Roof truss.

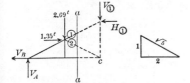

Fig. 2.27 Roof truss (left-hand portion).

2.5 PROPERTIES OF MATERIALS

(See Table 2.6 and Section 10.)

Intensity of stress is the stress per unit area, usually expressed in pounds per square inch. The simple term, **stress,** is normally used to indicate intensity of stress.

Ultimate stress is the greatest stress that can be produced in a body before rupture occurs.

Allowable stress or **working stress** is the intensity of stress that the material of a structure or a machine is designed to resist.

Factor of safety is a factor by which the ultimate stress is divided to obtain the allowable stress.

Elastic limit is the maximum intensity of stress to which a material may be subjected and return to its original shape upon the removal of the stress.

NOTE. For stresses below the elastic limit the deformations are directly proportional to the stresses producing them; that is, Hooke's law holds for stresses below the elastic limit.

Yield point is the intensity of stress beyond which the change in length increases rapidly with little if any increase in stress.

Modulus of elasticity is the ratio of stress to strain, for stresses below the elastic limit.

NOTE. Modulus of elasticity may also be defined as the stress that would produce a change of length of a bar equal to the original length of the bar, assuming the material to retain its elastic properties up to that point.

Poisson's ratio is the ratio of the relative change of diameter of a bar to its unit change of length under an axial load that does not stress it beyond the elastic limit.

NOTE. Poisson's ratio is usually denoted by $1/m$. It varies for different materials but is usually about ¼.

Intensity of stress (s) due to a force of P pounds producing tension, compression, or shear on an area of A square inches, over which it is uniformly distributed.

$$s = \frac{P}{A} \text{ pounds/inch}^2. \tag{2.128}$$

Table 2.6   Mechanical Properties of Some Engineering Materials[9]

Material	Equivalent	Ultimate strength, psi			Yield point, tension, psi	Modulus of elasticity, tension or compression, psi	Modulus of elasticity, shear, psi	Weight per cu in., lb
		Tension	Compression *	Shear				
Steel, forged-rolled:								
C, 0.10–0.20	SAE 1015	60,000	39,000	48,000	39,000	30,000,000	12,000,000	0.28
C, 0.20–0.30	SAE 1025	67,000	43,000	53,000	43,000	30,000,000	12,000,000	0.28
C, 0.30–0.40	SAE 1035	70,000	46,000	56,000	46,000	30,000,000	12,000,000	0.28
C, 0.60–0.80	....	125,000	65,000	75,000	65,000	30,000,000	12,000,000	0.28
Nickel	SAE 2330	115,000	....	92,000	......	30,000,000	12,000,000	0.28
Cast iron:								
Gray	ASTM 20	20,000	80,000	27,000	....	15,000,000	6,000,000	0.26
Gray	ASTM 35	35,000	125,000	44,000	....			0.26
Gray	ASTM 60	60,000	145,000	70,000	....	20,000,000	8,000,000	0.26
Malleable	SAE 32510	50,000	120,000	48,000		23,000,000	9,200,000	0.26
Wrought iron	....	48,000	25,000	38,000	25,000	27,000,000	....	0.28
Steel, cast:								
Low C	....	60,000	....	......				0.28
Medium C	....	70,000	....					0.28
High C	....	80,000	45,000	......	45,000			0.28
Aluminum alloy:								
Structural, No. 350	...	16,000	5,000	11,000	5,000	10,000,000	3,750,000	0.10
Structural, No. 17ST	..	58,000	35,000	35,000	35,000	10,000,000	3,750,000	0.10
Brass:								
Cast	....	40,000	....					0.30
Annealed	....	54,000	18,000	.....	18,000			0.30
Cold-drawn	....	96,700	49,000	.....	49,000	15,500,000	6,200,000	0.30
Bronze:								
Cast	....	22,000	....					0.31
Cold-drawn	....	85,000	....		....	15,000,000	6,000,000	0.31
Brick, clay:								
Grade SW	ASTM	....	3,000 (min) †	.....			....	0.072
Grade MW	ASTM	....	2,500 (min)					
Grade NW	ASTM	....	1,500 (min)					
Concrete, 1:2:4 (28 days)	....	....	2,000	....	....	3,000,000		0.087
Stone	....	....	8,000	.....				0.092
White oak:								
Parallel to grain	....		7,440	2,000	4,760 ‡	1,780,000		0.028
Across grain	....	800	....	....	1,320 ‡			
White pine:								
Parallel to grain	....		4,840	860	3,680 ‡	1,280,000	....	0.015
Across grain	....	300	....	.....	550 ‡			
Southern longleaf pine:								
Parallel to grain	....		8,440	1,500	6,150 ‡	1,999,000		0.024
Across grain	....	470		.....	1,199 ‡			

* The ultimate strength in compression for ductile materials is usually taken as the yield point. The bearing value for pins and rivets may be much higher, and for structural steel is taken as 90,000 psi.
† Average of five bricks.
‡ Proportional limit in compression.
Source: Urquhart, *Civil Engineering Handbook*, 4th Edition, 1959. Used with permission of McGraw-Hill Book Co.

Modulus of elasticity (E) of a bar of A square inches cross-sectional area and l inches length, which undergoes a change of length of d inches under an axial load of P pounds.

$$E = \frac{Pl}{Ad} \text{ pounds/inch}^2. \tag{2.129}$$

NOTE.   The load must be such as to produce an intensity of stress below the elastic limit. If s is the intensity of stress produced and e the ratio of change of length to total length, $E = s/e$ and $e = s/E$.

Change of length (d) of a bar of A square inches cross-sectional area, l inches length, and E pounds per square inch modulus of elasticity of material, due to an axial load of P pounds.

$$d = \frac{Pl}{AE} \text{ inches.} \tag{2.130}$$

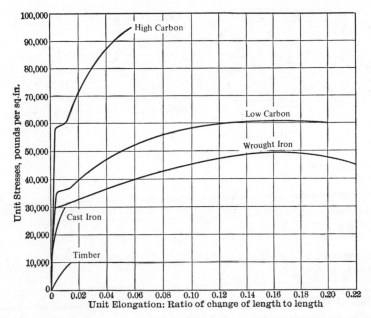

Fig. 2.28 Stress–strain diagram showing the relation of the intensities of stress of a material to the corresponding strains or deformations.

2.6 BEAMS

Vertical shear at any section of a beam is equal to the algebraic sum of all the vertical forces on one side of the section. The shear is positive when the part of the beam to the left of the section tends to move upward under the action of the resultant of the vertical forces.

NOTE.   In the study of beams, the reactions must be treated as applied loads and included in shear and moment. A section is always taken as cut by a plane normal to the axis of the beam. In all cases vertical means normal to the axis.

Bending moment at any section of a beam is equal to the algebraic sum of the moments, about the center of gravity of the section, of all the forces on one side of the section. Moment that causes compression in the upper fibers of a beam is positive.

NOTE.   The maximum moment occurs at a section where the shear is zero. A diagram of shears or of moments is a curve the ordinate to which at any section shows the value of the shear or moment at that section.

Neutral axis of a beam is the plane that undergoes no change in length due to the bending and along which the direct stress is zero. The fibers on one side of the neutral axis are stressed in tension and on the other side in compression and the intensities of these stresses in homogeneous beams are directly proportional to the distances of the fibers from the neutral axis.

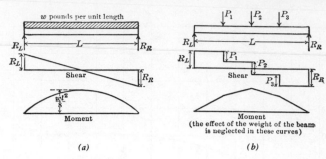

Fig. 2.29 Moment and shear diagram for a simple beam (*a*) with a uniformly distributed load, (*b*) with concentrated loads.

NOTE. The neutral axis at any section in a beam subject to bending only passes through the center of gravity of that section.

Elastic curve of a beam is the curve formed by the neutral plane when the beam deflects due to bending.

Equation of the elastic curve of a beam of J inches[4] moment of inertia and a modulus of elasticity of the material of E pounds per square inch, if x and y in inches are the abscissa and ordinate, respectively, of a point on the neutral axis referred to rectangular coordinates through the points of support, and M is the moment in inch pounds at that point.

$$M = EJ \frac{d^2y}{dx^2}. \tag{2.131}$$

NOTE. The equation of the elastic curve is used to find the slope and deflection of a beam under loading. A single integration gives the slope, integrating twice gives the deflection; in each case, however, the proper value of the constant of integration must be determined.

Three-moment equation gives the ratio between the moments M_a, M_b, and M_c at three consecutive points of support (a, b, and c) on a beam continuous over three or more supports.

CASE I. Concentrated loads. (See Fig. 2.30.)

$$M_a l_1 + 2M_b(l_1 + l_2) + M_c l_2$$
$$= P_1 l_1^2(k_1^3 - k_1) + P_2 l_2^2(3k_2^2 - k_2^3 - 2k_2). \tag{2.132}$$

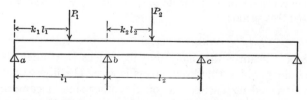

Fig. 2.30 Three-moment equation, concentrated loads.

Table 2.7   Beams under Various Loadings

Beam, loading and moment diagram	Reactions	Bending moment	Deflection
	$R_L = R_R = \dfrac{wl}{2}$	$M_x = \dfrac{wlx}{2} - \dfrac{wx^2}{2}$. $\quad M_{max} = \dfrac{wl^2}{8}$.	$d_{max} = \dfrac{5\,wl^4}{384\,EJ}$.
	$R_L = R_R = \dfrac{P}{2}$.	$M_x = \dfrac{Px}{2}$. $\quad M_{max} = \dfrac{Pl}{4}$.	$d_{max} = \dfrac{Pl^3}{48\,EJ}$.
	$R_L = \dfrac{Pb}{l}$. $\quad R_R = \dfrac{Pa}{l}$.	$M_{x_1} = \dfrac{Pbx_1}{l}$. $\quad M_{x_2} = \dfrac{Pax_2}{l}$. $\quad M_{max} = \dfrac{Pab}{l}$.	$d_{max} = \dfrac{Pab(2\,a+b)\sqrt{3\,b(2\,a+b)}}{27\,EJl}$,
	$R_L = R_R = P$.	$M_x = Px$. $\quad M_{max} = Pa$.	$d_{max} = \dfrac{Pa}{6\,EJ}\left(\dfrac{3}{4}\,l^2 - a^2\right)$.
	$R_L = \dfrac{wb\,(2\,c+b)}{2\,l}$. $\quad R_R = \dfrac{wb\,(2\,a+b)}{2\,l}$.	$M_x = R_L x - \dfrac{w\,(x-a)^2}{2}$. $\quad M_{max} = R_L\left[a + \dfrac{R_L}{2\,w}\right]$.	

117

Table 2.7 (*Continued*)

$R_L = \frac{1}{3} W.$ $\quad$ $R_R = \frac{2}{3} W.$	$M_x =$ $\frac{Wx}{3}\left(1 - \frac{x^2}{l^2}\right).$ $M_{max} = \frac{2\,Wl}{9\sqrt{3}}.$	$d_{max} = \frac{0.013044\,Wl^3}{EJ}.$
$R_L = R_R = \frac{W}{2}.$	$M_x =$ $W_x\left(\frac{1}{2} - \frac{2}{3}\frac{x^2}{l^2}\right).$ $M_{max} = \frac{Wl}{6}.$ (at center)	$d_{max} = \frac{Wl^3}{60\,EJ}.$
$R_L = P.$	$M_x = Px.$ $M_{max} = Pl.$	$d_{max} = \frac{Pl^3}{3\,EJ}.$
$R_L = wl.$	$M_x = \frac{wx^2}{2}.$ $M_{max} = \frac{wl^2}{2}.$	$d_{max} = \frac{wl^4}{8\,EJ}.$
$R_L = W.$	$M_x = \frac{Wx^3}{3\,l^2}.$ $M_{max} = \frac{Wl}{3}.$	$d_{max} = \frac{Wl^3}{15\,EJ}.$
$R_L = R_R = P.$	$M_x = Px.$ $M_{max} = Pa.$	$d_{end} = \frac{Pa^2(2\,a + 3\,l)}{6\,EJ}.$ $d_{center} = -\frac{Pal^2}{8\,EJ}.$

118

Table 2.7 (*Continued*)

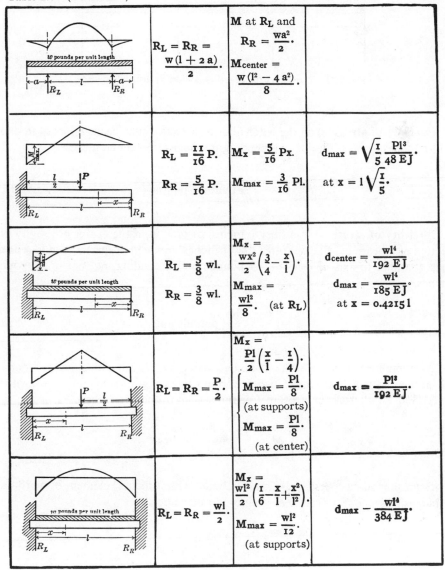

		M at R_L and $R_R = \dfrac{wa^2}{2}$. $M_{center} = \dfrac{w(l^2 - 4a^2)}{8}$.	
	$R_L = R_R = \dfrac{w(l + 2a)}{2}$.		
	$R_L = \dfrac{11}{16}P$. $R_R = \dfrac{5}{16}P$.	$M_x = \dfrac{5}{16}Px$. $M_{max} = \dfrac{3}{16}Pl$.	$d_{max} = \sqrt{\dfrac{1}{5}}\dfrac{Pl^3}{48\,EJ}$. at $x = l\sqrt{\dfrac{1}{5}}$.
	$R_L = \dfrac{5}{8}wl$. $R_R = \dfrac{3}{8}wl$.	$M_x = \dfrac{wx^2}{2}\left(\dfrac{3}{4} - \dfrac{x}{l}\right)$. $M_{max} = \dfrac{wl^2}{8}$. (at R_L)	$d_{center} = \dfrac{wl^4}{192\,EJ}$ $d_{max} = \dfrac{wl^4}{185\,EJ}$ at $x = 0.4215\,l$
	$R_L = R_R = \dfrac{P}{2}$.	$M_x = \dfrac{Pl}{2}\left(\dfrac{x}{l} - \dfrac{1}{4}\right)$. $M_{max} = \dfrac{Pl}{8}$. (at supports) $M_{max} = \dfrac{Pl}{8}$. (at center)	$d_{max} = \dfrac{Pl^3}{192\,EJ}$.
	$R_L = R_R = \dfrac{wl}{2}$.	$M_x = \dfrac{wl^2}{2}\left(\dfrac{1}{6} - \dfrac{x}{l} + \dfrac{x^2}{l^2}\right)$. $M_{max} = \dfrac{wl^2}{12}$. (at supports)	$d_{max} = \dfrac{wl^4}{384\,EJ}$.

CASE II. Uniformly distributed load. (See Fig. 2.31.)

$$M_a l_1 + 2M_b(l_1 + l_2) + M_c l_2 = -\tfrac{1}{4}w_1 l_1^3 - \tfrac{1}{4}w_2 l_2^3. \qquad (2.133)$$

Intensity of stress (s) in tension or compression on a fiber y inches distant from the center of gravity of a section of a beam with J inches^4 moment of inertia, due to a bending moment of M pound-inches.

$$s = \frac{My}{J} \text{ pounds/inch}^2. \qquad (2.134)$$

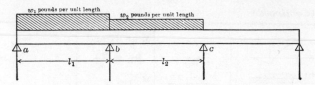

Fig. 2.31   Three-moment equation, uniformly distributed loads.

Intensity of stress (s) on the outer fiber of a rectangular beam h inches in depth and b inches in breadth, due to a bending moment of M pound-inches.

$$s = \frac{6M}{bh^2} \text{ pounds/inch}^2. \qquad (2.135)$$

Intensity of stress (s) in a fiber y inches distant from the center of gravity of a section of a beam of A inches^2 area and J inches^4 moment of inertia, due to a direct load (parallel to axis of beam) of P pounds and a bending moment of M pound-inches.

$$s = \frac{P}{A} \pm \frac{My}{J} \text{ pounds/inch}^2. \qquad (2.136)$$

Maximum moment (M) which can be carried by a beam with J inches^4 moment of inertia and y inches greatest distance from center of gravity to outer fiber, without exceeding an intensity of stress of f pounds/inch^2 in the outer fiber.

$$M = \frac{sJ}{y} \text{ pound-inches}. \qquad (2.137)$$

Section modulus (S) of a section of a beam with J inches^4 moment of inertia and y inches distance from center of gravity to outer fiber.

$$S = \frac{J}{y} \text{ inches}^3. \qquad (2.138)$$

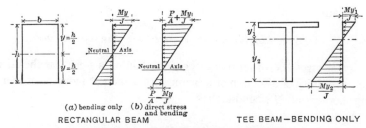

Fig. 2.32   Stress distribution in beam.

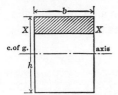

Fig. 2.33   Intensity of longitudinal shear.

Intensity of stress (s) on the outer fiber of a beam of section modulus of S inches³, due to a bending moment of M pound-inches.

$$s = \frac{M}{S} \text{ pounds/inch}^2. \tag{2.139}$$

Intensity of longitudinal shear (s) along a plane XX at the section of a beam where the total vertical shear is S pounds, if J inches⁴ is the moment of inertia of the total section about its center of gravity axis, b the width of the beam at plane XX, and Q inches³ the statical moment, taken about the center of gravity axis, of that portion of the section that lies outside of the axis XX.

$$s = \frac{SQ}{bJ} \text{ pounds/inch}^2. \tag{2.140}$$

NOTE.   The maximum intensity of shear always occurs at the center of gravity of the section of a beam.

Maximum intensity of shear (s) in a rectangular beam A inches² in area at a section where the total vertical shear is S pounds.

$$s = \frac{3}{2}\frac{S}{A} \text{ pounds/inch}^2. \tag{2.141}$$

NOTE.   The intensity of vertical shear is equal to that of the longitudinal shear acting at right angles to it. The intensity of vertical shear is obtained by the formula $s = SQ/bJ$.

2.7   COLUMNS

Euler's formula for the ultimate average intensity of stress (f) on a column l inches in length, with a least radius of gyration of r inches and of material of E pounds/inch² modulus of elasticity. f should not exceed the elastic limit.

Column with ends rounded, $\quad f = \pi^2 E \left(\frac{r}{l}\right)^2 \text{ pounds/inch}^2. \tag{2.142}$

Column with ends fixed, $\quad f = 4\pi^2 E \left(\frac{r}{l}\right)^2 \text{ pounds/inch}^2. \tag{2.143}$

Column with one end fixed $f = \dfrac{9}{4}\pi^2 E \left(\dfrac{r}{l}\right)^2 \text{ pounds/inch}^2. \tag{2.144}$
and one end rounded,

Gordon formula for allowable average intensity of stress (f) on a column l inches in length, with a least radius of gyration of r inches and a maximum allowable compression stress of f_c pounds/inch^2 on the material.

$$f = \frac{f_c}{1 + \dfrac{1}{c}\left(\dfrac{l}{r}\right)^2} \text{ pounds/inch}^2. \tag{2.145}$$

NOTE.   The following values of c are commonly used for steel columns.

Column with ends rounded	9,000
Column with ends fixed	20,000
Column with one end fixed and one end rounded	36,000

Pin-ended columns are generally considered to have ends rounded.

Straight-line formula for the allowable average intensity of stress (f) in a column l inches in length, with a least radius of gyration of r inches and a maximum allowable compression stress of f_c pounds/inch^2 on the material.

$$f = f_c - c\left(\frac{l}{r}\right) \text{ pounds/inch}^2. \tag{2.146}$$

Maximum intensity of stress (f) in a column of A square inches area of cross-section, l inches length, J inches^4 moment of inertia about the axis about which bending occurs and y inches distance from that axis to the most stressed fiber, due to a direct load of P pounds and a bending moment of M inch-pounds.

$$f = \frac{P}{A} + \frac{My}{J - \dfrac{Pl^2}{cE}} \text{ pounds/inch}^2 \text{ (approx.).} \tag{2.147}$$

NOTE.   The constant c for the common case of pin-ended columns subject to bending due to a uniformly distributed load may be taken as 10.

Maximum intensity of stress (f) in a short column of A square inches area of cross section, due to a load of P pounds applied a inches distant from the X axis of symmetry and b inches distant from the Y axis of symmetry, if J_x inches^4 is the moment of inertia about the X axis, y inches the distance from the X axis to the most

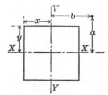

Fig. 2.34   Maximum intensity of stress.

stressed fiber, J_y inches^4 the moment of inertia about the Y axis, and x inches the distance from Y axis to the most stressed fiber.

$$f = \frac{P}{A} + \frac{Pay}{J_x} + \frac{Pbx}{J_y} \text{ pounds/inch}^2. \tag{2.148}$$

2.7.1 Timber Columns[8]

Maximum allowable unit compressive stress (F'_c) in axially loaded, square or rectangular, simple solid columns can be computed based on column length classification:

$$\text{Short columns,} \quad F'_c = F_c. \tag{2.149}$$

$$\text{Intermediate columns,} \quad F'_c = F_c \left[1 - \frac{1}{3} \left(\frac{l/d}{K} \right)^4 \right]. \tag{2.150}$$

$$\text{Long columns,} \quad F'_c = \frac{0.30E}{(l/d)^2}, \tag{2.151}$$

where l/d = slenderness ratio (unbraced height/least side, both in inches), E = modulus of elasticity, F'_c = allowable unit stress in compression parallel to grain, adjusted for l/d ratio (above), with the limiting magnitude of F'_c equal to F_c, the design value for compression parallel to grain for the species and grade of lumber used.

Short columns have an l/d ratio of 11 or less. Intermediate columns have an l/d ratio greater than 11 but less than K, where $K = 0.671(E/F_c)^{1/2}$. Long columns have an l/d ratio of K or greater.

Values of K for Selected Values of E (psi) and F_c (psi)

$E = 1,100,000$		$E = 1,300,000$		$E = 1,400,000$	
F_c	K	F_c	K	F_c	K
625	28.15	1050	23.61	625	31.76
700	26.60	1200	22.09	725	29.49
900	23.46	1650	18.83		

$E = 1,500,000$		$E = 1,600,000$		$E = 1,700,000$	
F_c	K	F_c	K	F_c	K
775	29.52	925	27.91	1200	25.26
		1000	26.84	1350	23.81
		1100	25.59		
		1150	25.03		
		1300	23.54		

Source: From Parker & Hauf, *Simplified Design of Structural Wood*, John Wiley & Sons, 1979.

These formulas apply to square-end, simple solid columns as well as to the pin-end condition from which they were derived. They are appropriate for wood columns subjected to normal loading and used in dry locations.

2.8 STRUCTURAL STEEL[1]

2.8.1 Product Availability

Table 2.8 shows groups of shapes, plates, and bars available in the various minimum yield stress and tensile levels afforded by seven steels. For complete information on each steel, reference should be made to the appropriate ASTM specification. A listing of the shape sizes included in each of the five groups follows in Table 2.9 corresponding to the groupings given in Table A of ASTM Specification A6.

Space does not permit inclusion of all rolled shapes, pipe, cold- and hot-formed tubing, and cold- and hot-rolled sheet, and strip or plates of greater thickness that are occasionally used in construction. For such products, reference should be made to the various producer's catalogs.

2.8.2 Selection of the Appropriate Structural Steel

ASTM A36 is the all-purpose grade steel widely used in building and bridge construction. ASTM A529 structural carbon steel, ASTM A441 and A572 high-strength low-alloy structural steels, ASTM A242 and A588 atmospheric corrosion-resistant high-strength low-alloy structural steels, and ASTM A514 quenched and tempered alloy structural steel plate may each have certain advantages over ASTM A36, depending on the application. These **high-strength steels** have proven to be economical choices where lighter members, resulting from use of higher allowable stresses, are not penalized because of instability, local buckling, deflection, or other similar reasons. They are frequently used in tension members, beams in continuous and composite construction where deflections can be minimized, and columns having low slenderness ratios. The reduction of dead load, and associated savings in shipping costs, can be significant factors. However, higher-strength steels are not to be used indiscriminately. Effective use of all steels depends on thorough cost and engineering analysis.

With suitable procedures and precautions, all steels listed in the AISC Specification are suitable for welded fabrication.

ASTM A242 and A588 **atmospheric corrosion-resistant** high-strength low-alloy steels are more expensive than the high-strength low-alloy steels. They are suitable for use in the bare (uncoated) condition, where exposure to normal atmosphere causes a tightly adherent oxide to form on the surface, protecting the steel from further oxidation. The reduction of maintenance resulting from use of these steels often offsets their higher initial cost. Designers should consult the steel producers on the corrosion-resistant properties and limitations of these steels prior to use in the bare (uncoated) condition.

When either A242 or A588 steel is exposed to a more corrosive atmospheric environment, its use in the coated condition provides longer coating life than with other structural steels. It should be noted that A588 steel, in addition to its ability to resist

Table 2.8 Availability of Shapes, Plates, and Bars According to ASTM Structural Steel Specifications

Steel Type	ASTM Designation	F_y Minimum Yield Stress (ksi)	F_u Tensile Stress[a] (ksi)	Shapes Group per ASTM A6 [b]1	2	3	4	5	To 1/2" Incl.	Over 1/2" to 3/4" Incl.	Over 3/4" to 1 1/4" Incl.	Over 1 1/4" to 1 1/2" Incl.	Over 1 1/2" to 2" Incl.	Over 2" to 2 1/2" Incl.	Over 2 1/2" to 4" Incl.	Over 4" to 5" Incl.	Over 5" to 6" Incl.	Over 6" to 8" Incl.	Over 8"
Carbon	A36	32	58-80															■	■
		36	58-80[c]	■	■	■	■	■	■	■	■	■	■	■	■	■	■	■	
	A529	42	60-85	■					■	■						■	■		
High-Strength Low-Alloy	A441	40	60											■	■	■	■	■	
		42	63			■	■					■	■						
		46	67		■					■	■								
		50	70	■					■	■									
	A572—Grade	42	60	■	■	■	■	■	■	■	■	■	■	■	■				
		50	65	■	■	■	■	■	■	■	■								
		60	75	■	■				■	■									
		65	80	■					■										
Corrosion-Resistant High-Strength Low-Alloy	A242	42	63									■	■	■	■				
		46	67		■	■			■	■	■								
		50	70	■	■				■	■									
	A588	42	63												■	■	■		
		46	67												■	■			
		50	70	■	■	■	■	■	■	■	■	■	■	■					
Quenched & Tempered Alloy	A514[d]	90	100-130						■	■	■	■	■	■	■				
		100	110-130						■	■	■	■							

[a] Minimum unless a range is shown.
[b] Includes bar-size shapes.
[c] For shapes over 426 lbs./ft., minimum of **58** ksi only applies.
[d] Plates only.
■ Available.
□ Not available.

Source: Reproduced by courtesy of American Institute of Steel Construction.

125

atmospheric corrosion, offers the advantage of being the only listed "as-rolled" structural steel available at 50 ksi minimum specified yield stress in thickness up to 4 inches (inclusive).

2.8.3 Structural Shapes—Designations, Dimensions, and Properties

The letter M designates shapes that cannot be classified as W, HP, or S shapes. Similarly, MC designates channels that cannot be classified as C shapes. Because many of the M and MC shapes are available only from a limited number of producers, or are infrequently rolled, their availability should be checked prior to specifying these shapes. They have various slopes on their inner flange surfaces, for which dimensions may be obtained from the respective producing mills. The flange thickness given in the tables is the average flange thickness.

In calculating the theoretical weights, properties, and dimensions of the rolled shapes, fillets, and roundings have been included for all shapes except angles. The properties of these rolled shapes are based on the smallest theoretical size fillets produced; dimensions for detailing are based on the largest theoretical size fillets produced. These properties and dimensions are either exact or slightly conservative for all producers who offer them, except as noted in the footnotes to Table 2.9.

Maximum lengths available vary widely with producers but a conservative range is from 60 to 75 feet.

2.8.4 Tables of Shapes

For convenience, larger and less commonly available shapes and tees have been omitted from the tables that follow. For complete listings refer to the *AISC Manual of Steel Construction*.

2.9 STRUCTURAL TIMBER

2.9.1 General[3]

Wooden structural members in large sizes have become increasingly costly and difficult to locate. As a result structural glued laminated timber members ("glulams") have been developed and have made possible the production of structural timbers in a wide variety of sizes and shapes.

Glulams are used as load-carrying structural framing for roofs and other structural portions of buildings, and for other construction, such as bridges, towers, and marine installations. The term "structural glued laminated timber" refers to an engineered, stress-rated product of a timber laminating plant, comprising assemblies of suitably selected and prepared wood laminations securely bonded together with adhesives. The grain of all laminations is approximately parallel longitudinally. The individual laminations do not exceed 2 inches (50 mm) in net thickness. Laminations may be comprised of pieces end-joined to form any lengths, pieces placed or glued edge to edge to make wider ones, or pieces bent to a curved form during gluing.

Table 2.9   Structural Shape Size Groupings for Tensile Property Classification[1]

Structural Shape	Group 1	Group 2	Group 3	Group 4	Group 5
W Shapes	W 24x55, 62	W 36x135 to 210 incl	W 36x230 to 300 incl	W 14x233 to 550 incl	W 14x605 to 730 incl
	W 21x44 to 57 incl	W 33x118 to 152 incl	W 33x201 to 241 incl	W 12x210 to 336 incl	
	W 18x35 to 71 incl	W 30x99 to 211 incl	W 14x145 to 211 incl		
	W 16x26 to 57 incl	W 27x84 to 178 incl	W 12x120 to 190 incl		
	W 14x22 to 53 incl	W 24x68 to 162 incl			
	W 12x14 to 58 incl	W 21x62 to 147 incl			
	W 10x12 to 45 incl	W 18x76 to 119 incl			
	W 8x10 to 48 incl	W 16x67 to 100 incl			
	W 6x9 to 25 incl	W 14x61 to 132 incl			
	W 5x16, 19	W 12x65 to 106 incl			
	W 4x13	W 10x49 to 112 incl			
		W 8x58, 67			
M Shapes	to 20 lb/ft incl				
S Shapes	to 35 lb/ft incl	over 35 lb/ft			
HP Shapes		to 102 lb/ft incl	over 102 lb/ft		
American Standard Channels (C)	to 20.7 lb/ft incl	over 20.7 lb/ft			
Miscellaneous Channels (MC)	to 28.5 lb/ft incl	over 28.5 lb/ft			
Angles (L), Structural & Bar-Size	to 1/2 in. incl	over 1/2 to 3/4 in. incl	over 3/4 in.		

Notes: Structural tees from W, M and S shapes fall in the same group as the structural shape from which they are cut.

Group 4 and Group 5 shapes are generally contemplated for application as compression members. When used in other applications or when subject to welding or thermal cutting, the material specification should be reviewed to determine if it adequately covers the properties and quality appropriate for the particular application. Where warranted, the use of killed steel or special metallurgical requirements should be considered.

Source: American Institute of Steel Construction.

Table 2.10 Selected Structural Shapes—Dimensions and Properties[1]

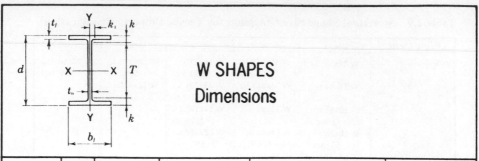

W SHAPES
Dimensions

Designation	Area A	Depth d		Web Thickness t_w	Web $\dfrac{t_w}{2}$	Flange Width b_f		Flange Thickness t_f		T	k	k_1
	In.²	In.		In.	In.	In.		In.		In.	In.	In.
W 18x119	35.1	18.97	19	0.655	5/8	5/16	11.265	11¼	1.060	1¹/₁₆	15½	1¾	15/16
x106	31.1	18.73	18¾	0.590	9/16	5/16	11.200	11¼	0.940	15/16	15½	1⅝	15/16
x 97	28.5	18.59	18⅝	0.535	9/16	5/16	11.145	11⅛	0.870	⅞	15½	1⁹/₁₆	⅞
x 86	25.3	18.39	18⅜	0.480	½	¼	11.090	11⅛	0.770	¾	15½	1⁷/₁₆	⅞
x 76	22.3	18.21	18¼	0.425	7/16	¼	11.035	11	0.680	11/16	15½	1⅜	13/16
W 18x 71	20.8	18.47	18½	0.495	½	¼	7.635	7⅝	0.810	13/16	15½	1½	⅞
x 65	19.1	18.35	18⅜	0.450	7/16	¼	7.590	7⅝	0.750	¾	15½	1⁷/₁₆	⅞
x 60	17.6	18.24	18¼	0.415	7/16	¼	7.555	7½	0.695	11/16	15½	1⅜	13/16
x 55	16.2	18.11	18⅛	0.390	⅜	3/16	7.530	7½	0.630	⅝	15½	1⁵/₁₆	13/16
x 50	14.7	17.99	18	0.355	⅜	3/16	7.495	7½	0.570	9/16	15½	1¼	13/16
W 16x100	29.4	16.97	17	0.585	9/16	5/16	10.425	10⅜	0.985	1	13⅝	1¹¹/₁₆	15/16
x 89	26.2	16.75	16¾	0.525	½	¼	10.365	10⅜	0.875	⅞	13⅝	1⁹/₁₆	⅞
x 77	22.6	16.52	16½	0.455	7/16	¼	10.295	10¼	0.760	¾	13⅝	1⁷/₁₆	⅞
x 67	19.7	16.33	16⅜	0.395	⅜	3/16	10.235	10¼	0.665	11/16	13⅝	1⅜	13/16
W 16x 57	16.8	16.43	16⅜	0.430	7/16	¼	7.120	7⅛	0.715	11/16	13⅝	1⅜	⅞
x 50	14.7	16.26	16¼	0.380	⅜	3/16	7.070	7⅛	0.630	⅝	13⅝	1⁵/₁₆	13/16
x 45	13.3	16.13	16⅛	0.345	⅜	3/16	7.035	7	0.565	9/16	13⅝	1¼	13/16
x 40	11.8	16.01	16	0.305	5/16	3/16	6.995	7	0.505	½	13⅝	1 3/16	13/16
x 36	10.6	15.86	15⅞	0.295	5/16	3/16	6.985	7	0.430	7/16	13⅝	1⅛	¾
W 14x426	125.0	18.67	18⅝	1.875	1⅞	15/16	16.695	16¾	3.035	3¹/₁₆	11¼	3¹¹/₁₆	1⁹/₁₆
x398	117.0	18.29	18¼	1.770	1¾	⅞	16.590	16⅝	2.845	2⅞	11¼	3½	1½
x370	109.0	17.92	17⅞	1.655	1⅝	13/16	16.475	16½	2.660	2¹¹/₁₆	11¼	3⁵/₁₆	1⁷/₁₆
x342	101.0	17.54	17½	1.540	1⁹/₁₆	13/16	16.360	16⅜	2.470	2½	11¼	3⅛	1⅜
x311	91.4	17.12	17⅛	1.410	1⁷/₁₆	¾	16.230	16¼	2.260	2¼	11¼	2¹⁵/₁₆	1⁵/₁₆
x283	83.3	16.74	16¾	1.290	1⁵/₁₆	11/16	16.110	16⅛	2.070	2¹/₁₆	11¼	2¾	1¼
x257	75.6	16.38	16⅜	1.175	1 3/16	⅝	15.995	16	1.890	1⅞	11¼	2⁹/₁₆	1 3/16
x233	68.5	16.04	16	1.070	1¹/₁₆	9/16	15.890	15⅞	1.720	1¾	11¼	2⅜	1 3/16
x211	62.0	15.72	15¾	0.980	1	½	15.800	15¾	1.560	1⁹/₁₆	11¼	2¼	1⅛
x193	56.8	15.48	15½	0.890	⅞	7/16	15.710	15¾	1.440	1⁷/₁₆	11¼	2⅛	1¹/₁₆
x176	51.8	15.22	15¼	0.830	13/16	7/16	15.650	15⅝	1.310	1⁵/₁₆	11¼	2	1¹/₁₆
x159	46.7	14.98	15	0.745	¾	⅜	15.565	15⅝	1.190	1 3/16	11¼	1⅞	1
x145	42.7	14.78	14¾	0.680	11/16	⅜	15.500	15½	1.090	1¹/₁₆	11¼	1¾	1

Source: American Institute of Steel Construction.

Table 2.10 (*Continued*)

W SHAPES
Properties

Nominal Wt. per Ft. Lb.	Compact Section Criteria $\frac{b_f}{2t_f}$	F_y' Ksi	$\frac{d}{t_w}$	F_y''' Ksi	r_T In.	$\frac{d}{A_f}$ In.	Axis X-X I In.⁴	S In.³	r In.	Axis Y-Y I In.⁴	S In.³	r In.	Torsional constant J In.⁴	Plastic Modulus Z_x In.³	Z_y In.³
119	5.3	—	29.0	—	3.02	1.59	2190	231	7.90	253	44.9	2.69	10.6	261	69.1
106	6.0	—	31.7	—	3.00	1.78	1910	204	7.84	220	39.4	2.66	7.48	230	60.5
97	6.4	—	34.7	54.7	2.99	1.92	1750	188	7.82	201	36.1	2.65	5.86	211	55.3
86	7.2	—	38.3	45.0	2.97	2.15	1530	166	7.77	175	31.6	2.63	4.10	186	48.4
76	8.1	64.2	42.8	36.0	2.95	2.43	1330	146	7.73	152	27.6	2.61	2.83	163	42.2
71	4.7	—	37.3	47.4	1.98	2.99	1170	127	7.50	60.3	15.8	1.70	3.48	145	24.7
65	5.1	—	40.8	39.7	1.97	3.22	1070	117	7.49	54.8	14.4	1.69	2.73	133	22.5
60	5.4	—	44.0	34.2	1.96	3.47	984	108	7.47	50.1	13.3	1.69	2.17	123	20.6
55	6.0	—	46.4	30.6	1.95	3.82	890	98.3	7.41	44.9	11.9	1.67	1.66	112	18.5
50	6.6	—	50.7	25.7	1.94	4.21	800	88.9	7.38	40.1	10.7	1.65	1.24	101	16.6
100	5.3	—	29.0	—	2.81	1.65	1490	175	7.10	186	35.7	2.51	7.73	198	54.9
89	5.9	—	31.9	64.9	2.79	1.85	1300	155	7.05	163	31.4	2.49	5.45	175	48.1
77	6.8	—	36.3	50.1	2.77	2.11	1110	134	7.00	138	26.9	2.47	3.57	150	41.1
67	7.7	—	41.3	38.6	2.75	2.40	954	117	6.96	119	23.2	2.46	2.39	130	35.5
57	5.0	—	38.2	45.2	1.86	3.23	758	92.2	6.72	43.1	12.1	1.60	2.22	105	18.9
50	5.6	—	42.8	36.1	1.84	3.65	659	81.0	6.68	37.2	10.5	1.59	1.52	92.0	16.3
45	6.2	—	46.8	30.2	1.83	4.06	586	72.7	6.65	32.8	9.34	1.57	1.11	82.3	14.5
40	6.9	—	52.5	24.0	1.82	4.53	518	64.7	6.63	28.9	8.25	1.57	0.79	72.9	12.7
36	8.1	64.0	53.8	22.9	1.79	5.28	448	56.5	6.51	24.5	7.00	1.52	0.54	64.0	10.8
426	2.8	—	10.0	—	4.64	0.37	6600	707	7.26	2360	283	4.34	331	869	434
398	2.9	—	10.3	—	4.61	0.39	6000	656	7.16	2170	262	4.31	273	801	402
370	3.1	—	10.8	—	4.57	0.41	5440	607	7.07	1990	241	4.27	222	736	370
342	3.3	—	11.4	—	4.54	0.43	4900	559	6.98	1810	221	4.24	178	672	338
311	3.6	—	12.1	—	4.50	0.47	4330	506	6.88	1610	199	4.20	136	603	304
283	3.9	—	13.0	—	4.46	0.50	3840	459	6.79	1440	179	4.17	104	542	274
257	4.2	—	13.9	—	4.43	0.54	3400	415	6.71	1290	161	4.13	79.1	487	246
233	4.6	—	15.0	—	4.40	0.59	3010	375	6.63	1150	145	4.10	59.5	436	221
211	5.1	—	16.0	—	4.37	0.64	2660	338	6.55	1030	130	4.07	44.6	390	198
193	5.5	—	17.4	—	4.35	0.68	2400	310	6.50	931	119	4.05	34.8	355	180
176	6.0	—	18.3	—	4.32	0.74	2140	281	6.43	838	107	4.02	26.5	320	163
159	6.5	—	20.1	—	4.30	0.81	1900	254	6.38	748	96.2	4.00	19.8	287	146
145	7.1	—	21.7	—	4.28	0.88	1710	232	6.33	677	87.3	3.98	15.2	260	133

AMERICAN INSTITUTE OF STEEL CONSTRUCTION

Table 2.10 (*Continued*)

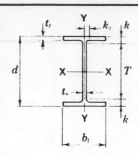

W SHAPES
Dimensions

Designation	Area A In.²	Depth d In.		Web Thickness t_w In.		$\frac{t_w}{2}$ In.	Flange Width b_f In.		Thickness t_f In.		Distance T In.	k In.	k_1 In.
W 14x132	38.8	14.66	14⅝	0.645	⅝	5/16	14.725	14¾	1.030	1	11¼	1 11/16	15/16
x120	35.3	14.48	14½	0.590	9/16	5/16	14.670	14⅝	0.940	15/16	11¼	1⅝	15/16
x109	32.0	14.32	14⅜	0.525	½	¼	14.605	14⅝	0.860	⅞	11¼	1 9/16	⅞
x 99	29.1	14.16	14⅛	0.485	½	¼	14.565	14⅝	0.780	¾	11¼	1 7/16	⅞
x 90	26.5	14.02	14	0.440	7/16	¼	14.520	14½	0.710	11/16	11¼	1⅜	⅞
W 14x 82	24.1	14.31	14¼	0.510	½	¼	10.130	10⅛	0.855	⅞	11	1⅝	1
x 74	21.8	14.17	14⅛	0.450	7/16	¼	10.070	10⅛	0.785	13/16	11	1 9/16	15/16
x 68	20.0	14.04	14	0.415	7/16	¼	10.035	10	0.720	¾	11	1½	15/16
x 61	17.9	13.89	13⅞	0.375	⅜	3/16	9.995	10	0.645	⅝	11	1 7/16	15/16
x190	55.8	14.38	14⅜	1.060	1 1/16	9/16	12.670	12⅝	1.735	1¾	9½	2 7/16	1 3/16
x170	50.0	14.03	14	0.960	15/16	½	12.570	12⅝	1.560	1 9/16	9½	2¼	1⅛
x152	44.7	13.71	13¾	0.870	⅞	7/16	12.480	12½	1.400	1⅜	9½	2⅛	1 1/16
x136	39.9	13.41	13⅜	0.790	13/16	7/16	12.400	12⅜	1.250	1¼	9½	1 15/16	1
x120	35.3	13.12	13⅛	0.710	11/16	⅜	12.320	12⅜	1.105	1⅛	9½	1 13/16	1
x106	31.2	12.89	12⅞	0.610	⅝	5/16	12.220	12¼	0.990	1	9½	1 11/16	15/16
x 96	28.2	12.71	12¾	0.550	9/16	5/16	12.160	12⅛	0.900	⅞	9½	1⅝	⅞
x 87	25.6	12.53	12½	0.515	½	¼	12.125	12⅛	0.810	13/16	9½	1½	⅞
x 79	23.2	12.38	12⅜	0.470	½	¼	12.080	12⅛	0.735	¾	9½	1 7/16	⅞
x 72	21.1	12.25	12¼	0.430	7/16	¼	12.040	12	0.670	11/16	9½	1⅜	⅞
x 65	19.1	12.12	12⅛	0.390	⅜	3/16	12.000	12	0.605	⅝	9½	1 5/16	13/16
W 12x 58	17.0	12.19	12¼	0.360	⅜	3/16	10.010	10	0.640	⅝	9½	1⅜	13/16
x 53	15.6	12.06	12	0.345	⅜	3/16	9.995	10	0.575	9/16	9½	1¼	13/16
W 12x 50	14.7	12.19	12¼	0.370	⅜	3/16	8.080	8⅛	0.640	⅝	9½	1⅜	13/16
x 45	13.2	12.06	12	0.335	5/16	3/16	8.045	8	0.575	9/16	9½	1¼	13/16
x 40	11.8	11.94	12	0.295	5/16	3/16	8.005	8	0.515	½	9½	1¼	¾
W 12x 35	10.3	12.50	12½	0.300	5/16	3/16	6.560	6½	0.520	½	10½	1	9/16
x 30	8.79	12.34	12⅜	0.260	¼	⅛	6.520	6½	0.440	7/16	10½	15/16	½
x 26	7.65	12.22	12¼	0.230	¼	⅛	6.490	6½	0.380	⅜	10½	⅞	½

Table 2.10 (*Continued*)

W SHAPES
Properties

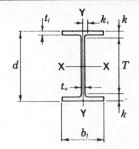

Nom-inal Wt. per Ft.	Compact Section Criteria				r_T	$\dfrac{d}{A_f}$	Elastic Properties						Tor-sional con-stant	Plastic Modulus	
	$\dfrac{b_f}{2t_f}$	F_y'	$\dfrac{d}{t_w}$	F_y'''			Axis X-X			Axis Y-Y				Z_x	Z_y
							I	S	r	I	S	r	J		
Lb.		Ksi		Ksi	In.		In.^{4}	In.^{3}	In.	In.^{4}	In.^{3}	In.	In.^{4}	In.^{3}	In.^{3}
132	7.1	—	22.7	—	4.05	0.97	1530	209	6.28	548	74.5	3.76	12.3	234	113
120	7.8	—	24.5	—	4.04	1.05	1380	190	6.24	495	67.5	3.74	9.37	212	102
109	8.5	58.6	27.3	—	4.02	1.14	1240	173	6.22	447	61.2	3.73	7.12	192	92.7
99	9.3	48.5	29.2	—	4.00	1.25	1110	157	6.17	402	55.2	3.71	5.37	173	83.6
90	10.2	40.4	31.9	—	3.99	1.36	999	143	6.14	362	49.9	3.70	4.06	157	75.6
82	5.9	—	28.1	—	2.74	1.65	882	123	6.05	148	29.3	2.48	5.08	139	44.8
74	6.4	—	31.5	—	2.72	1.79	796	112	6.04	134	26.6	2.48	3.88	126	40.6
68	7.0	—	33.8	57.7	2.71	1.94	723	103	6.01	121	24.2	2.46	3.02	115	36.9
61	7.7	—	37.0	48.1	2.70	2.15	640	92.2	5.98	107	21.5	2.45	2.20	102	32.8
190	3.7	—	13.6	—	3.50	0.65	1890	263	5.82	589	93.0	3.25	48.8	311	143
170	4.0	—	14.6	—	3.47	0.72	1650	235	5.74	517	82.3	3.22	35.6	275	126
152	4.5	—	15.8	—	3.44	0.79	1430	209	5.66	454	72.8	3.19	25.8	243	111
136	5.0	—	17.0	—	3.41	0.87	1240	186	5.58	398	64.2	3.16	18.5	214	98.0
120	5.6	—	18.5	—	3.38	0.96	1070	163	5.51	345	56.0	3.13	12.9	186	85.4
106	6.2	—	21.1	—	3.36	1.07	933	145	5.47	301	49.3	3.11	9.13	164	75.1
96	6.8	—	23.1	—	3.34	1.16	833	131	5.44	270	44.4	3.09	6.86	147	67.5
87	7.5	—	24.3	—	3.32	1.28	740	118	5.38	241	39.7	3.07	5.10	132	60.4
79	8.2	62.6	26.3	—	3.31	1.39	662	107	5.34	216	35.8	3.05	3.84	119	54.3
72	9.0	52.3	28.5	—	3.29	1.52	597	97.4	5.31	195	32.4	3.04	2.93	108	49.2
65	9.9	43.0	31.1	—	3.28	1.67	533	87.9	5.28	174	29.1	3.02	2.18	96.8	44.1
58	7.8	—	33.9	57.6	2.72	1.90	475	78.0	5.28	107	21.4	2.51	2.10	86.4	32.5
53	8.7	55.9	35.0	54.1	2.71	2.10	425	70.6	5.23	95.8	19.2	2.48	1.58	77.9	29.1
50	6.3	—	32.9	60.9	2.17	2.36	394	64.7	5.18	56.3	13.9	1.96	1.78	72.4	21.4
45	7.0	—	36.0	51.0	2.15	2.61	350	58.1	5.15	50.0	12.4	1.94	1.31	64.7	19.0
40	7.8	—	40.5	40.3	2.14	2.90	310	51.9	5.13	44.1	11.0	1.93	0.95	57.5	16.8
35	6.3	—	41.7	38.0	1.74	3.66	285	45.6	5.25	24.5	7.47	1.54	0.74	51.2	11.5
30	7.4	—	47.5	29.3	1.73	4.30	238	38.6	5.21	20.3	6.24	1.52	0.46	43.1	9.56
26	8.5	57.9	53.1	23.4	1.72	4.95	204	33.4	5.17	17.3	5.34	1.51	0.30	37.2	8.17

Table 2.10 (*Continued*)

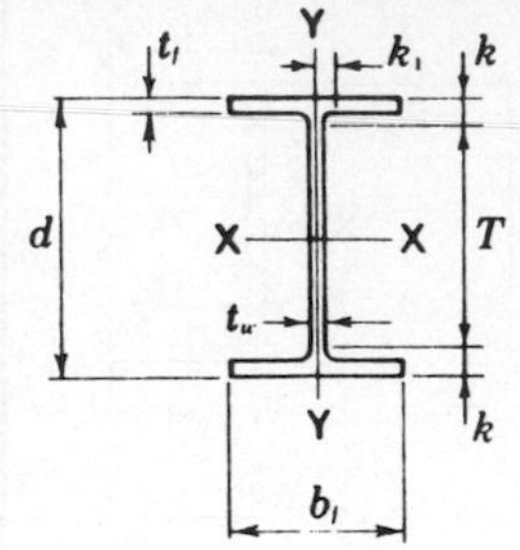

W SHAPES
Dimensions

Designation	Area A	Depth d		Web: Thickness t_w		Web: $\frac{t_w}{2}$	Flange: Width b_f		Flange: Thickness t_f		Distance: T	Distance: k	Distance: k_1
	In.^{2}	In.		In.		In.	In.		In.		In.	In.	In.
W 10x112	32.9	11.36	11 3/8	0.755	3/4	3/8	10.415	10 3/8	1.250	1 1/4	7 5/8	1 7/8	15/16
x100	29.4	11.10	11 1/8	0.680	11/16	3/8	10.340	10 3/8	1.120	1 1/8	7 5/8	1 3/4	7/8
x 88	25.9	10.84	10 7/8	0.605	5/8	5/16	10.265	10 1/4	0.990	1	7 5/8	1 5/8	13/16
x 77	22.6	10.60	10 5/8	0.530	1/2	1/4	10.190	10 1/4	0.870	7/8	7 5/8	1 1/2	13/16
x 68	20.0	10.40	10 3/8	0.470	1/2	1/4	10.130	10 1/8	0.770	3/4	7 5/8	1 3/8	3/4
x 60	17.6	10.22	10 1/4	0.420	7/16	1/4	10.080	10 1/8	0.680	11/16	7 5/8	1 5/16	3/4
x 54	15.8	10.09	10 1/8	0.370	3/8	3/16	10.030	10	0.615	5/8	7 5/8	1 1/4	11/16
x 49	14.4	9.98	10	0.340	5/16	3/16	10.000	10	0.560	9/16	7 5/8	1 3/16	11/16
W 10x 45	13.3	10.10	10 1/8	0.350	3/8	3/16	8.020	8	0.620	5/8	7 5/8	1 1/4	11/16
x 39	11.5	9.92	9 7/8	0.315	5/16	3/16	7.985	8	0.530	1/2	7 5/8	1 1/8	11/16
x 33	9.71	9.73	9 3/4	0.290	5/16	3/16	7.960	8	0.435	7/16	7 5/8	1 1/16	11/16
W 10x 30	8.84	10.47	10 1/2	0.300	5/16	3/16	5.810	5 3/4	0.510	1/2	8 5/8	15/16	1/2
x 26	7.61	10.33	10 3/8	0.260	1/4	1/8	5.770	5 3/4	0.440	7/16	8 5/8	7/8	1/2
x 22	6.49	10.17	10 1/8	0.240	1/4	1/8	5.750	5 3/4	0.360	3/8	8 5/8	3/4	1/2
W 8x67	19.7	9.00	9	0.570	9/16	5/16	8.280	8 1/4	0.935	15/16	6 1/8	1 7/16	11/16
x58	17.1	8.75	8 3/4	0.510	1/2	1/4	8.220	8 1/4	0.810	13/16	6 1/8	1 5/16	11/16
x48	14.1	8.50	8 1/2	0.400	3/8	3/16	8.110	8 1/8	0.685	11/16	6 1/8	1 3/16	5/8
x40	11.7	8.25	8 1/4	0.360	3/8	3/16	8.070	8 1/8	0.560	9/16	6 1/8	1 1/16	5/8
x35	10.3	8.12	8 1/8	0.310	5/16	3/16	8.020	8	0.495	1/2	6 1/8	1	9/16
x31	9.13	8.00	8	0.285	5/16	3/16	7.995	8	0.435	7/16	6 1/8	15/16	9/16
W 8x28	8.25	8.06	8	0.285	5/16	3/16	6.535	6 1/2	0.465	7/16	6 1/8	15/16	9/16
x24	7.08	7.93	7 7/8	0.245	1/4	1/8	6.495	6 1/2	0.400	3/8	6 1/8	7/8	9/16
W 8x21	6.16	8.28	8 1/4	0.250	1/4	1/8	5.270	5 1/4	0.400	3/8	6 5/8	13/16	1/2
x18	5.26	8.14	8 1/8	0.230	1/4	1/8	5.250	5 1/4	0.330	5/16	6 5/8	3/4	7/16
W 8x15	4.44	8.11	8 1/8	0.245	1/4	1/8	4.015	4	0.315	5/16	6 5/8	3/4	1/2
x13	3.84	7.99	8	0.230	1/4	1/8	4.000	4	0.255	1/4	6 5/8	11/16	7/16
x10	2.96	7.89	7 7/8	0.170	3/16	1/8	3.940	4	0.205	3/16	6 5/8	5/8	7/16
W 6x25	7.34	6.38	6 3/8	0.320	5/16	3/16	6.080	6 1/8	0.455	7/16	4 3/4	13/16	7/16
x20	5.87	6.20	6 1/4	0.260	1/4	1/8	6.020	6	0.365	3/8	4 3/4	3/4	7/16
x15	4.43	5.99	6	0.230	1/4	1/8	5.990	6	0.260	1/4	4 3/4	5/8	3/8
W 6x16	4.74	6.28	6 1/4	0.260	1/4	1/8	4.030	4	0.405	3/8	4 3/4	3/4	7/16
x12	3.55	6.03	6	0.230	1/4	1/8	4.000	4	0.280	1/4	4 3/4	5/8	3/8
x 9	2.68	5.90	5 7/8	0.170	3/16	1/8	3.940	4	0.215	3/16	4 3/4	9/16	3/8

Table 2.10 (*Continued*)

W SHAPES
Properties

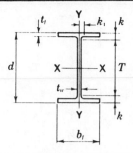

Nominal Wt. per Ft.	Compact Section Criteria				r_T	$\dfrac{d}{A_f}$	Elastic Properties						Torsional constant J	Plastic Modulus	
	$\dfrac{b_f}{2t_f}$	F_y'	$\dfrac{d}{t_w}$	F_y'''			Axis X-X			Axis Y-Y				Z_x	Z_y
							I	S	r	I	S	r			
Lb.		Ksi		Ksi	In.		In.⁴	In.³	In.	In.⁴	In.³	In.	In.⁴	In.³	In.³
112	4.2	—	15.0	—	2.88	0.87	716	126	4.66	236	45.3	2.68	15.1	147	69.2
100	4.6	—	16.3	—	2.85	0.96	623	112	4.60	207	40.0	2.65	10.9	130	61.0
88	5.2	—	17.9	—	2.83	1.07	534	98.5	4.54	179	34.8	2.63	7.53	113	53.1
77	5.9	—	20.0	—	2.80	1.20	455	85.9	4.49	154	30.1	2.60	5.11	97.6	45.9
68	6.6	—	22.1	—	2.79	1.33	394	75.7	4.44	134	26.4	2.59	3.56	85.3	40.1
60	7.4	—	24.3	—	2.77	1.49	341	66.7	4.39	116	23.0	2.57	2.48	74.6	35.0
54	8.2	63.5	27.3	—	2.75	1.64	303	60.0	4.37	103	20.6	2.56	1.82	66.6	31.3
49	8.9	53.0	29.4	—	2.74	1.78	272	54.6	4.35	93.4	18.7	2.54	1.39	60.4	28.3
45	6.5	—	28.9	—	2.18	2.03	248	49.1	4.32	53.4	13.3	2.01	1.51	54.9	20.3
39	7.5	—	31.5	—	2.16	2.34	209	42.1	4.27	45.0	11.3	1.98	0.98	46.8	17.2
33	9.1	50.5	33.6	58.7	2.14	2.81	170	35.0	4.19	36.6	9.20	1.94	0.58	38.8	14.0
30	5.7	—	34.9	54.2	1.55	3.53	170	32.4	4.38	16.7	5.75	1.37	0.62	36.6	8.84
26	6.6	—	39.7	41.8	1.54	4.07	144	27.9	4.35	14.1	4.89	1.36	0.40	31.3	7.50
22	8.0	—	42.4	36.8	1.51	4.91	118	23.2	4.27	11.4	3.97	1.33	0.24	26.0	6.10
67	4.4	—	15.8	—	2.28	1.16	272	60.4	3.72	88.6	21.4	2.12	5.06	70.2	32.7
58	5.1	—	17.2	—	2.26	1.31	228	52.0	3.65	75.1	18.3	2.10	3.34	59.8	27.9
48	5.9	—	21.3	—	2.23	1.53	184	43.3	3.61	60.9	15.0	2.08	1.96	49.0	22.9
40	7.2	—	22.9	—	2.21	1.83	146	35.5	3.53	49.1	12.2	2.04	1.12	39.8	18.5
35	8.1	64.4	26.2	—	2.20	2.05	127	31.2	3.51	42.6	10.6	2.03	0.77	34.7	16.1
31	9.2	50.0	28.1	—	2.18	2.30	110	27.5	3.47	37.1	9.27	2.02	0.54	30.4	14.1
28	7.0	—	28.3	—	1.77	2.65	98.0	24.3	3.45	21.7	6.63	1.62	0.54	27.2	10.1
24	8.1	64.1	32.4	63.0	1.76	3.05	82.8	20.9	3.42	18.3	5.63	1.61	0.35	23.2	8.57
21	6.6	—	33.1	60.2	1.41	3.93	75.3	18.2	3.49	9.77	3.71	1.26	0.28	20.4	5.69
18	8.0	—	35.4	52.7	1.39	4.70	61.9	15.2	3.43	7.97	3.04	1.23	0.17	17.0	4.66
15	6.4	—	33.1	60.3	1.03	6.41	48.0	11.8	3.29	3.41	1.70	0.876	0.14	13.6	2.67
13	7.8	—	34.7	54.7	1.01	7.83	39.6	9.91	3.21	2.73	1.37	0.843	0.09	11.4	2.15
10	9.6	45.8	46.4	30.7	0.99	9.77	30.8	7.81	3.22	2.09	1.06	0.841	0.04	8.87	1.66
25	6.7	—	19.9	—	1.66	2.31	53.4	16.7	2.70	17.1	5.61	1.52	0.46	18.9	8.56
20	8.2	62.1	23.8	—	1.64	2.82	41.4	13.4	2.66	13.3	4.41	1.50	0.24	14.9	6.72
15	11.5	31.8	26.0	—	1.61	3.85	29.1	9.72	2.56	9.32	3.11	1.46	0.10	10.8	4.75
16	5.0	—	24.2	—	1.08	3.85	32.1	10.2	2.60	4.43	2.20	0.966	0.22	11.7	3.39
12	7.1	—	26.2	—	1.05	5.38	22.1	7.31	2.49	2.99	1.50	0.918	0.09	8.30	2.32
9	9.2	50.3	34.7	54.8	1.03	6.96	16.4	5.56	2.47	2.19	1.11	0.905	0.04	6.23	1.72

Table 2.10 (*Continued*)

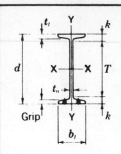

S SHAPES
Dimensions

Designation	Area A	Depth d		Web Thickness t_w		$\dfrac{t_w}{2}$	Flange Width b_f		Flange Thickness t_f		Distance T	k	Grip	Max. Flge. Fastener
	In.²	In.		In.		In.	In.		In.		In.	In.	In.	In.
S 18x70	20.6	18.00	18	0.711	11/16	3/8	6.251	6 1/4	0.691	11/16	15	1 1/2	11/16	7/8
x54.7	16.1	18.00	18	0.461	7/16	1/4	6.001	6	0.691	11/16	15	1 1/2	11/16	7/8
S 15x50	14.7	15.00	15	0.550	9/16	5/16	5.640	5 5/8	0.622	5/8	12 1/4	1 3/8	9/16	3/4
x42.9	12.6	15.00	15	0.411	7/16	1/4	5.501	5 1/2	0.622	5/8	12 1/4	1 3/8	9/16	3/4
S 12x50	14.7	12.00	12	0.687	11/16	3/8	5.477	5 1/2	0.659	11/16	9 1/8	1 7/16	11/16	3/4
x40.8	12.0	12.00	12	0.462	7/16	1/4	5.252	5 1/4	0.659	11/16	9 1/8	1 7/16	5/8	3/4
S 12x35	10.3	12.00	12	0.428	7/16	1/4	5.078	5 1/8	0.544	9/16	9 5/8	1 3/16	1/2	3/4
x31.8	9.35	12.00	12	0.350	3/8	3/16	5.000	5	0.544	9/16	9 5/8	1 3/16	1/2	3/4
S 10x35	10.3	10.00	10	0.594	5/8	5/16	4.944	5	0.491	1/2	7 3/4	1 1/8	1/2	3/4
x25.4	7.46	10.00	10	0.311	5/16	3/16	4.661	4 5/8	0.491	1/2	7 3/4	1 1/8	1/2	3/4
S 8x23	6.77	8.00	8	0.441	7/16	1/4	4.171	4 1/8	0.426	7/16	6	1	7/16	3/4
x18.4	5.41	8.00	8	0.271	1/4	1/8	4.001	4	0.426	7/16	6	1	7/16	3/4
x15.3	4.50	7.00	7	0.252	1/4	1/8	3.662	3 5/8	0.392	3/8	5 1/8	15/16	3/8	5/8
S 6x17.25	5.07	6.00	6	0.465	7/16	1/4	3.565	3 5/8	0.359	3/8	4 1/4	7/8	3/8	5/8
x12.5	3.67	6.00	6	0.232	1/4	1/8	3.332	3 3/8	0.359	3/8	4 1/4	7/8	3/8	—

Table 2.10   (*Continued*)

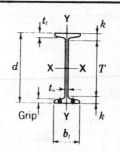

S SHAPES
Properties

Nominal Wt. per Ft	$\frac{b_f}{2t_f}$	F_y'	$\frac{d}{t_w}$	F_y'''	r_T	$\frac{d}{A_f}$	Axis X-X I	S	r	Axis Y-Y I	S	r	Torsional constant J	Z_x	Z_y
Lb.		Ksi		Ksi	In.		In.⁴	In.³	In.	In.⁴	In.³	In.	In.⁴	In.³	In.³
70	4.5	—	25.3	—	1.36	4.17	926	103	6.71	24.1	7.72	1.08	4.15	125	14.4
54.7	4.3	—	39.0	43.3	1.37	4.34	804	89.4	7.07	20.8	6.94	1.14	2.37	105	12.1
50	4.5	—	27.3	—	1.26	4.28	486	64.8	5.75	15.7	5.57	1.03	2.12	77.1	9.97
42.9	4.4	—	36.5	49.6	1.26	4.38	447	59.6	5.95	14.4	5.23	1.07	1.54	69.3	9.02
50	4.2	—	17.5	—	1.25	3.32	305	50.8	4.55	15.7	5.74	1.03	2.82	61.2	10.3
40.8	4.0	—	26.0	—	1.24	3.46	272	45.4	4.77	13.6	5.16	1.06	1.76	53.1	8.85
35	4.7	—	28.0	—	1.16	4.34	229	38.2	4.72	9.87	3.89	0.980	1.08	44.8	6.79
31.8	4.6	—	34.3	56.2	1.16	4.41	218	36.4	4.83	9.36	3.74	1.00	0.90	42.0	6.40
35	5.0	—	16.8	—	1.10	4.12	147	29.4	3.78	8.36	3.38	0.901	1.29	35.4	6.22
25.4	4.7	—	32.2	63.9	1.09	4.37	124	24.7	4.07	6.79	2.91	0.954	0.60	28.4	4.96
23	4.9	—	18.1	—	0.95	4.51	64.9	16.2	3.10	4.31	2.07	0.798	0.55	19.3	3.68
18.4	4.7	—	29.5	—	0.94	4.70	57.6	14.4	3.26	3.73	1.86	0.831	0.34	16.5	3.16
15.3	4.7	—	27.8	—	0.87	4.88	36.7	10.5	2.86	2.64	1.44	0.766	0.24	12.1	2.44
17.25	5.0	—	12.9	—	0.81	4.69	26.3	8.77	2.28	2.31	1.30	0.675	0.37	10.6	2.36
12.5	4.6	—	25.9	—	0.79	5.02	22.1	7.37	2.45	1.82	1.09	0.705	0.17	8.47	1.85

Table 2.10 (*Continued*)

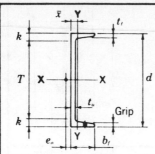

CHANNELS
AMERICAN STANDARD
Dimensions

Designation	Area A	Depth d	Web Thickness t_w	$\frac{t_w}{2}$	Flange Width b_f		Flange Average thickness t_f		Distance T	Distance k	Grip	Max. Flge. Fastener
	In.²	In.	In.	In.	In.		In.		In.	In.	In.	In.
C 12x30	8.82	12.00	0.510	½	¼	3.170	3⅛	0.501	½	9¾	1⅛	½	⅞
x25	7.35	12.00	0.387	⅜	³⁄₁₆	3.047	3	0.501	½	9¾	1⅛	½	⅞
x20.7	6.09	12.00	0.282	⁵⁄₁₆	⅛	2.942	3	0.501	½	9¾	1⅛	½	⅞
C 10x30	8.82	10.00	0.673	¹¹⁄₁₆	⁵⁄₁₆	3.033	3	0.436	⁷⁄₁₆	8	1	⁷⁄₁₆	¾
x25	7.35	10.00	0.526	½	¼	2.886	2⅞	0.436	⁷⁄₁₆	8	1	⁷⁄₁₆	¾
x20	5.88	10.00	0.379	⅜	³⁄₁₆	2.739	2¾	0.436	⁷⁄₁₆	8	1	⁷⁄₁₆	¾
x15.3	4.49	10.00	0.240	¼	⅛	2.600	2⅝	0.436	⁷⁄₁₆	8	1	⁷⁄₁₆	¾
C 9x20	5.88	9.00	0.448	⁷⁄₁₆	¼	2.648	2⅝	0.413	⁷⁄₁₆	7⅛	¹⁵⁄₁₆	⁷⁄₁₆	¾
x15	4.41	9.00	0.285	⁵⁄₁₆	⅛	2.485	2½	0.413	⁷⁄₁₆	7⅛	¹⁵⁄₁₆	⁷⁄₁₆	¾
x13.4	3.94	9.00	0.233	¼	⅛	2.433	2⅜	0.413	⁷⁄₁₆	7⅛	¹⁵⁄₁₆	⁷⁄₁₆	¾
C 8x18.75	5.51	8.00	0.487	½	¼	2.527	2½	0.390	⅜	6⅛	¹⁵⁄₁₆	⅜	¾
x13.75	4.04	8.00	0.303	⁵⁄₁₆	⅛	2.343	2⅜	0.390	⅜	6⅛	¹⁵⁄₁₆	⅜	¾
x11.5	3.38	8.00	0.220	¼	⅛	2.260	2¼	0.390	⅜	6⅛	¹⁵⁄₁₆	⅜	¾
C 6x13	3.83	6.00	0.437	⁷⁄₁₆	³⁄₁₆	2.157	2⅛	0.343	⁵⁄₁₆	4⅜	¹³⁄₁₆	⁵⁄₁₆	⅝
x10.5	3.09	6.00	0.314	⁵⁄₁₆	³⁄₁₆	2.034	2	0.343	⁵⁄₁₆	4⅜	¹³⁄₁₆	⅜	⅝
x 8.2	2.40	6.00	0.200	³⁄₁₆	⅛	1.920	1⅞	0.343	⁵⁄₁₆	4⅜	¹³⁄₁₆	⁵⁄₁₆	⅝
C 4x 7.25	2.13	4.00	0.321	⁵⁄₁₆	³⁄₁₆	1.721	1¾	0.296	⁵⁄₁₆	2⅝	¹¹⁄₁₆	⁵⁄₁₆	⅝
x 5.4	1.59	4.00	0.184	³⁄₁₆	¹⁄₁₆	1.584	1⅝	0.296	⁵⁄₁₆	2⅝	¹¹⁄₁₆	—	—
C 3x 6	1.76	3.00	0.356	⅜	³⁄₁₆	1.596	1⅝	0.273	¼	1⅝	¹¹⁄₁₆	—	—
x 5	1.47	3.00	0.258	¼	⅛	1.498	1½	0.273	¼	1⅝	¹¹⁄₁₆	—	—
x 4.1	1.21	3.00	0.170	³⁄₁₆	¹⁄₁₆	1.410	1⅜	0.273	¼	1⅝	¹¹⁄₁₆	—	—

Table 2.10 (*Continued*)

CHANNELS
AMERICAN STANDARD
Properties

Nominal Weight per Ft.	$\bar{x}$	Shear Center Location e_o	$\dfrac{d}{A_f}$	Axis X-X			Axis Y-Y		
				I	S	r	I	S	r
	In.	In.		In.^{4}	In.^{3}	In.	In.^{4}	In.^{3}	In.
30	0.674	0.618	7.55	162	27.0	4.29	5.14	2.06	0.763
25	0.674	0.746	7.85	144	24.1	4.43	4.47	1.88	0.780
20.7	0.698	0.870	8.13	129	21.5	4.61	3.88	1.73	0.799
30	0.649	0.369	7.55	103	20.7	3.42	3.94	1.65	0.669
25	0.617	0.494	7.94	91.2	18.2	3.52	3.36	1.48	0.676
20	0.606	0.637	8.36	78.9	15.8	3.66	2.81	1.32	0.692
15.3	0.634	0.796	8.81	67.4	13.5	3.87	2.28	1.16	0.713
20	0.583	0.515	8.22	60.9	13.5	3.22	2.42	1.17	0.642
15	0.586	0.682	8.76	51.0	11.3	3.40	1.93	1.01	0.661
13.4	0.601	0.743	8.95	47.9	10.6	3.48	1.76	0.962	0.669
18.75	0.565	0.431	8.12	44.0	11.0	2.82	1.98	1.01	0.599
13.75	0.553	0.604	8.75	36.1	9.03	2.99	1.53	0.854	0.615
11.5	0.571	0.697	9.08	32.6	8.14	3.11	1.32	0.781	0.625
13	0.514	0.380	8.10	17.4	5.80	2.13	1.05	0.642	0.525
10.5	0.499	0.486	8.59	15.2	5.06	2.22	0.866	0.564	0.529
8.2	0.511	0.599	9.10	13.1	4.38	2.34	0.693	0.492	0.537
7.25	0.459	0.386	7.84	4.59	2.29	1.47	0.433	0.343	0:450
5.4	0.457	0.502	8.52	3.85	1.93	1.56	0.319	0.283	0.449
6	0.455	0.322	6.87	2.07	1.38	1.08	0.305	0.268	0.416
5	0.438	0.392	7.32	1.85	1.24	1.12	0.247	0.233	0.410
4.1	0.436	0.461	7.78	1.66	1.10	1.17	0.197	0.202	0.404

Table 2.10 *(Continued)*

ANGLES
Equal legs and unequal legs
Properties for designing

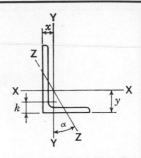

Size and Thickness	k	Weight per Foot	Area	AXIS X-X				AXIS Y-Y				AXIS Z-Z	
				I	S	r	y	I	S	r	x	r	Tan
In.	In.	Lb.	In.^{2}	In.^{4}	In.^{3}	In.	In.	In.^{4}	In.^{3}	In.	In.	In.	α
L 6 x6 x1	$1\frac{1}{2}$	37.4	11.0	35.5	8.57	1.80	1.86	35.5	8.57	1.80	1.86	1.17	1.000
$\frac{7}{8}$	$1\frac{3}{8}$	33.1	9.73	31.9	7.63	1.81	1.82	31.9	7.63	1.81	1.82	1.17	1.000
$\frac{3}{4}$	$1\frac{1}{4}$	28.7	8.44	28.2	6.66	1.83	1.78	28.2	6.66	1.83	1.78	1.17	1.000
$\frac{5}{8}$	$1\frac{1}{8}$	24.2	7.11	24.2	5.66	1.84	1.73	24.2	5.66	1.84	1.73	1.18	1.000
$\frac{1}{2}$	1	19.6	5.75	19.9	4.61	1.86	1.68	19.9	4.61	1.86	1.68	1.18	1.000
$\frac{3}{8}$	$\frac{7}{8}$	14.9	4.36	15.4	3.53	1.88	1.64	15.4	3.53	1.88	1.64	1.19	1.000
L 6 x4 x $\frac{3}{4}$	$1\frac{1}{4}$	23.6	6.94	24.5	6.25	1.88	2.08	8.68	2.97	1.12	1.08	0.860	0.428
$\frac{5}{8}$	$1\frac{1}{8}$	20.0	5.86	21.1	5.31	1.90	2.03	7.52	2.54	1.13	1.03	0.864	0.435
$\frac{1}{2}$	1	16.2	4.75	17.4	4.33	1.91	1.99	6.27	2.08	1.15	0.987	0.870	0.440
$\frac{3}{8}$	$\frac{7}{8}$	12.3	3.61	13.5	3.32	1.93	1.94	4.90	1.60	1.17	0.941	0.877	0.446
L 6 x3$\frac{1}{2}$x $\frac{3}{8}$	$\frac{7}{8}$	11.7	3.42	12.9	3.24	1.94	2.04	3.34	1.23	0.988	0.787	0.767	0.350
$\frac{5}{16}$	$\frac{13}{16}$	9.8	2.87	10.9	2.73	1.95	2.01	2.85	1.04	0.996	0.763	0.772	0.352
L 4 x4 x $\frac{3}{4}$	$1\frac{1}{8}$	18.5	5.44	7.67	2.81	1.19	1.27	7.67	2.81	1.19	1.27	0.778	1.000
$\frac{5}{8}$	1	15.7	4.61	6.66	2.40	1.20	1.23	6.66	2.40	1.20	1.23	0.779	1.000
$\frac{1}{2}$	$\frac{7}{8}$	12.8	3.75	5.56	1.97	1.22	1.18	5.56	1.97	1.22	1.18	0.782	1.000
$\frac{3}{8}$	$\frac{3}{4}$	9.8	2.86	4.36	1.52	1.23	1.14	4.36	1.52	1.23	1.14	0.788	1.000
$\frac{5}{16}$	$\frac{11}{16}$	8.2	2.40	3.71	1.29	1.24	1.12	3.71	1.29	1.24	1.12	0.791	1.000
$\frac{1}{4}$	$\frac{5}{8}$	6.6	1.94	3.04	1.05	1.25	1.09	3.04	1.05	1.25	1.09	0.795	1.000
L 4 x3$\frac{1}{2}$x $\frac{1}{2}$	$\frac{15}{16}$	11.9	3.50	5.32	1.94	1.23	1.25	3.79	1.52	1.04	1.00	0.722	0.750
$\frac{3}{8}$	$\frac{13}{16}$	9.1	2.67	4.18	1.49	1.25	1.21	2.95	1.17	1.06	0.955	0.727	0.755
$\frac{5}{16}$	$\frac{3}{4}$	7.7	2.25	3.56	1.26	1.26	1.18	2.55	0.994	1.07	0.932	0.730	0.757
$\frac{1}{4}$	$\frac{11}{16}$	6.2	1.81	2.91	1.03	1.27	1.16	2.09	0.808	1.07	0.909	0.734	0.759
L 4 x3 x $\frac{3}{8}$	$\frac{13}{16}$	8.5	2.48	3.96	1.46	1.26	1.28	1.92	0.866	0.879	0.782	0.644	0.551
$\frac{5}{16}$	$\frac{3}{4}$	7.2	2.09	3.38	1.23	1.27	1.26	1.65	0.734	0.887	0.759	0.647	0.554
$\frac{1}{4}$	$\frac{11}{16}$	5.8	1.69	2.77	1.00	1.28	1.24	1.36	0.599	0.896	0.736	0.651	0.558

Angles in shaded rows may not be readily available.   Availability is subject to rolling accumulation and geographical location, and should be checked with material suppliers.

Table 2.10 *(Continued)*

ANGLES
Equal legs and unequal legs
Properties for designing

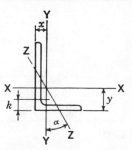

Size and Thickness	k	Weight per Foot	Area	AXIS X-X				AXIS Y-Y				AXIS Z-Z	
				I	S	r	y	I	S	r	x	r	Tan
In.	In.	Lb.	In.²	In.⁴	In.³	In.	In.	In.⁴	In.³	In.	In.	In.	α
L 3 x3 x ½	13/16	9.4	2.75	2.22	1.07	0.898	0.932	2.22	1.07	0.898	0.932	0.584	1.000
3/8	11/16	7.2	2:11	1.76	0.833	0.913	0.888	1.76	0.833	0.913	0.888	0.587	1.000
5/16	5/8	6.1	1.78	1.51	0.707	0.922	0.865	1.51	0.707	0.922	0.865	0.589	1.000
¼	9/16	4.9	1.44	1.24	0.577	0.930	0.842	1.24	0.577	0.930	0.842	0.592	1.000
3/16	½	3.71	1.09	0.962	0.441	0.939	0.820	0.962	0.441	0.939	0.820	0.596	1.000
L 3 x2½x 3/8	¾	6.6	1.92	1.66	0.810	0.928	0.956	1.04	0.581	0.736	0.706	0.522	0.676
¼	5/8	4.5	1.31	1.17	0.561	0.945	0.911	0.743	0.404	0.753	0.661	0.528	0.684
3/16	9/16	3.39	0.996	0.907	0.430	0.954	0.888	0.577	0.310	0.761	0.638	0.533	0.688
L 3 x2 x 3/8	11/16	5.9	1.73	1.53	0.781	0.940	1.04	0.543	0.371	0.559	0.539	0.430	0.428
5/16	5/8	5.0	1.46	1.32	0.664	0.948	1.02	0.470	0.317	0.567	0.516	0.432	0.435
¼	9/16	4.1	1.19	1.09	0.542	0.957	0.993	0.392	0.260	0.574	0.493	0.435	0.440
3/16	½	3.07	0.902	0.842	0.415	0.966	0.970	0.307	0.200	0.583	0.470	0.439	0.446
L 2½x2½x 3/8	11/16	5.9	1.73	0.984	0.566	0.753	0.762	0.984	0.566	0.753	0.762	0.487	1.000
5/16	5/8	5.0	1.46	0.849	0.482	0.761	0.740	0.849	0.482	0.761	0.740	0.489	1.000
¼	9/16	4.1	1.19	0.703	0.394	0.769	0.717	0.703	0.394	0.769	0.717	0.491	1.000
3/16	½	3.07	0.902	0.547	0.303	0.778	0.694	0.547	0.303	0.778	0.694	0.495	1.000
L 2½x2 x 3/8	11/16	5.3	1.55	0.912	0.547	0.768	0.831	0.514	0.363	0.577	0.581	0.420	0.614
5/16	5/8	4.5	1.31	0.788	0.466	0.776	0.809	0.446	0.310	0.584	0.559	0.422	0.620
¼	9/16	3.62	1.06	0.654	0.381	0.784	0.787	0.372	0.254	0.592	0.537	0.424	0.626
3/16	½	2.75	0.809	0.509	0.293	0.793	0.764	0.291	0.196	0.600	0.514	0.427	0.631
L 2 x2 x 3/8	11/16	4.7	1.36	0.479	0.351	0.594	0.636	0.479	0.351	0.594	0.636	0.389	1.000
5/16	5/8	3.92	1.15	0.416	0.300	0.601	0.614	0.416	0.300	0.601	0.614	0.390	1.000
¼	9/16	3.19	0.938	0.348	0.247	0.609	0.592	0.348	0.247	0.609	0.592	0.391	1.000
3/16	½	2.44	0.715	0.272	0.190	0.617	0.569	0.272	0.190	0.617	0.569	0.394	1.000
⅛	7/16	1.65	0.484	0.190	0.131	0.626	0.546	0.190	0.131	0.626	0.546	0.398	1.000

Angles in shaded rows may not be readily available. Availability is subject to rolling accumulation and geographical location, and should be checked with material suppliers.

AMERICAN INSTITUTE OF STEEL CONSTRUCTION

Table 2.11   Square and Round Bars[1]

	SQUARE AND ROUND BARS								
	Weight and area								
Size Inches	Weight Lb. per Foot ■	●	Area Square Inches ▨	◎	Size Inches	Weight Lb. per Foot ■	●	Area Square Inches ▨	◎
0					3	30.63	24.05	9.000	7.069
1/16	0.013	0.010	0.0039	0.0031	1/16	31.91	25.07	9.379	7.366
1/8	0.053	0.042	0.0156	0.0123	1/8	33.23	26.10	9.766	7.670
3/16	0.120	0.094	0.0352	0.0276	3/16	34.57	27.15	10.160	7.980
1/4	0.213	0.167	0.0625	0.0491	1/4	35.94	28.23	10.563	8.296
5/16	0.332	0.261	0.0977	0.0767	5/16	37.34	29.32	10.973	8.618
3/8	0.479	0.376	0.1406	0.1105	3/8	38.76	30.44	11.391	8.946
7/16	0.651	0.512	0.1914	0.1503	7/16	40.21	31.58	11.816	9.281
1/2	0.851	0.668	0.2500	0.1963	1/2	41.68	32.74	12.250	9.621
9/16	1.077	0.846	0.3164	0.2485	9/16	43.19	33.92	12.691	9.968
5/8	1.329	1.044	0.3906	0.3068	5/8	44.71	35.12	13.141	10.321
11/16	1.608	1.263	0.4727	0.3712	11/16	46.27	36.34	13.598	10.680
3/4	1.914	1.503	0.5625	0.4418	3/4	47.85	37.58	14.063	11.045
13/16	2.246	1.764	0.6602	0.5185	13/16	49.46	38.85	14.535	11.416
7/8	2.605	2.046	0.7656	0.6013	7/8	51.09	40.13	15.016	11.793
15/16	2.991	2.349	0.8789	0.6903	15/16	52.76	41.43	15.504	12.177
1	3.403	2.673	1.0000	0.7854	4	54.44	42.76	16.000	12.566
1/16	3.841	3.017	1.1289	0.8866	1/16	56.16	44.11	16.504	12.962
1/8	4.307	3.382	1.2656	0.9940	1/8	57.90	45.47	17.016	13.364
3/16	4.798	3.769	1.4102	1.1075	3/16	59.67	46.86	17.535	13.772
1/4	5.317	4.176	1.5625	1.2272	1/4	61.46	48.27	18.063	14.186
5/16	5.862	4.604	1.7227	1.3530	5/16	63.28	49.70	18.598	14.607
3/8	6.433	5.053	1.8906	1.4849	3/8	65.13	51.15	19.141	15.033
7/16	7.032	5.523	2.0664	1.6230	7/16	67.01	52.63	19.691	15.466
1/2	7.656	6.013	2.2500	1.7671	1/2	68.91	54.12	20.250	15.904
9/16	8.308	6.525	2.4414	1.9175	9/16	70.83	55.63	20.816	16.349
5/8	8.985	7.057	2.6406	2.0739	5/8	72.79	57.17	21.391	16.800
11/16	9.690	7.610	2.8477	2.2365	11/16	74.77	58.72	21.973	17.257
3/4	10.421	8.185	3.0625	2.4053	3/4	76.78	60.30	22.563	17.721
13/16	11.179	8.780	3.2852	2.5802	13/16	78.81	61.90	23.160	18.190
7/8	11.963	9.396	3.5156	2.7612	7/8	80.87	63.51	23.766	18.665
15/16	12.774	10.032	3.7539	2.9483	15/16	82.96	65.15	24.379	19.147
2	13.611	10.690	4.0000	3.1416	5	85.07	66.81	25.000	19.635
1/16	14.475	11.369	4.2539	3.3410	1/16	87.21	68.49	25.629	20.129
1/8	15.366	12.068	4.5156	3.5466	1/8	89.38	70.20	26.266	20.629
3/16	16.283	12.788	4.7852	3.7583	3/16	91.57	71.92	26.910	21.135
1/4	17.227	13.530	5.0625	3.9761	1/4	93.79	73.66	27.563	21.648
5/16	18.197	14.292	5.3477	4.2000	5/16	96.04	75.43	28.223	22.166
3/8	19.194	15.075	5.6406	4.4301	3/8	98.31	77.21	28.891	22.691
7/16	20.217	15.879	5.9414	4.6664	7/16	100.61	79.02	29.566	23.221
1/2	21.267	16.703	6.2500	4.9087	1/2	102.93	80.84	30.250	23.758
9/16	22.344	17.549	6.5664	5.1572	9/16	105.29	82.69	30.941	24.301
5/8	23.447	18.415	6.8906	5.4119	5/8	107.67	84.56	31.641	24.850
11/16	24.577	19.303	7.2227	5.6727	11/16	110.07	86.45	32.348	25.406
3/4	25.734	20.211	7.5625	5.9396	3/4	112.50	88.36	33.063	25.967
13/16	26.917	21.140	7.9102	6.2126	13/16	114.96	90.29	33.785	26.535
7/8	28.126	22.090	8.2656	6.4918	7/8	117.45	92.24	34.516	27.109
15/16	29.362	23.061	8.6289	6.7771	15/16	119.96	94.22	35.254	27.688
3	30.625	24.053	9.0000	7.0686	6	122.50	96.21	36.000	28.274

Source: American Institute of Steel Construction.

Table 2.11 (*Continued*)

SQUARE AND ROUND BARS
Weight and area

Size Inches	Weight Lb. per Foot ■	Weight Lb. per Foot ●	Area Square Inches ▨	Area Square Inches ⊘	Size Inches	Weight Lb. per Foot ■	Weight Lb. per Foot ●	Area Square Inches ▨	Area Square Inches ⊘
6	122.50	96.21	36.000	28.274	9	275.63	216.48	81.000	63.617
1/16	125.07	98.23	36.754	28.866	1/16	279.47	219.49	82.129	64.504
1/8	127.66	100.26	37.516	29.465	1/8	283.33	222.53	83.266	65.397
3/16	130.28	102.32	38.285	30.069	3/16	287.23	225.59	84.410	66.296
1/4	132.92	104.40	39.063	30.680	1/4	291.15	228.67	85.563	67.201
5/16	135.59	106.49	39.848	31.296	5/16	295.10	231.77	86.723	68.112
3/8	138.29	108.61	40.641	31.919	3/8	299.07	234.89	87.891	69.029
7/16	141.02	110.75	41.441	32.548	7/16	303.07	238.03	89.066	69.953
1/2	143.77	112.91	42.250	33.183	1/2	307.10	241.20	90.250	70.882
9/16	146.55	115.10	43.066	33.824	9/16	311.15	244.38	91.441	71.818
5/8	149.35	117.30	43.891	34.472	5/8	315.24	247.59	92.641	72.760
11/16	152.18	119.52	44.723	35.125	11/16	319.34	250.81	93.848	73.708
3/4	155.04	121.77	45.563	35.785	3/4	323.48	254.06	95.063	74.662
13/16	157.92	124.03	46.410	36.450	13/16	327.64	257.33	96.285	75.622
7/8	160.83	126.32	47.266	37.122	7/8	331.82	260.61	97.516	76.589
15/16	163.77	128.63	48.129	37.800	15/16	336.04	263.92	98.754	77.561
7	166.74	130.95	49.000	38.485	10	340.28	267.25	100.000	78.540
1/16	169.73	133.30	49.879	39.175	1/16	344.54	270.60	101.254	79.525
1/8	172.74	135.67	50.766	39.871	1/8	348.84	273.98	102.516	80.516
3/16	175.79	138.06	51.660	40.574	3/16	353.16	277.37	103.785	81.513
1/4	178.86	140.48	52.563	41.282	1/4	357.50	280.78	105.063	82.516
5/16	181.96	142.91	53.473	41.997	5/16	361.88	284.22	106.348	83.525
3/8	185.08	145.36	54.391	42.718	3/8	366.28	287.67	107.641	84.541
7/16	188.23	147.84	55.316	43.445	7/16	370.70	291.15	108.941	85.562
1/2	191.41	150.33	56.250	44.179	1/2	375.16	294.65	110.250	86.590
9/16	194.61	152.85	57.191	44.918	9/16	379.64	298.17	111.566	87.624
5/8	197.84	155.38	58.141	45.664	5/8	384.14	301.70	112.891	88.664
11/16	201.10	157.94	59.098	46.415	11/16	388.67	305.26	114.223	89.710
3/4	204.38	160.52	60.063	47.173	3/4	393.23	308.84	115.563	90.763
13/16	207.69	163.12	61.035	47.937	13/16	397.82	312.45	116.910	91.821
7/8	211.03	165.74	62.016	48.707	7/8	402.43	316.07	118.266	92.886
15/16	214.39	168.38	63.004	49.483	15/16	407.07	319.71	119.629	93.956
8	217.78	171.04	64.000	50.265	11	411.74	323.38	121.000	95.033
1/16	221.19	173.73	65.004	51.054	1/16	416.43	327.06	122.379	96.116
1/8	224.64	176.43	66.016	51.849	1/8	421.15	330.77	123.766	97.205
3/16	228.11	179.15	67.035	52.649	3/16	425.89	334.49	125.160	98.301
1/4	231.60	181.90	68.063	53.456	1/4	430.66	338.24	126.563	99.402
5/16	235.12	184.67	69.098	54.269	5/16	435.46	342.01	127.973	100.510
3/8	238.67	187.45	70.141	55.088	3/8	440.29	345.80	129.391	101.623
7/16	242.25	190.26	71.191	55.914	7/16	445.14	349.61	130.816	102.743
1/2	245.85	193.09	72.250	56.745	1/2	450.02	353.44	132.250	103.869
9/16	249.48	195.94	73.316	57.583	9/16	454.92	357.30	133.691	105.001
5/8	253.13	198.81	74.391	58.426	5/8	459.85	361.17	135.141	106.139
11/16	256.82	201.70	75.473	59.276	11/16	464.81	365.06	136.598	107.284
3/4	260.53	204.62	76.563	60.132	3/4	469.80	368.98	138.063	108.434
13/16	264.26	207.55	77.660	60.994	13/16	474.81	372.91	139.535	109.591
7/8	268.02	210.50	78.766	61.862	7/8	479.84	376.87	141.016	110.753
15/16	271.81	213.48	79.879	62.737	15/16	484.91	380.85	142.504	111.922
9	275.63	216.48	81.000	63.617	12	490.00	384.85	144.000	113.097

Adhesives for the manufacture of glulam must comply with Voluntary Product Standard PS 56-73. Wet-use adhesives must be used if the members are subject to occasional or continuous wetting, or for applications, either exterior or interior, where the moisture content of the wood will exceed 16%.

Timber construction has historically been recognized as an economical type of construction. Laminated wood does not require the added expense of false ceilings to cover or disguise the structural framework. Glulam members can be used to provide long clear spans eliminating interior walls and supports.

Since wood substance is relatively inert chemically, under normal conditions it is not subject to chemical change or deterioration. It is resistant to most acids, rust, and other corrosive agents. Thus, it is often used where chemical deterioration eliminates use of other structural materials.

Heavy timber sizes used in glulam construction are difficult to ignite. Glulam burns slowly and resists heat penetration by the formation of self-insulating char. In a large member subjected to fire, the uncharred inner portion maintains its strength.

Wood members change less in dimension with variations in temperature than do other materials. Usually, the effects of thermal expansion are negligible compared to dimensional changes (shrinkage or swelling) due to changes in moisture content, and the two changes tend to offset each other. Longitudinal dimensional changes can be neglected for most structural designs.

Glulams are available in a variety of appearance grades and finishes. Stress grades can be specified and range from 1600 to 2400 psi (11.2 to 16.6 MPa), extreme fiber tension when loaded in bending, with compression values, parallel to grain, from 1050 to 1700 psi (7.2 to 11.7 MPa).

Sizes of glulams most readily available range from 3⅛ to 10¾ inches (79 to 273 mm) in width with depths from 4½ to 60 inches (114 mm to 1.5 m).

Precalculated capacity tables and additional data are available from the American Institute of Timber Construction.

2.9.2   Standard Sizes for Finished Lumber

Standard sizes for finished lumber have recently been established as shown in Tables 2.12 and 2.13.

Machine stress-rated (MSR) lumber is also available. In the process, each piece of lumber is machine tested (nondestructively) to establish its modulus of elasticity. This lumber is also required to meet certain visual grading requirements.

2.9.3   Structural Lumber Design Values

Design values for stress-graded lumber are given in Table 2.14.

The allowable stresses to be used in design must, of course, conform to the requirements of the local building code. Many municipal codes are revised only infrequently and consequently may not be in agreement with current industry-recommended stress levels. The design values shown are those given in the *National Design Specification for Wood Construction* (1982) and recommended by the National Forest Products Association.

Table 2.12   Standard Sizes for Finished Dry^a Lumber

Nominal Thickness	Surfaced Thickness	Nominal Width	Surfaced Width
⅜″	5⁄16″	2″	1½″
½″	7⁄16″	3″	2½″
⅝″	9⁄16″	4″	3½″
¾″	⅝″	5″	4½″
1″	¾″	6″	5½″
1¼″	1″	7″	6½″
1½″	1¼″	8″ and wider ¾″	
1¾″	1⅜″	off nominal	
2″	1½″		
2½″	2″		
3″	2½″		
3½″	3″		
4″	3½″		

Standard lengths of finish are 3′ and longer in multiples of 1′. In the Superior grade, 3% of 3′ and 4′ and 7% of 3′ to 6′ are permitted. In the Prime grade, 20% of 3′ to 6′ is permitted.

^{a}Dry is defined as having a moisture content (by weight) of 19% or less.
Source: From Parker & Hauf, *Simplified Design of Structural Wood,* John Wiley & Sons, 1979.

2.10   REINFORCED CONCRETE

The American Concrete Institute (ACI) publishes material on all phases of concrete technology. Their *Manual of Concrete Practice* is a five-part compilation of current ACI standards and committee reports and is broadly accepted as the standard for concrete work of all types. Each standard of the institute bears a hyphenated number to identify it. The first three digits identify the committee originating the standard and the last two digits identify the year it was adopted. Thus standard ACI 214-77 was prepared by Committee 214 and was adopted as a standard in the year 1977.

The five groups of the standards and committee reports are:

Group	Example
100—Research and Administration	
This group contains all research and administration committees, including any committees not logically placed in other subdivisions.	ACI 116R-78, Cement and Concrete Terminology, Parts 1 and 2.
200—Materials and Properties of Concrete	
This group contains committees whose major concern is materials in concrete and properties of concrete.	ACI 211.1-81, Standard Practice for Selecting Proportions for Normal, Heavyweight, and Mass Concrete, Part 1.

Continued on p. 152.

Table 2.13 Properties of Structural Lumber—Standard Dressed Sizes*

Sizes

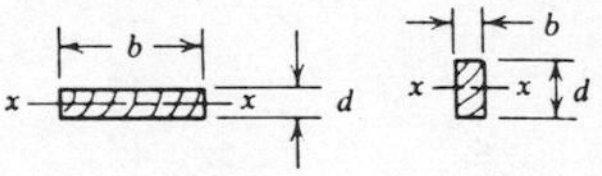

Nominal size (in.) b d	Dressed size (in.) b d	Area of section (sq in.) A	Moment of inertia (in.⁴) I	Section modulus (in.³) S	Weight per linear foot†
2 × 4	1½ × 3½	5.250	5.359	3.063	1.458
2 × 6	1½ × 5½	8.250	20.797	7.563	2.292
2 × 8	1½ × 7¼	10.875	47.635	13.141	3.021
2 × 10	1½ × 9¼	13.875	98.932	21.391	3.854
2 × 12	1½ × 11¼	16.875	177.979	31.641	4.688
2 × 14	1½ × 13¼	19.875	290.775	43.891	5.521
3 × 2	2½ × 1½	3.750	0.703	0.938	1.042
3 × 4	2½ × 3½	8.750	8.932	5.104	2.431
3 × 6	2½ × 5½	13.750	34.661	12.604	3.819
3 × 8	2½ × 7¼	18.125	79.391	21.901	5.035
3 × 10	2½ × 9¼	23.125	164.886	35.651	6.424
3 × 12	2½ × 11¼	28.125	296.631	52.734	7.813
3 × 14	2½ × 13¼	33.125	484.625	73.151	9.201
3 × 16	2½ × 15¼	38.125	738.870	96.901	10.590
4 × 2	3½ × 1½	5.250	0.984	1.313	1.458
4 × 3	3½ × 2½	8.750	4.557	3.646	2.431
4 × 4	3½ × 3½	12.250	12.505	7.146	3.403
4 × 6	3½ × 5½	19.250	48.526	17.646	5.347
4 × 8	3½ × 7¼	25.375	111.148	30.661	7.049
4 × 10	3½ × 9¼	32.375	230.840	49.911	8.933
4 × 12	3½ × 11¼	39.375	415.283	73.828	10.938
4 × 14	3½ × 13¼	46.375	678.475	102.411	12.877
4 × 16	3½ × 15¼	53.375	1,034.418	135.66	14.828
6 × 2	5½ × 1½	8.250	1.547	2.063	2.292
6 × 3	5½ × 2½	13.750	7.161	5.729	3.819
6 × 4	5½ × 3½	19.250	19.651	11.229	5.347
6 × 6	5½ × 5½	30.250	76.255	27.729	8.403
6 × 8	5½ × 7½	41.250	193.359	51.563	11.458
6 × 10	5½ × 9½	52.250	392.963	82.729	14.514
6 × 12	5½ × 11½	63.250	697.068	121.229	17.569
6 × 14	5½ × 13½	74.250	1,127.672	167.063	20.625
6 × 16	5½ × 15½	85.250	1,706.776	220.229	23.681

Source: Compiled from data in the 1982 edition of the *National Design Specification for Wood Construction*. Courtesy of National Forest Products Association.

*Based on an assumed average weight of 40 lb per cu ft.

Table 2.13 *(Continued)*

Sizes *(continued)*

Nominal size (in.) $b \quad d$	Dressed size (in.) $b \quad d$	Area of section (sq in.) A	Moment of inertia (in.^{4}) I	Section modulus (in.^{3}) S	Weight per linear foot†
8×2	$7\frac{1}{4} \times 1\frac{1}{2}$	10.875	2.039	2.719	3.021
8×3	$7\frac{1}{4} \times 2\frac{1}{2}$	18.125	9.440	7.552	5.035
8×4	$7\frac{1}{4} \times 3\frac{1}{2}$	25.375	25.904	14.802	7.049
8×6	$7\frac{1}{2} \times 5\frac{1}{2}$	41.250	103.984	37.813	11.458
8×8	$7\frac{1}{2} \times 7\frac{1}{2}$	56.250	263.672	70.313	15.625
8×10	$7\frac{1}{2} \times 9\frac{1}{2}$	71.250	535.859	112.813	19.792
8×12	$7\frac{1}{2} \times 11\frac{1}{2}$	86.250	950.547	165.313	23.958
8×14	$7\frac{1}{2} \times 13\frac{1}{2}$	101.250	1,537.734	227.813	28.125
8×16	$7\frac{1}{2} \times 15\frac{1}{2}$	116.250	2,327.422	300.313	32.292
10×2	$9\frac{1}{4} \times 1\frac{1}{2}$	13.875	2.602	3.469	3.854
10×3	$9\frac{1}{4} \times 2\frac{1}{2}$	23.125	12.044	9.635	6.424
10×4	$9\frac{1}{4} \times 3\frac{1}{2}$	32.375	33.049	18.885	8.993
10×6	$9\frac{1}{2} \times 5\frac{1}{2}$	52.250	131.714	47.896	14.514
10×8	$9\frac{1}{2} \times 7\frac{1}{2}$	71.250	333.984	89.063	19.792
10×10	$9\frac{1}{2} \times 9\frac{1}{2}$	90.250	678.755	142.896	25.069
10×12	$9\frac{1}{2} \times 11\frac{1}{2}$	109.250	1,204.026	209.396	30.347
10×14	$9\frac{1}{2} \times 13\frac{1}{2}$	128.250	1,947.797	288.563	35.625
10×16	$9\frac{1}{2} \times 15\frac{1}{2}$	147.250	2,948.068	380.396	40.903
10×18	$9\frac{1}{2} \times 17\frac{1}{2}$	166.250	4,242.836	484.896	46.181
12×2	$11\frac{1}{4} \times 1\frac{1}{2}$	16.875	3.164	4.219	4.688
12×3	$11\frac{1}{4} \times 2\frac{1}{2}$	28.125	14.648	11.719	7.813
12×4	$11\frac{1}{4} \times 3\frac{1}{2}$	39.375	40.195	22.969	10.938
12×6	$11\frac{1}{2} \times 5\frac{1}{2}$	63.250	159.443	57.979	17.569
12×8	$11\frac{1}{2} \times 7\frac{1}{2}$	86.250	404.297	107.813	23.958
12×10	$11\frac{1}{2} \times 9\frac{1}{2}$	109.250	821.651	172.979	30.347
12×12	$11\frac{1}{2} \times 11\frac{1}{2}$	132.250	1,457.505	253.479	36.736
12×14	$11\frac{1}{2} \times 13\frac{1}{2}$	155.250	2,357.859	349.313	43.125
12×16	$11\frac{1}{2} \times 15\frac{1}{2}$	178.250	3,568.713	460.479	49.514
14×16	$13\frac{1}{2} \times 15\frac{1}{2}$	209.250	4,189.359	540.563	58.125
14×18	$13\frac{1}{2} \times 17\frac{1}{2}$	236.250	6,029.297	689.063	65.625
14×20	$13\frac{1}{2} \times 19\frac{1}{2}$	263.250	8,341.734	855.563	73.125
14×22	$13\frac{1}{2} \times 21\frac{1}{2}$	290.250	11,180.672	1,040.063	80.625

† Based on an assumed average weight of 40 lb per cu ft.

Table 2.14 Design Values for Visually Graded Structural Lumber[5]

Species and commercial grade	Size classification	Extreme fiber in bending "F_b" Single-member uses	Extreme fiber in bending "F_b" Repetitive member uses	Tension parallel to grain "F_t"	Horizontal shear "F_v"	Compression perpendicular to grain "$F_{c\perp}$"	Compression parallel to grain "F_c"	Modulus of elasticity "E"	Grading rules agency
CALIFORNIA REDWOOD (Surfaced dry or surfaced green. Used at 19% max. m.c.)									
Clear Heart Structural	4" & less thick,	2300	2650	1550	145	650	2150	1,400,000	
Clear Structural	any width	2300	2650	1550	145	650	2150	1,400,000	
Select Structural		2050	2350	1200	80	650	1750	1,400,000	
Select Structural, Open grain		1600	1850	950	80	425	1300	1,100,000	
No. 1	2" to 4"	1700	1950	975	80	650	1400	1,400,000	
No. 1, Open grain	thick	1350	1550	775	80	425	1050	1,100,000	
No. 2	2" to 4"	1400	1600	800	80	650	1100	1,250,000	
No. 2, Open grain	wide	1100	1250	625	80	425	825	1,000,000	
No. 3		800	900	475	80	650	675	1,100,000	
No. 3, Open grain		625	725	375	80	425	500	900,000	
Stud		625	725	375	80	425	500	900,000	
Construction	2" to 4"	825	950	475	80	425	925	900,000	
Standard	thick	450	525	250	80	425	775	900,000	RIS
Utility	4" wide	225	250	125	80	425	500	900,000	
Select Structural		1750	2000	1150	80	650	1550	1,400,000	(see footnotes
Select Structural, Open grain		1400	1600	925	80	425	1150	1,100,000	1 through 6,
No. 1	2" to 4"	1500	1700	975	80	650	1400	1,400,000	8 through 10
No. 1, Open grain	thick	1150	1350	775	80	425	1050	1,100,000	and 12)
No. 2	5" and	1200	1400	660	80	650	1200	1,250,000	
No. 2, Open grain	wider	950	1100	500	80	425	875	1,000,000	
No. 3		700	800	375	80	650	725	1,100,000	
No. 3, Open grain		550	650	350	80	425	525	900,000	
Stud		700	800	375	80	650	725	1,100,000	
Clear Heart Structural or Clear Structural	5" by 5" and larger	1850	—	1250	135	650	1650	1,300,000	
Select Structural		1400	—	950	95	650	1200	1,300,000	
No. 1		1200	—	800	95	650	1050	1,300,000	
No. 2		975	—	650	95	650	900	1,100,000	
No. 3		550	—	375	95	650	550	1,000,000	
Select Decking, Close grain	Decking	1850	2150	—	—	—	—	1,400,000	RIS
Select Decking	2" thick	1450	1700	—	—	—	—	1,100,000	(see footnotes 1,
Commercial Decking	6" and wider	1200	1350	—	—	—	—	1,000,000	2, 8 and 9)
COAST SITKA SPRUCE (Surfaced dry or surfaced green. Used at 19% max. m.c.)									
Select Structural		1500	1700	875	65	455	1100	1,700,000	
No. 1	2" to 3"	1250	1450	750	65	455	875	1,700,000	
No. 2	thick	1050	1200	625	65	455	700	1,500,000	
No. 3	2" to 4"	575	675	350	65	455	425	1,300,000	
Appearance	wide	1250	1450	725	65	455	1050	1,700,000	
Stud		575	675	350	65	455	425	1,300,000	
Construction	2" to 4"	750	875	450	65	455	800	1,300,000	
Standard	thick	425	500	250	65	455	650	1,300,000	NLGA
Utility	4" wide	200	225	125	65	455	425	1,300,000	
Select Structural	2" to 4"	1300	1500	850	65	455	975	1,700,000	(A Canadian
No. 1	thick	1100	1250	725	65	455	875	1,700,000	agency. See
No. 2	5" and	900	1050	475	65	455	750	1,500,000	footnotes1
No. 3	wider	525	600	275	65	455	475	1,300,000	through 12
Appearance		1100	1250	725	65	455	1050	1,700,000	and 15
Stud		525	600	275	65	455	475	1,300,000	and 16)
Select Structural	Beams and	1150	—	675	60	455	775	1,500,000	
No. 1	Stringers	950	—	475	60	455	650	1,500,000	
Select Structural	Posts and	1100	—	725	60	455	825	1,500,000	
No. 1	Timbers	875	—	575	60	455	725	1,500,000	
Select	Decking	1250	1450	—	—	455	—	1,700,000	
Commercial		1050	1200	—	—	455	—	1,500,000	
COAST SPECIES (Surfaced dry or surfaced green. Used at 19% max. m.c.)									
Select Structural		1500	1700	875	65	370	1100	1,500,000	
No. 1	2" to 3"	1250	1450	750	65	370	875	1,500,000	
No. 2	thick	1050	1200	625	65	370	700	1,400,000	
No. 3	2" to 4"	575	675	350	65	370	425	1,200,000	
Appearance	wide	1250	1450	725	65	370	1050	1,500,000	
Stud		575	675	350	65	370	425	1,200,000	
Construction	2" to 4"	750	875	450	65	370	800	1,200,000	
Standard	thick	425	500	250	65	370	650	1,200,000	NLGA
Utility	4" wide	200	225	125	65	370	425	1,200,000	
Select Structural	2" to 4"	1300	1500	850	65	370	975	1,500,000	(A Canadian
No. 1	thick	1100	1250	725	65	370	875	1,500,000	agency. See
No. 2	5" and	900	1050	475	65	370	750	1,400,000	footnotes 1
No. 3	wider	525	600	275	65	370	475	1,200,000	through 12
Appearance		1100	1250	725	65	370	1050	1,500,000	and 15
Stud		525	600	275	65	370	475	1,200,000	and 16)
Select	Decking	1250	1450	—	—	370	—	1,500,000	
Commercial		1050	1200	—	—	370	—	1,400,000	

Source: Compiled from data in the 1982 edition of the *National Design Specification for Wood Construction.* Courtesy of National Forest Products Association.

Table 2.14 (Continued)

(Design values listed are for normal loading conditions. See other provisions in the footnotes and in the National Design Specification for adjustments of tabulated values.)

...es and commercial grade	Size classification	Extreme fiber in bending "F_b" Single-member uses	Extreme fiber in bending "F_b" Repetitive member uses	Tension parallel to grain "F_t"	Horizontal shear "F_v"	Compression perpendicular to grain "$F_{c\perp}$"	Compression parallel to grain "F_c"	Modulus of elasticity "E"	Grading rules agency
...TONWOOD (Surfaced dry or surfaced green. Used at 19% max. m.c.)									
Stud	2" to 3" thick 2" to 4" wide	525	600	300	65	320	350	1,000,000	NHPMA (see footnotes 1 through 12)
Construction	2" to 4" thick	675	775	400	65	320	650	1,000,000	
Standard		375	425	225	65	320	525	1,000,000	
Utility	4" wide	175	200	100	65	320	350	1,000,000	
...GLAS FIR-LARCH (Surfaced dry or surfaced green. Used at 19% max. m.c.)									
Dense Select Structural		2450	2800	1400	95	730	1850	1,900,000	
Select Structural		2100	2400	1200	95	625	1600	1,800,000	
Dense No. 1		2050	2400	1200	95	730	1450	1,900,000	
No. 1	2" to 4"	1750	2050	1050	95	625	1250	1,800,000	
Dense No. 2	thick	1700	1950	1000	95	730	1150	1,700,000	
No. 2	2" to 4"	1450	1650	850	95	625	1000	1,700,000	
No. 3	wide	800	925	475	95	625	600	1,500,000	
Appearance		1750	2050	1050	95	625	1500	1,800,000	WCLIB
Stud		800	925	475	95	625	600	1,500,000	WWPA
Construction	2" to 4"	1050	1200	625	95	625	1150	1,500,000	
Standard	thick	600	675	350	95	625	925	1,500,000	
Utility	4" wide	275	325	175	95	625	600	1,500,000	(see footnotes 1 through 12)
Dense Select Structural		2100	2400	1400	85	730	1650	1,900,000	
Select Structural		1800	2050	1200	95	625	1400	1,800,000	
Dense No. 1	2" to 4"	1800	2050	1200	95	730	1450	1,900,000	
No. 1	thick	1500	1750	1000	95	625	1250	1,800,000	
Dense No. 2	5" and	1450	1700	775	95	730	1250	1,700,000	
No. 2	wider	1250	1450	850	95	625	1050	1,700,000	
No. 3		725	850	375	95	625	675	1,500,000	
Appearance		1500	1750	1000	95	625	1500	1,800,000	
Stud		725	850	375	95	625	675	1,500,000	
Dense Select Structural		1900	---	1100	85	730	1300	1,700,000	
Select Structural	Beams and	1600	---	950	85	625	1100	1,600,000	
Dense No. 1	Stringers	1550	---	775	85	730	1100	1,700,000	
No. 1		1300	---	675	85	625	925	1,600,000	
Dense Select Structural		1750	---	1150	85	730	1350	1,700,000	WCLIB
Select Structural	Posts and	1500	---	1000	85	625	1150	1,600,000	
Dense No. 1	Timbers	1400	---	950	85	730	1200	1,700,000	(see footnotes 1 through 12)
No. 1		1200	---	825	85	625	1000	1,600,000	
Select Dex	Decking	1750	2000	---	---	625	---	1,800,000	
Commercial Dex		1450	1650	---	---	625	---	1,700,000	
Dense Select Structural		1900	---	1250	85	730	1300	1,700,000	
Select Structural	Beams and	1600	---	1050	85	625	1100	1,600,000	
Dense No. 1	Stringers	1550	---	1050	85	730	1100	1,700,000	
No. 1		1350	---	900	85	625	925	1,600,000	WWPA
Dense Select Structural		1750	---	1150	85	730	1350	1,700,000	
Select Structural	Posts and	1500	---	1000	85	625	1150	1,600,000	
Dense No. 1	Timbers	1400	---	950	85	730	1200	1,700,000	(see footnotes 1 through 13)
No. 1		1200	---	825	85	625	1000	1,600,000	
Selected Decking	Decking	---	2000	---	---	---	---	1,800,000	
Commercial Decking		---	1650	---	---	---	---	1,700,000	
Selected Decking	Decking	---	2150	(Surfaced at 15% max. m.c. and			---	1,900,000	
Commercial Decking		---	1800	used at 15% max. m.c.)			---	1,700,000	
...UGLAS FIR-LARCH (NORTH) (Surfaced dry or surfaced green. Used at 19% max. m.c.)									
Select Structural		2100	2400	1200	95	625	1550	1,800,000	
No. 1	2" to 3"	1750	2050	1050	95	625	1250	1,800,000	
No. 2	thick	1450	1650	850	95	625	1000	1,700,000	
No. 3	2" to 4"	800	925	475	95	625	600	1,500,000	
Appearance	wide	1750	2050	1050	95	625	1500	1,800,000	
Stud		800	925	475	95	625	600	1,500,000	
Construction	2" to 4"	1050	1200	625	95	625	1150	1,500,000	NLGA
Standard	thick	600	675	350	95	625	925	1,500,000	
Utility	4" wide	275	325	175	95	625	600	1,500,000	(A Canadian agency. See footnotes 1 through 12 and 15 and 16)
Select Structural	2" to 4"	1800	2050	1200	95	625	1400	1,800,000	
No. 1	thick	1500	1750	1000	95	625	1250	1,800,000	
No. 2	5" and	1250	1450	850	95	625	1050	1,700,000	
No. 3	wider	725	850	375	95	625	675	1,500,000	
Appearance		1500	1750	1000	95	625	1500	1,800,000	
Stud		725	850	375	85	625	675	1,500,000	
Select Structural	Beams and	1600	---	950	85	625	1100	1,600,000	
No. 1	Stringers	1300	---	675	85	625	925	1,600,000	
Select Structural	Posts and	1500	---	1000	85	625	1150	1,600,000	
No. 1	Timbers	1200	---	825	85	625	1000	1,600,000	
Select	Decking	1750	2000	---	---	625	---	1,800,000	
Commercial		1450	1650	---	---	625	---	1,700,000	

147

Table 2.14   (*Continued*)

Species and commercial grade	Size classification	Extreme fiber in bending "F_b" Single-member uses	Repetitive-member uses	Tension parallel to grain "F_t"	Horizontal shear "F_v"	Compression perpendicular to grain "F_{c⊥}"	Compression parallel to grain "F_c"	Modulus of elasticity "E"	Grading rules agency
ENGELMANN SPRUCE–ALPINE FIR (ENGELMANN SPRUCE–LODGEPOLE PINE) (Surfaced dry or surfaced green. Used at 19% max. m.c.)									
Select Structural		1350	1550	800	70	320	950	1,300,000	
No. 1	2" to 4"	1150	1350	675	70	320	750	1,300,000	
No. 2	thick	950	1100	550	70	320	600	1,100,000	
No. 3	2" to 4"	525	600	300	70	320	375	1,000,000	
Appearance	wide	1150	1350	675	70	320	900	1,300,000	
Stud		525	600	300	70	320	375	1,000,000	
Construction	2" to 4"	700	800	400	70	320	675	1,000,000	
Standard	thick	375	450	225	70	320	550	1,000,000	
Utility	4" wide	175	200	100	70	320	375	1,000,000	
Select Structural		1200	1350	775	70	320	850	1,300,000	
No. 1	2" to 4"	1000	1150	675	70	320	750	1,300,000	WWPA
No. 2	thick	825	950	425	70	320	625	1,100,000	(see footnotes
No. 3	5" and	475	550	250	70	320	400	1,000,000	1 through 13)
Appearance	wider	1000	1150	675	70	320	900	1,300,000	
Stud		475	550	250	70	320	400	1,000,000	
Select Structural	Beams and	1050	—	700	65	320	675	1,100,000	
No. 1	Stringers	875	—	600	65	320	550	1,100,000	
Select Structural	Posts and	975	—	650	65	320	700	1,100,000	
No. 1	Timbers	800	—	525	65	320	625	1,100,000	
Selected Decking	Decking	—	1300	—	—	—	—	1,300,000	
Commercial Decking		—	1100	—	—	—	—	1,100,000	
Selected Decking	Decking	—	1400	(Surfaced at 15% max. m.c. and				—	1,300,000	
Commercial Decking		—	1200	used at 15% max. m.c.)				—	1,200,000	
HEM-FIR (Surfaced dry or surfaced green. Used at 19% max. m.c.)									
Select Structural		1650	1900	975	75	405	1300	1,500,000	
No. 1	2" to 4"	1400	1600	825	75	405	1050	1,500,000	
No. 2	thick	1150	1350	675	75	405	825	1,400,000	
No. 3	2" to 4"	650	725	375	75	405	500	1,200,000	
Appearance	wide	1400	1600	825	75	405	1250	1,500,000	
Stud		650	725	375	75	405	500	1,200,000	
Construction	2" to 4"	825	975	500	75	405	925	1,200,000	WCLIB
Standard	thick	475	550	275	75	405	775	1,200,000	WWPA
Utility	4" wide	225	250	125	75	405	500	1,200,000	
Select Structural		1400	1650	950	75	405	1150	1,500,000	(see footnotes
No. 1	2" to 4"	1200	1400	800	75	405	1050	1,500,000	1 through 12)
No. 2	thick	1000	1150	625	75	405	875	1,400,000	
No. 3	5" and	575	675	300	75	405	550	1,200,000	
Appearance	wider	1200	1400	800	75	405	1250	1,500,000	
Stud		575	675	300	75	405	550	1,200,000	
Select Structural	Beams and	1300	—	750	70	405	925	1,300,000	
No. 1	Stringers	1050	—	525	70	405	750	1,300,000	
Select Structural	Posts and	1200	—	800	70	405	975	1,300,000	WCLIB
No. 1	Timbers	975	—	650	70	405	850	1,300,000	
Select Dex	Decking	1400	1600	—	—	405	—	1,500,000	(see footnotes
Commercial Dex		1150	1350	—	—	405	—	1,400,000	1 through 12)
Select Structural	Beams and	1250	—	850	70	405	925	1,300,000	
No. 1	Stringers	1050	—	725	70	405	775	1,300,000	
Select Structural	Posts and	1200	—	800	70	405	975	1,300,000	WWPA
No. 1	Timbers	950	—	650	70	405	850	1,300,000	
Selected Decking	Decking	—	1600	—	—	—	—	1,500,000	
Commercial Decking		—	1350	—	—	—	—	1,400,000	(see footnotes
Selected Decking	Decking	—	1700	(Surfaced at 15% max. m.c. and				—	1,600,000	1 through 1
Commercial Decking		—	1450	used at 15% max. m.c.)				—	1,400,000	

Table 2.14 (*Continued*)

es and commercial grade	Size classification	Design values in pounds per square inch: Extreme fiber in bending "F_b" — Single-member uses	Extreme fiber in bending "F_b" — Repetitive-member uses	Tension parallel to grain "F_t"	Horizontal shear "F_v"	Compression perpendicular to grain "$F_{c\perp}$"	Compression parallel to grain "F_c"	Modulus of elasticity "E"	Grading rules agency
HERN PINE (Surfaced dry. Used at 19% max. m.c.)									
elect Structural		2000	2300	1150	100	565	1550	1,700,000	
ense Select Structural		2350	2700	1350	100	660	1800	1,800,000	
o. 1		1700	1950	1000	100	565	1250	1,700,000	
o. 1 Dense	2" to 4" thick	2000	2300	1150	100	660	1450	1,800,000	
o. 2	2" to 4"	1400	1650	825	90	565	975	1,600,000	
o. 2 Dense	wide	1650	1900	975	90	660	1150	1,600,000	
o. 3		775	900	450	90	565	575	1,400,000	
o. 3 Dense		925	1050	525	90	660	675	1,500,000	
ud		775	900	450	90	565	575	1,400,000	
onstruction	2" to 4"	1000	1150	600	100	565	1100	1,400,000	
andard	thick	575	675	350	90	565	900	1,400,000	
tility	4" wide	275	300	150	90	565	575	1,400,000	
elect Structural		1750	2000	See Footnote 3 1150	90	565	1350	1,700,000	
ense Select Structural		2050	2350	1300	90	660	1600	1,800,000	
o. 1		1450	1700	975	90	565	1250	1,700,000	SPIB
o. 1 Dense	2" to 4" thick	1700	2000	1150	90	660	1450	1,800,000	
o. 2	5" and	1200	1400	625	90	565	1000	1,600,000	(see footnotes
o. 2 Dense	wider	1400	1650	725	90	660	1200	1,600,000	1, 3, 4, 5, 6,
o. 3		700	800	350	90	565	625	1,400,000	12, 17 18
o. 3 Dense		825	925	425	90	660	725	1,500,000	and 19)
ud		725	850	350	90	565	625	1,400,000	
ense Standard Decking	2" to 4"	2000	2300	—	—	660	—	1,800,000	
elect Decking	thick	1400	1650	—	—	565	—	1,600,000	
ense Select Decking	2" and	1650	1900	—	—	660	—	1,600,000	
ommercial Decking	wider	1400	1650	—	—	565	—	1,600,000	
ense Commercial Decking	Decking	1650	1900	—	—	660	—	1,600,000	
ense Structural 86	2" to 4"	2600	3000	1750	155	660	2000	1,800,000	
ense Structural 72	thick	2200	2550	1450	130	660	1650	1,800,000	
ense Structural 65		2000	2300	1300	115	660	1500	1,800,000	
HERN PINE (Surfaced green. Used any condition)									
elect Structural		1600	1850	925	95	375	1050	1,500,000	
ense Select Structural		1850	2150	1100	95	440	1200	1,600,000	
o. 1		1350	1550	800	95	375	825	1,500,000	
o. 1 Dense	2½" to 4" thick	1600	1800	925	95	440	950	1,600,000	
o. 2	2½" to 4" wide	1150	1300	675	85	375	650	1,400,000	
o. 2 Dense		1350	1500	775	85	440	750	1,400,000	
o. 3		625	725	375	85	375	400	1,200,000	
o. 3 Dense		725	850	425	85	440	450	1,300,000	
ud		625	725	375	85	375	400	1,200,000	
onstruction	2½" to 4"	825	925	475	95	375	725	1,200,000	
tandard	thick	475	525	275	85	375	600	1,200,000	
tility	4" wide	200	250	125	85	375	400	1,200,000	
elect Structural		1400	1600	See Footnote 3 900	85	375	900	1,500,000	
ense Select Structural		1600	1850	1050	85	440	1050	1,600,000	
o. 1		1200	1350	775	85	375	825	1,500,000	
o. 1 Dense	2½" to 4" thick	1400	1600	925	85	440	950	1,600,000	
o. 2	5" and	975	1100	500	85	375	675	1,400,000	SPIB
o. 2 Dense	wider	1150	1300	600	85	440	800	1,400,000	
o. 3		550	650	300	85	375	425	1,200,000	(see footnotes
o. 3 Dense		650	750	350	85	440	475	1,300,000	1,3,4,5,6,
tud		575	675	300	85	375	425	1,200,000	12, 17,18 and 19)
ense Standard Decking	2½" to 4"	1600	1800	—	—	440	—	1,600,000	
elect Decking	thick	1150	1300	—	—	375	—	1,400,000	
ense Select Decking	2" and	1350	1500	—	—	440	—	1,400,000	
ommercial Decking	wider	1150	1300	—	—	375	—	1,400,000	
ense Commercial Decking	Decking	1350	1500	—	—	440	—	1,400,000	
o. 1 SR		1350	—	875	110	375	775	1,500,000	
o. 1 Dense SR	5" and	1550	—	1050	110	440	925	1,600,000	
o. 2 SR	thicker	1100	—	725	95	375	625	1,400,000	
o. 2 Dense SR		1250	—	850	95	440	725	1,400,000	
ense Structural 86	2½" and	2100	2400	1400	145	440	1300	1,600,000	
ense Structural 72	thicker	1750	2050	1200	120	440	1100	1,600,000	
ense Structural 65		1600	1800	1050	110	440	1000	1,600,000	

Table 2.14 (*Continued*)

Applicable to Visually Graded Structural Lumber

1. Following is a list of agencies certified by the American Lumber Standards Committee Board of Review (as of 1982) for inspection and grading o untreated lumber under the rules indicated. For the most up-to-date list of certified agencies, write to:

American Lumber Standards Committee
Suite #204, 20010 Century Boulevard
Germantown, MD 20767

Rules Writing Agencies	Rules for which grading authorized
Northeastern Lumber Manufacturers Association (NELMA)	NELMA, NLGA
4 Fundy Road, Falmouth, Maine 04105	
Northern Hardwood and Pine Manufacturers Association (NHPMA)	NHPMA, WCLIB, WWPA, NLGA
Northern Bldg., Green Bay, Wisconsin 54301	
Redwood Inspection Service (RIS)	RIS, WCLIB, WWPA
One Lombard St., San Francisco, California 94111	
Southern Pine Inspection Bureau (SPIB)	SPIB, NELMA
4709 Scenic Highway, Pensacola, Florida 32504	
West Coast Lumber Inspection Bureau (WCLIB)	WCLIB, RIS, WWPA, NLGA
6980 SW Varnes Rd., PO Box 23145, Portland, Oregon 97223	
Western Wood Products Association (WWPA)	WWPA, WCLIB, NLGA, RIS
1500 Yeon Building, Portland, Oregon 97204	
National Lumber Grades Authority (NLGA)	
P.O. Box 97 Ganges, B.C., Canada VDS 1EO	

Non-Rules Writing Agencies	
California Lumber Inspection Service	RIS, WCLIB, WWPA, NLGA
Pacific Lumber Inspection Bureau, Inc.	RIS, WCLIB, WWPA, NLGA
Timber Products Inspection	RIS, SPIB, WCLIB, WWPA, NHPMA, NELMA, NLGA
Alberta Forest Products Association	NLGA
Canadian Lumbermans Association	NLGA
Cariboo Lumber Manufacturers Association	NLGA
Central Forest Products Association	NLGA
Council of Forest Industries of British Columbia	NLGA
Interior Lumber Manufacturers Association	NLGA
MacDonald Inspection	NLGA
Maritime Lumber Bureau	NLGA
Ontario Lumber Manufacturers Association	NLGA
Pacific Lumber Inspection Bureau	NLGA
Quebec Lumber Manufacturers Association	NLGA

2. The design values herein are applicable to lumber that will be used under dry conditions such as in most covered structures. For 2" to 4" thick lumber the DRY surfaced size shall be used. In calculating design values, the natural gain in strength and stiffness that occurs as lumber dries has been taken into consideration as well as the reduction in size that occurs when unseasoned lumber shrinks. The gain in load carrying capacity due to increased strength and stiffness resulting from drying more than offsets the design effect of size reductions due to shrinkage. For 5" and thicker lumber, the surfaced sizes also may be used because design values have been adjusted to compensate for any loss in size by shrinkage which may occur.

3. Tabulated tension parallel to grain values for all species for 5" and wider, 2" to 4" thick (and 2½" to 4" thick) size classifications apply to 5" and 6" widths only, for grades of Select Structural, No. 1, No. 2, No. 3, Appearance and Stud, (including dense grades). For lumber wider than 6" in these grades, the tabulated "F$_t$" values shall be multiplied by the following factors:

Grade (2" to 4" thick, 5" and wider) (2½" to 4" thick, 5" and wider) (Includes "Dense" grades)	Multiply tabulated "F$_t$" values by		
	5" & 6" wide	8" wide	10" and wider
Select Structural	1.00	0.90	0.80
No. 1, No. 2, No. 3 and Appearance	1.00	0.80	0.60
Stud	1.00	--	--

4. Design values for all species of Stud grade in 5" and wider size classifications apply to 5" and 6" widths only.

5. Values for "F$_b$", "F$_t$", and "F$_c$" for all species of the grades of Construction, Standard and Utility apply only to 4" widths. Design values for 2" and 3" widths of these grades are available from the grading rules agencies (see Note 1).

6. The values in Table 2.14 for dimension lumber 2" to 4" in thickness are based on edgewise use. When such lumber is used flatwise, the design value for extreme fiber in bending for all species may be multiplied by the following factors:

	Dimension lumber used flatwise		
Width	Thickness		
	2"	3"	4"
2" to 4"	1.10	1.04	1.00
5" and wider	1.22	1.16	1.11

7. The design values in Table 2.14 for extreme fiber in bending for decking may be increased by 10 percent for 2" thick decking and by 4 percent for 3" thick decking. (Not applicable to California Redwood.)

8. When 2" to 4" thick lumber is manufactured at a maximum moisture content of 15 percent and used in a condition where the moisture content does not exceed 15 percent, the design values for surfaced dry or surfaced green lumber shown in Table 2.14 may be multiplied by the following factors. (For Southern Pine and Virginia Pine-Pond Pine use tabulated design values without adjustment):

2" to 4" thick lumber manufactured and used at 15 percent maximum moisture content (MC 15)					
Extreme fiber in bending "F$_b$"	Tension parallel to grain "F$_t$"	Horizontal shear "F$_v$"	Compression perpendicular to grain "F$_{c\perp}$"	Compression* parallel to grain "F$_c$"	Modulus* of elasticity "E"
1.08	1.08	1.05	1.00	1.17	1.05
			*For Redwood use only	1.15	1.04

150

Table 2.14 *(Continued)*

9. When 2" to 4" thick lumber is designed for use where the moisture content will exceed 19 percent for an extended period of time, the design values shown herein shall be multiplied by the following factors, except that for Southern Pine and Virginia Pine-Pond Pine footnote 18 applies:

2" to 4" thick lumber used where moisture content will exceed 19%					
Extreme fiber in bending in "F_b"	Tension parallel to grain "F_t"	Horizontal shear "F_v"	Compression perpendicular to grain "$F_{c\perp}$"	Compression parallel to grain "F_c"	Modulus of elasticity "E"
0.86	0.84	0.97	0.67	0.70	0.97

10. When lumber 5" and thicker is designed for use where the moisture content will exceed 19 percent for an extended period of time, the design values shown in Table 2.14 (except those for Southern Pine and Virginia Pine-Pond Pine) shall be multiplied by the following factors:

5" and thicker lumber used where moisture content will exceed 19%					
Extreme fiber in bending in "F_b"	Tension parallel to grain "F_t"	Horizontal shear "F_v"	Compression perpendicular to grain "$F_{c\perp}$"	Compression parallel to grain "F_c"	Modulus of elasticity "E"
1.00	1.00	1.00	0.67	0.91	1.00

11. Specific horizontal shear values may be established by use of the following table when length of split, or size of check or shake is known and no increase in them is anticipated. For California Redwood, Southern Pine, Virginia Pine-Pond Pine, or Yellow-Poplar, the provisions in this Footnote apply only to the following F_v Values: 75 psi, California Redwood; 95 psi, Southern Pine (KD-15); 90 psi, Southern Pine (S-Dry); 85 psi, Southern Pine (S-Green); 95 psi, Virginia Pine-Pond Pine (KD-15); 90 psi, Virginia Pine-Pond Pine (S-Dry); 85 psi, Virginia Pine-Pond Pine (S-Green); and 75 psi, Yellow-Poplar.

Shear Stress Modification Factor					
When length of split on wide face is:	Multiply tabulated "F_v" value by: (nominal 2" lumber)	When length of split on wide face is:	Multiply tabulated "F_v" value by: (3" and thicker lumber)	When size of shake* is:	Multiply tabulated "F_v" value by: (3" and thicker lumber)
no split	2.00	no split	2.00	no shake	2.00
½ x wide face	1.67	½ x narrow face . . .	1.67	1/6 x narrow face . . .	1.67
¾ x wide face	1.50	1 x narrow face	1.33	1/3 x narrow face . . .	1.33
1 x wide face	1.33	1½ x narrow face or more .	1.00	1/2 x narrow face or more .	1.00
1½ x wide face or more .	1.00			*Shake is measured at the end between lines enclosing the shake and parallel to the wide face.	

12. Stress rated boards of nominal 1", 1¼" and 1½" thickness, 2" and wider, of most species, are permitted the design values shown for Select Structural, No. 1, No. 2, No. 3, Construction, Standard, Utility, Appearance, Clear Heart Structural and Clear Structural grades as shown in the 2" to 4" thick categories herein, when graded in accordance with the stress rated board provisions in the applicable grading rules. Information on stress rated board grades applicable to the various species is available from the respective grading rules agencies. Information on additional design values may also be available from the respective grading agencies

13. When Decking graded to WWPA rules is surfaced at 15 percent maximum moisture content and used where the moisture content will exceed 15 percent for an extended period of time, the tabulated design values for Decking surfaced at 15 percent maximum moisture content shall be multiplied by the following factors: Extreme Fiber in Bending "F_b", 0.79; Modulus of Elasticity "E", 0.92.

14. To obtain a recommended design value for Spruce Pine, multiply the appropriate design value for Virginia Pine-Pond Pine by the corresponding conversion factor shown below and round to the nearest 100,000 psi for modulus of elasticity; to the next lower multiple of 5 psi for horizontal shear and compression perpendicular to grain; to the next lower multiple of 50 psi for bending, tension parallel to grain and compression parallel to grain if 1000 psi or greater, 25 psi otherwise.

Conversion Factors for Determining Design Values for Spruce Pine							
Design Category	Extreme fiber in bending "F_b"		Tension parallel to grain "F_t"	Horizontal Shear "F_v"	Compression perpendicular to grain "$F_{c\perp}$"	Compression parallel to grain "F_c"	Modulus of elasticity "E"
	Single member uses	Repetitive member uses					
Conversion Factor	.784	.784	.784	.766	.965	.682	.807

15. National Lumber Grades Authority is the Canadian rules writing agency responsible for preparation, maintenance and dissemination of a uniform softwood lumber grading rule for all Canadian species.

16. For species graded to NLGA rules, values shown in Table 2.14 for Select Structural, No. 1, No. 2, No. 3 and Stud grades are not applicable to 3" x 4" and 4" x 4" sizes.

17. Repetitive member design values for extreme fiber in bending for Southern Pine grades of Dense Structural 86, 72 and 65 apply to 2" to 4" thicknesses only.

18. When 2" to 4" thick Southern Pine or Virginia Pine-Pond Pine lumber is surfaced dry or at 15 percent maximum moisture content (KD-15) and is designed for use where the moisture content will exceed 19 percent for an extended period of time, the design values in Table 4A for the corresponding grades of 2½" to 4" thick surfaced green Southern Pine lumber shall be used. The net green size may be used in such designs.

19. When 2" to 4" thick Southern Pine or Virginia Pine-Pond Pine lumber is surfaced dry or at 15 percent maximum moisture content (KD-15) and is designed for use under dry conditions, such as in most covered structures, the net DRY size shall be used in design. For other sizes and conditions of use, the net green size may be used in design.

Group	Example
300—Design and Construction Practices	
This group contains committees whose major concern is design and construction practices.	ACI 345-74, Recommended Practice for Concrete Highway Bridge Deck Construction, Part 2.
400—Structural Analysis	
This group contains committees whose major concern is analysis of structures or analysis of design practice.	ACI 435.IR-63, Reaffirmed 1979, Deflections of Prestressed Concrete Members, Part 4.
500—Special Products and Special Processes	
This group contains committees dealing with special products used with concrete or special processing of concrete.	ACI 523.2R-68, Guide for Low Density Precast Concrete Floor, Roof, and Wall Units, Part 5.

As with all other design activities, the authority having jurisdiction should be contacted to determine code applicability and special requirements which may exist.

2.10.1 Cement Types

Portland cement is produced in five types as defined by ASTM specification C-150.

Normal portland cement Type I—widely used for nonspecial applications where thick cross sections are not present.

Modified portland cement Type II—used for thick sections or where sulfate attack is a problem. Low heat of hydration avoids cracking.

High early strength portland cement Type III is used where early strength, shorter curing, or early form stripping is required.

Low-heat portland cement Type IV has a lower heat of hydration than Type II. Of particular use for mass concrete work such as dams. Fairly slow in gaining strength and not widely available.

Air-entraining portland cements are produced as Types IA, IIA, and IIIA and used where resistance to thawing and freezing are a significant concern.

2.10.2 Concrete Strength[7]

The **quantity** of cement and the **water/cement ratio** are two major variables affecting concrete strength. Mix designs normally establish minimum hardness (strength) of the aggregate and limit the quantity of sand used and thus avoid the effect of two other major variables. Concretes typically have compressive strengths (fully cured) that range from 3000 to 5000 psi (20 to 50 MPa), with tensile strengths usually from 10 to 15% and direct shear strengths about 20% of the compressive strength.

For concretes using well-proportioned mixes the characteristics in Tables 2.15 and 2.16 are typical.

The proportion of some typical mixes is shown in Table 2.17.

Table 2.15   Variation of Compressive Strength with Age[4]

(Strength at 28 days taken as 100)

Water-cement ratio by volume, gal per bag of cement	3 days	7 days	28 days	3 months	1 year
5	40	75	100	125	145
7	30	65	100	135	155
9	25	50	100	145	165

Source: Mark's Standard Handbook for Mechanical Engineers, Baumeister. © 1978. McGraw-Hill Book Company. Used with permission.

2.10.3   Reinforcing Steel (Rebar)[7]

Widely used as deformed bars providing greater bond to the concrete and shorter lap distances than plain bars, reinforcing steel is available in tensile ratings (minimum yield) of 40,000 to 75,000 psi (275 to 515 MPa). The most common are grade 40 and grade 60, with yield strengths of 40,000 and 60,000 psi (275 and 414 MPa), respectively.

Welding of rebar is difficult and uncertain, thus mechanical or thermite reaction filled sleeve splices are used where space for lap is a problem.

Rebar is rolled in the sizes shown in Table 2.18.

The **modulus of elasticity** of concrete will vary with its strength and weight. For normal 145-lb/ft^3 (2.324-kg/l) concrete:

$$E_c = 57,000 \sqrt{f'_c} \tag{2.152}$$

where: E_c = modulus of concrete (psi)
f'_c = compressive strength—ultimate @ 28 days (psi)

The **modular ratio** of reinforcing steel (rebar) to concrete is:

$$n = \frac{E_s}{E_c} \tag{2.153}$$

where E_s = modulus of steel usually taken as 29,000,000 psi (200 × 10^3 MPa).

This data for various strengths of concrete is shown in Table 2.19. The value of n used is the nearest whole number but should not be less than 6.

Table 2.16   Strength of Plain Concrete at 28 Days[4]

Max water content, gal per bag of cement	5	5.5	6	6.5	7.0
Compressive strength, lb/in^2	4,000	3,700	3,350	3,000	2,650
Modulus of rupture, lb/in^2	650	625	600	550	500
Tensile strength (split cyl. method), lb/in^2	350	325	300	275	250

Source: Mark's Standard Handbook for Mechanical Engineers, Baumeister. © 1978. McGraw-Hill Book Company. Used with permission.

Table 2.17   Comparison of Quantities and Properties of Some Typical Concrete Mixes[a,b,9]

	Mixtures		
	Rich	Medium	Lean
Cement-water ratio (by solid volume)	0.9	0.6	0.4
Type of mixture	Rich	Medium	Lean
Predicted 28-day strength, psi (w/c basis)	6,000	4,000	2,000
Workability (slump, in.)	3.0	3.0	3.0
Workability (texture)	Plastic	Plastic	Plastic
Quantity in 1 unit volume concrete by solid (absolute) volume: cement (c)	0.153	0.102	0.068
F.A. (a)	0.247	0.298	0.332
C.A. (b)	0.430	0.430	0.430
Water (w)	0.170	0.170	0.170
Air (assumed 0)	0	0	0
Total	1.000	1.000	1.000
Water-cement ratio: gal per bag cement	4.0	6.0	9.0
Weight	0.34	0.52	0.80
Bulk volume	0.54	0.80	1.20
Solid volume	1.11	1.67	2.50
Voids-cement ratio: solid volume	1.11	1.67	2.50
Proportions: by weight	1:1.4:2.4	1:2.5:3.6	1:4.1:5.3
Bulk volume	1:1.2:2.3	1:2.2:3.4	1:3.7:5.1
Solid volume	1:1.6:2.8	1:2.9:4.2	1:4.9:6.3
Ratio $b \div b_0$	0.70	0.70	0.70
Ratio C.A. to F.A., i.e., $b \div a$	1.74	1.44	1.30
Quantity in 1 cu yd: cement, bags	8.64	5.76	3.84
F.A., lb	1,103	1,331	1,487
C.A., lb	1,920	1,920	1,920
Water, gal	34.3	34.3	34.3
Yield, cu ft concrete per bag cement	3.1	4.7	7.0
Strength-economy index (psi per bag per cu yd)	695	695	469
Weight fresh concrete, lb per cu ft	153	151	150
Lb per cu yd	4,122	4,079	4,055

[a]Materials: F. A.—washed, sanded, graded, 0 to 4 (Standard Ottawa to Fine Sands); sp. gr., 2.65; bulk weight, 103.5 lb/ft³. C. A.—river gravel, graded, ⅜ to 1½ inch; sp. gr., 2.65; bulk weight, 98.3 lb/ft³. Cement—sp. gr., 3.15.

[b]Although quantities and relative proportions are representative of those for rich, medium, and lean mixtures, in general they are not identical with what would be obtained by some other (less direct) basis of adjustment, such as varying total aggregate at a constant ratio of coarse to fine, along with appropriate alterations in water, and/or cement.

Source: Urquhart, *Civil Engineering Handbook,* 4th Edition, 1959. Used with permission of McGraw-Hill Book Co.

Working stress design establishes conservative stress values for both the concrete and its steel reinforcement sufficiently below the yield stress to provide margins of safety to compensate for shrinking, minor cracking, and creep. However, for most major structures the **ultimate stress design** is used today and provides greater uniformity in safety factors among the different structural elements. The design methodology is covered in detail in the ACI (American Concrete Institute) Standards which has been adopted by most jurisdictions. It should be referred to for any extensive design work. The local code must be complied with and should be carefully reviewed as it may invoke only portions of the ACI standards, contain special provisions, and so on.

Table 2.18   Areas and Perimeters of Standard Deformed Bars

Bar designation	Nominal diameter, (in.)		
#3	0.375	**Area**	**0.11**
		Perimeter	1.178
#4	0.500	**Area**	**0.20**
		Perimeter	1.571
#5	0.625	**Area**	**0.31**
		Perimeter	1.963
#6	0.750	**Area**	**0.44**
		Perimeter	2.356
#7	0.875	**Area**	**0.60**
		Perimeter	2.749
#8	1.000	**Area**	**0.79**
		Perimeter	3.142
#9	i.128	**Area**	**1.00**
		Perimeter	3.544
#10	1.270	**Area**	**1.27**
		Perimeter	3.990
#11	1.410	**Area**	**1.56**
		Perimeter	4.430
#14	1.693	**Area**	**2.25**
		Perimeter	5.32
#18	2.257	**Area**	**4.00**
		Perimeter	7.09

Source: From Parker & Hauf, *Simplified Design of Reinforced Concrete,* John Wiley & Sons, 1976.

Table 2.19   Modulus of Elasticity of Normal Weight Concrete[7]

f'_c Ultimate Compressive Strength at 28-Day Period (psi)	E_c Modulus of Elasticity of Concrete (psi)	$n = \dfrac{E_s}{E_c}$
2,500	2,880,000	10
3,000	3,150,000	9
4,000	3,640,000	8
5,000	4,070,000	7

Source: From Parker & Hauf, *Simplified Design of Reinforced Concrete,* John Wiley & Sons, 1976.

155

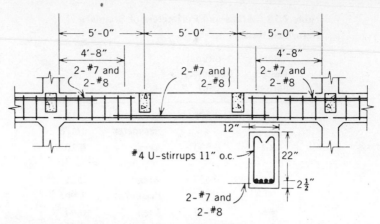

Fig. 2.35 Reinforcement locations. *Source:* Parker & Hauf, *Simplified Design of Reinforced Concrete,* John Wiley & Sons, 1976.

2.10.4   Reinforcement Locations[7]

Although theoretically all compression loads can be carried by the concrete, in fact many designs include reinforcement for compression as well as for tension and shear. For designs with rebar not handling compression, both positive and negative moments must be considered yielding designs with rebar both above and below the neutral axis, and in some cases added rebar at points of high shear loads (e.g., supports). Stirrups are frequently found in major beams to avoid the development of shear cracking at loading points. Economy of design, with the areas of concrete and rebar proportional to their load-carrying ability, including rebar cover and spacing

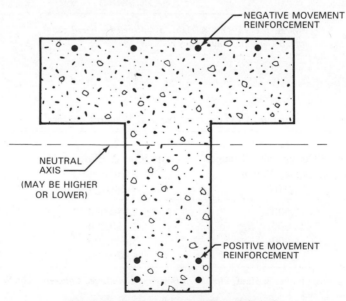

Fig. 2.36 Typical T beam (balanced design).

requirements, frequently results in a T configuration for beams with the bulk of the rebar found below the neutral axis (see Fig. 2.36).

2.11   FABRICATION

2.11.1   Riveted Joints

Shearing strength (r_s) of a rivet d inches in diameter, with an allowable stress in shear of f_s pounds/inch^2.

$$r_s = \frac{\pi d^2}{4} f_s \text{ pounds.} \tag{2.154}$$

Bearing strength (r_b) of a rivet d inches in diameter, with an allowable stress in bearing of f_b pounds/inch^2, against a plate t inches in thickness.

$$r_b = dt f_b \text{ pounds.} \tag{2.155}$$

Total stress (r) on each of n rivets resisting a pull or thrust of P pounds.

$$r = \frac{P}{n} \text{ pounds.} \tag{2.156}$$

Total stress (r_m) on the most stressed rivet of a group of rivets resisting the action of a couple of M inch-pounds, if y is the distance in inches from the center of gravity of the group of rivets to the outermost rivet and Σy^2 is the sum of the squares of the distances from the center of gravity of the group to each of the rivets (Fig. 2.37).

$$r_m = \frac{My}{\Sigma y^2} \text{ pounds.} \tag{2.157}$$

Resistance to moment (M) of a group of rivets, if the distance of the outermost rivet from the center of gravity of the group is y inches and the sum of the squares

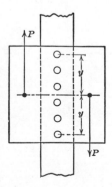

Fig. 2.37   Total stress (couple).

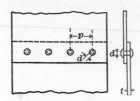

Fig. 2.38 Resistance to tearing.

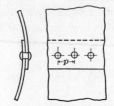

Fig. 2.39 Single-riveted lap joint.

of the distances from the center of gravity of the group to each of the rivets is Σy^2 and r is the total allowable stress on a rivet.

$$M = \frac{r\Sigma y^2}{y} \text{ inch-pounds.} \qquad (2.158)$$

Resistance to tearing (T) between rivets, of a plate t inches in thickness in which rivets of d inches diameter are placed with p inches pitch, if the allowable intensity of stress of the plate in tension is f_t pounds/inch^2.

$$T = t(p - d)f_t \text{ pounds.} \qquad (2.159)$$

Efficiency of a riveted joint is the ratio of the least strength of the joint to the tensile strength of the solid plate.

Single-riveted Lap Joint

Shearing one rivet $= \dfrac{\pi d^2}{4} f_s$.

Tearing plate between rivets $= (p - d)tf_t$. $\qquad (2.160)$

Crushing of rivet or plate $= dtf_b$.

where: f_s = allowable shearing stress in pounds/inch^2.
$\quad\quad f_b$ = allowable bearing stress in pounds/inch^2.
$\quad\quad f_t$ = allowable tension stress in pounds/inch^2.
$\quad\quad d$ = diameter of rivet in inches.
$\quad\quad t$ = thickness of plate in inches.

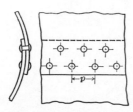

Fig. 2.40 Double-riveted lap joint.

Double-riveted Lap Joint

Shearing two rivets $= \dfrac{2\pi d^2}{4} f_s.$

Tearing between two rivets $= (p - d)tf_t.$ (2.161)

Crushing in front of rivets $= 2dtf_b.$

2.11.2 Welding Processes

Welding, as a method of joining metals, offers the opportunity to achieve a more efficient use of the materials and faster fabrication and erection. Welding also permits the designer to develop and use new and aesthetically appealing designs, and saves weight because connecting plates are not needed and allowances need not be made for reduced load-carrying ability due to holes for bolts, rivets, and so on.

The welding process joins two pieces of metal together by establishing a metallurgical bond between them. There are many different types of processes, most of which use a fusion technique. The two most widely used are **arc welding** and **gas welding.**

Arc Welding

The arc welding process obtains the intense heat needed to liquefy the metals to be joined from an electric arc that is developed between the workpiece to be welded and the electrode. The electrode may be either consumable or nonconsumable. An arc temperature of approximately 6500°F (3600°C) is created as a result of a continual electrical discharge between the workpiece and the electrode which, in the case of a consumable electrode, melts along with the surface of the workpiece. A weld bead is formed by the deposition of metal from the electrode by slowly moving the electrode along the workpiece, generally with a weaving motion. This process is repeated in layers, as necessary, with intermediate removal of surface slag after each welding pass, to create the necessary size weld (or thickness). The slag formed results from melting of the specially designed electrode coating or flux which **scavenges weld impurities** and produces an **inert gas cover** over the liquid weld metal to avoid its contamination and weakening by contacting the air. The arc welding process requires a continuous supply of electrical current having adequate voltage and amperage to maintain the arc, and may be either ac or dc. Typical voltage ranges are from 20 to 80 with amperage ranging from 50 to 500.

Principal Arc Welding Processes

Shielded Metal Arc. This process uses hand-held, "stick" type, coated electrodes. Advantages of this process are minimal cost, flexibility, and ease of use. The electrode coating is the flux. Weld quality is largely dependent on the skill of the operator.

Submerged Arc. A machine-controlled electrode (wire feed and travel speed) is submerged in a blanket of granular flux. High deposition rates and deep penetration are characteristic with this process, making it economical for large welds. Joint must be flat or horizontal and thus welding positioners may be necessary. The final weld surface is generally very smooth and blends well with the base metal.

Flux-Cored Arc. This process typically uses a hand-held wire gun with a continuous feed wire containing a flux internal to the wire (a hollow wire filled with a flux material). Higher deposition rates possible with the flux-cored arc process permit increased tolerance for poor joint fit. Wind shelters are not generally required, which permits increased ventilation around operators and the work.

Gas-Shielded Arc. This process is similar to the self-shielded flux-cored process except that shielding is accomplished by an external (annular) gas stream. Typical shielding gases are CO_2, **Helium,** or **Argon. Gas Metal Arc (MIG)** or **Gas Tungsten Arc (TIG)** welding are two popular processes using this technique. MIG filler metal is fed through the center of a welding gun wire guide. The TIG process creates the arc using a tungsten electrode that is not consumed with the desired bare filler metal rod separately added to the molten weld metal pool under the inert gas blanket. The TIG process may also be used where additional filler metal is not required. These processes are very flexible and are suitable for alloy or thin section work.

Other Frequently Used Arc Welding Processes

Electro Slag Welding. Used for welding very heavy and thick metallurgical sections.

Plasma Arc Welding. Sometimes called "plasma jet" welding. Frequently used for metal overlaying and may also be used for cutting.

Electron Beam Welding. Used for deep penetration welds requiring minimal distortion and metallurgical structure damage, usually accomplished in a vacuum chamber.

Resistance Welding. Spot, seam, and roll-spot welds created by passing a localized current through parts to be joined, with the parts held together under pressure by electrodes. Weld results from resistance heating of a localized area.

Stud Welding. Welding of a metal stud to another part by creating an arc between the two which produces a molten puddle. The stud is forced into the molten metal which is then permitted to solidify.

Gas Welding. Gas welding is a process whereby metal surfaces to be joined are melted, using a fuel gas–oxygen flame, and caused to flow together without the application of pressure to the parts being joined. A filler metal may or may not be used. The most widely used source of heat is a combination of **oxygen** and **acetylene gases,**

which are mixed in an oxyacetylene welding torch. The flame temperature of this mixture reaches approximately 5600° F (3100° C). Other fuel gases such as **propane, butane,** and **natural gas** can be used for welding and soldering of nonferrous metals, but these fuels do not provide adequate heat or proper inert atmosphere for welding of ferrous metals.

The gases used are stored in separate heavy metal cylinders, withdrawal being regulated by a pressure regulator on each cylinder. The gas passes through flexible rubber hoses to a welding torch where the combination of gases can be adjusted to obtain the desired characteristics.

The welder has considerable control over the temperature of the metal in the weld zone when using the gas welding technique. Deposition rates of weld filler metal are easily controlled because the source of heat and source of filler metal are separate. Heat can be applied to either the base metal or the filler metal while both are still within the flame envelope. These characteristics make gas welding well suited for joining thin metal sections and where fit-up is poor. Gas welding is not as economic a method of joining heavy sections as arc welding. Equipment required to perform gas welding is relatively inexpensive and portable. It can be used to preheat, postheat, weld, braze, and may be converted for oxygen cutting. The gas welding process is well adapted to short production runs and field repairs and alterations.

Distortion

Severe physical distortion can result from the numerous variables of the welding process and procedures used and from the physical design, sequence of welding operations, inadequate fixtures, improper welding parameters, and so on. Control of distortion as a result of shrinkage becomes extremely important. This is particularly true where welded structures consist of material having widely varying thicknesses because more massive weldments generally have a greater shrinkage than smaller weldments. The susceptibility to distortion increases, with welds that are not symmetrical, about the neutral axis of the section, where materials of greatly different thickness are welded together or where welds lie in different planes. Further, weld distortion may be increased where weld-deposited cross sections vary widely or where the rate of deposition of weld metal varies widely (e.g., where material is deposited at very high rates). The actual design of welded joints to reduce distortion is beyond the scope of this book, but references at the end of this chapter provide more specific guidance. If the sequence of welding is not carefully chosen, there may be cracking induced in initially deposited welds from subsequently deposited heavier welds. Further, if the welding sequence is not considered during the design of the weldment and not planned for in the course of fabrication, distortions in the positioning, angularity, and so on, of the various members making up the weldment will likely occur. For this reason, large carbon steel weldments are frequently **stress relieved** after welding using temperatures of 1100 to 1200° F (590 to 650° C) for a period of time approaching one hour per inch (25 mm) of thickness of the heaviest section, up to a maximum of 8 hours. This stress relief substantially reduces the amount of residual stress remaining in the welds and provides a way in which the welded structure can relax the thermally (weld-) induced stresses and provides a stable structure suitable for final machining.

2.11.3   Weld Strength

The strength of a welded structure is dependent on many factors, including: the load-carrying ability of the base metal, design of the weld joint and amount and configuration of deposited weld metal, type of weld metal, the variables of the individual welding procedures, and residual—after-welding—stresses. The load-carrying ability can be related to the weld strength by knowing the type of stresses involved and the weld metal cross-sectional area that is carrying the load. Assurance that welds perform as desired is obtained by: establishing welding procedures that define all essential variables, performing procedure qualification tests including destructive testing of the weld sample joints, assuring that welding of critical structures and components is accomplished by qualified welders, and, when welding is to be performed in accordance with code (e.g., ASME, AWS) requirements, verify that a quality control program is in effect to assure that code requirements are met.

The weld metal is as strong as and generally stronger than the adjacent base metal being welded. Thus, strength calculations of full penetration butt welds generally are not necessary, the critical requirements being selection of a weld filler material compatible with and as strong as the base metal, and selection of the proper weld procedure.

In application, fillet welds required may vary from 80% to less than 50%, the size of the thickness of the plate depending on whether the design is based on rigidity or actual weld strength.

Allowable Shear and Unit Forces

Table 2.20 presents the allowable shear values for various weld-metal strength levels and the more common fillet weld sizes. These values are for equal-leg fillet welds where the effective throat (t_e) equals 0.707 × leg size (ω).

$$F = 0.707 \, \omega \times \tau \qquad (2.162)$$

where $\tau = 0.30 \, (EXX)$.

2.12   POWER TRANSMISSION

2.12.1   Shafts

Maximum intensity of shear (s) in a shaft of r inches radius and J_0 inches^4 polar moment of inertia due to a torque (twisting moment) of M inch-pounds.

$$s = \frac{Mr}{J_0} \text{ pounds/inch}^2. \qquad (2.163)$$

NOTE.   For a solid round shaft $s = 2M/(\pi r^3)$.

Table 2.20 Allowable Load for Various Sizes of Fillet Welds[6]

	Strength Level of Weld Metal (EXX)					
	60	70	80	90	100	110
	Allowable Shear Stress on Throat of Fillet Weld or Partial-Penetration Groove Weld (1000 psi)					
$\tau =$	18.0	21.0	24.0	27.0	30.0	33.0
	Allowable Unit Force on Fillet Weld (1000 psi/linear in.)					
$f =$	12.73ω	14.85ω	16.97ω	19.09ω	21.21ω	23.33ω
Leg Size ω (in.)	Allowable Unit Force for Various Sizes of Fillet Welds (1000 Lbs./Linear in.)					
1	12.73	14.85	16.97	19.09	21.21	23.33
7/8	11.14	12.99	14.85	16.70	18.57	20.41
3/4	9.55	11.14	12.73	14.32	15.92	17.50
5/8	7.96	9.28	10.61	11.93	13.27	14.58
1/2	6.37	7.42	8.48	9.54	10.61	11.67
7/16	5.57	6.50	7.42	8.35	9.28	10.21
3/8	4.77	5.57	6.36	7.16	7.95	8.75
5/16	3.98	4.64	5.30	5.97	6.63	7.29
1/4	3.18	3.71	4.24	4.77	5.30	5.83
3/16	2.39	2.78	3.18	3.58	3.98	4.38
1/8	1.59	1.86	2.12	2.39	2.65	2.92
1/16	.795	.930	1.06	1.19	1.33	1.46

Source: Courtesy of the Lincoln Electric Company.

Angle (θ) of twist in a solid circular shaft, of r inches radius, l inches in length, and with E_s pounds/inch^2 modulus of elasticity in shear, due to a torque of M inch-pounds.

$$\theta = \frac{2Ml}{\pi r^4 E_s} \text{ radians.} \qquad (2.164)$$

NOTE. E_s for steel is commonly taken as 12,000,000.

Horsepower (P) transmitted by a shaft making n revolutions/minute under a torque of M inch-pounds.

$$P = \frac{2\pi n M}{33,000 \times 12} \text{ horsepower.} \qquad (2.165)$$

Diameter (d) of a solid circular shaft to transmit P horsepower at n revolutions/minute with a fiber stress in shear of s pounds/inch^2.

$$d = \sqrt[3]{\frac{321,000\, P}{ns}} \text{ inches.} \qquad (2.166)$$

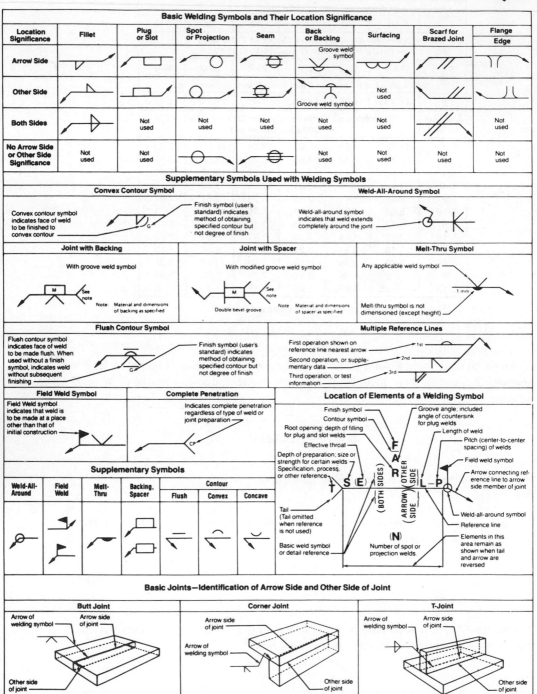

Fig. 2.41 American Welding Society standard welding symbols.

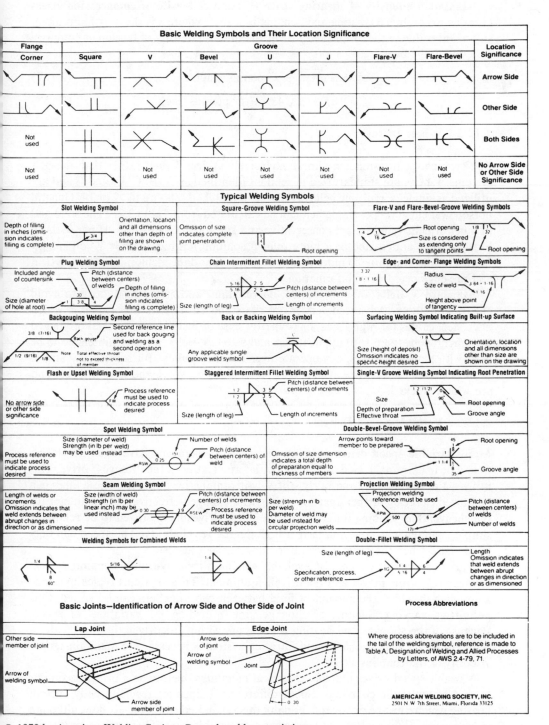

Basic Welding Symbols and Their Location Significance

Typical Welding Symbols

Basic Joints—Identification of Arrow Side and Other Side of Joint

Process Abbreviations

Where process abbreviations are to be included in the tail of the welding symbol, reference is made to Table A, Designation of Welding and Allied Processes by Letters, of AWS 2.4-79, 71.

AMERICAN WELDING SOCIETY, INC.
2501 N. W. 7th Street, Miami, Florida 33125

Maximum intensity of shearing stress (s') and of **tensile or compression stress** (f') due to combined twisting and bending in a shaft where s is the maximum intensity of shear due to the torque and f is the maximum intensity of tension or compression due to the bending.

$$s' = \tfrac{1}{2}\sqrt{4s^2 + f^2} \text{ pounds/inch}^2. \tag{2.167}$$

$$f' = \tfrac{1}{2}f + \tfrac{1}{2}\sqrt{4s^2 + f^2} \text{ pounds/inch}^2. \tag{2.168}$$

2.12.2   V-Belt Drives

Belt-type power transmission almost exclusively utilizes **V-belt** rather than flat-belt systems. V-belts are standardized in sections identified as A, B, C, D, and E and are utilized as single or multiple belts (available in matched sets) up to 12 or 14. General horsepower capacity per single belt can be calculated from Table 2.21. The sizes and number of belts required are conservative and suitable for most uses. If particularly heavy duty is anticipated or a more refined solution desired, consult V-belt manufacturers who can provide more detailed engineering data including both analytical and pre-engineered drive information.

Standard V-Belts are fully sealed, oil resistant, and static conductive to avoid electrical buildup. Advanced versions of the standard V-Belts are available which have steel cables for high loads, link-type for ease of installation, toothed or cogged for synchronous (zero slip) applications, and so on. V-Belts, although intended to be used with mating sheaves, can operate over flat-face pulleys for some applications. Both belts and sheaves are available in pre-engineered sizes, and matched sets and are widely stocked.

2.12.3   Variable-Speed Drives

Many **variable-speed drives** utilizing ingenious mechanisms have been developed. One type in extremely wide use for industrial application is the "Reeves" drive. This utilizes a relatively wide and fairly stiff V-belt running over sheaves whose faces can be moved closer or further apart, resulting in a changed effective pitch diameter for the sheave, thus yielding different speeds. These drives have proven to be relatively simple, maintenance free, and have a high degree of reliability. The speed can only be changed when the drive is operating, however, not at rest.

Variable-speed drives are also available which utilize a **hydraulic coupling** system between the prime mover and the driven equipment. These systems typically utilize a coupling in which the amount of fluid between two impellers can be varied, thus permitting varying amounts of slippage, and providing for variations of speed. Because of the slippage and heating that occurs within the driven fluid (typically a synthetic oil), a heat exchange and filtration system is necessary to maintain proper viscosity of the coupling fluid.

Variable-speed drives utilizing **electrical couplings** operate on a principle similar to that of hydraulic couplings. Speed differences (slippage) result in eddy current heating which must be dissipated. Usually this is handled by forced air flow, although in some cases fluid coolers can be used to remove this heat. There are several types of **variable-speed ac motors,** the choice of which is determined by the size, application, cost, and efficiency. The **wound rotor induction motor** can be controlled from

Table 2.21 V-Belt Drives

Recommended V-Belt Cross-Sections for Various Horsepowers and Speeds

Horsepower	MOTOR SPEED—RPM						
	1750	1160	870	690	575	490	435
½	A	A	A				
¾	A	A	A				
1	A	A	A				
1½	A	A	A				
2	A	A	A				
3	A	A	B (or A)				
5	B (or A)	B (or A)	B				
7½	B	B	B				
10	B	B	B or C				
15	B	B or C	C (or B)				
20	B or C	C (or B)	C	D	D		
25	C (or B)	C	C	D	D		
30	C	C	C	D	D		
40	C	C or D	C or D	D	D		
50	C	C or D	C or D	D	D	E	E
60	C	C or D	D (or C)	D	D	E	E
75	C	D (or C)	D	D	D (or E)	E	E
100	C	D	D	D or E	E (or D)	E	E
125		D	D	D or E	E (or D)	E	E
150		D	D	E (or D)	E	E	E
200		D	D	E	E	E	E
250		D	D	E	E	E	E
300 and above		D	D	E	E	E	E

Hp transmitted by V-Belts based on 180° arc of contact

Veloc in ft Per Min	Cross-Sect A	Cross-Sect B	Cross-Sect. C	Cross-Sect D	Cross-Sect E
	width ½" thick 5⁄16"	width 21⁄32" thick ½"	width 7⁄8" thick 5⁄8"	width 1¼" thick ¾"	width 1½" thick 1"
1000	.9	1.2	3.0	5.5	7.5
1100	1.0	1.3	3.2	6.0	8.2
1200	1.0	1.4	3.4	6.5	8.9
1300	1.1	1.5	3.6	7.0	9.6
1400	1.2	1.6	3.8	7.5	10.3
1500	1.3	1.7	4.0	8.0	11.0
1600	1.4	1.8	4.3	8.4	11.6
1700	1.5	1.9	4.6	8.8	12.2
1800	1.6	2.1	4.9	9.2	12.8
1900	1.6	2.2	5.2	9.6	13.4
2000	1.7	2.3	5.5	10.0	14.0
2100	1.8	2.4	5.7	10.5	14.8
2200	1.9	2.5	5.9	11.0	15.2
2300	1.9	2.6	6.1	11.5	15.8
2400	2.0	2.7	6.3	12.0	16.4
2500	2.1	2.8	6.5	12.5	17.0

Veloc in ft Per Min	Cross-Sect A	Cross-Sect B	Cross-Sect C	Cross-Sect D	Cross-Sect E
	width ½" thick 5⁄16"	width 21⁄32" thick ½"	width 7⁄8" thick 5⁄8"	width 1¼" thick ¾"	width 1½" thick 1"
2600	2.2	2.8	6.7	12.9	17.5
2700	2.2	2.9	6.9	13.3	18.0
2800	2.3	3.0	7.1	13.7	18.5
2900	2.3	3.1	7.3	14.1	19.3
3000	2.4	3.2	7.5	14.5	19.8
3100	2.5	3.3	7.7	14.8	20.0
3200	2.5	3.4	7.9	15.1	20.5
3300	2.5	3.5	8.1	15.4	21.0
3400	2.6	3.6	8.3	15.7	21.3
3500	2.6	3.7	8.5	16.0	21.8
3600	2.7	3.8	8.6	16.3	22.0
3700	2.7	3.9	8.7	16.6	22.8
3800	2.8	4.0	8.8	16.9	23.0
3900	2.8	4.1	8.9	17.2	23.3
4000	2.8	4.2	9.0	17.5	23.5
5000	2.8	4.2	9.0	17.5	23.5

$$\text{No of belts required} = \frac{\text{hp of drive}}{(\text{hp per belt}) \left(1 - \dfrac{.175\,(D-d)}{C}\right)}$$

D = pitch dia of large pulley, in
d = pitch dia of small pulley, in
C = center distance, in
For pump, compressor and blower drives 40% more belting than shown by above formula should be used.

Source: Courtesy of Dayco Corporation.

zero speed to full speed, but because of the high losses in the external rotor resistances, it is not often used in the larger ratings where operation at less than full speed is required for other than short periods. A more efficient variation of the wound rotor application is termed the **wound rotor slip recovery drive.** Here the rotor is connected to an ac/dc converter, which drives a dc/ac inverter. The inverter ac output is three-phase, 60 Hz, and may be coupled into the source supplying the motor by a

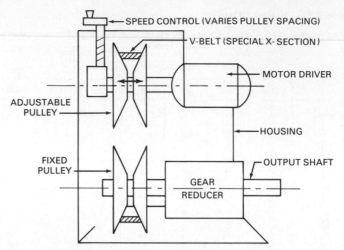

Fig. 2.42   Reeves-type variable-speed drive.

transformer. Another system with high efficiency is the **variable-speed synchronous motor.** The source at 60 Hz is converted to dc, which supplies an inverter, the output of which can be controlled to provide a variable frequency and voltage to supply the synchronous motor. Both of the latter two systems are quite efficient over their entire speed range which, in many applications, will offset their higher initial cost.

Dc motors of the series type are also used to provide speed variation, but can accelerate to infinite speed on loss of the driven load. Thus, for application, where loads may change suddenly, such as a belt drive which breaks, these motors may increase their speed very rapidly and the motor(s) may self-destruct due to centrifugal force.

2.13   BOILER AND PRESSURE VESSEL CODE

In most of the United States and many other parts of the world as well, the American Society of Mechanical Engineers (ASME) Boiler and Pressure Vessel Code has been adopted by jurisdictional bodies as the governing code. The code consists of the following sections:

1983 ASME
BOILER AND PRESSURE VESSEL CODE
An American National Standard

    I   Power Boilers
   II   Material Specifications
           Part A—Ferrous Materials
           Part B—Nonferrous Materials
           Part C—Welding Rods, Electrodes and Filler Metals

III   Subsection NCA—General Requirements for Division 1 and Division 2
III   Division 1
      Subsection NB—Class 1 Components
      Subsection NC—Class 2 Components
      Subsection ND—Class 3 Components
      Subsection NE—Class MC Components
      Subsection NF—Component Supports
      Subsection NG—Core Support Structures
      Appendices
III   Division 2—Code for Concrete Reactor Vessels and Containments
IV   Heating Boilers
 V   Nondestructive Examination
VI   Recommended Rules for Care and Operation of Heating Boilers
VII   Recommended Rules for Care of Power Boilers
VIII Pressure Vessels
      Division 1
      Division 2—Alternative Rules
 IX   Welding and Brazing Qualifications
  X   Fiberglass-Reinforced Plastic Pressure Vessels
 XI   Rules for Inservice Inspection of Nuclear Power Plant Components—Division 1

The code is intended to provide rules for the design, fabrication and operation of these equipment and components and is extensive and extremely detailed.

The code is revised periodically (typically triannually) and intermediate semiannual revisions are issued as addenda to maintain currency of the code. Before use of the code, the engineer should assure himself of the code applicability to the jurisdiction (usually state) in which the item will be assembled (as distinct from fabricated) and operated as well as specific additional or differing requirements particular to that jurisdiction.

2.14   UNIFORM BUILDING CODE

The Uniform Building Code covers the fire, life, and structural safety aspects of buildings and related structures. Published triannually by the International Conference of Building Officials, it is divided into the following sections:

Part I   Administrative
Part II   Definitions and Abbreviations
Part III   Requirements Based on Occupancy
Part IV   Requirements Based on Types of Construction
Part V   Engineering Regulations—Quality and Design of the Materials of Construction
Part VI   Detailed Regulations

Part VII   Fire-resistive Standards for Fire Protection
Part VIII   Regulations for Use of Public Streets and Projections Over Public
                Property
Part IX   Wall and Ceiling Coverings
Part X   Special Subjects
Part XI   Uniform Building Code Standards

In addition to the Uniform Building Code, this body issues plumbing, housing, security, fire, and other codes which supplement it.

The code has been adopted by many jurisdictions and the jurisdiction authority should be consulted regarding its applicability.

REFERENCES

1. *AISC Manual of Steel Construction,* 8th edition, American Institute of Steel Construction, Inc., Chicago, Illinois, 1980.
2. *American Welding Society Standard Welding Symbols,* American Welding Society Inc., Miami, Florida.
3. *Glulam Systems 1980,* American Institute of Timber Construction, Englewood, Colorado, 1980.
4. *Mark's Standard Handbook for Mechanical Engineers,* 8th edition, McGraw-Hill, New York, 1978.
5. *National Design Specification for Wood Construction,* National Forest Products Association, Washington, D.C., 1982.
6. *Procedure Handbook of Arc Welding,* 12th edition, Lincoln Electric Co., Cleveland, Ohio, 1973.
7. Harry Parker and Harold Hauf, *Simplified Design of Reinforced Concrete,* 4th edition, John Wiley & Sons, New York, 1976.
8. Harry Parker and Harold Hauf, *Simplified Design of Structural Wood,* John Wiley & Sons, New York, 1979.
9. L. Urquhart, *Civil Engineering Handbook,* 4th edition, McGraw-Hill, New York, 1959.

SECTION 3
FLUID MECHANICS

3.1 HYDROSTATICS

Pressure (p) due to a head of h feet in a liquid weighing w pounds/foot^3.

$$p = wh/144 \text{ pounds/inch}^2. \tag{3.1}$$

NOTE. In water, the pressure corresponding to a head of h feet is $0.434h$ pounds/inch^2 (3 kPa).

Head (*h*) corresponding to a pressure of p pounds/foot^2 in a liquid weighing w pounds/foot^3.

$$h = \frac{p}{w} \text{ feet.} \tag{3.2}$$

NOTE. In water, the head corresponding to pressure of p pounds/inch^2 is $2.3p$ feet.

Total normal pressure (P) on a plane or curved surface A square feet in area immersed in a liquid weighing w pounds/foot^3 with a head of h_0 feet on its center of gravity.

$$P = wAh_0 \text{ pounds.} \tag{3.3}$$

NOTE. The total pressure on a plane surface may be represented by a resultant force of P pounds acting normally to the area at its center of pressure.

Component of normal pressure (P_c) on a plane area of A square feet with h_0 feet head on its center of gravity and a projection of A_c square feet on a plane perpendicular to the component of pressure.

$$P_c = wA_ch_0 \text{ pounds.} \tag{3.4}$$

Vertical component of pressure (P_v) on a plane area of A square feet with h_0 feet head on its center of gravity and A_h square feet horizontal projection of area.

$$P_v = wA_hh_0 \text{ pounds.} \tag{3.5}$$

Table 3.1   Water Equivalents[3]

One U. S. Gallon						
One U. S. Gallon			=	.1337	Cubic Foot	
"	"	"	=	231.	Cubic Inches	
"	"	"	=	.833	British Imperial Gallon	
"	"	"	=	3.785	Liters	
"	"	"	=	3785.	Cubic Centimeters (Milliliters)	
"	"	" Water	=	8.33	Pounds (Lb.)	
One Cubic Foot			=	7.48	U. S. Gallons	
"	"	" Water	=	62.43	Pounds (Lb.) (at greatest density—39.2°F)	
One Acre Inch			=	27154.	U. S. Gallons	
One Acre Foot			=	325851.	U. S. Gallons	
One Second Foot			=	1.	Cu. Ft. per Second = 60. Cu. Ft. per min.	
"	"	"	=	7.48	U. S. Gals. per " = 448.8 U. S. gpm	
One Miners Inch			=	1.2 to 1.76	Cu. Ft. per Min. (Varies in different States)	
One Cubic Meter			=	1000.	Liters	
"	"	"	=	264.2	U. S. Gallons	
"	"	"	=	220.	British Imperial Gallons	
"	"	"	=	35.31	Cubic Feet	
One Boiler H.P. Hr.			=	4.	Gallons Water Evaporated per Hour	

Source: Reprinted from *The Permutit Water and Waste Treatment Data Book*
© 1953 The Permutit Co., Inc.

Horizontal component of pressure (P_h) on any area of A square feet with A_v square feet vertical projection of area and h_0 feet head on the center of gravity of the projected area.

$$P_h = wA_v h_0 \text{ pounds.} \qquad (3.6)$$

Resultant pressure (P_{bc}) on an area bc of A_{bc} square feet with a head above its base of h_1 feet on one side and h_2 feet on the other side, or a difference of head of h feet.

$$P_{bc} = wA_{bc}(h_1 - h_2) = wA_{bc}h \text{ pounds.} \qquad (3.7)$$

For an allowable design stress, usually established by the governing code (e.g., ASME, API, AWWA) the thickness t required for a pipe of d inches internal diameter to withstand a pressure of p pounds/inch^2 with a design (tensile) stress of f pounds/inch^2.

$$t = \frac{pd}{2f} \text{ inches.} \qquad (3.8)$$

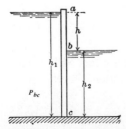

Fig. 3.1   Resultant pressure.

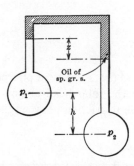

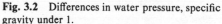

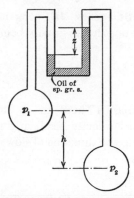

Fig. 3.2 Differences in water pressure, specific gravity under 1.

Fig. 3.3 Differences in water pressure, specific gravity over 1.

Difference in water pressure $(p_1 - p_2)$ in two pipes as indicated by a differential gauge with an oil of specific gravity s, when the difference in level of the surfaces of separation of the oil and water is z feet and the difference in level of the two pipes is h feet.

CASE I. When the oil has a specific gravity less than 1. (See Fig. 3.2.)

$$p_1 - p_2 = 0.434[z(1 - s) - h] \text{ pounds/inch}^2. \tag{3.9}$$

CASE II. When the oil has a specific gravity greater than 1. (See Fig. 3.3.)

$$p_1 - p_2 = 0.434[z(s - 1) - h] \text{ pounds/inch}^2. \tag{3.10}$$

3.2 HYDRODYNAMICS

Conservation of Energy. The law of conservation of energy states that with steady flow the total energy at any section is equal to the total energy at any further section in the direction of flow, plus the loss of energy due to friction in the distance between the two sections. Thus the various forms of energy are interchangeable and their sum is constant (less frictional losses).

Pressure energy (W_{pr}) per pound of water weighing w pounds/foot^3 due to a pressure of p pounds/foot^2.

$$W_{pr} = \frac{p}{w} = 0.016p \text{ foot-pounds.} \tag{3.11}$$

Potential energy (W_p) per pound of water due to a height of z feet of the center of gravity of the section above the datum level.

$$W_p = z \text{ foot-pounds.} \tag{3.12}$$

Kinetic energy *(W)* per pound of water due to a velocity of v feet/second, the acceleration due to gravity being g feet/second^2.

$$W = \frac{v^2}{2g} \text{ foot-pounds.} \tag{3.13}$$

Bernoulli's Theorem. In steady flow the total head (pressure head plus potential head plus velocity head) at any section is equal to the total head at any further section in the direction of flow, plus the lost head due to friction between these two sections.

$$\frac{p_1}{w} + z_1 + \frac{v_1^2}{2g} = \frac{p_2}{w} + z_2 + \frac{v_2^2}{2g} + \text{lost head.} \tag{3.14}$$

NOTE. This is also known as the conservation of energy equation.

Power *(P)* available at a section of A square feet area in a moving stream of water, due to a pressure of p pounds/foot^2, a velocity of v feet/second, and a height of z feet above the datum level.

$$P = wvA \left(\frac{p}{w} + z + \frac{v^2}{2g} \right) \text{ foot-pounds/second.} \tag{3.15}$$

Horsepower *(h.p.)* available at any section of a stream.

$$h.p. = \frac{wvA \left(\dfrac{p}{w} + z + \dfrac{v^2}{2g} \right)}{550} \text{ horsepower.} \tag{3.16}$$

Power *(P)* available in a jet A square feet in area discharging with a velocity of v feet/second.

$$P = \frac{wv^3 A}{2g} \text{ foot-pounds/second.} \tag{3.17}$$

3.2.1   Orifices

Theoretical velocity of discharge *(v)* through an orifice due to a head of h feet over the center of gravity of the orifice.

$$v = \sqrt{2gh} \text{ feet/second.} \tag{3.18}$$

Actual velocity of discharge *(v)* if the coefficient of velocity for the orifice is c_v.

$$v = c_v \sqrt{2gh} \text{ feet/second.} \tag{3.19}$$

Quantity of discharge (Q) through an orifice A square feet in area due to a head of h feet over the center of gravity of the orifice if the coefficient of discharge is c.

$$Q = cA\sqrt{2gh} \text{ feet}^3/\text{second}. \tag{3.20}$$

NOTE.   Orifice coefficients are given in Section 10.

Quantity of discharge (Q) through a submerged orifice A square feet in area due to a head of h_1 feet on one side of the orifice and h_2 feet on the other side, the coefficient of discharge being c.

$$Q = cA\sqrt{2g(h_1 - h_2)} \text{ feet}^3/\text{second}. \tag{3.21}$$

NOTE.   If $h = h_1 - h_2$, $Q = cA(2gh)^{1/2}$ feet^3/second.

Quantity of discharge (Q) through a large rectangular orifice b feet in width with a small head of h_1 feet above the top of the orifice and a head of h_2 feet above the bottom of the orifice, the coefficient of discharge being c.

$$Q = \tfrac{2}{3}cb\sqrt{2g}(h_2^{3/2} - h_1^{3/2}) \text{ feet}^3/\text{second}. \tag{3.22}$$

Velocity of discharge (v) and **quantity of discharge** (Q) through an orifice A_1 square feet in area, considering the velocity of approach in the approach channel of A_2 square feet area, due to a pressure head of h feet, if the coefficient of discharge is c and the coefficient of velocity is c_v.

$$v = c_v \sqrt{\frac{2gh}{1 - \left(\dfrac{A_1 c}{A_2}\right)^2}} \text{ feet/second}. \tag{3.23}$$

Time (t) to lower the water in a vessel of A_1 square feet constant cross section through an orifice A_2 square feet in area, from an original head of h_1 feet over the orifice to a final head of h_2 feet.

$$t = \frac{2A_1}{cA_2\sqrt{2g}}(\sqrt{h_1} - \sqrt{h_2}) \text{ seconds}. \tag{3.24}$$

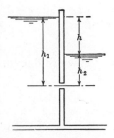

Fig. 3.4   Quantity of discharge, submerged orifice.

NOTE.   In general, problems involving the time required to lower the water in a reservoir of any cross section may be solved thus: Let A = cross-sectional area of the reservoir (this may be a variable in terms of h), Q = the rate of discharge through an orifice (or weir) as given by the ordinary formula, and h_1 and h_2 the initial and final heads.

$$t = \int_{h_2}^{h_1} \frac{A\,dh}{Q} \text{ seconds.}$$

For a suppressed weir this would be

$$t = \int_{h_2}^{h_1} \frac{A\,dh}{3.33bh^{3/2}} \text{ seconds.}$$

Mean velocity of discharge (v_m) in lowering water in a vessel of constant cross section, if the initial velocity of discharge is v_1 feet/second and the final velocity is v_2 feet/second.

$$v_m = \frac{v_1 + v_2}{2} \text{ feet/second.} \tag{3.25}$$

Constant head (h_m) that will produce the same mean velocity of discharge as is produced in lowering the water in a vessel of constant cross section from an initial head of h_1 feet over the orifice to a final head of h_2 feet.

$$h_m = \left(\frac{\sqrt{h_1} + \sqrt{h_2}}{2} \right)^2 \text{ feet.} \tag{3.26}$$

3.2.2   Weirs

Theoretical discharge (Q) over a rectangular weir b feet in width due to a head of H feet over the crest.

$$Q = \tfrac{2}{3}b\sqrt{2g}H^{3/2} \text{ feet}^3/\text{second.} \tag{3.27}$$

NOTE.   If the velocity head due to the velocity of approach v feet/second in the channel back of the weir is h feet: $Q = \tfrac{2}{3}b(2g)^{1/2}[(H + h)^{3/2} - h^{3/2}]$ feet^3/second. The actual discharge may be obtained by multiplying the theoretical discharge by a coefficient c which varies from 0.60 to 0.63 for contracted weirs (approach channel width $>$ weir width) and from 0.62 to 0.65 for suppressed weirs (approach channel width = weir width).

Francis formula for discharge (Q) over a rectangular weir b feet in width due to a head of H feet over the crest.
For a contracted weir.

$$Q = 3.33(b - 0.2H)H^{3/2} \text{ feet}^3/\text{second.} \tag{3.28}$$

Fig. 3.5 Discharge over a triangle weir.

For a contracted weir considering the velocity head h due to the velocity of approach.

$$Q = 3.33(b - 0.2H)[(H + h)^{3/2} - h^{3/2}] \text{ feet}^3/\text{second}. \qquad (3.29)$$

NOTE. In case contraction occurs on only one side of the weir, the term for width becomes $(b - 0.1H)$.

Bazin formula for discharge (Q) over a rectangular suppressed weir b feet in width due to a head of H feet over the crest and a height p feet of the crest above the bottom of the channel.

$$Q = (0.405 + \frac{0.00984}{H}) \left[1 + 0.55 \left(\frac{H}{p + H}\right)^2 \right] b\sqrt{2g}H^{3/2} \text{ feet}^3/\text{second}. \quad (3.30)$$

Fteley and Stearns' formula for discharge (Q) over a suppressed weir b feet in width due to a head of H feet over the crest.

$$Q = 3.31bH^{3/2} + 0.007b \text{ feet}^3/\text{second}. \qquad (3.31)$$

NOTE. Considering the velocity head h due to the velocity of approach. $Q = 3.31b(H + 1.5h)^{3/2} + 0.007b \text{ feet}^3/\text{second}.$

Discharge (Q) over a triangular weir, with the sides making an angle of α degrees with the vertical, due to a head of H feet over the crest (Fig. 3.5).

$$Q = c \tfrac{8}{15} \tan \alpha \sqrt{2g}H^{5/2} \text{ feet}^3/\text{second}. \qquad (3.32)$$

NOTE. If $\alpha = 45°$ (90° notch), $Q = 2.53H^{5/2} \text{ feet}^3/\text{second}.$

Discharge (Q) over a trapezoidal weir. Compute by adding the discharge over a suppressed weir b feet in width to that over the triangular weir formed by the sloping sides.

Fig. 3.6 Discharge over a trapezoidal weir.

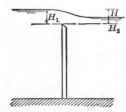

Fig. 3.7 Discharge over a trapezoidal weir.

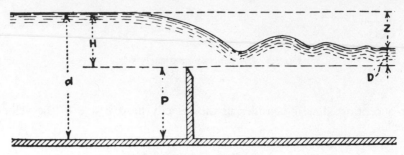

Fig. 3.8 Submerged weir.

Bazin formula[6] for discharge (Q) over a submerged sharp-crested weir b feet in width due to a head of H feet over the crest on the upstream side with a depth of submergence of D, a crest height of P above channel bottom, and a difference in water surface elevation above and below the weir of Z with $d = A/b$ A = cross-sectional area of the approach channel and b = effective width of the weir including the effect of end contractions = total width $- 0.1 \times$ number of contractions $\times H$ (or width $- 0.2H$, for contractions at both ends).

$$Q = bH^{3/2}\left(3.248 + \frac{0.079}{H}\right)\left(1 + 0.55\frac{H^2}{d^2}\right)\left(1.05 + 0.21\frac{D}{P}\right)\sqrt[3]{\frac{z}{h}}. \quad (3.33)$$

3.2.3   Flow in Open Channels

Chezy formula for quantity (Q) and velocity (v) of flow in an open channel A square feet in sectional area with p feet wetted perimeter, r feet hydraulic radius, h feet drop of water surface in distance l feet, and slope s of water surface ($s = h/l$).

$$r = \frac{A}{p} \text{ feet. } v = c\sqrt{rs} \text{ feet/second.} \quad (3.34)$$

$$Q = Av \text{ feet}^3/\text{second.} \quad (3.35)$$

NOTE. c is the coefficient and is usually found either by the Kutter formula or by the Bazin formula.

Kutter Formula

$$v = c\sqrt{rs} \text{ feet/second,} \quad (3.36)$$

where

$$c = \frac{41.6 + \dfrac{1.811}{n} + \dfrac{0.00281}{s}}{1 + \left(41.6 + \dfrac{0.00281}{s}\right)\dfrac{n}{\sqrt{r}}}.$$

NOTE.   Specific values of c are given in Table 10.26. n is the coefficient of roughness and has the following values:

Channel Lining	n
Smooth wooden flume	0.009
Neat cement and glazed pipe	0.010
Unplaned timber	0.012
Ashlar and brickwork	0.013
Rubble masonry	0.017
Very firm gravel	0.020
Earth free from stone and weeds	0.025
Earth with stone and weeds	0.030
Earth in bad condition	0.035

Bazin Formula

$$v = c\sqrt{rs} \text{ feet/second,} \tag{3.37}$$

where

$$c = \frac{87}{0.552 + \dfrac{m}{\sqrt{r}}}.$$

NOTE. Specific values of c are given in Table 10.27. m is the coefficient of roughness and has the following values:

Channel Lining	m
Smooth cement or matched boards	0.06
Planks and bricks	0.16
Masonry	0.46
Regular earth beds	0.85
Canals in good order	1.30
Canals in bad order	1.75

3.2.4 Jet Dynamics

Reaction of a jet (P) A square feet in area, the head on the orifice being h feet and the weight of the liquid w pounds/foot^3.

$$P = 2Awh \text{ pounds (theoretical).} \tag{3.38}$$

NOTE. P equals about $1.2AwH$ pounds (actual).

Energy of a jet (W) discharging with a velocity of v feet/second.

$$W = \frac{wv^3 A}{2g} \text{ foot-pounds.} \tag{3.39}$$

NOTE. If h_v is the velocity head and $Q\,(=Av)$ the quantity of flow in feet^3/second, $W = wQh_v$ foot-pounds.

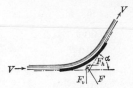

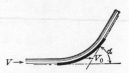

Fig. 3.9   Force exerted on fixed curve vane.   **Fig. 3.10**   Force exerted on moving curved vane.

Force (F) exerted on a fixed curve vane by a jet A square feet in area and v feet/second velocity.

$$F = \frac{Awv^2}{g} \sqrt{2(1 - \cos \alpha)} \text{ pounds.} \tag{3.40}$$

Vertical component of force (F_v) exerted by a jet on a fixed curved vane.

$$F_v = \frac{Awv^2}{g} \sin \alpha \text{ pounds.} \tag{3.41}$$

Horizontal component of force (F_h) exerted by a jet on a fixed curved vane.

$$F_h = \frac{Awv^2}{g} (1 - \cos \alpha) \text{ pounds.} \tag{3.42}$$

Force (F) exerted by a jet on a flat fixed plate perpendicular to the jet.

$$F = \frac{Awv^2}{g} \text{ pounds.} \tag{3.43}$$

Force (F) exerted on a moving curved vane by a jet A square feet in area with a velocity of v feet/second, the vane moving in the direction of the flow of the jet with a velocity of v_0 feet/second.

$$F = \frac{wA(v - v_0)^2}{g} \sqrt{2(1 - \cos \alpha)} \text{ pounds.} \tag{3.44}$$

Vertical component of force (F_v) exerted by a jet on a moving curved vane.

$$F_v = \frac{wA(v - v_0)^2}{g} \sin \alpha \text{ pounds.} \tag{3.45}$$

Horizontal component of force (F_h) exerted by a jet on a moving curved vane.

$$F_h = \frac{wA(v - v_0)^2}{g} (1 - \cos \alpha) \text{ pounds.} \tag{3.46}$$

NOTE.   If there is a series of vanes, $F_h = (wAv/g)(v - v_0)(1 - \cos \alpha)$ pounds. $F_v = (wAv/g)(v - v_0) \sin \alpha$ pounds.

Power (P) exerted on a (moving) vane.

$$P_h = F_h v_0 \text{ foot-pounds/second.} \tag{3.47}$$

NOTE. Maximum efficiency for a series of vanes occurs where $v_0 = v/2$ if there is no friction loss; then, $P = [wAv^3/(4g)](1 - \cos \alpha)$ foot-pounds/second.

3.2.5 Venturi Meter

Quantity of water (Q) flowing through a Venturi Meter with an area of A_1 square feet in the main pipe and an area A_2 square feet in the throat and a pressure head of h_1 feet in the main pipe and of h_2 feet in the throat, if the coefficient of the meter is c.

$$Q = c \frac{A_1 A_2}{\sqrt{A_1^2 - A_2^2}} \sqrt{2g(h_1 - h_2)} \text{ feet}^3/\text{second.} \tag{3.48}$$

3.3 FLOW THROUGH PIPES

Solution by Bernoulli's Theorem

If the total head at any point in the pipe system (preferably at the source) is known, the velocity of discharge at the end can be computed by applying Bernoulli's theorem between these two points, provided the losses of head can be determined. Following are expressions for the important losses of head which may occur.*

Darcy-Weisbach formula for friction loss (h_f) in a pipe of d feet internal diameter and l feet length with a velocity of v feet/second and a friction factor f.

$$h_f = f \frac{l}{d} \frac{v^2}{2g} \text{ feet.} \tag{3.49}$$

Reynolds number (R_e) is a dimensionless value which relates fluid viscosity and inertial factors. It is of particular use for predicting performance of full-size components from model tests and for establishing flow regimes in piping systems to predict fluid friction (head loss).

$$R_e = \frac{\rho v d}{\mu} \tag{3.50}$$

where: ρ = specific mass or density (lb/ft^3)
  d = length dimension (i.e., inside diameter for a round pipe ft)
  μ = absolute viscosity (lb-sec/ft^2 or slugs/ft-sec)
  v = velocity (ft/sec)

---

*These formulas apply to pipes flowing full under pressure, otherwise the pipe should be treated as an open channel.

There are three general cases of flow defined by R_e:

0–3000	**Laminar Flow**	viscous regime
2000–3000	**Transition Flow**	viscous regime
>3000	**Turbulent Flow**	inertia regime

Piping velocities for system designs that balance piping cost and friction loss are normally in the Turbulent Flow Regime (unless velocities are quite low or viscosities high); thus, where precalculated head loss tables for piping flow are utilized, they are computed on the turbulent flow basis.

Friction factor (F) can also be determined:

$$h_f = \frac{64}{R_e} \frac{l}{d} \frac{v^2}{2g} \text{ feet,} \tag{3.51}$$

thus, $F = \dfrac{64}{R_e}$. (Turbulent Flow)

NOTE.   A mean value for the friction factor for clean-cast iron pipe is 0.02. Tables 10.23 and 10.24 give values for various sizes of pipes and different velocities. In long pipelines, 1000 ft (300 m) or more, it is accurate enough to consider that the total head H is used up in overcoming friction in the pipe. Then:

$$H = f\frac{l}{d}\frac{v^2}{2g} \text{ feet,} \tag{3.52}$$

and $Q = Av$ feet^3/second.

Loss at entrance to a pipe (h_e) if the velocity of flow in the pipe is v feet/second and the entrance is sharp cornered.

$$h_e = 0.5\frac{v^2}{2g} \text{ feet.} \tag{3.53}$$

Loss due to sudden expansion (h_x) where one pipe is abruptly followed by a second pipe of larger diameter, if the velocity in the smaller pipe is v_1 feet/second and that in the larger pipe is v_2 feet/second.

$$h_x = \frac{(v_1 - v_2)^2}{2g} \text{ feet.} \tag{3.54}$$

Loss due to sudden contraction (h_c) where one pipe is abruptly followed by a second pipe of smaller diameter, if the velocity in the smaller pipe is v feet/second and c_c is a coefficient.

$$h_c = c_c\frac{v^2}{2g} \text{ feet.} \tag{3.55}$$

NOTE.   Values of c_c:

Ratio of areas	0.1	0.2	0.3	0.4	0.5	0.8	1.00
c_c	0.362	0.338	0.308	0.267	0.221	0.053	0.00

Loss due to bends (h_b).

$$h_b = c_b \frac{v^2}{2g} \text{ feet.} \qquad (3.56)$$

NOTE.   Values of c_b (d is the diameter of the pipe in feet and r is the radius of the bend in feet):

$\dfrac{d}{r}$	0.2	0.4	0.6	0.8	1.00
c_b	0.131	0.138	0.158	0.206	0.294

Nozzle loss (h_n) if the velocity of discharge is v feet/second and the velocity coefficient of the nozzle is c_v.

$$h_n = \left(\frac{1}{c_v^2} - 1 \right) \frac{v^2}{2g} \text{ feet.} \qquad (3.57)$$

Quantity of discharge (Q) in a pipe A square feet in area where the velocity is v feet/second.

$$Q = Av \text{ feet}^3/\text{second.} \qquad (3.58)$$

Diameter of pipe (d) required to deliver Q feet^3 of water/second under a head of h feet if the friction factor is f.

$$d = \sqrt[5]{\frac{fl}{2gh} \left(\frac{4Q}{\pi} \right)^2} \text{ feet.} \qquad (3.59)$$

Hydraulic gradient is a line the ordinates to which show the pressure heads at the different points in the pipe system. It may also be defined as the line to which water would rise in piezometer tubes placed at intervals along the pipe.

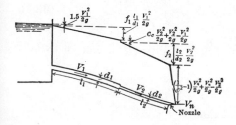

Fig. 3.11   Hydraulic gradient.

Solution by Chezy Formula

Quantity (Q) and velocity (v) of flow through a pipe when the hydraulic radius is r feet, and the slope of the hydraulic gradient is s, and the coefficient for the Chezy formula is c.

$$v = c\sqrt{rs} \text{ feet/second.} \tag{3.60}$$
$$Q = Av \text{ feet}^3/\text{second.} \tag{3.61}$$

NOTE. r equals the area in square feet divided by the wetted perimeter in feet, and s equals the head in feet divided by the length of the pipe in feet, or the slope of the hydraulic gradient.

3.4   FRICTION LOSS CALCULATIONS

An alternative and widely used method of calculating friction loss is to express all valves and fittings in the system in terms of **equivalent feet** of piping, adding this to the actual length of piping and establishing an equivalent length of piping for the entire system. This is then multiplied by values of head loss precalculated for that flow rate [usually expressed per 100 ft (30 m) of pipe length] to determine the piping system head loss. To this must be added head loss (or gain) for elevation differences as well as pressure drop (expressed in feet of head) for equipment such as strainers, heat exchangers, and vessels in the system, to determine the total system head loss. The method has been proven to be of suitable accuracy for the great majority of applications and has the added advantage of rapidity.

3.4.1   Friction Loss Graphs

Figure 3.12 and Table 3.2 express the resistance of valves and fittings in terms of length or pipe diameters. The **pipe diameter equivalent** method (Table 3.2) (which must be converted to equivalent feet) tends to be more accurate.

3.4.2   Friction Loss Tables[5]

Friction Losses in Pipes Carrying Water.   Among the many empirical formulas for friction losses that have been proposed, that of Williams and Hazen has been most widely used. In a convenient form it reads:

$$f = 0.2083 \left(\frac{100}{C}\right)^{1.85} \times \left(\frac{q^{1.85}}{d^{4.87}}\right) \tag{3.62}$$

where: f = friction head in feet of liquid per 100 ft of pipe (if desired in lb/in.^{2}, multiply $f \times 0.433 \times$ sp. gr.)
d = inside diameter of pipe (in.)
q = flow (gal/min)
C = constant accounting for surface roughness

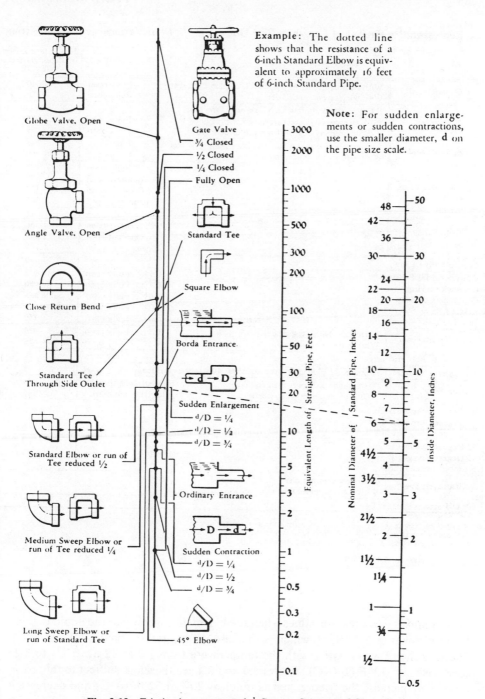

Example: The dotted line shows that the resistance of a 6-inch Standard Elbow is equivalent to approximately 16 feet of 6-inch Standard Pipe.

Note: For sudden enlargements or sudden contractions, use the smaller diameter, d on the pipe size scale.

Globe Valve, Open

Angle Valve, Open

Close Return Bend

Standard Tee Through Side Outlet

Standard Elbow or run of Tee reduced $\frac{1}{2}$

Medium Sweep Elbow or run of Tee reduced $\frac{1}{4}$

Long Sweep Elbow or run of Standard Tee

Gate Valve
$\frac{3}{4}$ Closed
$\frac{1}{2}$ Closed
$\frac{1}{4}$ Closed
Fully Open

Standard Tee

Square Elbow

Borda Entrance

Sudden Enlargement
$d/D = \frac{1}{4}$
$d/D = \frac{1}{2}$
$d/D = \frac{3}{4}$

Ordinary Entrance

Sudden Contraction
$d/D = \frac{1}{4}$
$d/D = \frac{1}{2}$
$d/D = \frac{3}{4}$

45° Elbow

Equivalent Length of Straight Pipe, Feet

Nominal Diameter of Standard Pipe, Inches

Inside Diameter, Inches

Fig. 3.12 Friction loss nomograph.[1] *Source:* Courtesy of Crane Co.

Table 3.2   Representative Equivalent Length in Pipe Diameters (L/D) of Various Valves and Fittings[1]

Description of Product				Equivalent Length In Pipe Diameters (L/D)
Globe Valves	Conventional	With no obstruction in flat, bevel, or plug type seat	Fully open	340
		With wing or pin guided disc	Fully open	450
	Y-Pattern	(No obstruction in flat, bevel, or plug type seat)		
		− With stem 60 degrees from run of pipe line	Fully open	175
		− With stem 45 degrees from run of pipe line	Fully open	145
Angle Valves	Conventional	With no obstruction in flat, bevel, or plug type seat	Fully open	145
		With wing or pin guided disc	Fully open	200
Gate Valves	Conventional Wedge Disc, Double Disc, or Plug Disc		Fully open	13
			Three-quarters open	35
			One-half open	160
			One-quarter open	900
	Pulp Stock		Fully open	17
			Three-quarters open	50
			One-half open	260
			One-quarter open	1200
	Conduit Pipe Line		Fully open	3**
Check Valves	Conventional Swing		0.5†...Fully open	135
	Clearway Swing		0.5†...Fully open	50
	Globe Lift or Stop		2.0†...Fully open	Same as Globe
	Angle Lift or Stop		2.0†...Fully open	Same as Angle
	In-Line Ball	2.5 vertical and 0.25 horizontal†...Fully open		150
Foot Valves with Strainer		With poppet lift-type disc	0.3†...Fully open	420
		With leather-hinged disc	0.4†...Fully open	75
Butterfly Valves (6-inch and larger)			Fully open	20
Cocks	Straight-Through	Rectangular plug port area equal to 100% of pipe area	Fully open	18
	Three-Way	Rectangular plug port area equal to 80% of pipe area (fully open)	Flow straight through	44
			Flow through branch	140
Fittings	90 Degree Standard Elbow			30
	45 Degree Standard Elbow			16
	90 Degree Long Radius Elbow			20
	90 Degree Street Elbow			50
	45 Degree Street Elbow			26
	Square Corner Elbow			57
	Standard Tee	With flow through run		20
		With flow through branch		60
	Close Pattern Return Bend			50

**Exact equivalent length is equal to the length between flange faces or welding ends.

†Minimum calculated pressure drop (psi) across valve to provide sufficient flow to lift disc fully.

Source: Courtesy of Crane Co.

This formula gives accurate values when the kinematic viscosity of the liquid is about 1.1 centistokes or 31.5 SSU, which is the case with water at about 60°F (15.6°C). The viscosity of water varies with the temperature from 1.8 at 32°F (0°C) to 0.29 centistokes at 212°F (100°C). Tables 3.3 and 3.4 are therefore subject to this error, which may increase the friction loss as much as 20% at 32°F (0°C) and decrease it as much as 20% at 212°F (100°C). Note that the tables may be used for any liquids having a viscosity of the same order as just indicated.

Values of C for various types of pipe are given in Table 3.3 together with the corresponding multiplier that should be applied to the tabulated values of the head loss in Table 3.4.

Table 3.3   Pipe C Values[5]

TYPE OF PIPE	VALUES OF C		
	Range — High = best, smooth, well laid — Low = poor or corroded	Average value for good, clean, new pipe	Commonly used value for design purposes
Cement—Asbestos....................................	160–140	150	140
Fibre...	—	150	140
Bitumastic-enamel-lined iron or steel centrifugally applied..	160–130	148	140
Cement-lined iron or steel centrifugally applied............	—	150	140
Copper, brass, lead, tin or glass pipe and tubing...........	150–120	140	130
Wood-stave...................................	145–110	120	110
Welded and seamless steel........................	150–80	140	100
Continuous-interior riveted steel (no projecting rivets or joints................................	—	139	100
Wrought-iron.................................	150–80	130	100
Cast-iron....................................	150–80	130	100
Tar-coated cast-iron..........................	145–80	130	100
Girth-riveted steel (projecting rivets in girth seams only)...	—	130	100
Concrete....................................	152–85	120	100
Full-riveted steel (projecting rivets in girth and horizontal seams)...................................	—	115	100
Vitrified.....................................	—	110	100
Spiral-riveted steel (flow with lap)...................	—	110	100
Spiral-riveted steel (flow against lap).................	—	100	90
Corrugated steel.............................	—	60	60

Value of C......................	150	140	130	120	110	100	90	80	70	60
Multiplier to correct tables........	.47	.54	.62	.71	.84	1.0	1.22	1.50	1.93	2.57

3.5   PUMPS

Pumping is a major requirement in many processes and can be carried out by a variety of devices. The most widely used pumps are of the centrifugal type.

3.5.1   Centrifugal Pumps

For centrifugal pumps pumping water:

$$\text{Brake h.p.} = \frac{\text{gpm} \times H \times s}{3960 \times \text{pump eff. (overall)}} \tag{3.63}$$

where: H = total head (in feet) of liquid pumped
s = specific gravity [water at 60°F (15.6°C) = 1.0]
gpm = gal/min

Pump efficiencies typically vary from 50 to 90%, depending on size, specific speed, and rotative speed. Close to the rating point a properly chosen pump and driver will have a combined efficiency of 80 to 90%.

Table 3.4 Friction Losses in Pipe: $C = 100$ (For Old Pipe)[5]

½ Inch

FLOW US gal per min	Standard Wt Steel .622" inside dia			Extra Strong Steel .546" inside dia		
	Velocity ft per sec	Velocity head ft	Head loss ft per 100 ft	Velocity ft per sec	Velocity head ft	Head loss ft per 100 ft
0.5	.528	.00	.582	.686	.01	1.10
1.0	1.06	.02	2.10	1.37	.03	3.96
1.5	1.58	.04	4.44	2.06	.07	8.38
2.0	2.11	.07	7.57	2.74	.12	14.3
2.5	2.64	.11	11.4	3.43	.18	21.6
3.0	3.17	.16	16.0	4.11	.26	30.2
3.5	3.70	.21	21.3	4.80	.36	40.2
4.0	4.23	.28	27.3	5.48	.47	51.4
4.5	4.75	.35	33.9	6.17	.59	64.0
5.0	5.28	.43	41.2	6.86	.73	77.7
5.5	5.81	.52	49.2	7.54	.88	92.7
6.0	6.34	.62	57.8	8.23	1.05	109
6.5	6.87	.73	67.0	8.91	1.23	126
7.0	7.39	.85	76.8	9.60	1.43	145
7.5	7.92	.97	87.3	10.3	1.6	165
8.0	8.45	1.11	98.3	11.0	1.9	185
8.5	8.98	1.25	110	11.6	2.1	207
9.0	9.51	1.4	122	12.3	2.4	231
9.5	10.0	1.6	135	13.0	2.6	255
10	10.6	1.7	149	13.7	2.9	280

¾ Inch

FLOW US gal per min	Standard Wt Steel .824" inside dia			Extra Strong Steel .742" inside dia		
	Velocity ft per sec	Velocity head ft	Head loss ft per 100 ft	Velocity ft per sec	Velocity head ft	Head loss ft per 100 ft
1.5	.903	.01	1.13	1.11	.02	1.88
2.0	1.20	.02	1.93	1.48	.03	3.21
2.5	1.51	.04	2.91	1.86	.05	4.85
3.0	1.81	.05	4.08	2.23	.08	6.79
3.5	2.11	.07	5.42	2.60	.11	9.03
4.0	2.41	.09	6.94	2.97	.14	11.6
4.5	2.71	.11	8.63	3.34	.17	14.4
5	3.01	.14	10.5	3.71	.21	17.5
6	3.61	.20	14.7	4.45	.31	24.5
7	4.21	.28	19.6	5.20	.42	32.6
8	4.82	.36	25.0	5.94	.55	41.7
9	5.42	.46	31.1	6.68	.69	51.8
10	6.02	.56	37.8	7.42	.86	63.0
11	6.62	.68	45.1	8.17	1.04	75.1
12	7.22	.81	53.0	8.91	1.23	88.3
13	7.82	.95	61.5	9.63	1.44	102
14	8.43	1.10	70.5	10.4	1.7	117
16	9.63	1.44	90.2	11.9	2.2	150
18	10.8	1.8	112	13.4	2.8	187
20	12.0	2.2	136	14.8	3.4	227

1 Inch

FLOW US gal per min	Standard Wt Steel 1.049" inside dia			Extra Strong Steel .957" inside dia		
	Velocity ft per sec	Velocity head ft	Head loss ft per 100 ft	Velocity ft per sec	Velocity head ft	Head loss ft per 100 ft
2	.742	.01	.595	.892	.01	.930
3	1.11	.02	1.26	1.34	.03	1.97
4	1.49	.03	2.14	1.79	.05	3.28
5	1.86	.05	3.24	2.23	.08	5.07
6	2.23	.08	4.54	2.68	.11	7.10
8	2.97	.14	7.73	3.57	.20	12.1
10	3.71	.21	11.7	4.45	.31	18.3
12	4.46	.31	16.4	5.36	.45	25.6
14	5.20	.42	21.8	6.25	.61	34.0
16	5.94	.55	27.9	7.14	.79	43.6
18	6.68	.69	34.7	8.03	1.00	54.2
20	7.43	.86	42.1	8.92	1.24	65.8
22	8.17	1.04	50.2	9.82	1.50	78.5
24	8.91	1.23	59.0	10.7	1.8	94.4
26	9.66	1.45	68.4	11.6	2.1	107

1½ Inch

FLOW US gal per min	Standard Wt Steel 1.610" inside dia			Extra Strong Steel 1.500" inside dia		
	Velocity ft per sec	Velocity head ft	Head loss ft per 100 ft	Velocity ft per sec	Velocity head ft	Head loss ft per 100 ft
4	.63	.01	.267	.73	.01	.376
5	.79	.01	.403	.91	.01	.569
6	.95	.01	.565	1.09	.02	.797
7	1.10	.02	.751	1.27	.03	1.06
8	1.26	.02	.962	1.45	.03	1.36
9	1.42	.03	1.20	1.63	.04	1.69
10	1.58	.04	1.45	1.82	.05	2.05
12	1.89	.06	2.04	2.18	.07	2.87
14	2.21	.08	2.71	2.54	.10	3.82
16	2.52	.10	3.47	2.90	.13	4.89
18	2.84	.13	4.31	3.27	.17	6.08
20	3.15	.15	5.24	3.63	.20	7.39
22	3.47	.19	6.26	3.99	.25	8.82
24	3.78	.22	7.34	4.36	.30	10.4
26	4.10	.26	8.51	4.72	.35	12.0
28	4.41	.30	9.76	5.08	.40	13.8
30	4.73	.35	11.1	5.45	.46	15.7
32	5.04	.39	12.5	5.81	.52	17.6
34	5.36	.45	14.0	6.17	.59	19.7
36	5.67	.50	15.5	6.54	.66	21.9
38	5.99	.56	17.2	6.90	.74	24.2
40	6.30	.62	18.9	7.26	.82	26.7
42	6.62	.68	20.7	7.63	.90	29.2
44	6.93	.75	22.5	7.99	.99	31.8
46	7.25	.82	24.5	8.35	1.08	34.5
48	7.57	.89	27.1	8.72	1.18	37.3
50	7.88	.97	28.5	9.08	1.28	40.3
55	8.67	1.17	34.0	9.99	1.55	49.0
60	9.46	1.39	40.0	10.9	1.8	56.4
65	10.2	1.6	46.4	11.8	2.2	65.4

Used with permission of Ingersoll-Rand Company, © 1965.

188

2 Inch

FLOW US gal per min	Standard Wt Steel 2.067" inside dia			Extra Strong Steel 1.939" inside dia		
	Velocity ft per sec	Velocity head ft	Head loss ft per 100 ft	Velocity ft per sec	Velocity head ft	Head loss ft per 100 ft
5	.48	.00	.120	.54	.00	.163
6	.57	.01	.167	.65	.01	.229
7	.67	.01	.223	.76	.01	.304
8	.77	.01	.285	.87	.01	.389
9	.86	.01	.355	.98	.01	.484
10	.96	.01	.431	1.09	.02	.588
12	1.15	.02	.604	1.30	.03	.824
14	1.34	.03	.803	1.52	.04	1.10
16	1.53	.04	1.03	1.74	.05	1.40
18	1.72	.05	1.28	1.96	.06	1.74
20	1.91	.06	1.55	2.17	.07	2.12
22	2.10	.07	1.85	2.39	.09	2.53
24	2.29	.08	2.18	2.61	.11	2.97
26	2.49	.10	2.52	2.83	.12	3.44
28	2.68	.11	2.89	3.04	.14	3.95
30	2.87	.13	3.29	3.26	.17	4.49
35	3.35	.17	4.37	3.80	.22	5.97
40	3.82	.23	5.60	4.35	.29	7.64
45	4.30	.29	6.96	4.89	.37	9.50
50	4.78	.36	8.46	5.43	.46	11.5
55	5.26	.43	10.1	5.98	.56	13.7
60	5.74	.51	11.9	6.52	.66	16.2
65	6.21	.60	13.7	7.06	.77	18.8
70	6.69	.70	15.8	7.61	.90	21.5
75	7.17	.80	17.9	8.15	1.03	24.5
80	7.65	.91	20.2	8.69	1.17	27.6
85	8.13	1.03	22.6	9.03	1.27	30.8
90	8.61	1.15	25.1	9.78	1.49	34.3
95	9.08	1.28	27.7	10.3	1.6	37.9
100	9.56	1.42	30.5	10.9	1.8	41.6
110	10.5	1.7	36.4	12.0	2.2	49.7
120	11.5	2.1	42.7	13.0	2.6	58.3
130	12.4	2.4	49.6	14.1	3.1	67.7
140	13.4	2.8	56.9	15.2	3.6	77.6
150	14.3	3.2	64.7	16.3	4.1	88.4

3 Inch

FLOW US gal per min	Cast Iron 3.0" inside dia			Std Wt Steel 3.068" inside dia			Extra Strong Steel 2.900" inside dia		
	Velocity ft per sec	Velocity head ft	Head loss ft per 100 ft	Velocity ft per sec	Velocity head ft	Head loss ft per 100 ft	Velocity ft per sec	Velocity head ft	Head loss ft per 100 ft
10	.45	.00	.070	.43	.00	.063	.49	.00	.083
15	.68	.01	.149	.65	.01	.134	.73	.01	.176
20	.91	.01	.254	.87	.01	.227	.97	.02	.299
25	1.13	.02	.383	1.09	.02	.344	1.21	.02	.452
30	1.36	.03	.537	1.30	.03	.481	1.45	.03	.633
35	1.59	.04	.714	1.52	.04	.640	1.70	.04	.842
40	1.82	.05	.914	1.74	.05	.820	1.94	.06	1.08
45	2.04	.06	1.14	1.95	.06	1.02	2.18	.07	1.34
50	2.27	.08	1.38	2.17	.07	1.24	2.43	.09	1.63
55	2.50	.10	1.64	2.39	.09	1.47	2.67	.11	1.94
60	2.72	.12	1.94	2.60	.11	1.74	2.91	.13	2.28
65	2.95	.14	2.24	2.82	.12	2.01	3.16	.15	2.65
70	3.18	.16	2.57	3.04	.14	2.31	3.40	.18	3.04
75	3.40	.18	2.92	3.25	.16	2.62	3.64	.21	3.45
80	3.63	.20	3.30	3.47	.19	2.96	3.88	.23	3.89
85	3.86	.23	3.69	3.69	.21	3.31	4.12	.26	4.35
90	4.09	.26	4.10	3.91	.24	3.67	4.37	.29	4.83
95	4.31	.29	4.53	4.12	.26	4.06	4.61	.33	5.34
100	4.54	.32	4.98	4.34	.29	4.47	4.85	.36	5.87
110	4.99	.39	5.94	4.77	.35	5.33	5.33	.44	7.01
120	5.45	.46	6.98	5.21	.42	6.26	5.81	.52	8.23
130	5.90	.54	8.09	5.64	.49	7.26	6.30	.62	9.54
140	6.35	.63	9.28	6.08	.57	8.32	6.79	.71	10.9
150	6.81	.72	10.6	6.51	.66	9.48	7.28	.82	12.5
160	7.26	.82	11.9	6.94	.75	10.7	7.76	.93	14.0
180	8.16	1.03	14.8	7.81	.95	13.2	8.72	1.01	17.4
200	9.08	1.28	18.0	8.68	1.17	16.1	9.70	1.46	21.2
220	9.99	1.55	21.4	9.55	1.42	19.2	10.7	1.78	25.3
240	10.9	1.8	25.2	10.4	1.7	22.6	11.6	2.07	29.7
260	11.8	2.2	29.2	11.3	2.0	26.2	12.6	2.46	34.4
280	12.7	2.5	33.5	12.2	2.3	30.0	13.6	2.88	39.5
300	13.6	2.9	38.0	13.0	2.6	34.1	14.5	3.26	44.8
320	14.5	3.3	42.8	13.9	3.0	38.4	15.5	3.77	50.5
340	15.4	3.7	47.9	14.8	3.4	43.0	16.5	4.22	56.5
360	16.3	4.1	53.3	15.6	3.8	47.8	17.5	4.73	62.8

4 Inch

FLOW US gal per min	Cast Iron 4.0" inside dia			Std Wt Steel 4.026" inside dia			Extra Strong Steel 3.826" inside dia		
	Velocity ft per sec	Velocity head ft	Head loss ft per 100 ft	Velocity ft per sec	Velocity head ft	Head loss ft per 100 ft	Velocity ft per sec	Velocity head ft	Head loss ft per 100 ft
70	1.79	.05	.635	1.76	.05	.615	1.95	.06	.789
80	2.04	.06	.813	2.02	.06	.788	2.23	.08	1.01
90	2.30	.08	1.01	2.27	.08	.980	2.51	.10	1.26
100	2.55	.10	1.23	2.52	.10	1.19	2.79	.12	1.53
110	2.81	.12	1.47	2.77	.12	1.42	3.07	.15	1.82
120	3.06	.15	1.72	3.02	.14	1.67	3.35	.17	2.14
130	3.32	.17	2.00	3.28	.17	1.93	3.63	.20	2.48
140	3.57	.20	2.29	3.53	.19	2.22	3.91	.24	2.84
150	3.83	.23	2.61	3.78	.22	2.53	4.19	.27	3.24
160	4.08	.26	2.93	4.03	.25	2.84	4.47	.31	3.64
170	4.34	.29	3.28	4.29	.29	3.18	4.75	.35	4.07
180	4.60	.33	3.64	4.54	.32	3.53	5.02	.39	4.52
190	4.86	.37	4.03	4.79	.36	3.90	5.30	.44	5.00
200	5.11	.41	4.43	5.05	.40	4.29	5.58	.48	5.50
220	5.62	.49	5.28	5.55	.48	5.12	6.14	.59	6.56
240	6.13	.58	6.21	6.05	.57	6.01	6.70	.70	7.70
260	6.64	.69	7.20	6.55	.67	6.97	7.26	.82	8.93
280	7.15	.79	8.25	7.06	.77	8.00	7.82	.95	10.2
300	7.66	.91	9.38	7.57	.89	9.09	8.38	1.09	11.6
320	8.17	1.04	10.6	8.07	1.01	10.2	8.94	1.24	13.1
340	8.68	1.17	11.8	8.58	1.14	11.5	9.50	1.4	14.7
360	9.19	1.31	13.1	9.08	1.28	12.7	10.0	1.6	16.3
380	9.70	1.46	14.5	9.59	1.43	14.1	10.6	1.7	18.0
400	10.2	1.6	16.0	10.1	1.6	15.5	11.2	1.9	19.8
420	10.7	1.8	17.5	10.6	1.7	17.0	11.7	2.1	21.7
440	11.2	1.9	19.0	11.1	1.9	18.5	12.3	2.3	23.6
460	11.7	2.1	20.7	11.6	2.1	20.0	12.8	2.5	25.7
480	12.3	2.3	22.4	12.1	2.3	21.7	13.4	2.8	27.8
500	12.8	2.5	24.1	12.6	2.5	23.4	14.0	3.0	30.0
550	14.0	3.0	28.8	13.9	3.0	27.9	15.3	3.6	35.7

6 Inch

FLOW US gal per min	Cast Iron 6.0" inside dia			Std Wt Steel 6.065" inside dia			Extra Strong Steel 5.761" inside dia		
	Velocity ft per sec	Velocity head ft	Head loss ft per 100 ft	Velocity ft per sec	Velocity head ft	Head loss ft per 100 ft	Velocity ft per sec	Velocity head ft	Head loss ft per 100 ft
100	1.13	.02	.171	1.11	.02	.162	1.23	.02	.208
120	1.36	.03	.239	1.33	.03	.227	1.48	.03	.292
140	1.59	.04	.318	1.56	.04	.302	1.72	.05	.388
160	1.82	.05	.408	1.78	.05	.387	1.97	.06	.497
180	2.04	.06	.507	2.00	.06	.481	2.22	.08	.618
200	2.27	.08	.616	2.22	.08	.584	2.46	.09	.751
220	2.50	.10	.735	2.44	.09	.697	2.71	.11	.895
240	2.72	.12	.863	2.67	.11	.819	2.96	.14	1.03
260	2.95	.14	1.00	2.89	.13	.950	3.20	.16	1.22
280	3.18	.16	1.15	3.11	.15	1.09	3.45	.19	1.40
300	3.40	.18	1.30	3.33	.17	1.24	3.69	.21	1.59
320	3.64	.21	1.47	3.56	.20	1.39	3.94	.24	1.79
340	3.86	.23	1.64	3.78	.22	1.56	4.19	.27	2.00
360	4.08	.26	1.83	4.00	.25	1.73	4.43	.31	2.23
380	4.31	.29	2.02	4.22	.28	1.92	4.68	.34	2.46
400	4.55	.32	2.22	4.44	.31	2.11	4.93	.38	2.71
450	5.11	.41	2.76	5.00	.39	2.62	5.54	.48	3.36
500	5.68	.50	3.36	5.56	.48	3.19	6.16	.59	4.09
550	6.25	.61	4.00	6.11	.58	3.80	6.77	.71	4.88
600	6.81	.72	4.70	6.66	.69	4.46	7.39	.85	5.73
650	7.38	.85	5.45	7.22	.81	5.17	8.00	.99	6.64
700	7.95	.98	6.25	7.78	.94	5.93	8.63	1.16	7.62
750	8.52	1.13	7.10	8.34	1.08	6.73	9.24	1.33	8.66
800	9.08	1.28	8.00	8.90	1.23	7.60	9.85	1.51	9.75
850	9.65	1.45	8.95	9.45	1.39	8.50	10.5	1.7	10.9
900	10.2	1.6	9.95	10.0	1.6	9.44	11.1	1.9	12.1
950	10.8	1.8	11.0	10.5	1.7	10.2	11.7	2.1	13.4
1000	11.4	2.0	12.1	11.1	1.9	11.5	12.3	2.4	14.7
1100	12.5	2.4	14.4	12.2	2.3	13.7	13.5	2.8	17.6
1200	13.6	2.9	16.9	13.3	2.7	16.1	14.8	3.4	20.7

Table 3.4   (*Continued*)

8 Inch

FLOW US gal per min	Cast Iron 8.0" inside dia			Std Wt Steel 7.981" inside dia			Extra Strong Steel 7.625" inside dia		
	Velocity ft per sec	Velocity head ft	Head loss ft per 100 ft	Velocity ft per sec	Velocity head ft	Head loss ft per 100 ft	Velocity ft per sec	Velocity head ft	Head loss ft per 100 ft
180	1.15	.02	.125	1.15	.02	.126	1.26	.02	.158
190	1.21	.02	.138	1.22	.02	.140	1.33	.03	.175
200	1.28	.03	.152	1.28	.03	.154	1.41	.03	.192
220	1.40	.03	.181	1.41	.03	.183	1.55	.04	.229
240	1.53	.04	.213	1.54	.04	.215	1.69	.04	.269
260	1.66	.04	.247	1.67	.04	.250	1.83	.05	.312
280	1.79	.05	.283	1.80	.05	.286	1.97	.06	.358
300	1.91	.06	.322	1.92	.06	.325	2.11	.07	.406
350	2.24	.08	.428	2.24	.08	.433	2.46	.09	.540
400	2.56	.10	.548	2.57	.10	.554	2.81	.12	.692
450	2.87	.13	.681	2.88	.13	.689	3.16	.15	.860
500	3.19	.16	.828	3.20	.16	.838	3.51	.19	1.05
550	3.51	.19	.987	3.52	.19	.999	3.86	.23	1.25
600	3.83	.23	1.16	3.85	.23	1.17	4.22	.28	1.46
650	4.15	.27	1.34	4.17	.27	1.36	4.57	.32	1.70
700	4.47	.31	1.54	4.49	.31	1.56	4.92	.38	1.95
750	4.79	.36	1.75	4.81	.36	1.77	5.27	.43	2.21
800	5.11	.41	1.97	5.13	.41	1.99	5.62	.49	2.49
850	5.43	.46	2.21	5.45	.46	2.23	5.97	.55	2.79
900	5.75	.51	2.46	5.77	.52	2.48	6.32	.62	3.10
950	6.06	.57	2.71	6.09	.58	2.74	6.67	.69	3.43
1000	6.38	.63	2.98	6.41	.64	3.02	7.03	.77	3.77
1100	7.03	.77	3.56	7.05	.77	3.60	7.83	.95	4.49
1200	7.66	.91	4.18	7.69	.92	4.23	8.43	1.10	5.28
1300	8.30	1.07	4.85	8.33	1.08	4.90	9.13	1.30	6.12
1400	8.95	1.24	5.56	8.97	1.25	5.62	9.83	1.50	7.02
1500	9.58	1.43	6.32	9.61	1.44	6.39	10.5	1.7	7.98
1600	10.2	1.6	7.12	10.3	1.7	7.20	11.2	2.0	8.99
1800	11.5	2.1	8.85	11.5	2.1	8.95	12.6	2.5	11.2
2000	12.8	2.6	10.8	12.8	2.5	10.9	14.1	3.1	13.6

10 Inch

FLOW US gal per min	Cast Iron 10.0" inside dia			Standard Wt Steel 10.02" inside dia		
	Velocity ft per sec	Velocity head ft	Head loss ft per 100 ft	Velocity ft per sec	Velocity head ft	Head loss ft per 100 ft
500	2.04	.06	.280	2.04	.06	.277
550	2.24	.08	.333	2.24	.08	.330
600	2.45	.09	.392	2.44	.09	.388
650	2.65	.11	.454	2.64	.11	.450
700	2.86	.13	.521	2.85	.13	.516
800	3.26	.17	.667	3.25	.16	.660
900	3.67	.21	.829	3.66	.21	.821
1000	4.08	.26	1.01	4.07	.26	.998
1100	4.49	.31	1.20	4.48	.31	1.19
1200	4.90	.37	1.41	4.89	.37	1.40
1300	5.31	.44	1.64	5.30	.44	1.62
1400	5.71	.51	1.88	5.70	.50	1.86
1500	6.12	.58	2.13	6.10	.58	2.11
1600	6.53	.66	2.40	6.51	.66	2.38
1700	6.94	.75	2.69	6.92	.74	2.66
1800	7.35	.84	2.99	7.32	.83	2.96
1900	7.76	.94	3.30	7.73	.93	3.27
2000	8.16	1.03	3.63	8.14	1.03	3.60
2200	8.98	1.25	4.33	8.95	1.24	4.29
2400	9.80	1.49	5.09	9.76	1.48	5.04
2600	10.6	1.7	5.90	10.6	1.7	5.84
2800	11.4	2.0	6.77	11.4	2.0	6.70
3000	12.2	2.3	7.69	12.2	2.3	7.61
3200	13.1	2.7	8.66	13.0	2.7	8.58
3400	13.9	3.0	9.69	13.8	3.0	9.60
3600	14.7	3.4	10.8	14.6	3.3	10.7
3800	15.5	3.7	11.9	15.5	3.7	11.8
4000	16.3	4.1	13.1	16.3	4.1	13.0
4500	18.4	5.3	16.3	18.3	5.2	16.1
5000	20.4	6.5	19.8	20.3	6.4	19.6

12 Inch

FLOW US gal per min	Cast Iron 12.0" inside dia			Standard Wt Steel 12.000" inside dia		
	Velocity ft per sec	Velocity head ft	Head loss ft per 100 ft	Velocity ft per sec	Velocity head ft	Head loss ft per 100 ft
800	2.27	.08	.275	2.27	.08	.275
900	2.56	.10	.341	2.56	.10	.341
1000	2.84	.13	.415	2.84	.13	.415
1100	3.12	.15	.495	3.12	.15	.495
1200	3.41	.18	.581	3.41	.18	.581
1300	3.69	.21	.674	3.69	.21	.674
1400	3.98	.25	.773	3.98	.25	.773
1500	4.26	.28	.878	4.26	.28	.878
1600	4.55	.32	.990	4.55	.32	.990
1800	5.11	.41	1.23	5.11	.41	1.23
2000	5.68	.50	1.50	5.68	.50	1.50
2200	6.25	.61	1.78	6.25	.61	1.78
2400	6.81	.72	2.10	6.81	.72	2.10
2600	7.38	.85	2.43	7.38	.85	2.43
2800	7.95	.98	2.78	7.95	.98	2.78
3000	8.52	1.13	3.17	8.52	1.13	3.17
3500	9.95	1.54	4.21	9.95	1.54	4.21
4000	11.4	2.0	5.39	11.4	2.0	5.39
4500	12.8	2.5	6.70	12.8	2.5	6.70
5000	14.2	3.1	8.15	14.2	3.1	8.15
5500	15.6	3.8	9.72	15.6	3.8	9.72
6000	17.0	4.5	11.4	17.0	4.5	11.4
6500	18.4	5.3	13.2	18.4	5.3	13.2
7000	19.9	6.2	15.2	19.9	6.2	15.2
7500	21.3	7.1	17.3	21.3	7.1	17.3
8000	22.7	8.0	19.4	22.7	8.0	19.4
8500	24.2	9.1	21.7	24.2	9.1	21.7
9000	25.6	10.2	24.2	25.6	10.2	24.2
9500	27.0	11.3	26.7	27.0	11.3	26.7
10000	28.4	12.5	29.4	28.4	12.5	29.4

18 Inch

FLOW US gal per min	Cast Iron 18.0" inside dia			Steel 17 18" inside dia		
	Velocity ft per sec	Velocity head ft	Head loss ft per 100 ft	Velocity ft per sec	Velocity head ft	Head loss ft per 100 ft
1000	1.26	.02	.058	1.38	.03	.072
1200	1.53	.04	.081	1.66	.04	.101
1400	1.78	.05	.108	1.94	.06	.135
1600	2.03	.06	.138	2.21	.08	.173
1800	2.27	.08	.171	2.49	.10	.215
2000	2.52	.10	.208	2.77	.12	.261
2500	3.15	.15	.314	3.46	.19	.394
3000	3.78	.22	.440	4.15	.27	.553
3500	4.41	.30	.586	4.85	.37	.735
4000	5.04	.39	.750	5.54	.48	.941
4500	5.67	.50	.932	6.23	.60	1.17
5000	6.30	.62	1.13	6.92	.74	1.42
6000	7.56	.89	1.59	8.31	1.1	1.99
7000	8.83	1.2	2.11	9.70	1.5	2.65
8000	10.1	1.6	2.70	11.1	1.9	3.39
9000	11.3	2.0	3.36	12.5	2.4	4.22
10000	12.6	2.5	4.08	13.8	3.0	5.12
12000	15.3	3.6	5.72	16.6	4.3	7.18
14000	17.8	4.9	7.61	19.4	5.8	9.55
16000	20.3	6.4	9.74	22.1	7.6	12.2
18000	22.7	8.0	12.1	24.9	9.6	15.2
20000	25.2	9.9	14.7	27.7	11.9	18.5
22000	27.7	11.9	17.6	30.3	14.3	22.0
24000	30.6	14.6	20.6	33.2	17.1	25.9
26000	32.8	16.7	23.9	36.0	20.1	30.0
28000	35.5	19.6	27.4	38.8	23.4	34.4
30000	37.8	22.2	31.2	41.5	26.8	39.1
32000	40.6	25.6	35.1	44.3	30.5	44.1
34000	42.8	28.5	39.4	47.1	34.5	49.4
36000	45.4	32.0	43.7	49.9	38.7	54.8

Other useful relationships:

$$\text{At constant rpm:} \frac{\text{Capacity } A}{\text{Capacity } B} = \frac{\text{Impeller dia. } A}{\text{Impeller dia. } B} = \frac{H_A}{H_B}. \qquad (3.64)$$

$$\text{At constant Impeller dia.:} \frac{\text{Capacity } A}{\text{Capacity } B} = \frac{N_A}{N_B} = \frac{H_A}{H_B} \qquad (3.65)$$

and

$$\frac{\text{Brake h.p.}_A}{\text{Brake h.p.}_B} = \frac{N_A^3}{N_B^3} \qquad (3.66)$$

where $N_{A,B}$ = rotative speed (rpm).
These relationships should only be applied to relatively small changes, that is, 10 to 15% maximum.

Specific speed is defined:

$$N_s = \frac{NQ}{H^{3/4}} \qquad (3.67)$$

where N_s = specific speed (dimensionless).

Centrifugal pumps are broadly classified into types based on specific speed (Fig. 3.13).

Centrifugal pumps have a head/flow relationship that typically has a rising characteristic. If the system head/flow curve is plotted with the pump curve, their intersection will be the operating point for the pump/pumping system (see Fig. 3.14).

While less of a problem today than formerly, care should still be taken to be certain the pump head curve is *continuously* rising to avoid a condition of instability and "hunting." In Figure 3.15, two flows, *a* and *b*, correspond to *h'*. The pump will "hunt" (oscillate) between them and its operation will be unstable.

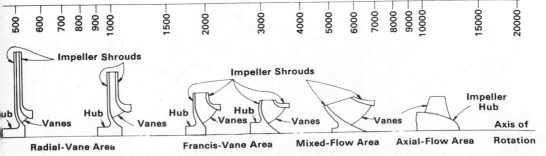

Values of Specific Speeds.

Fig. 3.13 Comparison of pump profiles. Courtesy of the Hydraulic Institute.[2]

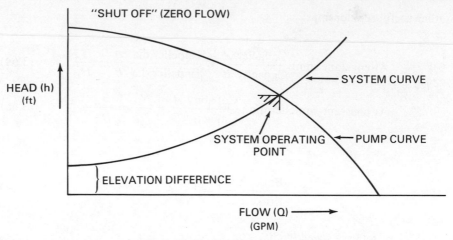

Fig. 3.14   Typical head/flow curve.

3.5.2   Other Pumps

Other types of pumps that are widely used are of the positive displacement type:

Screw type
Gear
Piston (Reciprocating)
Diaphragm
Vane

3.5.3   Pump Application

Capacity.   The theoretical cubic feet per minute displacement (V) of a pump that makes N pumping strokes of L feet forward per minute and has a piston of A square inches effective area is

$$V = \frac{ALN}{144} \text{ feet}^3/\text{minute.} \qquad (3.68)$$

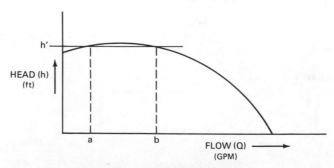

Fig. 3.15   Unstable head/flow curve.

NOTE. If the pump is double-acting, the total displacement is the sum of the displacements on the forward and return strokes, the effective area varies for the two sides of the piston. Due to clearance, slip, imperfect valve action, and so on, the actual displacement is reduced as much as 50% in some cases.

Water Horsepower. For a pump that discharges G pounds of water per minute through a total head H feet, the water horsepower (w.h.p.) is

$$\text{w.h.p.} = \frac{GH}{33,000} \text{ horsepower.} \tag{3.69}$$

NOTE. The total head must include the suction lift, the discharge lift, friction and velocity heads.

Overall Thermal Efficiency. For steam-driven pumping units it is customary to express the ratio of the heat actually converted into work in lifting the water to the heat supplied as overall thermal efficiency (ϵ_t), hence

$$\epsilon_t = \frac{2545}{w_a(h_1 - h_{f2})}, \tag{3.70}$$

where w_a is the actual steam consumption in pounds per water horsepower-hour.

Duty. The term "duty" is applied to steam-driven pumping units to indicate the foot-pounds of work done for every million Btu supplied. For a pump that discharges G pounds of water per minute through a total head of H feet while using M pounds of steam per minute with Q Btu available per pound, the duty (D) is

$$\begin{aligned} D &= \frac{GH}{MQ} \times 10^6 \text{ foot-pounds/million Btu} \\ &= \frac{\text{w.h.p.}}{MQ} \times 33 \times 10^9 \text{ foot-pounds/million Btu} \\ &= \epsilon_t \times 778 \times 10^6 \text{ foot-pounds/million Btu.} \end{aligned} \tag{3.71}$$

Pump drivers include the following principal types:

Electric motors
Engines—internal combustion and other
Compressed air, steam, and so forth

Engine drivers, particularly for positive development pump applications should include a clutch device to permit engine starting under a no-load condition.

Design life of pumps can range up to 40 years for carefully engineered and well-maintained installations; however, as with any machine, allowance for wear must be considered. Even pumping clear water without suspended material, pumps will suffer wear over an extended time.

Typically, pump requirements are stated for clear water service and often include at rated head, flow allowances of 7 to 10% for wear plus a 5% additional margin when fully worn. Centrifugal pumps are frequently furnished with replacement wear rings to permit restoration to original internal clearances over the pump life; other type pumps achieve this through replacement of sleeves, liners, and so on.

Pumps for abrasive service are available with a variety of types of replaceable liners (rubber, abrasion resistant, metallic alloys, etc.). Pumps for hazardous or sensitive materials can be obtained where the surfaces in contact with the fluid are either of compatible materials or isolated from the driver. Zero leakage ("canned") pumps have also been developed for particular services.

Net positive suction head (**NPSH**) defines the margin of absolute pressure required at the pump centerline to avoid flashing due to reductions in vapor pressure. Flashing (vaporization) can occur either in the suction or inlet piping or at the pump entrance, with loss of fluid flow into the pump or internally (usually at the impeller) causing cavitation which can be extremely destructive to the pump. Pump manufacturers normally provide data on required NPSH and recommended inlet piping to avoid the problem. NPSH requirements vary with pump sizes, types, and speeds.

3.6 PIPING SYSTEMS[1,4]

3.6.1 Design Considerations

The principal design considerations for fluid piping systems flowing full are: flow requirements; head loss (flow resistance); corrosion/erosion resistance; resistance to fouling; handling strength; temperature; operating, and test pressure requirements; and hanging and supports for the piping system.

Flow requirements are established by the process requirements and must consider maximum and in some cases minimum required flows as well as the normal flow for the system. **Economic sizing** of pressure piping systems would typically result in fluid flow velocities of 6 to 8 fps (1.8 to 2.4 mps) to 12-in. (300 mm) diameter lines and in larger lines above 12 in. (300 mm) of 10 to 15 fps (3 to 5 mps). For systems carrying steam the following values are typical: saturated, 100 to 175 fps (30 to 55 mps); superheated, 125 to 350 fps (40 to 110 mps). For complex or costly systems, detailed economic studies to evaluate pipe size versus system and pumping cost(s) are necessary. Caution must be exercised for high-velocity systems to avoid excessive erosion, noise, and vibration.

Head loss or flow resistance for the system is a function of both required flow(s) and the smoothness of the interior of the pipe. It is also a function of the head loss (pressure drop) through components such as heat exchangers, pressure vessels, valves, and fittings, as well as the elevation differences existing in the system. Excluding elevation differences that are constant, head losses will increase roughly as the square of the velocity (or flow).

Corrosion resistance can be achieved either by material selection or by the use of greater design allowances, that is, thicker wall piping. For conventional cold water service to 350°F (180°C), Carbon steel, ASTM A-106, Gr. B or ASTM A53, Gr. A or B are frequently specified. For other fluids, particularly those which are corrosive, considerable care should be taken in materials selection. A wide range of materials having high degrees of resistance to chemical attack are available, including plastic, high-nickel steel alloys, cast iron, glazed tile, and tempered glass. Each has its advantages and a proper selection should include consideration of each. Particular thought must be given to temperature requirements which might, for example, eliminate a highly corrosion-resistant plastic from consideration.

Erosion resistance can be achieved by providing greater design allowances or by selecting a piping material that is either very hard (such as high-nickel alloy piping) or relatively elastic (such as polyethylene or rubber-lined steel pipe). The plastics are relatively cheap and easy to assemble, but transitions and valve installations are sometimes awkward; further they are not usually able to accept elevated temperatures or high pressure because of their relatively low allowable (piping wall) tensile stress.

Handling strength should be considered where piping systems require long spans or have risk of mechanical damage. The risk of damage may be present during construction and installation as well as during operation, and this may require additional supports, barriers, or in some cases, the use of heavier wall piping. For some critical services **concentric piping systems** are used, with the inner line carrying the fluid under pressure and the outer line acting as a guard pipe or return line. Concentric piping can be used for steam jacketing (heating) of fluid in the inner line. It is expensive to fabricate and install, however; heating (tracing) of piping by a smaller line in contact with the larger line or by electric resistance heaters attached to the piping outer surface may be more economic.

Pressure boundary integrity of piping systems, including valves, instruments, and so forth, is assured through a **hydrostatic (hydro) test.** These hydro tests require venting and completely filling the system with water at a temperature above the NDT (nil ductility temperature—brittle failure temperature of the piping system material) and then pressurizing the system above maximum operating pressure. The hydro test pressures are usually a requirement of the applicable code (ANSI B31.1, ASME Boiler and Pressure Vessel Code, API Codes, etc.) and are typically on the order of 125 to 150% of the system maximum operating pressure. For systems not fully filled with water during operation, piping and equipment supports should be checked for adequacy during this flooded condition.

Pipe wall thickness is designated by scheduled numbers in which Schedule 40 is standard wall with Schedules 20 and 10 being progressively thinner and Schedules 80, 120, and 160 being thicker; ANSI standards B36.10 and B36.19 provide further information. For characteristics of pipe of various schedules up to 12 in (305 mm) in diameter, see Table 10.31.

3.6.2 Pipe Supports

Hangers or pipe supports are of three general types: **rigid, variable support,** and **constant support.** Rigid hangers, as their name implies, provide a rigid connection to the support structure, and although they may provide for horizontal linear expansion (through rollers, etc.) they do not provide for vertical motion. Vertical motion of systems or components having variable weight or large temperature ranges (causing changes in length) is provided for by either variable support or constant support hangers. Variable support hangers generally use a spring system to compensate for the load, but permit some movement, dependent upon the load and the spring constant. Constant support hangers generally have a linkage on a spring system to compensate for the load and maintain the supported element at a fixed location.

Hanger-type information and design guides are available from several major manufacturers. A basic requirement is to avoid excessive span between hangers and to arrange fluid lines to permit natural drainage. Provision for expansion and protection of insulation are also required.

3.6.3 Piping Materials

Material selection of piping systems requires consideration of the fluid carried, its properties, system life, maintenance requirements, and so forth. Table 3.5 is a guide to some typical applications. Where fluids are extremely corrosive or abrasive, more specialized materials may be required.

3.6.4 Valve and Fitting Material Selection

Piping equipment commonly used in industry falls into four basic material groups: bronze, cast iron, malleable iron, and steel. There are also one or more variants in each of these groups and each variant has individual service characteristics.

Bronze should not be used for temperatures exceeding 550°F (290°C). One manufacturer produces two grades of bronze. Normal bronze, an alloy of copper, tin, lead, and zinc, is widely used in valves and fittings for temperatures up to 450°F (230°C). Special bronze is a high-grade alloy used in piping equipment for higher pressures and for temperatures up to 550°F (290°C).

Cast iron is regularly made in two grades: Cast iron and High-tensile iron. It should not be used for temperatures exceeding 450°F (230°C). Cast iron is commonly used for small valves and fittings having light metal sections. High-tensile cast iron is a high-strength alloy cast iron used principally for castings for large size valves.

Malleable iron is particularly suited for use in screwed fittings, unions, and so on, and also is used to some extent for valves and flanges. It is characterized by pressure tightness, stiffness, and toughness, and is especially valuable for piping systems subject to expansion and contraction stresses and shock.

Steel is recommended for high pressure and temperatures and for services where working conditions, either internal or external, may be too severe for bronze or iron. Its superior strength and toughness and its resistance to piping strains, vibration, shock, low temperature, and damage by fire afford reliable protection when safety and utility are desired. Many different types of steel are both necessary and available because of the widely diversified services steel valves and fittings perform.

3.6.5 Piping Fabrication

The manner of fabrication of piping is of increasing importance. More and more welded piping systems are being used where threaded or flanged systems have been used in the past. The practice of welding pipe joints is now used for all sizes of high-pressure, high-temperature lines and for general service installations.

Butt-welded joints are designed to combine serviceability with ease of installation. Butt-welding consists of beveling the two ends, lining up the two openings, and then making the circumferential butt-welds.

Table 3.5 Typical Piping Materials[a]

Material	Typical Temperature Limit	Material Type	Comments
Low-alloy steel 2¼ Cr, 1 Mo	1015°F (546°C)	ASTM A-335, Gr. P22 (seamless) or ASTM A-369, Gr. FP22 (seamless)	Steam to 24″ (610 mm) To 2650 psig (18,300 kPa)
Low-alloy steel 2¼ Cr, 1 Mo	975°F (524°C)	ASTM A-155, Gr. 2¼ Cr Class 1 (seam-welded)	Steam/water to 24″ (610 mm) To 900 psig (6200 kPa)
Low-alloy steel 1¼ Cr	850°F (454°C)	ASTM A-335, Gr. P-11 or seam-welded ASTM A-155, Class 2 Gr., 1¼ Cr	General-purpose low-alloy steel
Stainless steel	1100°F (590°C)	ASTM A-376, Gr. TP 304 or A-312 Gr. TP 304 or 304L seamless (seam-welded frequently used for lower temperature and pressures)	General-purpose stainless steel, typically 8″ (200 mm) and smaller
Carbon steel	850°F (454°C)	ASTM SA-106, Gr. B (seamless)	Steam/water To 24″ (610 mm)
Carbon steel	350°F (177°C)	ASTM A-53, Gr. A or B seamless (or ASTM A-120 for gravity drains)	General-purpose carbon steel; water, air, steam To 24″ (610 mm)
Red brass	250°F (120°C)	ASTM B-43	Instrument air to 4″ (100 mm)
Copper tubing	400°F[b] (204°C)	ASTM B-88 and ANSI H23.1	Water to 4″ (100 mm)
Cast iron	150°F (67°C)	ANSI A21.11 and ANSI A21.6	Water, sewage underground service mechanical or push-on joints—to 24″ (610 mm)
Ductile cast iron	150°F (67°C)	ANSI A21.51	Water, sewage
Concrete	150°F (67°C)	ASTM C-76, Class III Wall A or Class IV Wall B	Wastewater, nonpressure drainage
Vitrified clay	150°F (67°C)	Extra strength ASTM C-700	Nonpressure sewer
Polyvinyl chloride	140°F (60°C)	ASTM D1784, Class 12454-B	Chemical drains To 6″ (150 mm)

[a]Pressure requirements will establish pipe wall thickness; see applicable design code.
[b]May depend on joint type, i.e., mechanical, solder,, etc.

Socket-welded joints simplify the welder's tasks. Socket-welding fittings have deep sockets with ample "come and go," hence pipe need not be cut to precise lengths.

The socket-type weld has advantages over the butt-weld that recommend it strongly for smaller size piping. Pipe doesn't have to be cut accurately unless it must be butted against the fitting shoulder. Since the pipe end slips into, and is supported by, the socket, the joint is self-aligning. Tack welding, special clamps to line up and

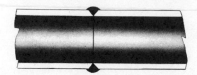

Fig. 3.16 Butt-welded pipe joint. Courtesy of   **Fig. 3.17**   Butt-welded tee. Courtesy of Crane Co.[4]
Crane Co.[4]

hold the joint, and backing rings are unnecessary. Pipe used in socket-welding does not require beveling. "Icicles" and weld spatter cannot enter the pipe.

Welded flanged joints are also used in joining flanges to piping. These welded joints are superior to screwed flanged joints because the possibility of leakage through the thread is eliminated. The full thickness of the pipe wall is maintained, and the welded flange becomes an integral part of the pipe. Two typical methods for making such joints are shown in Fig. 3.19.

Flanges are available in flat face, raised face, or various standard ring joint types to suit different pressure and gasket requirements.

Valves and in-line instruments are available in either threaded, butt-weld, or flanged-end types. Fittings are available in threaded, butt-weld, flanged, and socket-weld types. Flanges are available as weld neck (weld end) or slip-on types.

Valves and flanges are available in standardized ratings for carbon or alloy steel materials and can be applied with confidence to low-temperature systems, under 600°F (320°C), for the stated pressure. Higher temperatures or nonsteel valves require temperature derating. For large-diameter piping systems such as those 48 inches (1200 mm) and larger, valves and flanges carrying other pressure ratings such as 25, 50, 75, and 125 pounds (11, 23, 34, and 57 kg) are also available, usually on special order. Manufacturers should be contacted for further information.

3.6.6   Valve Types

The selection of valves for hydraulic systems is of fundamental importance to proper system operation. The following must be considered:

1.   Is the installation high- or low-pressure/temperature?
2.   What fluids will be sent through them?

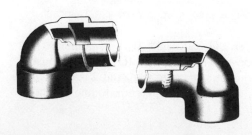

Fig. 3.18 Socket-welded ell. Courtesy of Crane Co.[4]

welding neck   slip-on   Cranelap

Fig. 3.19   Pipe flanges. Courtesy of Crane Co.[4]

3. Will operating conditions be moderate or severe (i.e., frequency of operation, high pressure drop, hostile environment, etc.)?
4. Is throttling control or on-off control required?
5. How much headroom must be allowed for valve stems?
6. What size is the line?
7. Will the valves have to be dismantled frequently for inspection and servicing?
8. Is the installation relatively permanent, or must the piping be broken into frequently?

The principal types of valves are **gate, globe, angle, check, butterfly,** and **ball.**

Gate Valves

Fluids flow through gate valves in a straight line. This construction offers little resistance to flow and reduces pressure drop to a minimum. A gatelike disc, actuated by a stem screw and handwheel, moves up and down at right angles to the path of flow, and seats against two seat faces to shut off flow.

Gate valves are best for services that require infrequent valve operation, and where the disc is kept either fully opened or closed. They are not practical for throttling. With the usual type of gate valve, close regulation is impossible. Velocity of flow against a partly opened disc may cause vibration and chattering, and result in damage to the seating surfaces. Also, when throttled, the disc is subjected to severe wire-drawing erosive effects.

Globe and Angle Valves

Fluids change direction when flowing through a globe valve. The seat construction increases resistance to and permits close regulation of fluid flow. Disc and seat can

Table 3.6   Standardized Valve and Flange Ratings

ANSI Class	PN (Metric) Designation
150	20
300	50
400	68
600	100
900	150
1500	250
2500	420

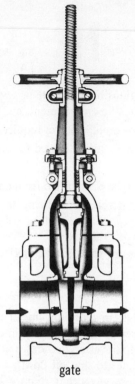

gate

Fig. 3.20   Gate valve. Courtesy of Crane Co.[4]

be quickly and conveniently reseated or replaced. This feature makes them ideal for services that require frequent valve maintenance. Shorter disc travel saves operators' time when valves must be operated frequently.

Angle valves have the same operating characteristics as globe valves. Used when making a 90° turn in a line, an angle valve reduces the number of joints and saves make-up time. It also gives less restriction to flow than the elbow and globe valve it displaces.

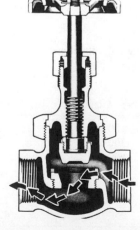

Fig. 3.21   Globe valve. Courtesy of Crane Co.[4]

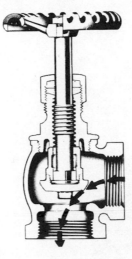

Fig. 3.22 Angle valve. Courtesy of Crane Co.[4]

Check Valves

Check valves are used to prevent back flow in lines. Check valves conform to one of the two basic patterns. Flow moves through swing check valves (Fig. 3.23) in approximately a straight line comparable to that in gate valves. In lift check valves (Fig. 3.24), flow moves through the body in a changing course as in globe and angle valves. In both swing and lift types, flow keeps the valve open while gravity and reversal of flow close it automatically.

Butterfly Valves

Butterfly valves are of the "quarter-turn" family and are so designated because a 90° turn of their operator fully opens or closes the valve. The valves utilize elastomer seals and their popularity can be attributed to the improvements made in elastomer materials. They are well suited to wide-open or fully closed position and in some cases may be used for noncritical throttling applications. They are generally lighter in weight than conventional valves. The position of the lever indicates whether they are

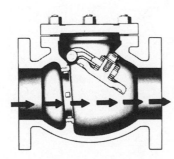

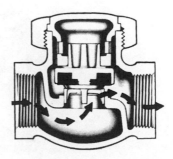

Fig. 3.23 Swing check valve. Courtesy of Crane Co.[4]

Fig. 3.24 Lift check valve. Reproduced by courtesy of the Crane Co.[4]

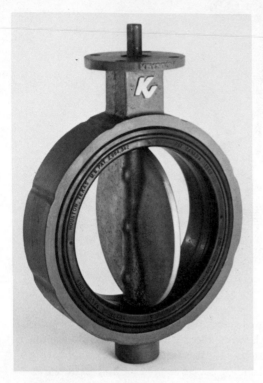

Fig. 3.25   Butterfly valve. Courtesy of Keystone Valve Co.

wide open, partially open, or fully closed. They are easily adapted to lever, manual, gear, cylinder, or motor operation.

Ball Valves

The advantages of quarter-turn ball valves include straight-through flow, minimum turbulence, low torque, tight closure, and compactness. Reliable operation, easy maintenance, and long-life economy justify their extensive application. Industrial,

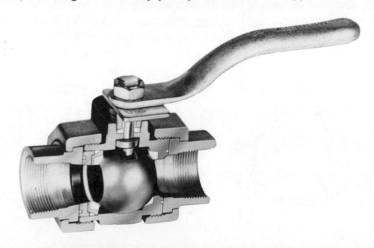

Fig. 3.26   Ball valve. Courtesy of Pacific Valves.

chemical, petrochemical, refinery, pulp and paper, gas transmission, water works and sewage, and power plants are utilizing ball valves where other types of valves have proven inadequate.

Relief Devices

Relief valves are intended to open when the system pressure reaches a desired value. They are frequently of the angle body design with a spring whose compression can be varied, thus establishing the system discharge pressure. Normally they are provided with a lever that permits periodic exercising (testing) to assure operation when needed. Some larger relief valves utilize an electric solenoid "pilot" which initiates operation of the main disc. For certain critical safety applications, valve performance testing may be required. Principal codes such as that of the American Petroleum Institute and the ASME Boiler and Pressure Vessel Code require their use.

Rupture diaphragms are used in lieu of relief valves for applications where usage is rare and reseating is not necessary. Both relief valves and rupture diaphragms can

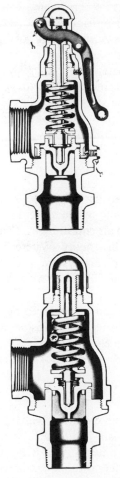

Fig. 3.27 Relief valve. Courtesy of Crane Co.[4]

be installed to protect against either **excess pressure** or potential collapse due to **vacuum.**

3.6.7   Seating Materials

The seat and disc constitute the heart of a valve and do most of its work. Valve manufacturers provide a wide choice of seating materials for increasing pressure or temperatures or for more rigorous service.

For relatively low pressures and temperatures, say 500 psi (3450 kPa) or 350°F (180°C) and for ordinary fluids, seating materials are not a particularly difficult problem. Bronze and iron valves usually have bronze or bronze-faced seating surfaces, or iron valves may be all iron. Nonmetallic "composition" discs are available for tight seating on hard-to-hold fluids such as air or gasoline. Valves are available with **backseating capability** to permit use in vacuum service or for maintenance of packing while under pressure.

As pressures and temperatures increase or as the service becomes more severe, careful consideration must be given to many factors, no one of which can be over-emphasized to the detriment of others. Long, trouble-free life requires the proper combination of hardness, wear-resistance, and resistance to corrosion, erosion, galling, seizing, and temperature. A successful combination in one instance may not serve equally well in all others.

The selection of seating materials for corrosive fluids, regardless of pressure or temperature, is almost endless. Included are many types of alloys as well as linings or coatings of many kinds. The safest policy in specifying seating materials is close adherence to valve manufacturer's recommendations, usually found in catalogs, or supplied on request.

3.6.8   Vents and Drains

Venting and drainage capability should be included in all piping systems. Vents are needed to permit complete filling of the system and avoid stoppage of flow or flow surges and interruptions due to air entrainment. Drains are necessary to permit ready removal of the working fluid by gravity. Where drainage is a process requirement some slight inclination of the piping system, usually 1 in 12, is provided. For other systems, typically large diameter, no slope may be necessary if drain provisions are provided at low points.

Both venting and drainage can be achieved by automatic valves or, for less frequent use, by manual valves. Vents and drains are normally provided at the high and low points of a system, respectively, and a typical system may require several of each. For certain flammable fluids, **flame arrestors** are incorporated into the vents.

3.6.9   Water Hammer

Water hammer caused by rapid closure of valving or rapid change in flow rate can be extremely destructive and is a potential problem in any filled, noncompressible, fluid dynamic system. The hammer is caused by the conversion of the fluid momentum forces into shock waves due to the incompressibility of the fluid. The potential for water hammer increases with the momentum (either mass or velocity) of the fluid flowing in the system. To overcome this, **surge chambers** or **shock suppressors** are used where rapid valve closing may occur. (Surge chambers may also be required on the discharge of reciprocating pumps to dampen pulsations.) Where practical, hammer can be avoided by providing slower-closing valves, surge chambers, or **pressure relief devices.**

REFERENCES

1. *Flow of Fluids Through Valves, Fittings and Pipe,* Technical Paper No. 410, Crane Co., New York, 1980.
2. *Hydraulic Institute Standards for Centrifugal, Rotary and Reciprocating Pumps,* Hydraulic Institute, Cleveland, Ohio.
3. *Permutit Water Conditioning Data Book,* Permutit Co., Inc., Paramas, New Jersey, 1953.
4. *Piping Pointers—Application and Maintenance of Valves and Piping Equipment,* Publication No. VC-1013A, Crane Co., New York.
5. Shaw and Loomis, *Cameron Hydraulic Data,* Compressed Air Magazine Co., Phillipsburg, New Jersey, 1962.
6. L. Urquhart, *Civil Engineering Handbook,* 4th edition, McGraw-Hill, New York, 1959.

SECTION 4
THERMODYNAMICS/HEAT TRANSFER

4.1 HEAT

In the following formulas, when specific units are not stated, any units may be used provided identical properties are expressed in the same units. Absolute pressure is indicated by p, total volume by V, specific volume by v, absolute temperature by T, and thermometer temperature by t. In all formulas containing indicated units, the temperature is measured in Fahrenheit degrees.

Measurement of Heat

Heat is the transient form of energy transmitted from one body to another when the two bodies are not at the same temperature. The ratio of the quantity of heat required to increase the temperature of a body in a specified state to that required to increase the temperature of an equal mass of water through the same temperature is called the specific heat of the body.

The unit of energy commonly used in the measurement of heat is the British thermal unit (Btu) which equals 2.930×10^{-4} kilowatt-hours or substantially $\frac{1}{180}$th of the quantity of heat required to raise one pound of water from 32 to 212°F at standard atmospheric pressure.

The quantity of heat (Q) added to M pounds of a substance having a constant specific heat (c) and causing the temperature to increase from t_1 to t_2 is

$$Q = Mkc(t_2 - t_1) \text{ units.} \tag{4.1}$$

NOTE.   See Table 4.1 for definite values of c. The constant k depends on the units of measurement, as follows:

Q	M	$t_2 - t_1$	k
gram-calories	grams	Cent.	1
kilogram-calories	kilograms	Cent.	1
British thermal units	pounds	Cent.	1.8
British thermal units	pounds	Fahr.	1
joules	grams	Cent.	4.18
joules	pounds	Fahr.	1054
kilowatt-hours	kilograms	Cent.	1.16×10^{-3}
kilowatt-hours	pounds	Fahr.	2.93×10^{-4}

If the specific heat varies with the temperature (as it usually does) according to the relation

$$c = a + bt + ft^2 \qquad (4.2)$$

where a, b, and f are constants which are determined by experiment, then

$$Q = Mk\left[a(t_2 - t_1) + \frac{b}{2}(t_2^2 - t_1^2) + \frac{f}{3}(t_2^3 - t_1^3) \right] \text{ units.} \qquad (4.3)$$

The quantity of heat added to a substance can also be determined by applying the First Law of Thermodynamics, which states that energy can be neither created nor destroyed. This principle may be expressed in an equation as follows:

$$Q = 778(W + \Delta E) \text{ Btu,} \qquad (4.4)$$

where Q is the heat interchange in Btu, W is the external work done in foot-pounds, and ΔE is the change in the internal energy in foot-pounds.

NOTE.   Although the accepted value for the mechanical equivalent of heat is 778.26 foot-pounds, it is sufficient to use 778 foot-pounds in the solution of most engineering problems.

Influence of Heat on the Length of a Solid Body

If heat is applied to a solid body which has a length l_0 at a temperature of 0 degrees Centigrade, the length l_t at a temperature of t degrees Centigrade is

$$l_t = l_0(1 + at). \qquad (4.5)$$

NOTE: See Section 10 for definite values of a (the Centigrade mean coefficient of linear expansion). The mean coefficient of cubical expansion equals $3a$, approximately. When the temperature is expressed in Fahrenheit degrees, Equation (4.5) becomes $l_t = l_{32}\{1 + a[(t - 32)/1.8]\}$.

Measurement of External Work

External work is the result of a force acting through a distance to overcome external resistances. For mechanical processes this work may be expressed by

$$W = 144 \int_{V_1}^{V_2} p\,dV \text{ foot-pounds,} \qquad (4.6)$$

where p is the intensity of pressure in pounds absolute per square inch and V is the total volume expressed in cubic feet.

If the specific heat varies with the temperature according to the relation

$$c = a + bt + ft^2, \qquad (4.7)$$

then

$$\Delta s = M \int_{T_1}^{T_2} \frac{c\, dT}{T} = M \left[(a - 460b + 211{,}600f) \ln \frac{T_2}{T_1} \right. \tag{4.8}$$

$$\left. + (b - 920f)(T_2 - T_1) + \frac{f}{2}(T_2^2 - T_1^2) \right] \text{ units of entropy.}$$

If the temperature remains constant, then

$$\Delta s = M \left(\frac{Q}{T} \right) \text{ units of entropy.} \tag{4.9}$$

If no heat is added to or rejected from the substance and the expansion or compression is frictionless (i.e., reversible), then

$$\Delta s = 0 \text{ units of entropy.}$$

Steady Flow

When the same quantity of working fluid progresses continuously and uniformly in one direction, the process is termed the steady flow condition. If the conservation of energy principle is applied to such a process between two sections 1 and 2, the equation for each pound of working fluid, in its simplest form for engineering applications, is

$$E_1 + 144p_1v_1 + \frac{U_1^2}{64.4} = E_2 + 144p_2v_2 + \frac{U_2^2}{64.4} + W + 778Q_{\text{loss}} \text{ foot-pounds.}$$

$$\tag{4.10}$$

where E is the internal energy in foot-pounds, $144pv$ is the flow work in foot-pounds, $\frac{U^2}{64.4}$ is the kinetic energy in foot-pounds, W is the external work done in foot-pounds, and Q_{loss} is the heat lost to (negative) or gained from (positive) the surroundings in Btu.

NOTE. In the expression $U^2/64.4$, U is the velocity of the fluid flowing in feet per second and 64.4 is equal to $2g$ [where g is the acceleration due to gravity and is assumed to equal 32.2 feet/second². (9.81 m/s²)].

Enthalpy

The combination $E + 144pv$ occurs so often that the special term "enthalpy" has been universally adopted. This property is represented by the symbol h when the combination is expressed in Btu; that is, $(E + 144pv)/778$.

4.2   PERFECT GASES

Characteristic Equation

The relation between pressure, volume, and absolute temperature can be determined by combining the two experimental gas laws, those of Boyle and Charles, or Gay-Lussac. This relation for any two conditions 1 and 2 may be expressed by

$$\frac{p_1 V_1}{T_1} = \frac{p_2 V_2}{T_2} = MR. \tag{4.11}$$

Values of the gas constant (R) for various gases are given, as follows: air, 53.3; carbon dioxide, 35.1; carbon monoxide, 55.1; helium, 386; hydrogen, 767; nitrogen, 55.1; oxygen, 48.3.

Fundamental Equations

The dual relation between pressure and volume, temperature and pressure, or temperature and volume for many changes met in practice may be represented by exponential equations. These equations are

$$pV^n = \text{constant}, \quad Tp^{(1-n)/n} = \text{constant}, \quad TV^{n-1} = \text{constant}. \tag{4.12}$$

The exponent (n) may be determined from the following relations

$$n = \frac{\log p_1 - \log p_2}{\log V_2 - \log V_1}, \tag{4.13}$$

or

$$n = \frac{c - c_p}{c - c_v} = \frac{c - kc_v}{c - c_v} \tag{4.14}$$

where c is the specific heat for a polytropic change (pV^n = constant), c_p is the specific heat at constant pressure, and c_v is the specific heat at constant volume and k is c_p/c_v (Table 4.1).

NOTE.   Another useful relation between c_p and c_v is $778(c_p - c_v) = R$.

The change in the internal energy (ΔE) is independent of the path and may be determined by the following equations:

$$\Delta E = 778 M c_v (T_2 - T_1) \text{ foot-pounds} \tag{4.15}$$

$$= M(778 c_p - R)(T_2 - T_1) \text{ foot-pounds} \tag{4.16}$$

$$= \frac{MR(T_2 - T_1)}{k - 1} \text{ foot-pounds} \tag{4.17}$$

$$= \frac{144(p_2 V_2 - p_1 V_1)}{k - 1} \text{ foot-pounds}. \tag{4.18}$$

Table 4.1 Mean Specific Heat of Gases at Constant Pressure (from 32°F to *t*°F in Btu per lb per °F)[a]

Gas	Temperature (°F)					
	100	300	500	1000	1500	2000
Air	0.240	0.241	0.243	0.249	0.257	0.263
Oxygen, O_2	0.218	0.222	0.225	0.235	0.243	0.249
Nitrogen, N_2	0.248	0.249	0.251	0.256	0.262	0.270
Hydrogen, H_2	3.41	3.44	3.45	3.47	3.51	3.55
Water vapor, H_2O	0.444	0.449	0.454	0.472	0.493	0.516
Carbon monoxide, CO	0.248	0.249	0.251	0.258	0.265	0.273
Carbon dioxide, CO_2	0.200	0.213	0.224	0.246	0.262	0.274
Typical flue gas of average bituminous coal. Based on 20% excess air and 5% H_2O	0.243	0.247	0.250	0.260	0.269	0.277

[a] The volume of 1 lb of a gas at any given temperature and pressure may be found from

$$V = \frac{t + 460}{W \times P \times 35.38}$$

where: V = Volume in cubic feet
t = Temperature in °F
W = Weight of gas in lb per cubic foot from Table 10.18
P = Absolute pressure in lb per square inch.

The change of entropy (Δs) is also independent of the path and may be determined by the following equations:

$$\Delta s = M \left[c_v \ln \frac{T_2}{T_1} + (c_p - c_v) \ln \frac{V_2}{V_1} \right] \text{ units of entropy} \qquad (4.19)$$

$$= M \left[c_p \ln \frac{T_2}{T_1} - (c_p - c_v) \ln \frac{P_2}{P_1} \right] \text{ units of entropy} \qquad (4.20)$$

$$= M \left[cv \ln \frac{P_2}{P_1} + c_p \ln \frac{V_2}{V_1} \right] \text{ units of entropy.} \qquad (4.21)$$

The heat interchange and the external work done are dependent on the path or the character of the change that takes place between two conditions 1 and 2. Of the innumerable possible changes, the only ones of importance to the engineer are: constant pressure changes, during which the pressure remains constant; constant volume changes, during which the volume remains constant; isothermal changes, during which the temperature remains constant; adiabatic changes, during which no heat is received from or rejected to external bodies; polytropic changes, during which the heat supplied to or withdrawn from the gas by external bodies is directly proportional to the change in temperature.

A summary of the convenient formulas for these paths is given in Table 4.2.

Table 4.2   Summary of Convenient Formulas for Perfect Gases Between Conditions 1 and 2

Path	pV-Relation	Heat Interchange (B.t.u.)	External Work Done (ft.-lbs.)	Change of Entropy (units of entropy)	Specific Heat
Constant Pressure	$p_1V_1^0 = p_2V_2^0$ $p = \text{const.}$	$Mc_p(T_2 - T_1)$ $\frac{MR}{778}\left(\frac{k}{k-1}\right)(T_2 - T_1)$	$144\,p(V_2 - V_1)$ $MR(T_2 - T_1)$	$Mc_p \ln\frac{T_2}{T_1}$ $Mc_p \ln\frac{V_2}{V_1}$	c_p
Constant Volume	$p_1V_1^\infty = p_2V_2^\infty$ $V = \text{const.}$	$Mc_v(T_2 - T_1)$ $\frac{MR}{778}\left(\frac{1}{k-1}\right)(T_2 - T_1)$	0	$Mc_v \ln\frac{T_2}{T_1}$ $Mc_v \ln\frac{p_2}{p_1}$	c_v
Isothermal or Isodynamic	$p_1V_1 = p_2V_2$ $pV = \text{const.}$	$MT(s_2 - s_1)$ $0.1851\,p_1V_1 \ln\frac{V_2}{V_1}$ $\frac{MRT}{778}\ln\frac{V_2}{V_1}$	$144\,p_1V_1 \ln\frac{V_2}{V_1}$ $MRT \ln\frac{V_2}{V_1}$	$\frac{Q}{T}$	∞
Reversible Adiabatic or Isentropic	$p_1V_1^k = p_2V_2^k$ $pV^k = \text{const.}$	0	$\frac{144(p_1V_1 - p_2V_2)}{k-1}$ $\frac{MR(T_1 - T_2)}{k-1}$	0	0
Polytropic	$p_1V_1^n = p_2V_2^n$ $pV^n = \text{const.}$	$Mc_v\left(\frac{n-k}{n-1}\right)(T_2 - T_1)$	$\frac{144(p_1V_1 - p_2V_2)}{n-1}$ $\frac{MR(T_1 - T_2)}{n-1}$	$Mc_v\left(\frac{n-k}{n-1}\right)\ln\frac{T_2}{T_1}$	$c_v\left(\frac{n-k}{n-1}\right)$

4.3   LIQUIDS AND VAPORS

Physical Conditions

A liquid and its vapor may exist in either of the following six conditions:

1. Compressed or subcooled liquid is a liquid at a temperature less than the saturation temperature corresponding to the pressure.

2. Saturated liquid is a liquid which under its given pressure will begin to vaporize when heat is added to it.

3. Saturated vapor is a vapor which under its given pressure will start changing to the liquid form when heat is removed.

4. Wet vapor is a physical mixture of saturated liquid and saturated vapor. In each pound of mixture, the fractional part by weight which is saturated vapor is designated by the symbol x.

5. Superheated vapor is a vapor the temperature of which is greater than the saturation temperature corresponding to the pressure imposed on it.

6. Supersaturated vapor is vapor the temperature and specific volume of which are less than those corresponding to the saturated condition for the pressure imposed on it. This condition occurs only during rapid expansion as in nozzles and is of special importance to turbine designers.

Properties of Liquids and Vapors

The various properties of the more important liquids and vapors are available in tabulated form. In general the properties given are pressure (p), temperature (t), specific volume (v), enthalpy (h), internal energy (E), and entropy (s).

Properties of Steam

The properties of saturated liquid, saturated vapor, and superheated vapor are given in Section 10.

The properties of compressed liquid can be computed from Table 4 in Keenan and Keyes, *Thermodynamic Properties of Steam*.

The properties of wet vapor can be computed from the saturated properties as follows:

$$v = v_f + xv_{fg}, \; h = h_f + xh_{fg}, \; s = s_f + xs_{fg}. \qquad (4.22)$$
$$E = u_f + xu_{fg} = h - 0.1851pv.$$

Thermodynamic Processes

The heat interchange (Q), work done (W), change in the internal energy (ΔE), and change of entropy (Δs) for processes most frequently met in engineering practice are given in Table 4.3.

Table 4.3   Summary of Convenient Formulas for Steam Between States 1 and 2

Path	Heat Interchange (B.t.u.)	Work Done (B.t.u.)	Change in the Internal Energy (B.t.u.)	Change of Entropy (units of entropy)
Constant Pressure p = const.	$M(h_2 - h_1)$	$0.1851\,p\,(V_2 - V_1)$	$M(E_2 - E_1)$	$M(s_2 - s_1)$
Constant Volume V = const.	$M(E_2 - E_1)$	0	$M(E_2 - E_1)$	$M(s_2 - s_1)$
Reversible Adiabatic s = const.	0	$M(E_1 - E_2)$	$M(E_2 - E_1)$	0
Isothermal T = const.	$MT(s_2 - s_1)$	$M[T(s_2 - s_1) - (E_2 - E_1)]$	$M(E_2 - E_1)$	$M(s_2 - s_1)$
Isodynamic E = const.	$\dfrac{M(T_1 + T_2)}{2}(s_2 - s_1)$	$\dfrac{M(T_1 + T_2)}{2}(s_2 - s_1)$	0	$M(s_2 - s_1)$
Exponential pV^n = const.	$W + \Delta E$	$\dfrac{0.1851\,(p_1V_1 - p_2V_2)}{n - 1}$	$M(E_2 - E_1)$	$M(s_2 - s_1)$

4.4   FLOW OF GASES AND VAPORS

Steady Flow Equation

Equation (4.10) for the steady flow of the working fluid when applied to nozzles and orifices, provided there is no loss or gain of heat, no friction, and no external work is done, reduces to

$$E_1 + 144p_1v_1 + \frac{U_1^2}{64.4} = E_2 + 144p_2v_2 + \frac{U_2^2}{64.4} \text{ foot-pounds.} \quad (4.23)$$

Velocity

Since h may be substituted for $E + 144pv$ divided by 778, the change in kinetic energy is

$$\frac{U_2^2 - U_1^2}{64.4} = 778(h_1 - h_2) \text{ foot-pounds.} \quad (4.24)$$

If the initial velocity U_1 is small, it may be neglected, giving

$$\frac{U_2^2}{64.4} = 778(h_1 - h_2) \text{ foot-pounds, or} \quad (4.25)$$

$$U_2 = 223.8 \sqrt{h_1 - h_2} \text{ feet/second.} \quad (4.26)$$

Weight

The weight of working fluid flowing must be constant throughout the process. If G represents this weight in pounds/second, a the area in square feet, U the velocity in feet/second, and v the specific volume in feet^3/pound at any pressure, then

$$G = \frac{a_1 U_1}{v_1} = \frac{a_2 U_2}{v_2} \text{ pounds/second.} \quad (4.27)$$

This weight is a maximum when the absolute pressure P_t at the throat is

$$P_t = p_1 \left(\frac{2}{n+1} \right)^{n/(n-1)} \text{ pounds/inch}^2. \quad (4.28)$$

For dry saturated steam $n = 1.135$, then

$$p_t = 0.58p_1 \text{ pounds/inch}^2. \quad (4.29)$$

For superheated steam $n = 1.30$, then

$$p_t = 0.55p_1 \text{ pounds/inch}^2. \quad (4.30)$$

For diatomic (two atoms per molecule) gases $n = 1.40$, then

$$p_t = 0.53p_1 \text{ pounds/inch}^2. \tag{4.31}$$

The pressure (p_t) that makes the weight of working fluid flowing a maximum is called **critical pressure**. When the final absolute pressure p_2 is less than the critical pressure, then the weight discharged remains constant and the term applied for such a condition is unretarded flow. When the final absolute pressure is greater than the critical pressure, then the weight discharged decreases as p_2 increases, and the flow is said to be retarded.

4.4.1 Flow of Gases

Velocity

For a perfect gas, c_pT may be substituted for h, and k for n. Equation (4.26) may be modified to give

$$U_2 = 223.8\sqrt{c_p(T_1 - T_2)} \text{ feet/second} \tag{4.32}$$

or

$$U_2 = \sqrt{\frac{2gkp_1v_1}{k-1}\left[1 - \left(\frac{p_2}{p_1}\right)^{(k-1)/k}\right]} \text{ feet/second.} \tag{4.33}$$

Weight

Substituting Equation (4.33) in Equation (4.27) gives

$$G = a_2\sqrt{\frac{2gk}{k-1}\left(\frac{p_1}{v_1}\right)\left[\left(\frac{p_2}{p_1}\right)^{2/k} - \left(\frac{p_2}{p_1}\right)^{(k+1)/k}\right]} \text{ pounds/second.} \tag{4.34}$$

It is often more convenient to use the throat area a_t. The equations used must be classified depending on the type of flow, retarded or unretarded. All units in feet.

For retarded flow of diatomic gases ($k = 1.40$), Equation (4.34) reduces to

$$G = \frac{15.03a_tp_1}{\sqrt{RT_1}}\sqrt{\left(\frac{p_2}{p_1}\right)^{1.43} - \left(\frac{p_2}{p_1}\right)^{1.71}} \text{ pounds/second,} \tag{4.35}$$

and for air ($R = 53.34$)

$$G = \frac{2.056a_tp_1}{\sqrt{T_1}}\sqrt{\left(\frac{p_2}{p_1}\right)^{1.43} - \left(\frac{p_2}{p_1}\right)^{1.71}} \text{ pounds/second.} \tag{4.36}$$

Fliegner's empirical formula for (inch units) retarded flow of air is

$$G = \frac{1.06a_t}{\sqrt{T_1}} \sqrt{p_2(p_1 - p_2)} \text{ pounds/second.} \tag{4.37}$$

For unretarded flow of diatomic gases ($k = 1.40$), Equation (4.34) reduces (inch or foot units) to

$$G = \frac{3.885a_t p_1}{\sqrt{RT_1}} \text{ pounds/second,} \tag{4.38}$$

and for air ($R = 53.34$)

$$G = \frac{0.532a_t p_1}{\sqrt{T_1}} \text{ pounds/second.} \tag{4.39}$$

Fliegner's empirical formula for unretarded flow of air is

$$G = \frac{0.53a_t p_1}{\sqrt{T_1}} \text{ pounds/second.} \tag{4.40}$$

4.4.2   Flow of Steam

Velocity

For steam, since the enthalpy (h) may be determined from the tables in Section 10, the velocity is

$$U_2 = 223.8 \sqrt{h_1 - h_2} \text{ feet/second.} \tag{4.41}$$

Weight

Although it is possible to deduce equations involving the exponent n, the most convenient form (foot units) is

$$G = \frac{a_2 U_2}{v_2} \text{ pounds/second.} \tag{4.42}$$

If the throat area a_t is used,

$$G = \frac{a_t U_t}{v_t} \text{ pounds/second.} \tag{4.43}$$

The throat pressure equals p_2 for retarded flow and equals the critical pressure ($0.58p_1$ for wet or dry saturated vapor and $0.55p_1$ for superheated vapor) for unretarded flow.

Rankine's empirical formula for retarded flow of dry saturated steam (inch units) is

$$G = 0.0292 a_t \sqrt{p_2(p_1 - p_2)} \text{ pounds/second.} \tag{4.44}$$

Rankine's empirical formula for unretarded flow of dry saturated vapor (inch or foot units) is

$$G = \frac{a_t p_1}{70} \text{ pounds/second.} \tag{4.45}$$

Grashof's empirical formula for unretarded flow of dry saturated vapor (inch units) is

$$G = \frac{a_t p_1^{0.97}}{60} \text{ pounds/second.} \tag{4.46}$$

Equations (4.44), (4.45), and (4.46) should be divided by $(x_1)^{1/2}$ for wet steam and by $1 + 0.00065\,\Delta t$ for superheated steam.

NOTE. Δt is the number of Fahrenheit degrees of superheat.

4.4.3   Steam Calorimeters

Throttling Calorimeter

Equation (4.23) for the steady flow of the working fluid when applied to a throttling calorimeter, provided there is no loss or gain of heat, no external work is done, and the initial and final velocities are equal or negligible, reduces to

$$h_1 = h_2 \text{ Btu.} \tag{4.47}$$

This calorimeter is limited in its use because the steam in the calorimeter must be superheated. For reliable results, at least 10° of superheat must be available. The quality (x) can be determined from

$$x = \frac{h_2 - h_{f_1}}{h_{fg_1}}. \tag{4.48}$$

NOTE. The percentage priming equals $(1 - x)100$.

Separating Calorimeters

This type of calorimeter is designed to separate the moisture from the steam. The drip (G_m) is collected and weighed. The saturated steam (G_s), for the same time interval, may be condensed and weighed or discharged through an orifice. The priming $(1 - x)$ can be determined from

$$1 - x = \frac{G_m}{G_s + G_m}. \tag{4.49}$$

4.5   THERMAL EQUIPMENT

4.5.1   Steam Engines

Mean Effective Pressure

The indicator card shows the pressure distribution in the cylinder of a steam engine at every point of the working and exhaust strokes. The mean effective pressure (M.E.P. or P) in pounds per square inch is

$$\text{M.E.P.} = \frac{aS}{l} \text{ pounds per square inch} \tag{4.50}$$

where a is the area of the card in square inches, l is the length of the card in inches, and S is the scale of the indicator spring in pounds per square inch per inch of height, or

$$\begin{aligned}
\text{M.E.P.} = \Big[& p_1 \times C + p_1(C + Cl) \ln \frac{R + Cl}{C + Cl} - p_2(1 - K) \\
& - p_2(K + Cl) \ln \frac{K + Cl}{Cl} \Big] \times \text{D.F. pounds/inch}^2, \tag{4.51}
\end{aligned}$$

where p_1 is the admission pressure in pounds absolute per square inch, p_2 is the exhaust pressure in pounds absolute per square inch, C is the percent cut-off, R is the percent release, K is the percent compression, and Cl is the percent clearance. The diagram factor (D.F.) is usually between 0.85 and 0.95.

NOTE.   The percent events are expressed as decimal fractions.

In compound engines it is customary to neglect the clearance in the conventional card. If E represents the expansion ratio, then

$$\text{M.E.P.} = (p_1 \times \frac{1}{E} + p_1 \times \frac{1}{E} \ln E - p_2) \times \text{D.F. pounds/inch}^2. \tag{4.52}$$

Rankine Efficiency

The Rankine cycle for a steam engine consists of the reception of heat energy at constant pressure, a frictionless adiabatic expansion to the exhaust pressure and the rejection of heat energy at constant pressure. The approximate efficiency for such a cycle (ϵ_R) is

$$\epsilon_R = \frac{Q_1 - Q_2}{Q_1} = \frac{h_1 - h_2}{h_1 - h_{f2}} = \frac{2545}{w_R(h_1 - h_{f2})} \qquad (4.53)$$

where h_1 is the enthalpy at the initial conditions, h_2 is the enthalpy after isentropic expansion to the exhaust pressure, h_{f2} is the enthalpy of the saturated liquid at the exhaust pressure, and w_R is the ideal steam consumption in pounds per horsepower-hour.

Thermal Efficiency

The actual cycle for a steam engine takes into account the fact that heat losses occur in the actual steam engine. The efficiency for such a cycle (ϵ_T) is

$$\epsilon_T = \frac{Q_1 - Q_2 - Q_{\text{losses}}}{Q_1} = \frac{2545}{w_A(h_1 - h_{f2})} \qquad (4.54)$$

where w_A is the actual steam consumption in pounds per horsepower-hour.

Engine Efficiency or Rankine Cycle Ratio

The ratio of the ideal steam consumption per horsepower-hour (w_R) to the actual steam consumption in pounds per horsepower-hour (w_A) is termed engine efficiency (ϵ_E), cylinder efficiency, or Rankine cycle ratio, hence

$$E = \frac{w_R}{w_A} = \frac{2545}{w_A(h_1 - h_2)} = \frac{\epsilon_T}{\epsilon_R}. \qquad (4.55)$$

Brake Horsepower

The output of a steam engine may be determined by means of a friction brake. If F represents the net force in pounds acting at a distance R feet from the center of the shaft rotating at N revolutions per minute, then the brake horsepower (B.H.P.) is

$$\text{B.H.P.} = \frac{2\pi RNF}{33,000} \text{ horsepower.} \qquad (4.56)$$

Mechanical Efficiency

The ratio of the brake horsepower (B.H.P.) to the indicated horsepower (I.H.P.) is termed mechanical efficiency (ϵ_M), hence

$$\epsilon_M = \frac{\text{B.H.P.}}{\text{I.H.P.}}. \qquad (4.57)$$

Carnot Efficiency

The Carnot cycle consists of the reception and rejection of heat energy, each at constant temperature together with frictionless adiabatic expansion and compression. The efficiency for such a cycle (ϵ_c) is

$$\epsilon_c = \frac{Q_1 - Q_2}{Q_1} = \frac{T_1 - T_2}{T_1}. \tag{4.58}$$

(This formula is extremely widely used and is applicable to any heat cycle.)

4.5.2 Steam Turbines

Steam flow required (F) can be determined:

$$F = \frac{R_T}{E_E} \times h_p \tag{4.59}$$

where: F = flow required (lb/hr)
R_T = steam rate—theoretical (lb/hr)
h_p = power output (h.p. or kw)
E_E = engine efficiency (%)

Theoretical steam rates and **efficiencies** vary widely depending on machine size and application, but can be obtained from the manufacturers. As a guide some approximate rates and efficiencies are shown in Tables 4.4 and 4.5. For large-capacity turbine applications **regenerative feedwater heating** is used to reduce the overall heat rate for the turbine/feedwater cycle. As the number of stages of heating increase, the incremental increase in efficiency declines (Table 4.6).

4.5.3 Steam Boilers

Maximum Allowable Working Pressure.   For a steam boiler drum with a shell of r inches radius, t inches thick, an ultimate tensile strength of f pounds/inch^2,

Table 4.4   Typical Theoretical Steam Rates (lb/kw-hr)

Exhaust Pressure	Throttle Conditions (psi/°F)				
	150/Sat.	400/650	600/825	1200/825	1800/1000 Reheat
2.5 in Hg abs.	10.9	8.04	6.92	6.58	5.77
4.0 in Hg abs.	11.8	8.52	7.28	6.90	6.01
5 psig	21.7	13.0	10.4	9.40	7.87
50 psig	46.0	19.4	14.3	12.2	9.77
100 psig	—	26.5	18.1	14.4	11.2
200 psig	—	48.2	27.0	18.5	13.6

Table 4.5   Typical Turbine Efficiencies (%)

Turbine Size (kw)	Throttle Conditions (psi/°F)		
	150/Sat	600/825	1800/1000 Reheat
1,000	64	64	—
10,000	78	77	72
100,000	—	82	81

factor of safety F.S., and an efficiency of the longitudinal joint ϵ percent, the maximum allowable working pressure (p) in pounds/inch^2 gauge is

$$p = \frac{ft\epsilon}{\text{F.S.}r} \text{ pounds/inch}^2. \tag{4.60}$$

Thickness of Bumped Head.   For a bumped head of bumped radius r inches, working pressure of p pounds/inch^2 gauge, an ultimate tensile strength of f pounds/inch^2, and a factor of safety F.S., the thickness (t) in inches is

$$t = \frac{\text{F.S.}rp}{Kf} \text{ inches.} \tag{4.61}$$

NOTE. $K = 1$ for convex heads and $K = 0.6$ for concave heads. The factor of safety (F.S.) is usually taken as 5.

NOTE.   Equations (4.60) and (4.61) may be superceded by the requirements of a local code or the ASME Boiler and Pressure Vessel Code. The local licensing authority should be consulted.

Boiler Horsepower.   One boiler horsepower is the evaporation of 34.5 pounds (15.7 kg) of water/hour at 212°F (100°C) and atmospheric pressure. If G_a pounds of water/hour enter the boiler with an enthalpy h_{f2} and leaves as steam with an enthalpy h_1, then the boiler horsepower (P_B) is

$$P_B = \frac{G_a(h_1 - h_{f2})}{33,475} \text{ horsepower.} \tag{4.62}$$

Table 4.6   Approximate Reduction in Heat Rate (%), Cumulative

Number of Stages	Throttle Conditions (psi/°F)	
	400/600	1800/1000 Reheat
1	5.0	7.0
2	7.5	9.5
4	9.0	11.0
6	10.0	12.0

Equivalent Evaporation. The numerator of Equation (4.62) represents the actual heat absorbed per hour. Because 970.3 Btu (245 kcal) are required to evaporate one pound (0.454 kg) of water at 212°F (100°C), the equivalent evaporation (G_e) in pounds per hour is

$$G_e = \frac{G_a(h_1 - h_{f2})}{970.3} \text{ pounds/hour.} \qquad (4.63)$$

Factor of Evaporation. In order to determine the equivalent evaporation per hour (G_e) it is necessary to multiply the actual evaporation per hour (G_a) by a factor. This factor is termed factor of evaporation (F) and is

$$F = \frac{h_1 - h_{f2}}{970.3}. \qquad (4.64)$$

Boiler Efficiency. The ratio of the heat absorbed in the boiler to the heat supplied by the fuel is termed boiler efficiency ϵ_B and is

$$\epsilon_B = \frac{G_a(h_1 - h_{f2})}{G_f(H)} \qquad (4.65)$$

where G_f is the weight of fuel in pounds per hour and H is the calorific heating value in Btu per pound of fuel.

4.5.4 Chimneys and Drafts

Intensity of Draft. For a chimney H feet high whose gases have an absolute temperature of T_1 degrees, and an outside absolute temperature of T_2 degrees, the intensity of the draft (D) in inches of water is

$$D = 7.64H \left(\frac{1}{T_2} - \frac{1}{T_1} \right) \text{ inches of water.} \qquad (4.66)$$

NOTE. This formula neglects the effect of friction. For a chimney with a friction factor f, H feet high, C feet circumference, A square feet passage area, and discharging G pounds of gases per second, the draft loss (d) in inches of water is

$$d = \frac{fG^2CH}{A^3} \text{ inches of water,} \qquad (4.67)$$

where $f = 0.0015$ for steel stacks with gases at 600°F (316°C), 0.0011 at 810°F (432°C); and 0.0020 for brick or brick-lined stacks with gases at 650°F (343°C), 0.0015 at 810°F (432°C).

Effective Area of Chimney. The retardation of ascending gases by friction within the stack has the effect of decreasing the inside cross-sectional area, or of lining the chimney with a layer of gas with no velocity. If the thickness of this lining is assumed to be 2 inches (5 cm) for all chimneys, then the effective area (E) for square or round chimneys with A square feet of passage area is approximately

$$E = A - 0.6\sqrt{A} \text{ feet}^2. \tag{4.68}$$

Boiler Horsepower. For a chimney H feet high and A square feet of passage area, **Kent's** empirical formula for the boiler horsepower (P_B) is

$$P_B = 3.33(A - 0.6\sqrt{A})\sqrt{H} \text{ horsepower.} \tag{4.69}$$

NOTE. This formula is based on the assumptions that the boiler horsepower capacity varies as the effective area (E) and the available draft is sufficient to effect combustion of 5 pounds (2.27 kg) of coal per hour per rated horsepower (the water-heating surface divided by 10).

4.5.5 Internal Combustion Engines

Compression Ratio. The ratio of the total volume at the beginning of compression (V_1) and the volume at the end of compression or clearance volume (V_c) is a significant ratio for the various internal combustion engine cycles. This ratio is known as compression ratio (r_k) and may also be expressed in terms of the piston displacement (P.D.), hence

$$r_k = \frac{V_1}{V_c} = \frac{\text{P.D.} + V_c}{V_c}. \tag{4.70}$$

Otto Efficiency. The Otto cycle consists of a frictionless adiabatic compression, constant-volume burning, frictionless adiabatic expansion, and rejection of heat energy at constant volume. If 1 and 2 are used to designate the states at the beginning and end of compression, respectively, then the efficiency for such a cycle (ϵ_0) is

$$\epsilon_0 = 1 - \frac{T_1}{T_2} = 1 - \left(\frac{p_1}{p_2}\right)^{(k-1)/k} = 1 - \left(\frac{V_2}{V_1}\right)^{k-1}$$
$$= 1 - \left(\frac{V_c}{\text{P.D.} + V_c}\right)^{k-1} = 1 - \left(\frac{1}{r_k}\right)^{k-1}. \tag{4.71}$$

NOTE. For explanation of the exponent k, see section on perfect gases.

Joule or Brayton Efficiency. The Joule cycle consists of a frictionless adiabatic compression, constant-pressure burning, frictionless adiabatic expansion, and rejection of heat energy at constant pressure. If 1 and 2 are used to designate the states at the beginning and end of compression, respectively, then the efficiency for such a cycle (ϵ_J) is

$$\epsilon_J = 1 - \frac{T_1}{T_2} = 1 - \left(\frac{p_1}{p_2}\right)^{(k-1)/k} = 1 - \left(\frac{V_2}{V_1}\right)^{k-1} = 1 - \left(\frac{1}{r_k}\right)^{k-1}. \tag{4.72}$$

Diesel Efficiency.   The Diesel cycle consists of a frictionless adiabatic compression, constant-pressure burning, frictionless adiabatic expansion, and the rejection of heat energy at constant volume. If 1 and 2 are used to designate the states at the beginning and end of compression, respectively, and 3 and 4 are used to designate the states at the beginning and end of expansion, respectively, then the efficiency for such a cycle (ϵ_D) is

$$\epsilon_D = 1 - \frac{1}{k}\left(\frac{T_4 - T_1}{T_3 - T_2}\right) = 1 - \left(\frac{1}{r_k}\right)^{k-1}\left[\frac{r_c^k - 1}{k(r_c - 1)}\right], \qquad (4.73)$$

where r_c represents the ratio of the total volume at the end of burning (V_3) to the volume at the start of burning (V_2). This ratio is termed cut-off ratio.

Thermal Efficiency.   The actual cycle for an internal combustion engine takes into account the fact that heat losses occur in the actual engine. The efficiency for such a cycle (ϵ_T) is

$$\epsilon_T = \frac{Q_1 - Q_2 - Q_{\text{losses}}}{Q_1} = \frac{2545}{w_A(H_1)}, \qquad (4.74)$$

where w_A is the actual fuel consumption in pounds per horsepower-hour and H_1 is the calorific heating value of the fuel per pound.

Engine Efficiency.   The ratio of the ideal fuel consumption per horsepower-hour (w_I) to the actual fuel consumption per horsepower-hour (w_A) is termed engine efficiency (ϵ_E), hence

$$\epsilon_E = \frac{w_I}{w_A} = \frac{\epsilon_T}{\epsilon_I}. \qquad (4.75)$$

NOTE.   The ideal efficiency (ϵ_I) can be either the Otto, Joule, or Diesel cycle, depending on which one of these cycles the actual engine is operating.

Brake Horsepower.   The output of an internal combustion engine may be determined by means of a friction brake. Equation (4.56) is applicable in this case.

The empirical equation for determining the brake horsepower (B.H.P.) for an engine with n cylinders of d inches diameter and L inches stroke at N revolutions/minute, the clearance being m percent of the stroke, is

$$\text{B.H.P.} = \frac{d^2LnN}{14,000}\left(0.48 - \frac{1}{10m}\right) \text{ horsepower.} \qquad (4.76)$$

Diameter.   If an engine cylinder is designed for maximum obtainable indicated horsepower (I.H.P.) with a mean effective pressure (M.E.P.) pounds/inch^2, the number of explosions per minute at full load being y and the stroke L in feet being x times the diameter (d) in feet, then

$$d = \sqrt[3]{\frac{300(\text{I.H.P.})}{(\text{M.E.P.})xy}} \text{ feet.} \qquad (4.77)$$

4.5.6 Vacuum Pumps

Vacuum pumps are frequently used for processes where the presence of air is detrimental, where simulation of conditions in outer space is necessary, and so on. Vacuum pumps are typically integrated into a system in which the pump is one component in a process train that may consist of the system or test chamber, the pump **"backing" system,** instrumentation, and control components. Some selected data on vacuum pumps are listed in Table 4.7.

Lobe-type pumps are typically high capacity and generally used as the first stage on "rough" (i.e., low-vacuum) systems; they would normally discharge into a **Stokes pump** (positive displacement), which in turn discharges to the atmosphere.

Oil diffusion pumps are of moderate capacity and need to discharge into a "backing train" typically consisting of one or more lobe-type pumps, discharging into a Stokes-type pump; but care in the design of the installation is required to assure matching of capacities. Also **cold traps** and **baffles** must be provided to avoid oil **backstreaming** out of the pump and into the vacuum chamber/system with a potential for contamination of the system.

Table 4.7 Typical Vacuum Pump Data

Type	Typical Vacuum Limit	Capacity	Remarks
Ejector (steam jets)	$\frac{1}{2}''$ Hg	Moderate	Efficient, low cost, no moving parts
Rotary			
Nash	$3''$ Hg $0.7''$ Hg (multistage)	Can be high	Can have high water requirements
Stokes	to 10^{-2} Torr[a]	Limited	
Lobe type	to 10^{-1} Torr[a]	High	
Oil diffusion	to 10^{-9} Torr[a]	Moderate	Backing train required
Cryo pumps	to 10^{-12} Torr[a]	Very high	Depends on Cryo fluid temp. (generally $20°\text{K} \pm$) and pumping area available

[a]Typical values with care can be 1 to 2 orders of magnitude lower. 1 Torr = $\frac{1}{760}$ of 1 atmosphere (approx. 1 mm Hg).

Cryopumping provides very high capacity but because of limited heat absorption capability, shielding from radiant heat sources usually by liquid nitrogen-cooled panels at $-320°F$ ($-196°C$) is required.

4.5.7   Air Compressors

Capacity.   The capacity of an air compressor is usually measured in terms of the cubic feet of "free air" handled per minute (V_a), which is air at atmospheric pressure (p_a) and atmospheric temperature (t_a). If $P.D.$ represents the piston displacement in cubic feet per minute of a compressor operating with suction pressure p_a pounds absolute per square inch, a discharge pressure p_2 pounds absolute per square inch, compression according to the law pV^n = a constant, and m percent clearance (expressed as a decimal), then

$$V_a = P.D. \left[1 + m - m \left(\frac{p_2}{p_a} \right)^{1/n} \right] \text{feet}^3/\text{minute.} \qquad (4.78)$$

NOTE.   For general expressions relating pressure, volume, and temperature, see section on perfect gases. For air compressors n = 1.20 to 1.35.

Volumetric Efficiency.   The ratio of the volume of "free air" (V_a) to the piston displacement ($P.D.$) is termed displacement or volumetric efficiency (ϵ_v).
For single-stage compression

$$\epsilon_v = \frac{V_a}{P.D.} = 1 + m - m \left(\frac{p_2}{p_a} \right)^{1/n}. \qquad (4.79)$$

For multistage air compressors

$$\epsilon_v = \frac{V_a}{P.D.} = 1 + m - m \left(\frac{p_x}{p_a} \right)^{1/n} \qquad (4.80)$$

where p_x is the discharge pressure in pounds absolute per square inch leaving the first-stage cylinder.
For two-stage compression

$$p_x = \sqrt{p_a p_2} \text{ pounds/inch}^2. \qquad (4.81)$$

For three-stage compression

$$p_x = \sqrt[3]{p_a^2 p_2} \quad \text{and} \quad p_y = \sqrt[3]{p_a p_2^2} \text{ pounds/inch}^2, \qquad (4.82)$$

where p_y is the discharge pressure in pounds absolute per square inch leaving the second-stage cylinder.

Power.   The power (P) required in foot-pounds per minute to compress V_a cubic feet of "free air" per minute polytropically according to the law pV^n = a constant, from atmospheric pressure p_a to a discharge pressure p_2 for single-stage compression is

$$P = 144 p_a V_a \frac{n}{n-1} \left[\left(\frac{p_2}{p_a} \right)^{(n-1)/n} - 1 \right] \text{ foot-pounds/minute.} \qquad (4.83)$$

For two-stage compression

$$P = 288 p_a V_a \frac{n}{n-1} \left[\left(\frac{p_2}{p_a} \right)^{(n-1)/(2n)} - 1 \right] \text{ foot-pounds/minute.} \qquad (4.84)$$

For three-stage compression

$$P = 432 p_a V_a \frac{n}{n-1} \left[\left(\frac{p_2}{p_a} \right)^{(n-1)/(3n)} - 1 \right] \text{ foot-pounds/minute.} \qquad (4.85)$$

The power (P) required in foot-pounds per minute to compress V_a cubic feet of "free air" per minute isothermally according to the law $pV^n = $ a constant, from atmospheric pressure (p_a) to a discharge pressure (p_2) is

$$P = 144 p_a V_a \ln \frac{p_2}{p_a} \text{ foot-pounds/minute.} \qquad (4.86)$$

Efficiency of Compression. The ratio of the isothermal power to the polytropic power is termed efficiency of compression (ϵ_c).

For single-stage compression

$$\epsilon_c = \frac{\ln \dfrac{p_2}{p_a}}{\dfrac{n}{n-1} \left[\left(\dfrac{p_2}{p_a} \right)^{(n-1)/n} - 1 \right]}. \qquad (4.87)$$

For two-stage compression

$$\epsilon_c = \frac{\ln \dfrac{p_2}{p_a}}{\dfrac{2n}{n-1} \left[\left(\dfrac{p_2}{p_a} \right)^{(n-1)/(2n)} - 1 \right]}. \qquad (4.88)$$

For three-stage compression

$$\epsilon_c = \frac{\ln \dfrac{p_2}{p_a}}{\dfrac{3n}{n-1} \left[\left(\dfrac{p_2}{p_a} \right)^{(n-1)/(3n)} - 1 \right]}. \qquad (4.89)$$

4.6   REFRIGERATION

Ideal Compression Refrigeration Cycle.   In order to simplify the references to conditions in the ideal compression refrigeration cycle, a complete description of the cycle is given. The compressor draws the vapor (usually saturated or slightly super-heated) from the evaporator at condition 1, compresses it adiabatically and without friction to condition 2 in the superheated region, and then discharges the vapor to the condenser. The cooling water condenses the vapor to a saturated liquid at condition 3. The liquid is drawn off and then passes through an expansion valve to condition 4. This partially vaporized liquid now enters the evaporator where further evaporation takes place before entering the compressor.

Refrigerating Effect.   The refrigerant enters the evaporator with an enthalpy of h_{f_3} Btu/pound and leaves with an enthalpy of h_1 Btu/pound. If G_r pounds of refrigerant are circulated per minute and the refrigerating effect per minute is represented by R, then

$$R = G_r(h_1 - h_{f_3}) \text{ Btu/minute.} \tag{4.90}$$

Capacity.   The cubic feet per minute (V_1) handled by a compressor operating at N revolutions per minute with a piston displacement ($P.D.$) per revolution, and drawing in G_r pounds of refrigerant per minute, each pound having a specific volume v_1 cubic feet is

$$V_1 = N \times P.D. = G_r v_1 \text{ feet}^3/\text{minute.} \tag{4.91}$$

NOTE.   This formula assumes no clearance. If the refrigerant is compressed from a suction pressure p_1 pounds absolute per square inch to a discharge pressure p_2 pounds absolute per square inch according to the law $pV^n = $ a constant and a clearance of m percent expressed as a decimal, then

$$V_1 = N \times P.D. \left[1 + m - m \left(\frac{p_2}{p_1}\right)^{1/n} \right] \text{ feet}^3/\text{minute.} \tag{4.92}$$

Tonnage.   One ton of refrigeration is the heat equivalent to the melting of one ton (2000 pounds, 907.2 kg) of ice at 32°F (0°C) in 24 hours. Since one pound of ice melting at 32°F (0°C) will absorb approximately 144 Btu (36.3 kcal), then a ton of refrigeration will absorb 288,000 Btu (72.6 × 10³ kcal) per day or 200 Btu (50.4 kcal) per minute, hence

$$\text{Tonnage} = \frac{R}{200} = \frac{G_r(h_1 - h_{f_3})}{200} \text{ tons.} \tag{4.93}$$

Power.   The refrigerant enters the compressor with an enthalpy h_1 and leaves with an enthalpy h_2 Btu/pound. If G_r pounds of refrigerant are circulated per minute, the power (P) expressed in foot-pounds per minute for adiabatic compression is

$$P = 778 G_r(h_2 - h_1) \text{ foot-pounds/minute.} \tag{4.94}$$

If V_1 cubic feet of refrigerant per minute enter the compressor with a suction pressure p_1 pounds absolute per square inch and is compressed to a discharge pressure p_2 pounds absolute per square inch according to the law pV^n = a constant, then the power (P) is

$$P = 144p_1V_1 \frac{n}{n-1}\left[\left(\frac{p_2}{p_1}\right)^{(n-1)/n} - 1\right] \text{ foot-pounds/minute.} \qquad (4.95)$$

Coefficient of Performance. The ratio of the refrigerating effect R to the power ($P/778$) is called the coefficient of performance (c. of p.). Hence, for adiabatic compression

$$\text{c. of p.} = \frac{778R}{P} = \frac{h_1 - h_{f_3}}{h_2 - h_1}, \qquad (4.96)$$

and, for polytropic compression

$$\text{c. of p.} = \frac{5.40G_r(h_1 - h_{f_3})}{p_1V_1 \dfrac{n}{n-1}\left[\left(\dfrac{p_2}{p_1}\right)^{(n-1)/n} - 1\right]}. \qquad (4.97)$$

Heat Removed in the Condenser. If G pounds of refrigerant per minute enter the condenser with an enthalpy h_2 and leave with an enthalpy h_{f_3} Btu/pound, then the heat removed per minute in the condenser (Q_c) is

$$Q_c = \frac{W}{778} + R = G(h_2 - h_{f_3}) \text{ Btu/minute.} \qquad (4.98)$$

Weight of Cooling Water Required. If Q_c Btu/minute are to be removed in the condenser and the temperatures of the cooling water entering and leaving the condenser are t_c and t_h, respectively, then the pounds of cooling water per minute (G_w) required are

$$G_w = \frac{Q_c}{h_{f_h} - h_{f_c}} = \frac{G(h_2 - h_{f_3})}{h_{f_h} - h_{f_c}} \text{ pounds/minute.} \qquad (4.99)$$

4.7 HEAT TRANSMISSION

Fundamental Equation. The heat transmitted in engineering apparatus is effected by a combination of the heat transferred by conduction, convection, and radiation. If the temperature is low and the rate of flow of the fluid over the surface is high, the radiation factor is ignored. The fundamental equation for the heat (Q) conducted in time (t), through a material having a thermal conductivity (k) and a surface area (S) which is normal to the flow of heat and of thickness (x) in the direction of the flow of heat with a temperature difference (θ) between its surfaces, is

$$Q = \frac{ckS\theta t}{x} \text{ units in time } (t). \qquad (4.100)$$

NOTE. Average values of k for various engineering materials expressed in gram-calories per second per square centimeter per centimeter per degree Centigrade, are given in Section 10. The constant c depends on the units of measurement as follows:

Q	S	x	θ	t	c	c^a
gram-calories	cm^2	cm	Cent.	seconds	1	0.000344
kilogram-calories	meters^2	cm	Cent.	hours	36,000	12.4
British thermal units	feet^2	inches	Fahr.	hours	2,903	1
joules	cm^2	cm	Cent.	seconds	4.18	0.00144
joules	feet^2	inches	Fahr.	seconds	851	0.293
kilowatt-hours	meters^2	cm	Cent.	hours	41.8	0.0144
kilowatt-hours	feet^2	inches	Fahr.	hours	0.851	0.000293

^{a}Values of c if k is expressed in Btu per hour per square foot per inch per degree Fahrenheit.

If the heat is transmitted through a body composed of an inside film, two materials, and an outside film, Equation (4.100), when expressed in English units, becomes

$$Q = \frac{\theta_m}{\dfrac{1}{a_1 S_{mf1}} + \dfrac{x_1}{k_1 S_{m1}} + \dfrac{x_2}{k_2 S_{m2}} + \dfrac{1}{a_2 S_{mf2}}} \text{ Btu/hour,} \qquad (4.101)$$

where θ_m is the mean temperature difference in degrees Fahrenheit between the two fluids while passing over the body, a_1 and a_2 are the inside and outside film coefficients in Btu per hour per square foot per degree Fahrenheit, S_{mf1} and S_{mf2} are the areas of the inside and outside films in square feet, x_1 and x_2 are the thicknesses of the materials in inches, k_1 and k_2 are the conductivities of the materials in Btu per hour per square foot per inch of thickness per degree Fahrenheit, and S_{m1} and S_{m2} are the mean surface areas of the materials.

Mean Temperature Difference. For building walls, roofs, partitions, and so forth, steam and refrigerating pipes carrying wet or saturated vapor and surrounded by atmospheric air, the heat is assumed to be transmitted from the hot fluid at a uniform temperature (t_1) to a cold fluid at a uniform temperature (t_2). For these cases the mean temperature difference (θ_m) is

$$\theta_m = t_1 - t_2 \text{ degrees Fahrenheit.} \qquad (4.102)$$

In heat exchangers such as boilers, superheaters, condensers, economizers, liquid and gas heaters or coolers, the temperature of either one or both fluids changes. If the hot fluid enters the apparatus at a temperature t_1 and leaves at t_2 and the contiguous cold fluid temperatures are t_a and t_b, respectively, then

$$\theta_m = \frac{(t_1 - t_a) - (t_2 - t_b)}{\ln \dfrac{t_1 - t_a}{t_2 - t_b}} \text{ degrees Fahrenheit.} \tag{4.103}$$

NOTE.   Equation (4.103) gives the logarithmic mean temperature difference and is applicable only when the overall coefficient of heat transfer (K), the weight (W) of the hot fluid and the weight (w) of the cold fluid and their specific heat (C) and (c), respectively, are approximately constant during the transfer of heat (see also Fig. 4.1).

Mean Surface Area.   The most important surfaces encountered in engineering practice are the plane or uniform cross-sectional surface, cylindrical and spherical surfaces. If S_1 and S_2 represent the inside and outside surface areas, respectively, then the mean surface area (S_m) for the plane surface is

$$S_m = \frac{S_2 + S_1}{2} = S \text{ feet}^2. \tag{4.104}$$

For the cylindrical surface, the mean surface area (S_m) is

$$S_m = \frac{S_2 - S_1}{\ln \dfrac{S_2}{S_1}} = \frac{2\pi L(r_2 - r_1)}{\ln \dfrac{r_2}{r_1}} \text{ feet}^2, \tag{4.105}$$

where L is the length in feet and r_1 and r_2 are the inside and outside radii in feet, respectively.

For the spherical surface, the mean surface area (S_m) is

$$S_m = \sqrt{S_2 S_1} = 4\pi r_1 r_2 \text{ feet}^2. \tag{4.106}$$

4.7.1   Heat Exchangers

Heat exchangers are an important type of process equipment widely used in most industries. **Fired heat exchangers** are generally classified as boilers, and typically are governed by the ASME Boiler and Pressure Vessel Code (Section 2); they are not considered in the following discussion. **Unfired heat exchangers** are often of the **counter flow design** in which the fluid flows are counter to each other, thus assuring that the fluid leaving the exchanger is close in temperature (approach) to the entering fluid. Heat exchangers often use a gas or vapor, frequently steam, as one medium with a liquid as the other. In the process, the vapor is condensed on the exterior of the exchange surface, typically tubing, and the resultant liquid may be further cooled in a drain cooler section. Noncondensing heat exchangers are used for process applications where the temperature exchange occurs above the dew point of the vapor, or where the exchange is between noncondensing media. Heat exchangers range from

General Heat Transfer Cases

Nomenclature:

$$LMTD = \text{logarithmic mean temperature difference} = \frac{GTD - LTD}{Log_e \dfrac{GTD}{LTD}}$$

TR = temperature rise of cold substance.
TF = temperature fall of hot substance.
ITD = initial temperature difference between hot and cold substances.
FTD = final temperature difference between hot and cold substances.
GTD = greatest temperature difference.
LTD = least temperature difference.

Case I—Hot substance giving up heat with temperature remaining constant to a cold substance absorbing heat with rising temperature.

GTD = ITD

LTD = ITD–TR

This case includes:—steam condensers, ammonia condensers, and boiler feed water heaters.

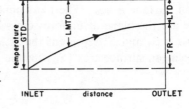

Case II—Hot substance giving up heat with falling temperature to a cold substance absorbing heat with temperature remaining constant.

GTD = ITD

LTD = ITD–TF

This case includes:—steam boilers, ammonia brine coolers, ammonia direct expansion coils in cold storage rooms, evaporator with hot liquid coils or jacket.

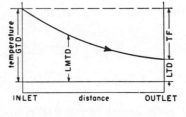

Case III—Two substances both changing temperature, one giving up heat with falling temperature to the other absorbing heat with rising temperature, **parallel flow.**

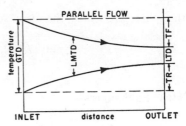

Case IV—Two substances both changing temperature, one giving up heat with falling temperature to the other absorbing heat with rising temperature, **countercurrent flow.**

Cases III & IV include:—steam superheaters and economizers, brine coils in cold storage rooms, compressor intercoolers, cylinder jackets and oil coolers.

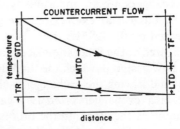

GENERAL HEAT TRANSFER CASES

Fig. 4.1 General heat transfer cases. Copyrighted by Ingersoll-Rand Company, 1962. Published with permission.[1]

simple single-pass, straight-tube exchangers to those having multiple passes with U-tube, spiral, or other more complex configurations (Fig. 4.2).

Provision for shell expansion is necessary where heaters operate with significant thermal differences from ambient. For exchangers with the **fluid heads (channel)** at one end only, this is normally accomplished by permitting shell motion away from the channel on rollers or sliding supports. Straight-tube exchangers with channels at each end normally allow one end of the exchanger to float by means of expansion or packed joints in the shell or at the shell-to-channel connection.

Since pressure can be more easily withstood by smaller diameters, the higher-pressure fluid is usually contained on the interior of the internal tubing, with the larger-diameter shell retaining the lower-pressure fluid and thus permitting more economical design. The tubing system is an extension of the channel, which is also constructed to resist the higher pressures.

As a matter of design convenience, manufacturers typically assign pressure ratings to both the channel and shell side of exchangers. These typically are:

Channel	Shell
0–25 psig (172 kPa)	Normally shell thicknesses
150 psig (1,034 kPa)	are tailored to the
300 psig (2,069 kPa)	operating pressure
600 psig (4,137 kPa)	conditions.
1,250 psig (8,619 kPa)	
3,000 psig (20,700 kPa)	

For higher pressures, frequently a special type of locking system is used (in lieu of a conventional blind flange) to provide access to the interior of the channel. These take the form of interrupted bayonet threads, split locking rings, and so forth.

Tubing is connected to the tube sheet normally by **flaring (rolling).** This sealing may also be accomplished by a system of packing, or by tube-to-tube sheet seal welding. Tubing is usually of circular cross section, relatively thin wall (14 to 22 BWG), seamless material. Copper bearing, stainless, or alloy materials are common because of their resistance to corrosion and thus loss of heat transfer capability due to fouling (corrosion product buildup). In some types of exchangers, special tube cross sections with fins or twists are used to promote either internal or external turbulence and thus improve heat transfer coefficients or increase surface area.

Maintenance of a low "U" factor is important to heat-transfer equipment operation and can be achieved by provision for mechanical or chemical cleaning. During the course of the life of a heat exchanger, some tube leakage (at the tube sheet) can occur. This tube leakage is often repaired by plugging the tube and removing it from service; thus, design of heat exchangers should allow for excess tubing capacity, on the order of 5% or more, depending upon the duty anticipated.

TEMA (Tubular Exchanger Manufacturers Association) Standards are widely applied throughout industry. The TEMA Standards include Thermal and Mechanical Standards, Nomenclature, Fabrication, Installation, Operation and Maintenance, and Materials Specification data, as well as Recommended Practices.

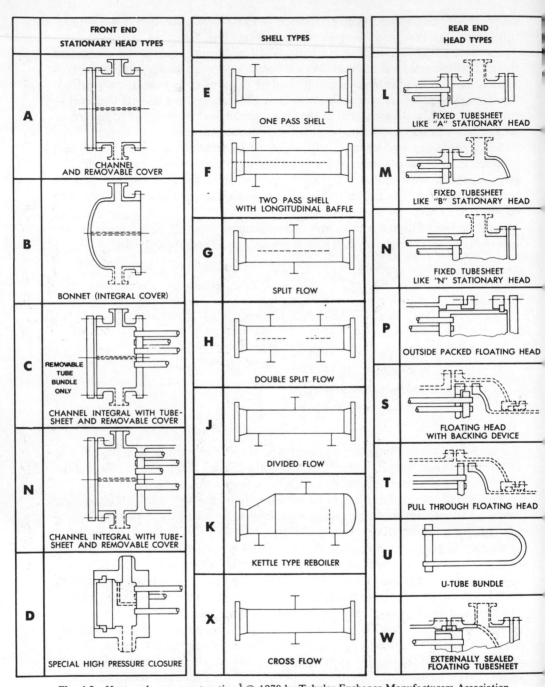

Fig. 4.2 Heat exchanger construction.[3] © 1978 by Tubular Exchange Manufacturers Association.

4.7.2 Building Heat Transfer

Overall Coefficient of Heat Transfer. It is desirable in the solution of engineering problems involving the transfer of heat through typical walls, roofs, partitions, floors, and so on, to use a coefficient of heat transmission that will take into account the effects of conduction, convection, and radiation, together with the type, thickness, and position of the materials, and which may be used with the difference of the temperatures of the fluid temperatures on each side of the composite section. This quantity is termed overall coefficient of heat transfer (U) and is expressed in Btu per hour per square foot of surface area per degree Fahrenheit. The heat transmitted per hour (Q) becomes

$$Q = US\theta_m \text{ Btu/hour.} \tag{4.107}$$

NOTE. Average values of U for the usual building structures are given below.

Overall Coefficients of Heat Transfer U for Building Structures[a] (Expressed in Btu per Hour per Square Foot per Degree Fahrenheit)

Structure	Thickness (inches)		
	8	12	16
Walls			
Brick, without interior plaster	0.50	0.36	0.28
Brick, with interior plaster	0.46	0.34	0.27
Concrete, without interior plaster	0.69	0.54	0.48
Concrete, with interior plaster	0.62	0.49	0.44
Haydite, without interior plaster	0.36	0.26	0.21
Haydite, with interior plaster	0.34	0.24	0.20
Hollow tile, without interior plaster	0.40	0.30	0.25
Hollow tile, with interior plaster	0.38	0.29	0.24
Limestone, without interior plaster	0.71	0.49	0.37
Limestone, with interior plaster	0.64	0.45	0.35
Wood, shingled or clapboarded, with interior plaster			0.25
Stucco, with interior plaster			0.30
Brick veneer, with interior plaster			0.27
Partitions			
4-inch hollow clay tile, plaster both sides			0.40
4-inch common brick, plaster both sides			0.43
4-inch hollow gypsum tile, plaster both sides			0.27
Wood lath and plaster on one side of studding			0.62
Wood lath and plaster on both sides of studding			0.34
Metal lath and plaster on one side of studding			0.69

[a]Correction for exposure:

	North	East	South	West
Multiply U by	1.3	1.1	1.0	1.2

Overall Coefficients of Heat Transfer U for Building Structures (*continued*)

Structure	Thickness (inches)		
	8	12	16

Partitions			
Metal lath and plaster on both sides of studding			0.39
Plasterboard and plaster on one side of studding			0.61
Plasterboard and plaster on both sides of studding			0.34
2-inch corkboard and plaster on one side of studding			0.12
2-inch corkboard and plaster on both sides of studding			0.063

	4	6	8	10
Floors				
Concrete, no ceiling and no flooring	0.65	0.59	0.53	0.49
Concrete, plastered ceiling and no flooring	0.59	0.54	0.50	0.45
Concrete, no ceiling and terrazzo flooring	0.61	0.56	0.51	0.47
Concrete, plastered ceiling and terrazzo flooring	0.56	0.52	0.47	0.44
Concrete, on ground and no flooring	1.07	0.90	0.79	0.70
Concrete, on ground and terrazzo flooring	0.98	0.84	0.74	0.66
Frame construction, no ceiling, maple or oak flooring on yellow pine subflooring on joists				0.34
Frame construction, metal lath and plaster ceiling, maple or oak flooring on yellow pine subflooring on joists				0.35
Frame construction, wood lath and plaster ceiling, maple or oak flooring on yellow pine subflooring on joists				0.24
Frame construction, plasterboard ceiling, maple or oak flooring on yellow pine subflooring on joists				0.24

	2	4	6
Roofs, Tar, and Gravel			
Concrete, no ceiling and no insulation	0.82	0.72	0.64
Concrete, no ceiling and 1-inch rigid insulation	0.24	0.23	0.22
Concrete, metal lath and plaster ceiling and no insulation	0.42	0.40	0.37
Concrete, metal lath and plaster ceiling and 1-inch rigid insulation	0.19	0.18	0.18
1-inch wood, no ceiling and no insulation			0.49
1-inch wood, no ceiling and 1-inch rigid insulation			0.20
1-inch wood, metal lath and plaster ceiling and no insulation			0.32
1-inch wood, metal lath and plaster ceiling and 1-inch rigid insulation			0.16
Metal, no ceiling and no insulation			0.95
Metal, no ceiling and 1-inch rigid insulation			0.25
Metal, metal lath and plaster ceiling and no insulation			0.46

		Thickness (inches)	
Structure	8	12	16

Roofs, tar, and gravel

Metal, metal lath and plaster ceiling and 1-inch rigid insulation	0.19
Wood shingles, rafters exposed	0.46
Wood shingles, metal lath and plaster	0.30
Wood shingles, wood lath and plaster	0.29
Wood shingles, plasterboard and plaster	0.29
Asphalt shingles, rafters exposed	0.56
Asphalt shingles, metal lath and plaster	0.34
Asphalt shingles, wood lath and plaster	0.32
Asphalt shingles, plasterboard and plaster	0.32

Glass

Single windows and skylights	1.13
Double windows and skylights	0.45
Triple windows and skylights	0.281
Hollow glass tile wall, 6 × 6 × 2 inch thick blocks	
Wind velocity 15 mph, outside surface; still air, inside surface	0.60
Still air outside and inside surfaces	0.48

Doors

	1	1¼	1½	1¾	2	2½	3
Wood	0.69	0.59	0.52	0.51	0.46	0.38	0.33

4.8 AIR AND VAPOR MIXTURES

Specific or Absolute Humidity. The weight of water vapor per unit volume of space occupied, expressed in grains or pounds per cubic foot, is termed absolute humidity. In order to simplify the solution of problems involving air and vapor mixtures, it is convenient to express the weight of water per cubic foot (d_s) in terms of the weight of dry air per cubic foot (d_a). This ratio has no specific name although the term absolute humidity (ϕ) is often applied to this ratio.

$$\phi = \frac{d_s}{d_a} = \frac{v_a}{v_s} = 0.622 \left(\frac{p_s}{B - p_s} \right) \text{ pounds.} \qquad (4.108)$$

NOTE. In Equation (4.108) the perfect gas laws are assumed to hold for both the water vapor and the dry air present in the moisture-laden air. The total pressure of the moisture-laden air (B) expressed in pounds absolute per square inch is assumed to be equal to the sum of the partial pressure exerted by the water vapor (p_s) and the partial pressure exerted by the dry air (p_a), both expressed in pounds absolute per square inch.

Relative Humidity. The ratio of the actual density of the water vapor in the moisture-laden air (d_s) to the density of saturated vapor (d_{sat}) at the same temperature is termed relative humidity (H). Assuming the perfect gas laws to satisfy this low pressure vapor, then

$$H = \frac{d_s}{d_{sat}} = \frac{v_{sat}}{v_s} = \frac{p_s}{p_{sat}}.$$ (4.109)

NOTE. Although p_s may be determined from Ferrell's or Carrier's equation in terms of the barometric reading, the wet-bulb, and dry-bulb temperatures, it is customary to use Equation (4.109) for determining the partial pressure (p_s) and the specific volume (v_s) of the water vapor. In engineering practice the psychrometric tables are used for determining the relative humidity (H).

Temperature, Dry Bulb	Difference Between Wet and Dry Bulb									
	2°	4°	6°	8°	10°	12°	14°	16°	18°	20°
32°F	79	59	39	20	2	—	—	—	—	—
40°F	84	68	52	37	22	8	—	—	—	—
45°F	86	71	59	44	32	19	6	—	—	—
50°F	87	74	62	50	38	26	16	5	—	—
55°F	88	76	65	54	43	33	24	14	5	—
60°F	89	78	68	58	48	39	30	22	13	5
65°F	90	80	70	61	52	44	35	28	20	12
70°F	90	81	72	64	56	48	40	32	26	19
75°F	91	82	74	66	58	51	44	37	30	24
80°F	92	83	75	68	61	54	47	41	34	29
85°F	92	84	77	70	63	56	50	44	38	33
90°F	92	85	78	71	65	58	52	47	41	36
95°F	93	86	79	72	66	60	54	49	44	39
100°F	93	86	80	74	68	62	57	52	46	42

NOTE. The relative humidity should range between 35 and 45.

Dew Point. When moisture-laden air is cooled until the temperature reaches that corresponding to the saturation temperature for the partial pressure of the water vapor, condensation or precipitation begins. This temperature is called the dew point.

Determination of Weight of Moisture Precipitated. In order to precipitate moisture from V cubic feet of moisture-laden air at condition 1 it is necessary to cool the air to the dew point temperature (t_3) for the final condition 2 desired. The pounds of moisture precipitated (M_p) is

$$M_p = \frac{V}{v_{a_1}}(\phi_1 - \phi_2) = \frac{V}{v_{a_1}}\left(\frac{v_{a_1}}{v_{s_1}} - \frac{v_{a_2}}{v_{s_2}}\right) \text{ pounds.}$$ (4.110)

NOTE. The specific volume of the dry air (v_a) may be determined from

$$v_a = \frac{53.34(t + 460)}{144(B - p_s)} \text{ feet}^3/\text{pound},$$ (4.111)

and the specific volume of the water vapor (v_s) may be determined from

$$v_s = \frac{v_{\text{sat}}}{H} \text{ feet}^3/\text{pound}.$$ (4.112)

Determination of the Quantity of Heat Removed from the Moisture-Laden Air. In order to remove the moisture (M_p), as given in Equation (4.110), it is necessary to supply refrigeration. This refrigeration must cool the dry air and the water vapor in addition to precipitating the moisture, thus the total amount of heat removed (R) in Btu is

$$R = \frac{V}{v_{a1}} \left[0.241(t_1 - t_3) + \frac{v_{a2}}{v_{s2}} (h_{s_1} - h_{s_3}) \right.$$

$$\left. + \left(\frac{v_{a1}}{v_{s1}} - \frac{v_{a2}}{v_{s2}} \right) (h_{s_1} - h_{f_3}) \right] \text{ Btu.}$$ (4.113)

NOTE. h_s may be assumed to be the same as the enthalpy (h) for the saturated vapor at the same temperature.

Determination of the Heat Added. In order to precipitate the required moisture from the air at condition 1, it was necessary to cool the air to the dew point temperature (t_3) for condition 2. The saturated air must now be heated to obtain the desired temperature (t_2). This heat must be supplied to the dry air and the water vapor; thus the total amount of heat added (Q) in Btu is

$$Q = \frac{V}{v_{a1}} \left[0.241(t_3 - t_2) + \frac{v_{a2}}{v_{s2}} (h_{s2} - h_{s3}) \right] \text{ Btu.}$$ (4.114)

REFERENCES

1. *Cameron Hydraulic Data,* 13th edition, Compressed Air Magazine Co., Phillipsburg, New Jersey, 1962.
2. *Standards of Tubular Exchanger Manufacturers Association,* 6th edition, Tubular Exchanger Manufacturers Association, Inc., Tarrytown, New York, 1978.

BIBLIOGRAPHY

S. Dushman, *Scientific Foundations of Vacuum Technique,* 2nd edition, John Wiley & Sons, New York, 1962.

Phillip J. Potter, *Steam Power Plants,* The Ronald Press Co., New York, 1949.

B. G. A. Skrotzki and W. A. Vopat, *Power Station Engineering and Economy,* McGraw-Hill, New York, 1960.

SECTION 5
ELECTRICITY AND ELECTRONICS

J. M. SHULMAN

Fellow Engineer (Retired)
Westinghouse Electric Corporation

5.1 ELECTRICAL ENGINEERING TERMS, UNITS, AND SYMBOLS

5.1.1 Standards

In the United States, standards covering design, manufacture, and use of electrical and electronic equipment are published by the Institute of Electrical and Electronic Engineers (IEEE), American National Standards Institute (ANSI), and National Electrical Manufacturers Association (NEMA). Each of these organizations issues a catalog of its standards and updates it at least once a year.[1,2,3] International standards are published by the International Electro-technical Commission (IEC), located at One Rue de Varembe, Geneva, Switzerland.

IEEE publishes a computer-compiled index of all its standards and all ANSI electrical standards.[4] In its first section the standards are listed by an alphabetical index of key words. The second section is a numerical listing by ANSI and IEEE standard numbers, giving for each the latest date of publication, information for ordering from IEEE, and a complete table of contents. The table of contents enables a reader to quickly determine the possible application of that particular standard.

5.1.2 Terms and Units

One of the IEEE standards is a dictionary of electrical and electronic terms.[5] In addition to terms and definitions previously standardized by IEEE, it now includes many in the ANSI standards and those of the IEC.

The dictionary points out a distinction between use of the terms "electric" and "electrical" as adjectives. "Electric" means containing, producing, arising from, actuated by or carrying electricity, or designed to carry electricity and capable of doing so. "Electrical" means related to, pertaining to or associated with electricity, but not having its properties or characteristics. Examples: electric current, electric motor, electrical engineer, electrical handbook. A note included in the definition of both words states that, although some dictionaries indicate they are synonymous, usage in the electrical engineering field has in general been restricted to the meanings

241

in the foregoing definitions. Also, it is recognized that there are borderline cases wherein the usage determines the selection.

The **International System of Units (SI)** is an updated version of the meter-kilogram-second-ampere (MKSA) metric system (which has long been the standard of many countries) and is now being adopted throughout the world.

SI units are divided into three classes—base units, supplementary units, and derived units. Base units are those that are regarded as dimensionally independent. They are as follows:

Quantity	Unit	Symbol
Length	meter	m
Mass	kilogram	kg
Time	second	s
Electric current	ampere	A
Thermodynamic temperature	kelvin	K
Amount of substance	mole	mol
Luminous intensity	candela	cd

Supplementary units are those which may be regarded as base units or derived units, as follows:

Quantity	Unit	Symbol
Plane angle	radian	rad
Solid angle	steradian	sr

Derived units are those formed by combining base units, supplementary units, and other derived units according to the algebraic relations linking the corresponding quantities.

In some countries, particularly English-speaking ones, much of the present engineering practice and technical literature uses non-SI units. To encourage and facilitate orderly changeover to metric practice and the use of SI units, a standard on metric practice has been published[6] which contains, in addition to information on base, supplementary, and derived units, complete tables of conversion factors from English system and obsolete metric system units to SI units.

5.1.3   Symbols, Graphics, and Alphabets

A unit symbol is a letter or group of letters that may be used in place of the name of the unit. A standard is available[7] that contains a complete alphabetical listing of all SI units and their unit symbols for use in electrical engineering.

Quantity symbols are those used in formulas. A quantity symbol is normally a single letter modified when appropriate by one or more subscripts or superscripts. The standard on quantity symbols[8] lists both quantity and unit symbols used in electrical science and electrical engineering in tables by the following categories: (1) Space and time, (2) Mechanics, (3) Heat, (4) Radiation and light, (5) Fields and

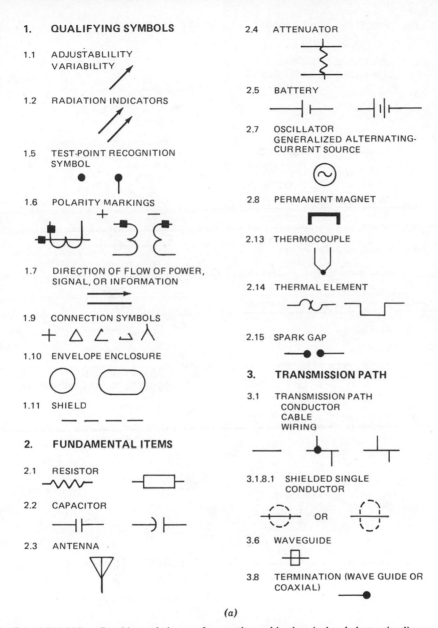

1. QUALIFYING SYMBOLS

1.1 ADJUSTABLILITY
VARIABILITY

1.2 RADIATION INDICATORS

1.5 TEST-POINT RECOGNITION
SYMBOL

1.6 POLARITY MARKINGS

1.7 DIRECTION OF FLOW OF POWER,
SIGNAL, OR INFORMATION

1.9 CONNECTION SYMBOLS

1.10 ENVELOPE ENCLOSURE

1.11 SHIELD

2. FUNDAMENTAL ITEMS

2.1 RESISTOR

2.2 CAPACITOR

2.3 ANTENNA

2.4 ATTENUATOR

2.5 BATTERY

2.7 OSCILLATOR
GENERALIZED ALTERNATING-
CURRENT SOURCE

2.8 PERMANENT MAGNET

2.13 THERMOCOUPLE

2.14 THERMAL ELEMENT

2.15 SPARK GAP

3. TRANSMISSION PATH

3.1 TRANSMISSION PATH
CONDUCTOR
CABLE
WIRING

3.1.8.1 SHIELDED SINGLE
CONDUCTOR

OR

3.6 WAVEGUIDE

3.8 TERMINATION (WAVE GUIDE OR
COAXIAL)

(a)

Fig. 5.1 (a)(b)(c)(d)   Graphic symbols most frequently used in electrical and electronics diagrams.

circuits, (6) Electronics and telecommunication, (7) Machines and power engineering. Table 8 is a list of symbols for physical constants, Table 9 a list of selected mathematical symbols, and Table 10 an alphabetical list of all quantity symbols.

Device and wiring graphic symbols for use in schematic wiring diagrams of electrical and electronic circuits and in single-line diagrams of electric power circuits have been standardized.[9] Those most frequently used in practice are shown in Fig. 5.1. Figure 5.2 illustrates three types of diagrams that use graphic symbols.

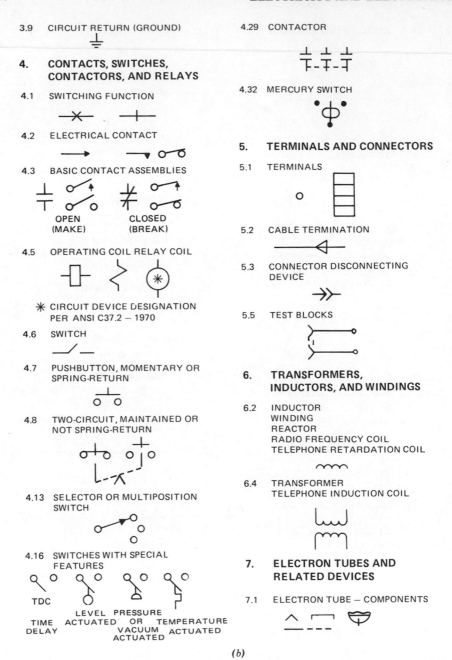

3.9   CIRCUIT RETURN (GROUND)

4.   **CONTACTS, SWITCHES, CONTACTORS, AND RELAYS**

4.1   SWITCHING FUNCTION

4.2   ELECTRICAL CONTACT

4.3   BASIC CONTACT ASSEMBLIES

OPEN   CLOSED
(MAKE)   (BREAK)

4.5   OPERATING COIL RELAY COIL

⁕ CIRCUIT DEVICE DESIGNATION
   PER ANSI C37.2 – 1970

4.6   SWITCH

4.7   PUSHBUTTON, MOMENTARY OR SPRING-RETURN

4.8   TWO-CIRCUIT, MAINTAINED OR NOT SPRING-RETURN

4.13 SELECTOR OR MULTIPOSITION SWITCH

4.16 SWITCHES WITH SPECIAL FEATURES

TDC

TIME   LEVEL PRESSURE
DELAY   ACTUATED   OR   TEMPERATURE
                   VACUUM   ACTUATED
                  ACTUATED

4.29 CONTACTOR

4.32 MERCURY SWITCH

5.   **TERMINALS AND CONNECTORS**

5.1   TERMINALS

5.2   CABLE TERMINATION

5.3   CONNECTOR DISCONNECTING DEVICE

5.5   TEST BLOCKS

6.   **TRANSFORMERS, INDUCTORS, AND WINDINGS**

6.2   INDUCTOR
      WINDING
      REACTOR
      RADIO FREQUENCY COIL
      TELEPHONE RETARDATION COIL

6.4   TRANSFORMER
      TELEPHONE INDUCTION COIL

7.   **ELECTRON TUBES AND RELATED DEVICES**

7.1   ELECTRON TUBE – COMPONENTS

(b)

Fig. 5.1   (*Continued*)

Prefixes and symbols for decimal multiples and submultiples of SI units are standardized from 10^{18} to 10^{-18}. Those most frequently used in electrical and electronic engineering practice are found in Table 10.36.

SI quantities most frequently used in practice are listed alphabetically in Table 5.1, with their quantity symbols, unit names, and unit symbols.

Letter symbols are mainly restricted to the English and Greek alphabets. In print

7.3   TYPICAL APPLICATIONS

7.7   NUCLEAR-RADIATION DETECTOR
        IONIZATION CHAMBER
        PROPORTIONAL COUNTER TUBE
        GEIGER-MÜLLER COUNTER TUBE

8.   **SEMICONDUCTOR DEVICES**

8.2   ELEMENT SYMBOLS

8.5   TYPICAL APPLICATIONS: TWO-
        TERMINAL DEVICES

8.6   TYPICAL APPLICATIONS: THREE-
        (OR MORE) TERMINAL DEVICES

8.7   PHOTOSENSITIVE CELL

8.8   SEMICONDUCTOR THERMO-
        COUPLE

8.9   HALL ELEMENT
        HALL GENERATOR

9.   **CIRCUIT PROTECTORS**

9.1   FUSE

9.3   LIGHTNING ARRESTER
        ARRESTER
        GAP

9.4   CIRCUIT BREAKER

9.5   PROTECTIVE RELAY

10.   **ACOUSTIC DEVICES**

10.1   AUDIBLE-SIGNALING DEVICE

11.   **LAMPS AND VISUAL-
          SIGNALING DEVICES**

11.2   VISUAL-SIGNALING DEVICE

    11.2.1   ANNUNCIATOR

    11.2.6   INDICATING LAMP

12.   **READOUT DEVICES**

12.1   METER INSTRUMENT

A	DB	I	OP	RF	VA
AH	DBM	INT	OSCG	SY	VAR
C	DM	μA	PH	TLM	VARH
CMA	DTR	UA	PI	t^0	VI
CMC	F	MA	PH	THC	VU
CMV	G	NM	RD	TT	W
CRO	GD	OHM	REC	V	WH
			ETC.		

(c)

Fig. 5.1 (*Continued*)

the quantity symbols appear in italic (sloping) type and the unit symbols in roman (upright) type. Quantity symbols are used in both capital and lower-case letter versions. Because of typography variations and use of both capital and lower-case letters, good engineering practice dictates that whenever a possibility of ambiguity or misunderstanding exists in the use of either a quantity or unit symbol, the symbol should be fully identified at the location where it is used.

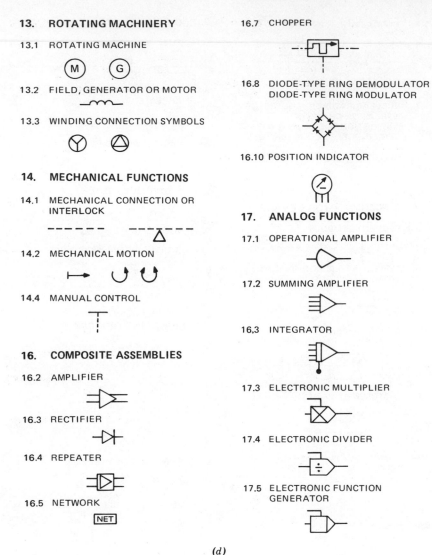

13.   ROTATING MACHINERY

13.1   ROTATING MACHINE

13.2   FIELD, GENERATOR OR MOTOR

13.3   WINDING CONNECTION SYMBOLS

14.   MECHANICAL FUNCTIONS

14.1   MECHANICAL CONNECTION OR
       INTERLOCK

14.2   MECHANICAL MOTION

14.4   MANUAL CONTROL

16.   COMPOSITE ASSEMBLIES

16.2   AMPLIFIER

16.3   RECTIFIER

16.4   REPEATER

16.5   NETWORK

16.7   CHOPPER

16.8   DIODE-TYPE RING DEMODULATOR
       DIODE-TYPE RING MODULATOR

16.10 POSITION INDICATOR

17.   ANALOG FUNCTIONS

17.1   OPERATIONAL AMPLIFIER

17.2   SUMMING AMPLIFIER

16.3   INTEGRATOR

17.3   ELECTRONIC MULTIPLIER

17.4   ELECTRONIC DIVIDER

17.5   ELECTRONIC FUNCTION
       GENERATOR

(d)

Fig. 5.1   (*Continued*)

5.2   CIRCUIT ELEMENTS

5.2.1   Circuit Definition

All useful applications of electricity and electronics involve conductors or systems of conductors through which current is intended to flow called circuits. In order for current to flow, one or more voltage sources must be present in the circuit. Voltage sources have many different forms, and circuit currents have magnitudes, directions, and forms depending on voltages. Voltages and currents may be of constant magnitude and direction (continuous dc), varying magnitude and constant direction (vary-

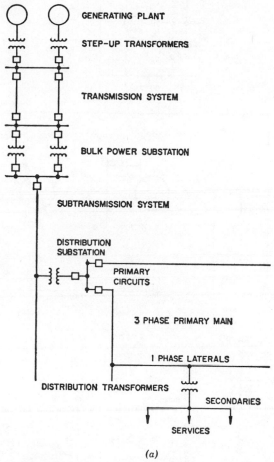

GENERATING PLANT

STEP-UP TRANSFORMERS

TRANSMISSION SYSTEM

BULK POWER SUBSTATION

SUBTRANSMISSION SYSTEM

DISTRIBUTION SUBSTATION

PRIMARY CIRCUITS

3 PHASE PRIMARY MAIN

I PHASE LATERALS

DISTRIBUTION TRANSFORMERS

SECONDARIES

SERVICES

(a)

Fig. 5.2 Typical diagrams with graphic symbols. (a) Single-line diagram of an electric power system, (b) electronic circuit with transistors, (c) electronic circuit with integrated circuits. (a used with the permission of the Westinghouse Electric Corporation. b and c, used with permission of The American Radio Relay League, Inc.)

ing or pulsating dc), cyclic reversal of direction (ac) or transient (of short duration compared to the steady-state condition). All circuits contain three linear and bilateral properties that determine the nature of current resulting from voltage, resistance, inductance, and capacitance.

5.2.2 Resistance

The physical properties of a conductor that tend to oppose the flow of current through it, and the conductor dimensions determine its resistance.

$$R = \frac{\rho L}{A},$$

(5.1)

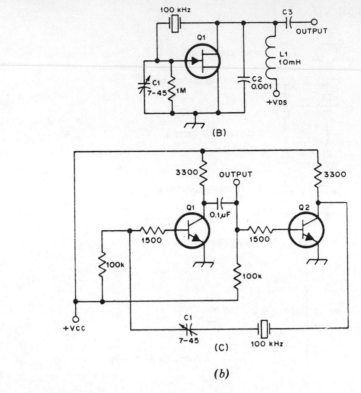

(B)

(C)

(b)

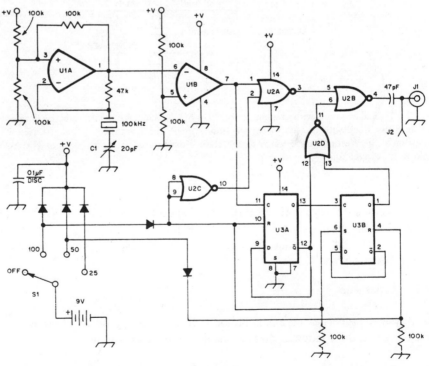

(c)

Fig. 5.2 *(Continued)*

Table 5.1   Quantities and SI Units Most Frequently Used in Electrical Engineering

Quantity	Quantity Symbol	Unit	Unit Symbol
Admittance	Y	siemens	S
Angular velocity	ω	radian per second	rad/s
Capacitance	C	farad	F
Conductance	G	siemens	S
Conductivity	σ	siemens per meter	S/m
Current	I	ampere	A
Current density	J	ampere per square meter	A/m^2
Electric charge	Q	coulomb	C
Electric field strength	E	volt per meter	V/m
Electric flux	Ψ	coulomb	C
Electric flux density	D	coulomb per square meter	C/m^2
Electromotive force	E	volt	V
Energy	W	joule	J
Force	F	newton	N
Frequency	f	hertz	Hz
Impedance	Z	ohm	Ω
Inductance	L	henry	H
Inductance, mutual	M	henry	H
Length	L	meter	m
Magnetic field strength (also called magnetizing force)	H	ampere per meter	A/m
Magnetic flux	Φ	weber	Wb
Magnetic flux density	B	tesla	T
Magnetomotive force	$\mathcal{F}$	ampere	A
Mass	m	kilogram	kg
Permeance	$\mathcal{P}$	henry	H
Permeability	μ	henry per meter	H/m
Permittivity	ϵ	farad per meter	F/m
Phase angle	Θ	radian	rad
Potential, potential difference	V	volt	V
Power, real	P	watt	W
Power, reactive	Q	var	var
Pressure	p	newton per square meter	N/m^2
Reactance	X	ohm	Ω
Reluctance	$\mathcal{R}$	reciprocal henry	H^{-1}
Resistance	R	ohm	Ω
Resistivity	ρ	ohm meter	$\Omega \cdot m$
Susceptance	B	siemens	S
Temperature	T	degrees Celsius	°C
Time	t	second	s
Voltage	V	volt	V
Wavelength	λ	meter	m
Work	W	joule	J

where L is length in the direction of current flow, A is cross-sectional area perpendicular to flow, and ρ is a constant for the material, called resistivity or specific resistance. Rho is numerically equal to R when L and A are equal to one. For L in meters (m) and A in square meters (m²), the unit for ρ is ohm meters.

Resistivity of metallic conductors varies linearly between approximately 100 and $-200°C$ (Fig. 5.3). It approaches zero nonlinearly between $-200°C$ and absolute zero temperature $-273.13°C$. Since most practical work is in the linear range, extending the linear portion of the curve to the abscissa yields a value of temperature T, on which the rate of change of resistivity α_1 at any temperature t_1 higher than T can be calculated

$$\alpha_1 = \frac{1}{T + t_1}. \qquad (5.2)$$

T is 234.5 for standard annealed copper. If t_1 is $+20°C$, $\alpha_1 = 1/254.5 = 0.00393$. α_1 is called the temperature coefficient of resistivity (or resistance) at temperature t_1, which must be stated. Table 5.2 lists values of resistivity, temperature coefficient of resistivity, and reference temperature T for copper and other commonly used conductors.

If R_1 and α_1 are known at any temperature t_1 in the linear range, R_2 at any temperature t_2 also in the linear range can be found from the formula

$$R_2 = R_1[1 + \alpha_1(t_2 - t_1)]. \qquad (5.3)$$

If reference temperature T is known instead of α_1, R_2 can be found from the formula

$$R_2 = R_1\left(\frac{T + t_2}{T + t_1}\right). \qquad (5.4)$$

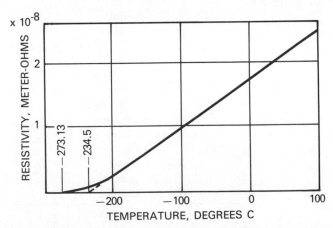

Fig. 5.3 Variation of resistivity of copper with temperature. From *Electrical Engineering Circuits* by H. H. Skilling. © 1965 by John Wiley & Sons. Reproduced by permission.

Table 5.2   Resistivity of Metals and Temperature Coefficients of Resistivity

Metal	Resistivity at 20°C (ohm meters)	Conductivity (% of Standard Annealed Copper)	Temperature Coefficient of Resistivity per Degree Celsius at 20°C	Reference Temperature (T) in Equations (5.2) and (5.4)
Copper, standard annealed	1.7241×10^{-8}	100	0.00393	234.5
Copper, soft-drawn	1.76×10^{-8}	98	0.00385	240
Copper, medium hard-drawn	$1.76–1.78 \times 10^{-8}$	97–98	0.00383 av	241
Copper, hard-drawn	1.77×10^{-8}	97.3	0.00382	242
Aluminum, 99.97% pure	2.66×10^{-8}	64.6	0.00427	214
Aluminum, average commercial hard-drawn	2.83×10^{-8}	61.0	0.00403	228
Brass, typical	$7–8 \times 10^{-8}$	21–25	0.002	480
Gold	2.44×10^{-8}	70.8	0.0034	274
Iron, pure	10×10^{-8}	17	0.005	180
Platinum	10×10^{-8}	17	0.003	313
Silver	1.63×10^{-8}	106	0.0038	243
Steel, carbon, medium to hard	$20–50 \times 10^{-8}$	3–9	0.002–0.005	480–180
Tin	11.5×10^{-8}	15	0.0042	218
Tungsten	5.51×10^{-8}	31	0.0045	202
Zinc	5.8×10^{-8}	30	0.0037	250

Source: Electrical Engineering Circuits, 2nd edition, H. H. Skilling, © 1965 by John Wiley & Sons. Reproduced by permission.

Most resistance–temperature calculations in practice are for copper, for which $T = 234.5$, a figure easily memorized, so Equation (5.4) is more frequently used.

American Wire Gauge (B & S) dimensional and resistance data for standard annealed copper and for hard-drawn aluminum wire are given in Tables 5.3 and 5.4.

5.2.3   Inductance and Ferromagnetic Circuits

A varying current flowing in a conductor creates a magnetic field around the conductor, which in turn creates an electromotive force (voltage) in the conductor and also in any other conductor close enough to be in the magnetic field. **Self-inductance** is the property of the conductor that produces an induced voltage in itself with changing current. The term "inductance" alone means self-inductance. When a varying current in one conductor induces a voltage in a neighboring conductor the effect is called **mutual inductance.**

Table 5.3 Wire Table, Standard Annealed Copper, American Wire Gauge (B & S), English Units

Gauge No. (AWG)	Diameter in Mils at 20°C	Cross Section at 20°C		Ohms per 1000 ft[a] at 20°C (= 68°F)	Pounds per 1000 ft	Feet per Pound	Feet per Ohm[b] at 20°C (= 68°F)	Ohms per Pound at 20°C (= 68°F)	Pounds per Ohm at 20°C (= 68°F)
		Circular Mils	Square Inches						
0000	460.0	211,600.	0.1662	0.049 01	640.5	1.561	20,400.	0.000 076 52	13,070.
000	409.6	167,800.	0.1318	0.061 80	507.9	1.968	16,180.	0.000 1217	8,219
00	364.8	133,100.	0.1045	0.077 93	402.8	2.482	12,830.	0.000 1935	5,169.
0	324.9	105,500.	0.082 89	0.098 27	319.5	3.130	10,180.	0.000 3076	3,251.
1	289.3	83,690.	0.065 73	0.1239	253.3	3.947	8,070.	0.000 4891	2,044.
2	257.6	66,370.	0.052 13	0.1563	200.9	4.977	6,400.	0.000 7778	1,286.
3	229.4	52,640.	0.041 34	0.1970	159.3	6.276	5,075.	0.001 237	808.6
4	204.3	41,740.	0.032 78	0.2485	126.4	7.914	4,025.	0.001 966	508.5
5	181.9	33,100.	0.026 00	0.3133	100.2	9.980	3,192.	0.003 127	319.8
6	162.0	26,250.	0.020 62	0.3951	79.46	12.58	2,531.	0.004 972	201.1
7	144.3	20,820.	0.016 35	0.4982	63.02	15.87	2,007.	0.007 905	126.5
8	128.5	16,510.	0.012 97	0.6282	49.98	20.01	1,592.	0.012 57	79.55
9	114.4	13,090.	0.010 28	0.7921	39.63	25.23	1,262.	0.019 99	50.03
10	101.9	10,380.	0.008 155	0.9989	31.43	31.82	1,001.	0.031 78	31.47
11	90.74	8,234.	0.006 467	1.260	24.92	40.12	794.0	0.050 53	19.79
12	80.81	6,530.	0.005 129	1.588	19.77	50.59	629.6	0.080 35	12.45
13	71.96	5,178.	0.004 067	2.003	15.68	63.80	499.3	0.1278	7.827
14	64.08	4,107.	0.003 225	2.525	12.43	80.44	396.0	0.2032	4.922
15	57.07	3,257.	0.002 558	3.184	9.858	101.4	314.0	0.3230	3.096
16	50.82	2,583.	0.002 028	4.016	7.818	127.9	249.0	0.5136	1.947
17	45.26	2,048.	0.001 609	5.064	6.200	161.3	197.5	0.8167	1.224

18	40.30	1,624.	0.001 276	6.385	4.917	203.4	156.6	1.299	0.7700
19	35.89	1,288.	0.001 012	8.051	3.899	256.5	124.2	2.065	0.4843
20	31.96	1,022.	0.000 802 3	10.15	3.092	323.4	98.50	3.283	0.3046
21	28.46	810.1	0.000 636 3	12.80	2.452	407.8	78.11	5.221	0.1915
22	25.35	642.4	0.000 504 6	16.14	1.945	514.2	61.95	8.301	0.1205
23	22.57	509.5	0.000 400 2	20.36	1.542	648.4	49.13	13.20	0.075 76
24	20.10	404.0	0.000 317 3	25.67	1.223	817.7	38.96	20.99	0.047 65
25	17.90	320.4	0.000 251 7	32.37	0.9699	1,031.	30.90	33.37	0.029 97
26	15.94	254.1	0.000 199 6	40.81	0.7692	1,300.	24.50	53.06	0.018 85
27	14.20	201.5	0.000 158 3	51.47	0.6100	1,639.	19.43	84.37	0.011 85
28	12.64	159.8	0.000 125 5	64.90	0.4837	2,067.	15.41	134.2	0.007 454
29	11.26	126.7	0.000 099 53	81.83	0.3836	2,607.	12.22	213.3	0.004 688
30	10.03	100.5	0.000 078 94	103.2	0.3042	3,287.	9.691	339.2	0.002 948
31	8.928	79.70	0.000 062 60	130.1	0.2413	4,145.	7.685	539.3	0.001 854
32	7.950	63.21	0.000 049 64	164.1	0.1913	5,227.	6.095	857.6	0.001 166
33	7.080	50.13	0.000 039 37	206.9	0.1517	6,591.	4.833	1,364.	0.000 7333
34	6.305	39.75	0.000 031 22	260.9	0.1203	8,310.	3.833	2,168.	0.000 4612
35	5.615	31.52	0.000 024 76	329.0	0.095 42	10,480.	3.040	3,448.	0.000 2901
36	5.000	25.00	0.000 019 64	414.8	0.075 68	13,210.	2.411	5,482.	0.000 1824
37	4.453	19.83	0.000 015 57	523.1	0.060 01	16,660.	1.912	8,717.	0.000 1147
38	3.965	15.72	0.000 012 35	659.6	0.047 59	21,010.	1.516	13,860.	0.000 072 15
39	3.531	12.47	0.000 009 793	831.8	0.037 74	26,500.	1.202	22,040.	0.000 045 38
40	3.145	9.888	0.000 007 766	1,049.	0.029 93	33,410.	0.9534	35,040.	0.000 028 54

Source: *Handbook of Engineering Fundamentals*, 3rd edition, O. W. Eshbach and M. Souders, © 1975 by John Wiley & Sons. Reproduced with permission.

[a]Resistance at the stated temperatures of a wire whose length is 1000 ft at 20°C.

[b]Length at 20°C of a wire whose resistance is 1 ohm at the stated temperatures.

m-g donna 11/4/83

Table 5.4 Wire Table, Aluminum
Hard-drawn Aluminum Wire at 20°C (68°F), American Wire Gauge (B & S), English Units

| Gauge No. | Diameter, Mils | Cross Section | | Ohms per 1000 ft | Pounds per 1000 ft | Pounds per Ohm | Feet per Ohm |
		Circular Mils	Square Inches				
0000	460.	212,000.	0.166	0.0804	195.	2,420.	12,400.
000	410.	168,000.	0.132	0.101	154.	1,520.	9,860.
00	365.	133,000.	0.105	0.128	122.	957.	7,820.
0	325.	106,000.	0.0829	0.161	97.0	602.	6,200.
1	289.	83,700.	0.0657	0.203	76.9	379.	4,920.
2	258.	66,400.	0.0521	0.256	61.0	238.	3,900.
3	229.	52,600.	0.0413	0.323	48.4	150.	3,090.
4	204.	41,700.	0.0328	0.408	38.4	94.2	2,450.
5	182.	33,100.	0.0260	0.514	30.4	59.2	1,950.
6	162.	26,300.	0.0206	0.648	24.1	37.2	1,540.
7	144.	20,800.	0.0164	0.817	19.1	23.4	1,220.
8	128.	16,500.	0.0130	1.03	15.2	14.7	970.
9	114.	13,100.	0.0103	1.30	12.0	9.26	770.
10	102.	10,400.	0.008 15	1.64	9.55	5.83	610.
11	91.	8,230.	0.006 47	2.07	7.57	3.66	484.
12	81.	6,530.	0.005 13	2.61	6.00	2.30	384.
13	72.	5,180.	0.004 07	3.29	4.76	1.45	304.
14	64.	4,110.	0.003 23	4.14	3.78	0.911	241.
15	57.	3,260.	0.002 56	5.22	2.99	0.573	191.
16	51.	2,580.	0.002 03	6.59	2.37	0.360	152.

17	45.	2,050.	0.001 61	8.31	1.88	0.227	120.
18	40.	1,620.	0.001 28	10.5	1.49	0.143	95.5
19	36.	1,290.	0.001 01	13.2	1.18	0.0897	75.7
20	32.	1,020.	0.000 802	16.7	0.939	0.0564	60.0
21	28.5	810.	0.000 636	21.0	0.745	0.0355	47.6
22	25.3	642.	0.000 505	26.5	0.591	0.0223	37.8
23	22.6	509.	0.000 400	33.4	0.468	0.0140	29.9
24	20.1	404.	0.000 317	42.1	0.371	0.008 82	23.7
25	17.9	320.	0.000 252	53.1	0.295	0.005 55	18.8
26	15.9	254.	0.000 200	67.0	0.234	0.003 49	14.9
27	14.2	202.	0.000 158	84.4	0.185	0.002 19	11.8
28	12.6	160.	0.000 126	106.	0.147	0.001 38	9.39
29	11.3	127.	0.000 099 5	134.	0.117	0.000 868	7.45
30	10.0	101.	0.000 078 9	169.	0.0924	0.000 546	5.91
31	8.9	79.7	0.000 062 6	213.	0.0733	0.000 343	4.68
32	8.0	63.2	0.000 049 6	269.	0.0581	0.000 216	3.72
33	7.1	50.1	0.000 039 4	339.	0.0461	0.000 136	2.95
34	6.3	39.8	0.000 031 2	428.	0.0365	0.000 085 4	2.34
35	5.6	31.5	0.000 024 8	540.	0.0290	0.000 053 7	1.85
36	5.0	25.0	0.000 019 6	681.	0.0230	0.000 033 8	1.47
37	4.5	19.8	0.000 015 6	858.	0.0182	0.000 021 2	1.17
38	4.0	15.7	0.000 012 3	1,080.	0.0145	0.000 013 4	0.924
39	3.5	12.5	0.000 009 79	1,360.	0.0115	0.000 008 40	0.733
40	3.1	9.9	0.000 007 77	1,720.	0.0091	0.000 005 28	0.581

Source: Handbook of Engineering Fundamentals, 3rd edition, O. W. Eshbach and M. Souders, © 1975 by John Wiley & Sons. Reproduced with permission.

Amount of inductance is a function of conductor configuration, dimensions, and magnetic characteristics of the medium surrounding it. Inductance is increased when a conductor is wound into the form of a coil. It is further increased by a large magnitude when the medium for the magnetic field inside the coil is magnetic material instead of free space or air.

A coil of N turns has an inductance of 1 henry if a current of 1 ampere in it produces a flux of 1 weber

$$L = \frac{N\Phi}{I} \text{ henrys.} \tag{5.5}$$

The **magnetic permeability** of free space is $4\pi \times 10^{-7}$. One weber is equal to 10^8 flux lines. Air-core inductors therefore have inductances very low compared with iron-core inductors. The latter, however, are not linear, so air-core inductors must be used when linearity is a requirement.

The following formulas give inductances for coils and other conductor configurations in air most frequently encountered in engineering practice.

Self-inductance (L) per centimeter axial length of the turns near the center of an air solenoid, A square centimeters in sectional area, wound uniformly with n turns per centimeter length (dimensions of sectional area negligible compared with the axial length).

$$L = 12.6n^2A \times 10^{-9} \text{ henrys.} \tag{5.6}$$

NOTE.   If the solenoid is filled completely with a medium of constant permeability μ, the self-inductance per centimeter length is $12.6n^2\mu A \times 10^{-9}$ henrys, and if filled partially throughout its length with a medium of constant permeability μ and B square centimeters in constant sectional area, the self-inductance per centimeter length is $12.6n^2(\mu B + A - B) \times 10^{-9}$ henrys.

Self-inductance (L) of a single-layer short solenoid of N turns, l centimeters in axial length, and r centimeters in radius (l small compared with r).

$$L = 12.6rN^2 \left[\ln \frac{8r}{l} - \frac{1}{2} + \frac{l^2}{32r^2} \left(\ln \frac{8r}{l} + \frac{1}{4} \right) \right] \times 10^{-9} \text{ henrys.} \tag{5.7}$$

Self-inductance (L) of a multiple-layer short solenoid of N turns, l centimeters in axial length, R centimeters in external radius, and r centimeters in internal radius (l small compared with R or r).

$$L = 12.6aN^2 \left[\ln \frac{8a}{b} \left(1 + \frac{3b^2}{16a^2} \right) - \left(2 + \frac{b^2}{16a^2} \right) \right] \times 10^{-9} \text{ henrys.} \tag{5.8}$$

NOTE. $a = (R + r)/2$ and $b = 0.2235(l + R - r)$. ln equals $\log_e$.

Self-inductance (L) of a toroidal coil wound uniformly with a single layer of N turns on a surface generated by the revolution of a circle r centimeters in radius about an axis R centimeters from the center of the circle.

$$L = 12.6N^2(R - \sqrt{R^2 - r^2}) \times 10^{-9} \text{ henrys.} \tag{5.9}$$

Self-inductance (L) of a toroidal coil of rectangular section, r and R centimeters in internal and external radius, respectively, sides of section $(R - r)$ centimeters and l centimeters, respectively, and wound uniformly with a single layer of N turns.

$$L = 2N^2l \ln \frac{R}{r} \times 10^{-9} \text{ henrys.} \tag{5.10}$$

Self-inductance (L) per centimeter length of one of two parallel straight cylindrical wires, each r centimeters in radius, their axes d centimeters apart, and conducting the same current in opposite directions (distance d small compared with the length of the wires).

$$L = \left(2 \ln \frac{d}{r} + 0.5\right) \times 10^{-9} \text{ henrys.} \tag{5.11}$$

NOTE. The self-inductance of each wire per mile is $0.08047 + 0.7411 \log (d/r)$ millihenrys, and for two wires is twice as great. Equation (5.11) also gives the self-inductance per centimeter length of one of three wires, located at the vertices of an equilateral triangle (d is the distance between the axes of any two wires, provided the algebraic sum of the instantaneous currents conducted respectively by the three wires in the same direction equals zero).

Self-inductance (L) per mile length of one of three unsymmetrically spaced but completely transposed wires, each r inches in radius, with axial spacings of d_{12}, d_{23}, and d_{13} inches, and with the algebraic sum of the instantaneous currents conducted respectively by the three wires in the same direction equal to zero.

$$L = 0.08047 + 0.7411 \log \frac{\sqrt[3]{d_{12}d_{23}d_{13}}}{r} \text{ millihenrys.} \tag{5.12}$$

NOTE. A completely transposed three-wire circuit is one in which each wire occupies each position for one-third of the distance.

Self-inductance (L) per centimeter length of two straight cylindrical concentric wires of equal section conducting the same current in opposite directions, the inner radius of the outer conductor being b centimeters and the radius of the solid inner conductor being c centimeters.

$$L = \left(2 \ln \frac{b}{c} + \frac{1}{2} + \frac{c^2}{3b^2} - \frac{c^4}{12b^4} + \frac{c^6}{30b^6} - \dots\right) \times 10^{-9} \text{ henrys.} \tag{5.13}$$

Self-inductance (L) of a single circular turn of wire of circular section, the mean radius of the turn being R centimeters and the radius of the section r centimeters.

$$L = 12.6R \left[\left(1 + \frac{r^2}{8R^2} \right) \ln \frac{8R}{r} + \frac{r^2}{24R^2} - 1.75 \right] \times 10^{-9} \text{ henrys.} \quad (5.14)$$

Mutual inductance (M) of two coils in which a current of I_1 amperes in one establishes a flux of Φ_2 webers through the N_2 turns of the other.

$$M = \frac{N_2\Phi_2}{I_1} \text{ henrys.} \quad (5.15)$$

Mutual inductance (M) of two parallel circular coaxial turns, each r centimeters in radius and their planes d centimeters apart (d small compared with r).

$$M = 12.6r \left[\ln \frac{8r}{d} \left(1 + \frac{3d^2}{16r^2} \right) - \left(2 + \frac{d^2}{16r^2} \right) \right] \times 10^{-9} \text{ henrys.} \quad (5.16)$$

Mutual inductance (M) of two concentric solenoids, the exterior of N_1 turns and length l centimeters and the interior of N_2 turns and sectional area A_2 square centimeters (the axial length of the interior solenoid small compared with the axial length of the exterior solenoid).

$$M = \frac{12.6N_1N_2A_2}{l} \times 10^{-9} \text{ henrys.} \quad (5.17)$$

Self-inductance (L) of two series connections of self-inductance L_1 and L_2 henrys, respectively, and **mutual inductance** M_{12} henrys.

$$L = L_1 + L_2 \pm 2M_{12} \text{ henrys.} \quad (5.18)$$

NOTE.   The sign is $+$ when the mutual fluxes are in conjunction and is $-$ when the mutual fluxes are in opposition. The mutual inductance (M) of two series connections of self-inductance L_1 and L_2 henrys, respectively, with K percent coupling is given by $M = K(L_1L_2)^{1/2}$ henrys. The coefficient of coupling K is expressed as a decimal fraction representing the ratio of the flux caused by one coil to that which links the other.

Self-inductance (L) of several coils of self-inductances L_1, L_2, L_3, wound on the same core with 100% coupling in conjunction ($+$ sign) or opposition ($-$ sign).

$$L = (\sqrt{L_1} \pm \sqrt{L_2} \pm \sqrt{L_3} \pm \cdots)^2 \text{ henrys.} \quad (5.19)$$

Relative to free-space permeability, ferromagnetic substances, including iron, steel, nickel, cobalt, and magnetic alloys, have very high permeability. For example, the relative permeability of transformer-quality steel may exceed 2000. This means a given coil with a closed-loop magnetic core of steel instead of an air core will have an inductance approximately 2000 times that given by the air-core formula. However, relative permeabilities of ferromagnetic substances vary greatly with flux density and may also be affected by heat treatment, previous magnetic history, and impurities introduced to reduce losses.

Magnetization curves of commercially used forms of iron and steel for magnetic circuits are shown in Fig. 5.4. Relative permeabilities of cast steel, cast iron, transformer hot-rolled steel, and some specially developed high-permeability alloys as a function of flux density are shown in Fig. 5.5.

Toroidal cores with ferromagnetic properties are manufactured of powdered iron and ferrite materials with a range of relative permeabilities from 1 to 8000. Although designed and used primarily for high-frequency and very-high-frequency applications, toroidal cores are also being used in power transformers at 50 and 60 Hz.

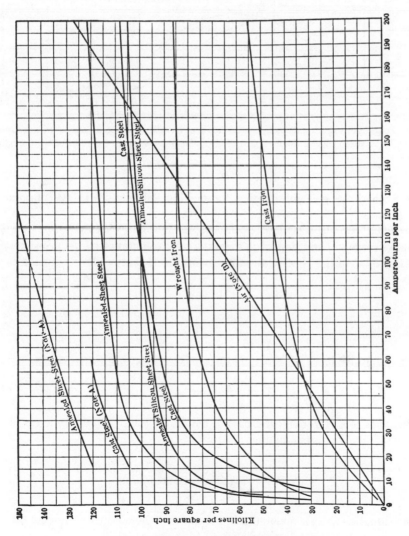

NOTE A.   Multiply abscissa scale by 10.
NOTE B.   Multiply abscissa scale by 200.

Fig. 5.4   Magnetization curves of iron and steel.

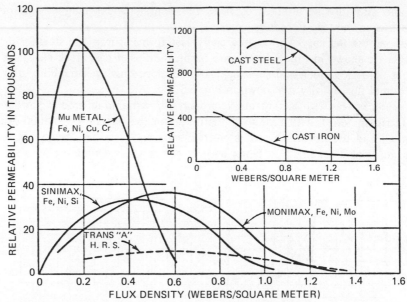

Fig. 5.5 Permeability of ferromagnetic materials. From *Handbook of Engineering Fundamentals*, 3rd ed., by O. W. Eshbach and M. Souders. © 1975 by John Wiley & Sons. Reproduced with permission.

Tables 5.5 and 5.6 give information for making inductances using commercially available cores.

In a closed loop ferromagnetic circuit magnetic flux (Φ) is established by a **magnetomotive force** ($\mathcal{F}$) and limited in magnitude by **reluctance** ($\mathcal{R}$), a property of the ferromagnetic material. The relationship is similar to that of Ohm's law for electric circuits

$$\Phi = \frac{\mathcal{F}}{\mathcal{R}}. \tag{5.20}$$

Magnetomotive force ($\mathcal{F}$) due to N turns of conductor each conducting I amperes

$$\mathcal{F} = NI \text{ ampere turns.} \tag{5.21}$$

Reluctance of a **ferromagnetic path** L meters long and A square meters in cross section with relative permeability

$$\mathcal{R} = \frac{L}{\mu A} \text{ reciprocal henrys.} \tag{5.22}$$

From Equation (5.20), reluctance is also equal to the number of ampere turns per weber.

$$\mathcal{R} = \frac{\mathcal{F}}{\Phi}. \tag{5.23}$$

Table 5.5 Powdered Iron Toroidal Core Data

Powdered-Iron Toroidal Cores—A_L Values ($\mu H/100$ Turns)[a]

Core Size	41-Mix Green $\mu = 75$	3-Mix Gray $\mu = 35$ 0.05–0.5 MHz	15-Mix Rd & Wh $\mu = 25$ 0.1–2 MHz	1-Mix Blue $\mu = 20$ 0.5–5 MHz	2-Mix Red $\mu = 10$ 1–30 MHz	6-Mix Yellow $\mu = 8$ 10–90 MHz	10-Mix Black $\mu = 6$ 60–150 MHz	12-Mix Gn & Wh $\mu = 3$ 100–200 MHz	0-Mix Tan $\mu = 1$ 150–300 MHz
T-200	755	360	NA[b]	250[c]	120	100[c]	NA	NA	NA
T-184	1640	720	NA	500[c]	240	195	NA	NA	NA
T-157	970	420	360[c]	320[c]	140	115	NA	NA	NA
T-130	785	330	250[c]	200	110	96	NA	NA	15.0
T-106	900	405	345[c]	325[c]	135	116	NA	NA	19.0[c]
T-94	590	248	200[c]	160	84	70	58	32	10.6
T-80	450	180	170	115	55	45	32[c]	22	8.5
T-68	420	195	180	115	57	47	32	21	7.5
T-50	320	175	135	100	49[c]	40	31	18	6.4
T-44	229	180	160	105	52[c]	42	33	NA	6.5
T-37	308	120[c]	90	80	40[c]	30	25	15	4.9
T-30	375	140[c]	93	85	43	36	25	16	6.0
T-25	225	100	85	70	34	27	19	13	4.5
T-20	175	90	65	52	27	22	16	10	3.5
T-16	130	61	NA	44	22	19	13	8	3.0
T-12	112	60	50[c]	48	20[c]	17[c]	7.5	3.0	

Table 5.5 (Continued)
Number of Turns versus Wire Size and Core Size
(Approximate Maximum of Turns—Single-Layer, Wound Enameled Wire)

Wire Size	T-200	T-130	T-106	T-94	T-80	T-68	T-50	T-37	T-25	T-12
10	33	20	12	12	10	6	4	1		
12	43	25	16	16	14	9	6	3		
14	54	32	21	21	18	13	8	5	1	
16	69	41	28	28	24	17	13	7	2	
18	88	53	37	37	32	23	18	10	4	1
20	111	67	47	47	41	29	23	14	6	1
22	140	86	60	60	53	38	30	19	9	2
24	177	109	77	77	67	49	39	25	13	4
26	223	137	97	97	85	63	50	33	17	7
28	281	173	123	123	108	80	64	42	23	9
30	355	217	154	154	136	101	81	54	29	13
32	439	272	194	194	171	127	103	68	38	17
34	557	346	247	247	218	162	132	88	49	23
36	683	424	304	304	268	199	162	108	62	30
38	875	544	389	389	344	256	209	140	80	39
40	1103	687	492	492	434	324	264	178	102	51

Physical Dimensions[e]

Core Size	Outer Dia. (in.)[d]	Inner Dia. (in.)	Height (in.)	Cross-Sect. Area (cm²)	Mean Length (cm)	Core Size	Outer Dia. (in.)	Inner Dia. (in.)	Height (in.)	Cross-Sect. Area (cm²)	Mean Length (cm)
T-200	2.000	1.250	0.550	1.330	12.97	T-50	0.500	0.303	0.190	0.121	3.20
T-184	1.840	0.950	0.710	2.040	11.12	T-44	0.440	0.229	0.159	0.107	2.67
T-157	1.570	0.950	0.570	1.140	10.05	T-37	0.375	0.205	0.128	0.070	2.32
T-130	1.300	0.780	0.437	0.733	8.29	T-30	0.307	0.151	0.128	0.065	1.83
T-106	1.060	0.560	0.437	0.706	6.47	T-25	0.255	0.120	0.096	0.042	1.50
T-94	0.942	0.560	0.312	0.385	6.00	T-20	0.200	0.088	0.067	0.034	1.15
T-80	0.795	0.495	0.250	0.242	5.15	T-16	0.160	0.078	0.060	0.016	0.75
T-68	0.690	0.370	0.190	0.196	4.24	T-12	0.125	0.062	0.050	0.010	0.74

Source: *The Radio Amateur's Handbook*, 60th edition, American Radio Relay League, Newington, Conn., 1983.

[a] Turns = 100 $(L_{\mu H}/A_L$ value$)^{1/2}$. All frequency figures optimum.

[b] NA = not available in that size.

[c] Updated values (1979) from Micrometals, Inc.

[d] Courtesy of Amidon Assoc., North Hollywood, Calif., and Micrometals, Inc.

[e] Inches × 25.4 = mm.

Table 5.6   Ferrite Toroidal Coil Data

Ferrite Toroids—A_L Values (mH/1000 Turns)[a] Enameled Wire

Core Size	63-Mix $\mu = 40$	61-Mix $\mu = 125$	43-Mix $\mu = 950$	72-Mix $\mu = 2000$	75-Mix $\mu = 5000$
FT-23	7.9	24.8	189.0	396.0	990.0
FT-37	17.7	55.3	420.0	884.0	2210.0
FT-50	22.0	68.0	523.0	1100.0	2750.0
FT-82	23.4	73.3	557.0	1172.0	2930.0
FT-114	25.4	79.3	603.0	1268.0	3170.0

Number turns = 1000 $\sqrt{\text{desired } L(\text{mH}) \div A_L \text{ value (above)}}$

Ferrite Magnetic Properties

Property	Unit	63-Mix	61-Mix	43-Mix	72-Mix	75-Mix
Initial perm. (μ_1)		40	125	950	2000	5000
Maximum perm.		125	450	3000	3500	8000
Saturation flux density at 13 oer	Gauss	1850	2350	2750	3500	3900
Residual flux density	Gauss	750	1200	1200	1500	1250
Curie temp.	°C	500	300	130	150	160
Vol. resistivity	ohm/cm	1×10^8	1×10^8	1×10^5	1×10^2	5×10^2
Opt. freq. range	MHz	15–25	0.2–10	0.01–1	0.001–1	0.001–1
Specific gravity		4.7	4.7	4.5	4.8	4.8
Loss factor		9.0×10^{-5} @ 25 MHz	2.2×10^{-5} @ 2.5 MHz	2.5×10^{-5} @ 0.2 MHz	9.0×10^{-6} @ 0.1 MHz	5.0×10^{-6} @ 0.1 MHz
$\dfrac{1}{uO}$						
Coercive force	Oer.	2.40	1.60	0.30	0.18	0.18
Temp. coeff. of initial perm.	%/°C 20–70°C	0.10	0.10	0.20	0.60	

Ferrite Toroids—Physical Properties[b]

Core Size	OD	ID	Hgt	A_e	l_e	V_e	A_s	A_w
FT-23	0.230	0.120	0.060	0.00330	0.529	0.00174	0.1264	0.01121
FT-37	0.375	0.187	0.125	0.01175	0.846	0.00994	0.3860	0.02750
FT-50	0.500	0.281	0.188	0.02060	1.190	0.02450	0.7300	0.06200
FT-82	0.825	0.520	0.250	0.03810	2.070	0.07890	1.7000	0.21200
FT-114	1.142	0.748	0.295	0.05810	2.920	0.16950	2.9200	0.43900

OD = Outer diameter (inches)[c] A_e = Effective magnetic cross-sectional area (inches²)
ID = Inner diameter (inches) l_e = Effective magnetic path length (inches)
Hgt = Height (inches) V_e = Effective magnetic volume (inches³)
A_w = Total window area (inches²) A_s = Surface area exposed for cooling (inches²)

Source: *The Radio Amateur's Handbook*, 60th edition, American Radio Relay League, Newington, Conn., 1983.

[a]Number of turns = 1000 $(L_{(\text{mH})}/A_L \text{ value})^{1/2}$.
[b]Courtesy of Amidon Assoc., North Hollywood, Calif.
[c]Inches × 25.4 = mm.

Permeance ($\mathcal{P}$) of the magnetic circuit is the reciprocal of reluctance

$$\mathcal{P} = \frac{\mu A}{L} = \frac{\Phi}{\mathcal{F}} \text{ henrys.} \tag{5.24}$$

Because μ varies with flux density in ferromagnetic circuits, data on ferromagnetic materials are always given in terms of flux density (B) as a function of magnetic field strength (also called magnetizing force) (H). From Equations (5.22) and (5.23), $B = \mu H$ teslas.

Much current technical literature and data on ferromagnetic materials uses CGS metric units or English units. Conversion factors to SI units for magnetic flux (Φ), magnetomotive force ($\mathcal{F}$), magnetic flux density (B), and magnetic field strength (H) are

To Convert from	To	Multiply by
(Φ) line (English)	weber	10^{-8}
(Φ) maxwell (CGS)	weber	10^{-8}
($\mathcal{F}$) ampere-turn (English)	ampere-turn	1.0
($\mathcal{F}$) gilbert (CGS)	ampere-turn	0.796
(B) line per square inch (English)	tesla	1.55×10^{-5}
(B) gauss (CGS)	tesla	10^{-4}
(H) ampere-turn per inch (English)	amp.-turn per meter	39.4
(H) oersted (CGS)	amp.-turn per meter	79.6

The **energy of a magnetic field** (W) established by a circuit with inductance L henrys conducting a current I amperes

$$W = \tfrac{1}{2}LI^2 \text{ joules.} \tag{5.25}$$

When flux is caused by changing current i, an increase of i converts electric to magnetic energy and a decrease of i converts magnetic to electric energy

$$W = N \int i\,d\Phi \times 10^{-8} \text{ joules.} \tag{5.26}$$

5.2.4 Capacitance

Two conductors across which a voltage is applied, separated by an insulating medium, form a capacitor. Quantitatively the capacitance is the amount of charge per volt which accumulates at the boundaries between conductors and insulating medium. With charge (Q) in coulombs and V in volts

$$C = \frac{Q}{V} \text{ farads.} \tag{5.27}$$

The **dielectric constant** (k) of the insulating medium in a capacitor is the ratio of electric flux density D to electric field strength E, relative to the ratio for an insulating medium of free space equal to one. Table 5.7 lists relative dielectric constants of insulating mediums (dielectrics) used in capacitors. Table 5.8 lists the resistivity and temperature coefficient of resistance of conductors in addition to those in Table 5.2. Table 5.9 lists the resistivity and dielectric constant of insulating materials in addition to those in Table 5.7.

Table 5.7　Relative Dielectric Constant (Permittivity)[a]

Air	1.0006	Glass, crown (window)*	6
Hydrogen	1.0003	Mica, good quality*	7
Water, pure	78	Porcelain*	6.5
Alcohol, ethyl	25.7	Rubber, vulcanized*	3
Oil, petroleum*	2.1	Teflon (all frequencies)	2.1
Paraffin wax*	2.25	Polystyrene (all freq.)	2.55
Paper, dry*	3.5	Polyethylene (all freq.)	2.26
Paper, oiled*	3.5	Titanium oxides and titanates*	10 to 10,000

Source: Electrical Engineering Circuits, 2nd edition, H. H. Skilling, © 1965 by John Wiley & Sons. Reproduced with permission.

[a]$K = \epsilon/\epsilon_0$; values given for 20°C and atmospheric pressure; frequency less than 1 megahertz unless otherwise indicated. Substances marked * are quite variable, and K may differ between samples by 10 to 20% or even more.

Equations for calculating common capacitor configurations and lines most used in practice follow.

Capacitance (C) of a parallel plate capacitor in which the positive and negative charges are each distributed uniformly over a surface area of A square centimeters, the uniform distance between the oppositely charged surfaces is d centimeters, and the medium between the oppositely charged surfaces is of dielectric constant k (d is assumed to be small compared with all other dimensions).

$$C = \frac{kA}{36\pi d \times 10^5} \text{ microfarads.} \tag{5.28}$$

Table 5.8　Resistivity (ρ) and Temperature Coefficient of Resistance (α) of Certain Conductors[a]

Material	ρ ($\mu\Omega/cm^3$)	α	Material	ρ ($\mu\Omega/cm^3$)	α
Aluminum	2.688	0.00403	Mercury	95.8	0.00089
Antimony	39.1 (0°C)	0.0036	Molybdenum	5.08 (0°C)	0.0047 (0–100°C)
Barium	9.8	0.0033	Monel metal	42	—
Beryllium	10.1	—	Nickel	7.8	0.00537 (20–100°C)
Bismuth	120	0.004	Osmium	9.5	0.0033
Carbon	3500 (0°C)	−0.0009	Palladium	11	—
Calcium	4.59	0.00364 (0–600°C)	Platinum	9.83 (0°C)	0.003
Cerium	78	—	Potassium	6.1 (0°C)	0.0055 (0°C)
Cesium	19 (0°C)	—	Rhodium	5.11 (0°C)	0.0043 (0°C)
Chromium	2.6 (0°C)	—	Silver	1.629 (18°C)	0.0038
Cobalt	9.7	0.00658 (0–100°C)	Sodium	4.3 (0°C)	0.0054
Copper	1.724	0.00393	Strontium	24.8	—
Gold	2.44	0.0034	Tantalum	15.5	0.0031
Graphite	800 (0°C)	—	Tellurium	2×10^5	—
Iron	9.8	0.0065 (0–100°C)	Thallium	17.6 (0°C)	0.0040 (0°C)
Iron, cast	79–104	—	Thorium	18	0.0021 (20–1800°C)
Lead	22.0	0.0039	Tin	11.5	0.0042
Lithium	8.55 (0°C)	0.0047 (0°C)	Titanium	3.0	—
Magnesium	4.46	0.0040	Tungsten	5.5	0.0047 (0–100°C)
Manganese	5	—	Zinc	5.75 (0°C)	0.0037

[a]Temperature is 20°C unless otherwise specified.

Table 5.9 Resistivity (ρ) and Dielectric Constant (k) of Certain Insulators at Room Temperature

Material	ρ (MΩ/cm^3)	k	Material	ρ (MΩ/cm^3)	k
Alcohol, ethyl	0.3	5.0–54.6	Oil, olive	5×10^6	3.11
Alcohol, methyl	0.14	31.2–35.0	Oil, paraffin	10^{10}	—
Amber	5×10^{10}	—	Oil, petroleum	2×10^{10}	2.13
Amylacetate	—	4.81	Paper	10^4–10^9	1.7–3.8
Asbestos paper	1.6×10^5	2.7	Paraffin	5×10^{10}–5×10^{12}	1.9–2.3
Asphalt	—	2.7	Porcelain	3×10^8	4.4
Bakelite	10^5–10^{10}	4.5–5.5	Quartz	10^8–5×10^{12}	4.7–5.1
Beeswax	6×10^8	—	Rosin	7×10^9–5×10^{10}	2.5
Cellophane	—	8	Rubber, hard	3×10^{10}–10^{12}	2.0–3.5
Celluloid	2×10^4	13.3	Sealing wax	10^9–8×10^9	—
Cellulose acetate	—	5	Selenium	0.06	6.1–7.4
Glass	5×10^5–10^{10}	5.5–9.1	Shellac	10^{10}	3.0–3.7
Glycerine	—	56.2	Silica, fused	10^8–10^{13}	3.5–3.6
Gutta percha	3×10^4	2.9	Slate	10^2–10^4	6.6–7.4
Ice	720	86	Sulphur	8×10^9–10^{11}	2.9–3.2
Ivory	200	—	Turpentine	—	2.23
Marble	10^3–10^5	8.3	Water, dist.	0.5	81
Mica	4×10^7–2×10^{11}	5–7	Wood, paraffined	3×10^4–4×10^7	4.1

Capacitance (C) of two concentric spheres, the inner, r_1 centimeters in external radius, the outer, r_2 centimeters in internal radius, and separated by a medium of dielectric constant k.

$$C = \frac{r_1 r_2 k}{9(r_2 - r_1) \times 10^5} \text{ microfarads.} \qquad (5.29)$$

Capacitance (C) of two coaxial cylinders per centimeter axial length, the inner, r_1 centimeters in external radius, the outer, r_2 centimeters in internal radius, and separated by a medium of dielectric constant k (ln equals log$_e$).

$$C = \frac{k}{18 \ln \dfrac{r_2}{r_1} \times 10^5} \text{ microfarads.} \qquad (5.30)$$

NOTE.   The capacitance per mile is $0.03882k/\log (r_2/r_1)$ microfarads.

Capacitance (C) of two parallel cylinders per centimeter length, each cylinder r centimeters in radius, their centers separated by a distance of d centimeters, and immersed in a medium of dielectric constant k (r small compared with d and all dimensions small compared with distance to surrounding objects).

$$C = \frac{k}{36 \ln \dfrac{d}{r} \times 10^5} \text{ microfarads.} \qquad (5.31)$$

NOTE.   The capacitance per mile is $(1.941k \times 10^{-2})/\log (d/r)$ microfarads. The capacitance per conductor (to neutral) of a balanced three-phase transmission line with conductors located at the vertices of an equilateral triangle equals $(3.882k \times 10^{-2})/\log (d/r)$ micro-farads per mile.

Capacitance (C) to neutral per mile of one conductor of a balanced three-phase transmission line with unsymmetrical spacing but completely transposed, d_{12}, d_{23}, and d_{13} being the axial spacings in inches and r being the conductor radius in inches.

$$C = \frac{3.882 \times 10^{-2}}{\log \dfrac{\sqrt[3]{d_{12}d_{23}d_{13}}}{r}} \text{ microfarads.} \tag{5.32}$$

NOTE.   See note following Equation (5.12).

Total capacitance (C_0) of several parallel capacitors of capacitance C_1, C_2, and C_3 farads, respectively.

$$C_0 = C_1 + C_2 + C_3 \text{ farads.} \tag{5.33}$$

Total capacitance (C_0) of several series capacitors of capacitance C_1, C_2, and C_3 farads, respectively.

$$C_0 = \frac{1}{\dfrac{1}{C_1} + \dfrac{1}{C_2} + \dfrac{1}{C_3}} \text{ farads.} \tag{5.34}$$

Capacitors used in electric power applications are rated in vars (volt amperes reactive) or kilovars instead of farads or microfarads because their useful output is reactive volt amperes at the alternating current frequency of the power system. If C is capacitance in microfarads, V is system RMS volts and f is system frequency in hertz

$$\text{KVAR} = 2\pi f C V^2 \times 10^{-9}. \tag{5.35}$$

Energy (W) stored in a capacitor of C farads at potential V volts.

$$W = \tfrac{1}{2}CV^2 \text{ joules.} \tag{5.36}$$

5.3   DIRECT-CURRENT CIRCUITS

5.3.1   Ohm's Law

In a circuit where a source of constant voltage V and a load resistance R_L are connected by two conductors (Fig. 5.6), the current is proportional to voltage and inversely proportional to resistance. When any two of the three quantities are known the third can be found from the equations

$$I = V/R \text{ amperes, } R = V/I \text{ ohms, } V = IR \text{ volts.} \tag{5.37}$$

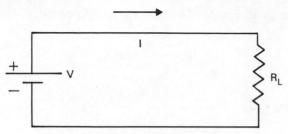

Fig. 5.6 Simplest series electric circuit.

A practical circuit always contains resistance in the source R_S and in the conductors R_C in addition to the load resistance R_L (Fig. 5.7). If R_S and R_C are significant in magnitude compared with R_L,

$$I = \frac{V}{R_L + R_S + R_C} \text{ amperes.} \tag{5.38}$$

5.3.2 Series and Parallel Connections

Figures 5.6 and 5.7 are the simplest possible form of a circuit with a single series path. If a series path has more than one voltage source, the total voltage in the path is their algebraic sum; if more than one resistance, the total resistance is their numerical sum.

Equivalent resistance R_p of a parallel circuit the respective branches of which have resistances R_1, R_2, R_3, and so on, and contain no voltage, each resistance in ohms.

$$\frac{1}{R_p} = \frac{1}{R_1} + \frac{1}{R_2} + \frac{1}{R_3}, \text{ etc., siemens.} \tag{5.39}$$

When there are only two branches, Equation (5.39) reduces to

$$R_p = \frac{R_1 R_2}{R_1 + R_2} \text{ ohms.} \tag{5.40}$$

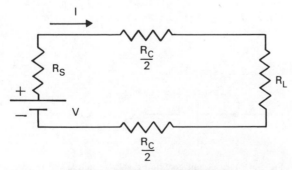

Fig. 5.7 Simplest practical circuit.

When there are n parallel paths of equal resistance R_1 ohms,

$$R_p = \frac{R_1}{n} \text{ ohms.} \tag{5.41}$$

5.3.3 Kirchhoff's Laws

A circuit containing three or more parallel paths is called a network. It is possible to determine the currents in each branch of simple networks by changing parallel paths to equivalent series paths, but a more systematic approach applicable to all networks is to set up equations based on Kirchhoff's two laws:

1. The sum of voltage sources (rises) and voltage drops around any closed series path of a network equals zero.
2. At any current junction in a network the total of individual currents leaving the junction is equal to the total of currents flowing toward the junction.

Figure 5.8 shows a part-circuit with voltage source V, resistance R, and current I_{AB} flowing in the direction A to B. In using Kirchhoff's first law it is necessary to consistently distinguish a voltage rise from a voltage drop by algebraic sign

$$V_{AB} = +V - I_{AB}R_{AB} \text{ volts.} \tag{5.42}$$

The sign of the voltage source or current is positive when acting in the direction shown in Fig. 5.8, and is negative when acting in the opposite direction. When V_{AB} is positive it is called a potential rise from A to B, and when V_{AB} is negative it is called a potential drop from A to B. If V is zero, $V_{AB} = -I_{AB}R_{AB}$ volts, and if either I_{AB} or R_{AB} is zero, $V_{AB} = +V$ volts.

Current I_{AB} in Fig. 5.8 flowing from A toward B

$$I_{AB} = \frac{+V - V_{AB}}{R_{AB}} \text{ amperes.} \tag{5.43}$$

The direction of current is determined by its sign, a positive sign indicating flow from A to B and a negative sign a flow from B to A. When V_{AB} = zero, $I_{AB} = +V/R_{AB}$, and when V = zero, $I_{AB}R_{AB} = -V_{AB}$.

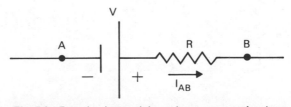

Fig. 5.8 Part-circuit containing voltage source and resistance.

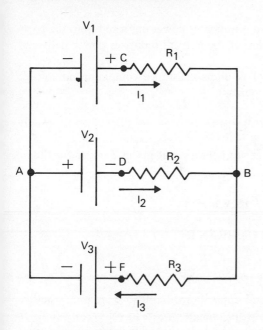

Fig. 5.9 Network with three branches.

Total current I_0 flowing toward a junction from which currents I_1, I_2, I_3, and so on, flow away, all currents being measured in amperes.

$$I_0 = I_1 + I_2 + I_3, \cdots \text{amperes.} \tag{5.44}$$

In the network of Fig. 5.9 the magnitudes of currents, voltage sources and resistances are indicated by the symbols I, V, and R, respectively, and directions of currents and voltage sources are indicated by arrows and $+/-$ signs, respectively, any unknown direction being assumed arbitrarily. The Kirchhoff's law equations are

$$\textbf{1.} +V_1 - I_1R_1 + I_2R_2 + V_2 = 0, \tag{5.45}$$
$$\textbf{2.} +V_1 - I_1R_1 - I_3R_3 - V_3 = 0, \tag{5.46}$$
$$\textbf{3.} \ I_1 + I_2 - I_3 = 0. \tag{5.47}$$

The magnitude and direction of each current may be determined by solving the simultaneous equations, a positive value of current indicating the same direction and a negative value indicating the opposite direction to that assumed in the figure.

5.3.4 Power and Energy in a Direct-Current Circuit

Power P delivered to or from a part-circuit conducting a current of I amperes and across which the voltage is V volts.

$$P = VI \text{ watts.} \tag{5.48}$$

A voltage rise in the direction of the current indicates power delivered from the part-circuit, and a voltage drop in the direction of the current indicates power delivered to the part-circuit.

Power P delivered to a part-circuit of R ohms resistance containing no voltage source and conducting a current of I amperes.

$$P = I^2R \text{ watts.} \tag{5.49}$$

Energy W delivered to or used by a circuit operating at power P watts for time t hours.

$$W = VIt = I^2Rt \text{ watt-hours.} \tag{5.50}$$

5.4   TRANSIENT VOLTAGES AND CURRENTS

5.4.1   Definition of Transient State

Because of the presence of inductance and capacitance in electric circuits, whether intended or not intended, whenever an abrupt change occurs in voltage, current, or impedance, transient conditions exist in one or more of these during the time interval between the initial change and the establishment of a steady-state condition. Figures 5.10 and 5.11 illustrate the transient waveforms that occur, for example, when a switch is closed, applying voltage to various combinations of resistance, capacitance, and inductance. Transients also occur when current flow in a circuit is interrupted by the opening of a switch. In the terminology of power systems, large transient increases in voltage or current during switching operations or accidental short circuits are called **surges.**

5.4.2   R-L Circuits

Current (i) flowing in a series circuit of R ohms resistance and L henrys self-inductance t seconds after a constant emf of E volts is impressed upon the circuit.*

$$i = \frac{E}{R}[1 - \epsilon^{-(Rt)/L}] + I\epsilon^{-(Rt)/L} \text{ amperes.} \tag{5.51}$$

NOTE. I is the current in amperes flowing in the circuit at the instant before the emf is impressed. It is a positive quantity if flowing in conjunction with and is a negative quantity if flowing in opposition to the emf.

Current (i) flowing in a series circuit of R ohms resistance and L henrys self-inductance t seconds after its source of emf is short-circuited, the current flowing in the circuit at the instant before the short-circuit being I amperes.

$$i = I\epsilon^{-(Rt)/L} \text{ amperes.} \tag{5.52}$$

*The value of ϵ throughout is 2.718.

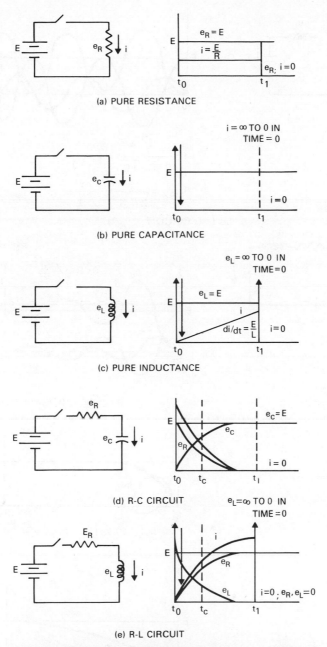

(a) PURE RESISTANCE

(b) PURE CAPACITANCE

(c) PURE INDUCTANCE

(d) R-C CIRCUIT

(e) R-L CIRCUIT

TRANSIENTS ON CIRCUIT CLOSE AT TIME T_0 AND OPEN AT TIME T_1 FOR R, C, L, R-C AND R-L CIRCUITS

Fig. 5.10 Transients on circuit close at time t_0 and open at time t_1, for R, C, L, R-C, and R-L circuits. Used with permission of Westinghouse Electric Corporation.

273

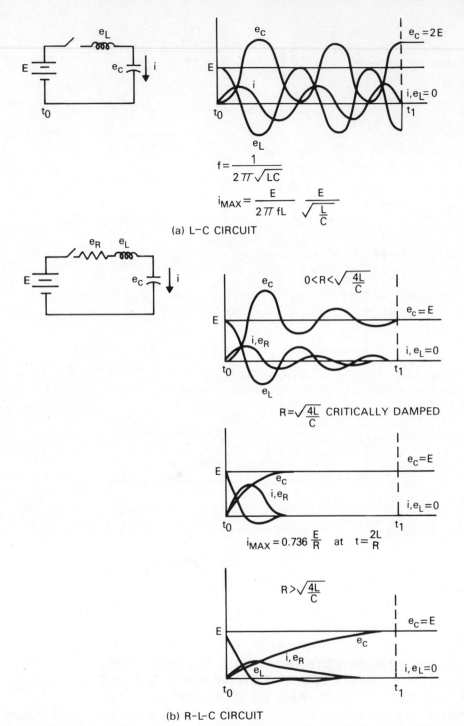

$$f = \frac{1}{2\pi\sqrt{LC}}$$

$$i_{MAX} = \frac{E}{2\pi fL} \qquad \frac{E}{\sqrt{\frac{L}{C}}}$$

(a) L–C CIRCUIT

$$0 < R < \sqrt{\frac{4L}{C}}$$

$$R = \sqrt{\frac{4L}{C}} \quad \text{CRITICALLY DAMPED}$$

$$i_{MAX} = 0.736\,\frac{E}{R} \quad \text{at} \quad t = \frac{2L}{R}$$

$$R > \sqrt{\frac{4L}{C}}$$

(b) R–L–C CIRCUIT

Fig. 5.11 Transients on circuit close at time t_0 and open at time t_1, for *L-C* and *R-L-C* circuits. Used with permission of Westinghouse Electric Corporation.

5.4.3 R-C Circuits

Current (i) flowing in a circuit of R ohms resistance and C farads series capacitance t seconds after a constant emf of E volts is impressed upon the circuit.

$$i = \left(\frac{E - V}{R}\right)\epsilon^{-t/(RC)} \text{ amperes.} \qquad (5.53)$$

NOTE. V is the potential across the capacitor at the instant before the emf is impressed. It is a positive quantity if acting in opposition to, and a negative quantity if acting in conjunction with, the impressed emf.

Potential (v) across the capacitor at any time t under the conditions stated in Equation (5.53).

$$v = E[1 - \epsilon^{-t/(RC)}] + V\epsilon^{-t/(RC)} \text{ volts.} \qquad (5.54)$$

Current (i) flowing in a circuit of R ohms resistance and C farads series capacitance t seconds after its source of emf is short-circuited, the potential across the condenser at the instant before the short-circuit being V volts.

$$i = \frac{V}{R}\epsilon^{-t/(RC)} \text{ amperes.} \qquad (5.55)$$

Potential (v) across the capacitor at any time t under the conditions stated in Equation (5.55).

$$v = V\epsilon^{-t/(RC)} \text{ volts.} \qquad (5.56)$$

5.4.4 R-L-C Circuits

Current (i) flowing in a circuit of R ohms resistance, L henrys self-inductance, and C farads series capacitance t seconds after a constant emf of E volts is impressed upon the circuit, the potential across the capacitor and the current flowing in the circuit at the instant before the emf is impressed being V volts and I amperes, respectively.

CASE I. $R^2C > 4L$.

$$i = \left[\frac{E - V - aLI}{(b - a)L}\right]\epsilon^{-at} - \left[\frac{E - V - bLI}{(b - a)L}\right]\epsilon^{-bt} \text{ amperes.} \qquad (5.57)$$

CASE II. $R^2C = 4L$.

$$i = \left\{I + \left[\frac{2(E - V) - RI}{2L}\right]t\right\}\epsilon^{-Rt/(2L)} \text{ amperes.} \qquad (5.58)$$

CASE III. $R^2C < 4L.$

$$i = \left\{ \left[\frac{2(E - V) - RI}{2\omega_1 L} \right] \sin \omega_1 t + I \cos \omega_1 t \right\} \epsilon^{-Rt/(2L)} \text{ amperes.} \quad (5.59)$$

NOTE. $a = \dfrac{RC - (R^2C^2 - 4LC)^{1/2}}{2LC}$, $b = \dfrac{RC + (R^2C^2 - 4LC)^{1/2}}{2LC}$, and $\omega1 = $
$\dfrac{(4LC - R^2C^2)^{1/2}}{2LC}$. The current (I) is positive when flowing in the same direction as the impressed emf and the sign of the potential (V) is obtained as in Equation (5.52).

Current (*i*) flowing in a circuit of R ohms resistance, L henrys self-inductance, and C farads series capacitance t seconds after its source of emf is short-circuited, the potential across the capacitor and the current flowing in the circuit at the instant before the short-circuit being V volts and I amperes, respectively.

NOTE.   Write Equation (5.57), (5.58), or (5.59), making E zero in each case.

Current (*i*) flowing in a circuit of R ohms resistance, L henrys self-inductance, and C farads series capacitance t seconds after a sinusoidal emf, $e = E_m \sin (\omega t + \alpha)$ volts, is impressed upon the circuit, the potential across the capacitor and the current flowing in the circuit at the instant before the emf is impressed being V volts and I amperes, respectively.

CASE I. $R^2C > 4L.$

$$i = G\epsilon^{-at} - H\epsilon^{-bt} + \frac{E_m}{Z} \sin (\omega t + \alpha - \theta) \text{ amperes.} \quad (5.60)$$

CASE II. $R^2C = 4L.$

$$i = (J + Kt)\epsilon^{-Rt/(2L)} + \frac{E_m}{Z} \sin (\omega t + \alpha - \theta) \text{ amperes.} \quad (5.61)$$

CASE III. $R^2C < 4L.$

$$i = (M \sin \omega_1 t + N \cos \omega_1 t)\epsilon^{-Rt/(2L)} + \frac{E_m}{Z} \sin (\omega t + \alpha - \theta) \text{ amperes.} \quad (5.62)$$

NOTE. $G = \dfrac{E_m \sin \alpha - V - aLI - \dfrac{E_m L}{Z}[b \sin (\alpha - \theta) + \omega \cos (\alpha - \theta)]}{(b - a)L}$.

$H = \dfrac{E_m \sin \alpha - V - bLI - \dfrac{E_m L}{Z}[a \sin (\alpha - \theta) + \omega \cos (\alpha - \theta)]}{(b - a)L}$.

$J = I - \dfrac{E_m}{Z} \sin (\alpha - \theta)$ ω_1, a, and b, as in Equation (5.59).

$$K = \frac{1}{L}\left\{ E_m \sin \alpha - V - \frac{RI}{2} - \frac{E_m}{Z}\left[\frac{R}{2}\sin(\alpha - \theta) + L\omega \cos(\alpha - \theta)\right]\right\}.$$

$$M = \frac{K}{\omega_1}. \qquad N = J. \qquad \omega = 2\pi f. \qquad \theta = \cos^{-1}\frac{R}{Z}.$$

$$Z = \sqrt{R^2 + \left(\omega L - \frac{1}{\omega C}\right)^2}.$$

$\alpha = \sin^{-1}(e/E_m)$, where e equals the algebraic value of the sinusoidal emf at the instant that it is impressed on the circuit. The current (I) is positive if flowing in the same direction as the impressed emf and the potential (V) is positive if acting in opposition to the impressed emf, both at time (t) equals zero. If the circuit contains no capacitor, the series capacitance is infinite and $a = 0$, $b = R/L$, and $Z = (R^2 + \omega^2 L^2)^{1/2}$.

5.5 ALTERNATING-CURRENT CIRCUITS

5.5.1 Types of Alternating Currents

The term alternating current refers to a current that reverses at regularly recurring intervals of time and that has alternately positive and negative values. An alternating current may be either sinusoidal, in which case it has only one frequency component, called the fundamental; or nonsinusoidal, in which case it has a fundamental frequency plus one or more harmonics. A harmonic is a sinusoidal component having a frequency an integral multiple of the fundamental.

5.5.2 Sinusoidal Voltages and Currents

Electromotive force (e) of an emf of maximum value E_m volts and angular velocity ω radians per second at any harmonic time t seconds.

$$e = E_m \sin \omega t \text{ volts.} \tag{5.63}$$

NOTE. A cycle is a single sequence of values from zero to positive maximum to zero to negative maximum to zero. The frequency (f) is the sequence rate in cycles per second (H_2). The angular velocity (ω) in radians per second equals 2π times the frequency (f). The time (t) is the time in seconds measured from the instant when the value is zero and is increasing to a positive maximum. When an emf is indicated by the expression $e = E_m \sin(\omega t + \alpha)$, time is measured from the instant when $e = E_m \sin \alpha$.

Current (i) flowing at any emf time t seconds in a circuit of R ohms resistance, L henrys self-inductance, and C farads series capacitance, upon which an emf, $e = E_m \sin \omega t$, is impressed.

$$i = \frac{E_m}{\sqrt{R^2 + \left(L\omega - \frac{1}{C\omega}\right)}} \sin\left[\omega t - \tan^{-1}\left(\frac{L\omega - \frac{1}{C\omega}}{R}\right)\right] \text{ amperes.} \tag{5.64}$$

NOTE.   It is assumed that the emf has been impressed upon the circuit long enough to produce a current.

Maximum current (I_m) flowing in a circuit under the conditions stated in Equation (5.64).

$$I_m = \frac{E_m}{\sqrt{R^2 + \left(L\omega - \dfrac{1}{C\omega}\right)^2}} \text{ amperes.} \qquad (5.65)$$

Effective or root-mean-square emf (E) of an emf, $e = E_m \sin \omega t$ volts.

$$E = \frac{E_m}{\sqrt{2}} \text{ volts.} \qquad (5.66)$$

NOTE.   The effective current (I) of a current equals $I_m/2^{1/2}$ amperes.

Average emf (E_a) of an emf, $e = E_m \sin \omega t$ volts.

$$E_a = \frac{2E_m}{\pi} \text{ volts.} \qquad (5.67)$$

NOTE.   The average current (I_a) of a current equals $2I_m/\pi$ amperes.

Form factor ($f.f.$) and **amplitude factor ($a.f.$)**, respectively, of an emf, $e = E_m \sin \omega t$ volts.

$$f.f. = \frac{E}{E_a} = 1.11. \qquad (5.68)$$

$$a.f. = \frac{E_m}{E} = 1.414. \qquad (5.69)$$

NOTE.   The form factor ($f.f.$) and amplitude factor ($a.f.$), respectively, of a current are $I/I_a = 1.11$ and $I_m/I = 1.414$.

5.5.3   Reactance and Impedance

Reactance (X) of a circuit of L henrys self-inductance and C farads series capacitance when conducting a current of ω radians per second angular velocity.

$$X = L\omega - \frac{1}{C\omega} \text{ ohms.} \qquad (5.70)$$

NOTE. $L\omega$ is called the inductive reactance and $1/(C\omega)$ the capacitive reactance of the circuit, each measured in ohms.

Impedance (Z) of a circuit of R ohms resistance and X ohms series reactance.

$$Z = \sqrt{R^2 + X^2} \text{ ohms.} \qquad (5.71)$$

Phase angle (θ) of a circuit of R ohms resistance and X ohms series reactance

$$\theta = \tan^{-1} \frac{X}{R}. \tag{5.72}$$

NOTE. The phase angle (θ) of a circuit in radians divided by the angular velocity (ω) of the conducted current in radians per second equals the time t in seconds by which the current lags or leads the emf. A positive value of X/R indicates a lagging current, and a negative value a leading current.

Power factor (*p.f.*) of a part-circuit of R ohms resistance, Z ohms impedance, and phase angle θ containing no generated emf.

$$p.f. = \frac{R}{Z} = \cos \theta. \tag{5.73}$$

Total reactance (X_s) of a series circuit the respective parts of which have reactances of X_1, X_2, X_3, and so on, ohms.

$$X_s = X_1 + X_2 + X_3 + \cdots \text{ ohms.} \tag{5.74}$$

NOTE. The addition is algebraic, inductive reactance being positive and capacitive reactance negative.

Total impedance (Z_s) of a series circuit of R_s ohms total resistance and X_s ohms total reactance. See Equation (5.71).

NOTE. The total impedance of a series circuit does not equal the sum of the impedances of its respective parts unless the ratio of reactance to resistance in each part is the same and the net reactances are of the same sign.

5.5.4 Power and Reactive Power

Power (P) delivered to or from a part-circuit conducting an effective current of I amperes across which the effective potential rise in the direction of the current is V volts, the phase angle between the current and the potential rise being θ.

$$P = VI \cos \theta \text{ watts.} \tag{5.75}$$

NOTE. Positive power indicates net power delivered from, and negative power indicates net power delivered to the part-circuit.

Reactive power (Q) under the conditions stated in Equation (5.75).

$$Q = VI \sin \theta \text{ vars.} \tag{5.76}$$

NOTE. Leading reactive power is considered by convention to be positive. Lagging reactive power is considered negative.

Volt-amperes (*VA*), or apparent power, under the conditions stated in Equation (5.75) is the product of V volts and I amperes (*VI*).

Power (*P*) delivered to a part-circuit of R ohms resistance conducting an effective current of I amperes and containing no generated emf.

$$P = I^2 R \text{ watts.} \tag{5.77}$$

NOTE.   The net power delivered to a reactance is zero.

Reactive power (*Q*) delivered to a part-circuit of X ohms reactance, conducting an effective current of I amperes, and containing no generated emf.

$$Q = I^2 X \text{ vars.} \tag{5.78}$$

NOTE. Q is positive for a capacitive reactance and negative for an inductive reactance. The reactive power delivered to a resistance is zero.

Volt-amperes (*V-A*) delivered to a part-circuit of Z ohms impedance conducting an effective current of I amperes and containing no generated emf.

$$V\text{-}A = I^2 Z \text{ volt-amperes.} \tag{5.79}$$

Volt-amperes (*V-A*) corresponding to a power of P watts and a reactive power of Q vars.

$$V\text{-}A = \sqrt{P^2 + Q^2}. \tag{5.80}$$

Effective vector expression (*E*) and (*I*) for an emf, $e = E_m \sin(\omega t + \alpha)$ volts, and a current, $i = I_m \sin(\omega t - \beta)$ amperes.

$$\mathbf{E} = \left(\frac{E_m}{\sqrt{2}} \cos \alpha + \mathbf{j} \frac{E_m}{\sqrt{2}} \sin \alpha \right) = \frac{E_m}{\sqrt{2}} \underline{/\alpha} \text{ volts.} \tag{5.81}$$

$$\mathbf{I} = \left(\frac{I_m}{\sqrt{2}} \cos \beta - \mathbf{j} \frac{I_m}{\sqrt{2}} \sin \beta \right) = \frac{I_m}{\sqrt{2}} \underline{/-\beta} \text{ amperes.} \tag{5.82}$$

NOTE.   In symbolic notation, the horizontal component of a vector is without prefix and its sign is $+$ to the right and $-$ to the left of the Y axis; the vertical component is designated by the prefix $\mathbf{j}$ and its sign is $+$ above and $-$ below the X axis. In some mathematical operations the symbol $\mathbf{j}$ has the value $(-1)^{1/2}$. The symbols $\underline{/\alpha}$ and $\underline{/-\beta}$ indicate vectors making the angles $+\alpha$ and $-\beta$, respectively, with the X axis and having magnitudes given by the quantity preceding the symbols. This is known as the polar form of the vector expression.

5.5.5   Vector and Symbolic Expressions

Vector electromotive force (*E*$_{AD}$) in a circuit the constituent parts of which contain the vector emf's $\mathbf{E}_{AB}$, $\mathbf{E}_{BC}$, and $\mathbf{E}_{CD}$ volts.

$$\mathbf{E}_{AD} = \mathbf{E}_{AB} + \mathbf{E}_{BC} + \mathbf{E}_{CD} \text{ volts.} \tag{5.83}$$

NOTE.   Each vector emf must be referred to the same axis of reference. The subscripts in each case indicate the direction of emf rise.

Vector current (I_{BA}) flowing from B toward a junction A from which the vector currents I_{AC}, I_{AD}, and I_{AF} amperes flow away.

$$I_{BA} = I_{AC} + I_{AD} + I_{AF} \text{ amperes.} \tag{5.84}$$

Electromotive force equivalent (E) of a vector emf, $E = a + jb$ volts.

$$E = \sqrt{a^2 + b^2} \text{ volts.} \tag{5.85}$$

NOTE.   In polar form, $E = (a^2 + b^2)^{1/2}\underline{/\tan^{-1}(b/a)}$. The current equivalent (I) of a vector current $I = (c + jd)$ amperes is $(c^2 + d^2)^{1/2}$ amperes, and in polar form $I = (c^2 + d^2)^{1/2}\underline{/\tan^{-1}(d/c)}$.

Symbolic expression (Z) for the impedance of a circuit of R ohms resistance and X ohms reactance.

$$Z = (R + jX) \text{ ohms.} \tag{5.86}$$

NOTE.   The resistance component has no prefix and is always $+$; the reactance component has the prefix j, a $+$ sign indicating net inductive reactance and a $-$ sign net capacitive reactance. In polar form, $Z = (R^2 + X^2)^{1/2}\underline{/\tan^{-1}(X/R)}$.

Symbolic impedance (Z_{AD}) between the ends A and D of a part-circuit containing several series parts of symbolic impedance Z_{AB}, Z_{BC}, and Z_{CD} ohms, respectively.

$$\begin{aligned} Z_{AD} &= Z_{AB} + Z_{BC} + Z_{CD} \text{ ohms} \\ &= (R_{AB} + R_{BC} + R_{CD}) + j(X_{AB} + X_{BC} + X_{CD}) \text{ ohms.} \end{aligned} \tag{5.87}$$

Vector current (I) flowing in the direction of an emf rise, $E = (a + jb)$ volts acting in a circuit of symbolic impedance $Z = (r + jx)$ ohms.

$$I = \left(\frac{a + jb}{r + jx}\right) \text{ amperes.} \tag{5.88}$$

NOTE.   To rationalize, multiply both numerator and denominator by the denominator with the sign of its j term reversed. We then have

$$I = \frac{(a + jb)(r - jx)}{(r + jx)(r - jx)} = \frac{ar - j^2bx + jbr - jax}{r^2 - j^2x^2}.$$

In this operation, $j = (-1)^{1/2}$ or $j^2 = -1$. Hence

$$I = \frac{(ar + bx) + j(br - ax)}{r^2 + x^2} = \left(\frac{ar + bx}{r^2 + x^2}\right) + j\left(\frac{br - ax}{r^2 + x^2}\right).$$

Alternately, in polar form,

$$I = \sqrt{a^2 + b^2} \frac{\Big/ \tan^{-1} \dfrac{b}{a}}{\sqrt{r^2 + x^2} \Big/ \tan^{-1} \dfrac{x}{r}} = \frac{\sqrt{a^2 + b^2}}{\sqrt{r^2 + x^2}} \Big/ \tan^{-1} \frac{b}{a} - \tan^{-1} \frac{x}{r}.$$

Vector potential rise (V_{AB}) between the ends A and B of a part-circuit of symbolic impedance Z_{AB} ohms conducting a current of vector value I_{AB} amperes and containing an emf rise of vector value E_{AB} volts.

$$V_{AB} = +E_{AB} - I_{AB}Z_{AB} \text{ volts.} \tag{5.89}$$

NOTE.   If $E_{AB} = a + jb$, $I_{AB} = c + jd$, and $Z_{AB} = r + jx$,

$$V_{AB} = a + jb - (c + jd)(r + jx)$$
$$= a + jb - cr - j^2dx - jcx - jdr$$

and since $j^2 = -1$,

$$V_{AB} = (a - cr + dx) + j(b - cx - dr).$$

Vector potential rise (V_{AD}) between the ends A and D of a part-circuit containing several series parts across which the respective vector potential rises are V_{AB}, V_{BC}, and V_{CD} volts.

$$V_{AD} = V_{AB} + V_{BC} + V_{CD} \text{ volts.} \tag{5.90}$$

Power (P) delivered to or from a part-circuit conducting a vector current $I = (c + jd)$ amperes and across which the vector potential rise in the direction of the current is $V = (a + jb)$ volts.

$$P = (ac + bd) \text{ watts.} \tag{5.91}$$

NOTE.   The signs of a, b, c, and d are preserved. Positive power indicates power delivered from, and negative power indicates power delivered to, the part-circuit. The power does not equal $(a + jb)(c + jd)$.

Reactive power (Q) under the conditions stated in Equation (5.91).

$$Q = (ad - bc) \text{ vars.} \tag{5.92}$$

NOTE.   The signs of a, b, c, and d should be preserved. Leading reactive power is positive; lagging reactive power is negative.

Conductance (G) and **susceptance (B)** of a branch of R ohms resistance, X ohms series reactance, and Z ohms impedance.

$$G = \frac{R}{R^2 + X^2} = \frac{R}{Z^2} \text{ siemens.} \tag{5.93}$$

$$B = \frac{X}{R^2 + X^2} = \frac{X}{Z^2} \text{ siemens.} \tag{5.94}$$

Admittance (Y) of a branch of Z ohms impedance, G siemens conductance, and B siemens susceptance.

$$Y = \frac{1}{Z} = \sqrt{G^2 + B^2} \text{ siemens.} \tag{5.95}$$

Total conductance (G_0) of several parallel branches of G_1, G_2, and G_3 siemens conductance, respectively.

$$G_0 = G_1 + G_2 + G_3 \text{ siemens.} \tag{5.96}$$

Total susceptance (B_0) of several parallel branches of B_1, B_2, and B_3 siemens susceptance, respectively.

$$B_0 = B_1 + B_2 + B_3 \text{ siemens.} \tag{5.97}$$

NOTE. The addition is algebraic, inductive susceptance being positive and capacitive susceptance negative.

Total admittance (Y_0) of several parallel branches of total conductance G_0 siemens and total susceptance B_0 siemens. See Equation (5.95).

NOTE. The total admittance of a parallel circuit does not equal the sum of the admittances of the respective branches unless the ratio of susceptance to conductance in each branch is the same and the net susceptances are of the same sign.

Phase angle (θ) of a circuit of G siemens conductance and B siemens susceptance.

$$\theta = \tan^{-1} \frac{B}{G}. \tag{5.98}$$

Power factor ($p.f.$) of a part-circuit of G siemens conductance and Y siemens admittance, containing no generated emf.

$$p.f. = \frac{G}{Y}. \tag{5.99}$$

Resistance (R) and **reactance (X)** of a circuit of G siemens conductance, B siemens susceptance, and Y siemens admittance.

$$R = \frac{G}{G^2 + B^2} = \frac{G}{Y^2} \text{ ohms.}$$

$$X = \frac{B}{G^2 + B^2} = \frac{B}{Y^2} \text{ ohms.} \tag{5.100}$$

Impedance (Z) of a circuit of Y siemens admittance.

$$Z = \frac{1}{Y} \text{ ohms.} \tag{5.101}$$

Symbolic expression (Y) for the admittance of a circuit of G siemens conductance and B siemens susceptance.

$$\mathbf{Y} = (G - \mathbf{j}B) \text{ siemens.} \tag{5.102}$$

NOTE.   In polar form, $\mathbf{Y} = \sqrt{G^2 + B^2} \Big/ \tan^{-1}\left(-\dfrac{B}{G}\right)$.

Symbolic admittance ($\mathbf{Y_0}$) of a parallel circuit containing several branches of symbolic admittance $\mathbf{Y_1}$, $\mathbf{Y_2}$, and $\mathbf{Y_3}$ siemens, respectively.

$$\mathbf{Y_0} = \mathbf{Y_1} + \mathbf{Y_2} + \mathbf{Y_3} \text{ siemens.} \tag{5.103}$$

Vector current (I) flowing in the direction of an emf rise **E** acting in a circuit of **Y** siemens symbolic admittance.

$$\mathbf{I} = \mathbf{EY} \text{ amperes.} \tag{5.104}$$

5.5.6   Nomenclature for Three-phase Circuits

Line emf or voltage, E_1 volts; phase emf or voltage, E_p volts; line current, I_1 amperes; phase current, I_p amperes; phase angle between phase voltage and phase current, Θ_p. In Equations (5.109) and (5.111), $\mathbf{E}_a$, $\mathbf{E}_b$, and $\mathbf{E}_c$ are any three voltage vectors that may exist in a three-phase system, such as voltages to neutral or to ground, line-

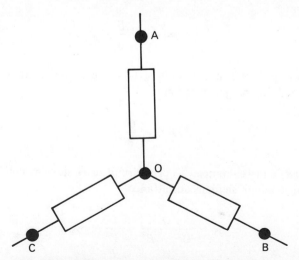

Fig. 5.12   Balanced wye-connected three-phase circuit.

to-line voltages, and induced voltages. Likewise, I_a, I_b, and I_c in Equations (5.110) and (5.112) may be line currents, phase currents, the currents in a Δ-connected winding, and so on. The subscripts 1, 2, and 0 in Equations (5.109) to (5.112) denote, respectively, positive-, negative-, and zero-sequence components. Positive phase rotation, ABC in counterclockwise direction.

Conditions for balanced three-phase circuit: all phase currents, phase emf's, and phase voltages, respectively, equal and differing in phase by 120°. Conditions for unbalanced three-phase circuit: phase currents, phase emf's, or phase voltages, respectively, unequal or not differing in phase by 120°.

5.5.7 Three-phase Circuits

Balanced Y-connected branches (Fig. 5.12).

$$E_1 = \sqrt{3}\, E_p.$$
$$I_1 = I_p.$$
$$E_{OA} + E_{OB} + E_{OC} = 0.$$
$$E_{AB} + E_{BC} + E_{CA} = 0. \tag{5.105}$$
$$E_{AB} = E_{OB} - E_{OA} =$$
$$\sqrt{3}\, E_{OB}\underline{/30°} = \sqrt{3}\, E_{OC}\underline{/90°}.$$
$$I_{OA} + I_{OB} + I_{OC} = 0.$$

Balanced Δ-connected branches (Fig. 5.13).

$$E_1 = E_p.$$
$$I_1 = \sqrt{3}\, I_p.$$
$$E_{AB} + E_{BC} + E_{CA} = 0.$$
$$I_{AB} + I_{BC} + I_{CA} = 0. \tag{5.106}$$

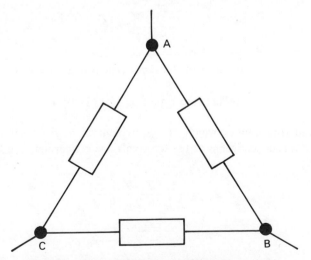

Fig. 5.13 Balanced delta-connected three-phase circuit.

$$I'_{AA} = I_{CA} - I_{AB} =$$
$$\sqrt{3}\,I_{CA}\underline{/30°} = \sqrt{3}\,I_{BC}\underline{/90°}.$$ (5.106a)
$$I'_{AA} = I'_{BB} + I'_{CC} = 0.$$

Unbalanced Y-connected branches.

$$E_{AB} + E_{BC} + E_{CA} = 0.$$
$$E_{AB} = E_{OB} - E_{OA}.$$ (5.107)
$$I_{OA} + I_{OB} + I_{OC} = 0.$$

NOTE.   If there is a current flowing out of the point O in a neutral connection, it must be added vectorially to the left-hand side of the last equation.

Unbalanced Δ-connected branches.

$$E_{AB} + E_{BC} + E_{CA} = 0.$$ (5.108)
$$I'_{AA} = I_{CA} - I_{AB}.$$

Symmetrical components of voltage in an unbalanced three-phase system.

$$E_{a1} = \tfrac{1}{3}(E_a + E_b\underline{/120°} + E_c\underline{/120°}).$$
$$E_{a2} = \tfrac{1}{3}(E_a + E_b\overline{/120°} + E_c\underline{/120°}).$$ (5.109)
$$E_0 = \tfrac{1}{3}(E_a + E_b + E_c).$$

Symmetrical components of current in an unbalanced three-phase system.

Replace E in Equation (5.109) by **I**.   (5.110)

Three-phase voltages in terms of the symmetrical components of voltage.

$$E_a = E_{a1} + E_{a2} + E_0.$$
$$E_b = E_{a1}\overline{/120°} + E_{a2}\underline{/120°} + E_0.$$ (5.111)
$$E_c = E_{a1}\underline{/120°} + E_{a2}\overline{/120°} + E_0.$$

Three-phase currents in terms of the symmetrical components of current.

Replace E in Equation (5.111) by **I**.   (5.112)

Y-connected impedances which are equivalent to a given set of Δ-connected impedances so far as conditions at the terminals are concerned. See Fig. 5.14.

$$Z_A = \frac{Z_{CA}Z_{AB}}{Z_{AB} + Z_{BC} + Z_{CA}}.$$

$$Z_B = \frac{Z_{AB}Z_{BC}}{Z_{AB} + Z_{BC} + Z_{CA}}.$$ (5.113)

$$Z_C = \frac{Z_{BC}Z_{CA}}{Z_{AB} + Z_{BC} + Z_{CA}}.$$

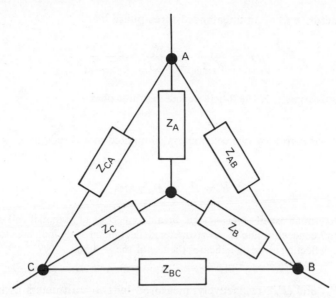

Fig. 5.14 Equivalent wye and delta impedances.

NOTE. If the impedances are balanced, the Y-connected impedances are ⅓ of the Δ-connected impedances.

Δ-connected impedances that are equivalent to a given set of Y-connected impedances so far as conditions at the terminals are concerned. See Fig. 5.14.

$$Z_{AB} = \frac{Z_A Z_B + Z_B Z_C + Z_C Z_A}{Z_C}.$$

$$Z_{BC} = \frac{Z_A Z_B + Z_B Z_C + Z_C Z_A}{Z_A}. \qquad (5.114)$$

$$Z_{CA} = \frac{Z_A Z_B + Z_B Z_C + Z_C Z_A}{Z_B}.$$

NOTE. If the impedances are balanced, the Δ-connected impedances are three times the Y-connected impedances.

Power (P) delivered to or from a balanced three-phase line.

$$P = \sqrt{3} E_1 I_1 \cos \theta_p \text{ watts.} \qquad (5.115)$$

Power factor (p.f.) of a balanced three-phase load.

$$p.f. = \cos \theta_p. \qquad (5.116)$$

Power (P) delivered to or from an unbalanced three-phase line.

$$P = E_{p1} I_{p1} \cos \theta_{p1} + E_{p2} I_{p2} \cos \theta_{p2} + E_{p3} I_{p3} \cos \theta_{p3} \text{ watts.} \qquad (5.117)$$

NOTE. The subscripts 1, 2, and 3 here denote the three phases.

Power factor (*p.f.*) of an unbalanced three-phase load.

$$p.f. = \frac{P}{E_{p_1}I_{p_1} + E_{p_2}I_{p_2} + E_{p_3}I_{p_3}}. \tag{5.118}$$

NOTE.   The subscripts 1, 2, and 3 here denote the three phases.

Power (*P*) measured by two wattmeters connected in a three-phase line as shown in Fig. 5.15.

$$P = P_A \pm P_B \text{ watts.} \tag{5.119}$$

NOTE.   To determine use of + or − sign, break connection of potential coil of wattmeter A at line C and connect to line B. A wattmeter deflection on scale indicates the use of the + sign and a deflection off scale indicates the use of the − sign.

Power (*P_A*) **and** (*P_B*), respectively, measured by two wattmeters connected in a three-phase balanced line as shown in Fig. 5.15.

$$P_A = E_1 I_1 \cos (30° - \theta_p) \text{ watts.} \tag{5.120}$$
$$P_B = E_1 I_1 \cos (30° + \theta_p) \text{ watts.}$$

Phase angle (*θ*) of a balanced three-phase load when two wattmeters connected as shown in Fig. 5.15 measure P_A and P_B watts, respectively.

$$\theta = \tan^{-1} \sqrt{3}\, \frac{P_A - P_B}{P_A + P_B}. \tag{5.121}$$

NOTE.   The additions and subtractions are algebraic.

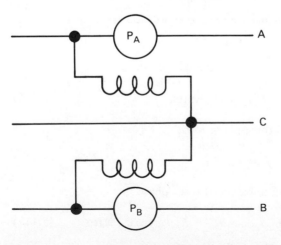

Fig. 5.15 Three-phase power measured by two wattmeters.

5.5.8   Nonsinusoidal Alternating Voltages and Currents

Any nonsinusoidal voltage or current waveform can be resolved into a fundamental frequency sine wave and its harmonics by a Fourier's series. If y is the amplitude of a nonsinusoidal wave at time t; $Y_1, Y_2, \cdots, Y_n$ the maximum amplitude of the 1st, 2nd, $\cdots$, nth harmonics; $\theta_1, \theta_2, \cdots, \theta_n$ the angles that determine relative phase of the harmonics; and $\omega(2\pi f)$ the angular velocity of the fundamental wave corresponding to f hertz,

$$y = Y_1 \sin(\omega t + \theta_1) + Y_2 \sin(2\omega t + \theta_2) + \cdots + Y_n \sin(n\omega t + \theta_n). \quad (5.122)$$

Effective or root-mean-square value (E) of a periodically time-varying emf or voltage $e = f(t)$ volts having a period of T seconds.

$$E = \sqrt{\frac{1}{T} \int_0^T e^2 dt} \text{ volts.} \quad (5.123)$$

Effective or root-mean-square value (I) of a periodically time-varying current $i = f(t)$ amperes having a period of T seconds.

$$I = \sqrt{\frac{1}{T} \int_0^T i^2 dt} \text{ amperes.} \quad (5.124)$$

Electromotive force (e) of a nonsinusoidal emf or voltage at any time t seconds.

$$e = E_{m_1} \sin(\omega t + \theta_1) + E_{m_3} \sin(3\omega t + \theta_3) + E_{m_5} \sin(5\omega t + \theta_5) + \cdots \text{ volts.} \quad (5.125)$$

NOTE. $E_{m_1}, E_{m_3}, E_{m_5}, \cdots$ represent the maximum values of the first, third, fifth, etc., harmonics, and θ_1, θ_2, and $\theta_3, \cdots$ their respective phase angles with a common axis of reference. The angular velocity ω is that of the fundamental or first harmonic. Alternators do not normally generate even harmonics of voltage. Some nonsinusoidal voltages, such as the outputs of rectifiers, however, may contain odd or even harmonics and average or dc components. In such cases the terms $E_0 + E_{m_2} \sin(2\omega t + \theta_2) + E_{m_4} \sin(4\omega t + \theta_4) + \cdots$ should be added to the right-hand member of Equation (5.125).

Current (i) at any time t seconds flowing in a circuit of R ohms resistance, L henrys self-inductance, and C farads series capacitance upon which a nonsinusoidal emf of the form stated in Equation (5.125) is impressed.

$$i = I_{m_1} \sin(\omega t + \theta_1') + I_{m_3} \sin(3\omega t + \theta_3') + I_{m_5} \sin(5\omega t + \theta_5') + \cdots \text{ amperes.} \quad (5.126)$$

NOTE.

$$I_{m1} = \frac{E_{m1}}{\sqrt{R^2 + \left(L\omega - \dfrac{1}{C\omega}\right)^2}}, \qquad \theta_1' = \theta_1 - \tan^{-1}\left(\frac{L\omega - \dfrac{1}{C\omega}}{R}\right),$$

$$I_{m3} = \frac{E_{m3}}{\sqrt{R^2 + \left(3L\omega - \dfrac{1}{3C\omega}\right)^2}}, \qquad \theta_3' = \theta_3 - \tan^{-1}\left(\frac{3L\omega - \dfrac{1}{3C\omega}}{R}\right),$$

$$I_{m5} = \frac{E_{m5}}{\sqrt{R^2 + \left(5L\omega - \dfrac{1}{5C\omega}\right)^2}}, \qquad \theta_5' = \theta_5 - \tan^{-1}\left(\frac{5L\omega - \dfrac{1}{5C\omega}}{R}\right).$$

Effective emf (E) of a nonsinusoidal emf of the form stated in Equation (5.125).

$$E = \sqrt{\frac{(E_{m1})^2 + (E_{m3})^2 + (E_{m5})^2}{2}} \text{ volts.} \qquad (5.127)$$

NOTE.   The effective value of a nonsinusoidal potential or current is obtained in the same manner.

Power (P) delivered to a part-circuit conducting a nonsinusoidal current of the form stated in Equation (5.126) and upon which is impressed a nonharmonic emf of the form stated in Equation (5.125).

$$P = \frac{E_{m1}I_{m1}}{2} \cos(\theta_1 - \theta_1') + \frac{E_{m3}I_{m3}}{2} \cos(\theta_3 - \theta_3')$$

$$+ \frac{E_{m5}I_{m5}}{2} \cos(\theta_5 - \theta_5') + \cdots \text{ watts.} \quad (5.128)$$

Power factor ($p.f.$) of a part-circuit conducting a nonsinusoidal current of effective value I amperes and absorbing energy at a rate of P watts, the effective value of the nonsinusoidal potential between its ends being V volts.

$$p.f. = \frac{P}{VI}. \qquad (5.129)$$

Sinusoidal emf and current equivalent to the nonsinusoidal forms stated in Equations (5.125) and (5.126).

$$e = \sqrt{2}\,E \sin \omega t \text{ volts;} \qquad (5.130)$$
$$i = \sqrt{2}\,I \sin(t \pm \cos^{-1} p.f.) \text{ amperes.}$$

NOTE. $p.f.$ is the power factor of the circuit upon which the nonharmonic emf is impressed.

Resistance (R) of a part-circuit containing no source of generated emf, conducting an effective current of I amperes, and absorbing energy at a rate of P watts.

$$R = \frac{P}{I^2} \text{ ohms.} \qquad (5.131)$$

Impedance (Z) of a part-circuit containing no source of generated emf, conducting an effective current of I amperes, and across which the potential is V volts.

$$Z = \frac{V}{I} \text{ ohms.} \qquad (5.132)$$

Reactance (X) of a part-circuit of R ohms resistance and Z ohms impedance.

$$X = \sqrt{Z^2 - R^2} \text{ ohms.} \qquad (5.133)$$

NOTE.   The reactance of a part-circuit to a nonsinusoidal current does not equal $[L\omega - (1/C\omega)]$ ohms.

5.6   DIRECT-CURRENT GENERATORS AND MOTORS

5.6.1   Rotating Masses

A generator is accelerated to its rated speed by applying mechanical torque (T) to its shaft; a motor is accelerated by creating mechanical torque at its shaft. For either machine, where I is total inertia of the rotating mass and α is angular acceleration,

$$\alpha = \frac{T}{I}. \qquad (5.134)$$

Units for α, T, and I in the English system, which is still in common use, can be converted to SI system units.

	(From) English Unit	(To) SI Unit	Multiply by
α	rad/s	rad/s^2	1.0
T	lbf·ft	newton meter (N·m)	1.356
I	lb·ft^2	kilogram meter^2 (kg·m^2)	0.0421

Time t to accelerate a rotating mass from standstill to rated speed n revolutions per minute when it has inertia I and average torque T_A during the accelerating interval.

$$t = \frac{0.105 \cdot I \cdot n}{g \cdot T_A} \text{ seconds.} \qquad (5.135)$$

I, g, and T_A can be in either metric or English units:

	I	g	T_A
Metric	$kg \cdot m^2$	9.807 m/s²	N·m
English	$lb \cdot ft^2$	32.17 ft/s²	lbf·ft

5.6.2   Mechanical Shaft Power

Mechanical shaft power, applied to a shaft in the case of a generator, or taken from the shaft of a motor at rated speed n revolutions per minute.

$$\text{For } T \text{ in N} \cdot \text{m,} \quad P = \frac{nT}{5190} \text{ kilowatts.} \tag{5.136}$$

$$\text{For } T \text{ in lbf} \cdot \text{ft,} \quad P = \frac{nT}{5250} \text{ horsepower.} \tag{5.137}$$

All the equations in Subsections (5.6), and in subsections 5.7 and 5.8 on alternating current rotating machines, relate to the conversion of mechanical shaft power to electric power or vice versa, and are applicable to either a generator or a motor.

5.6.3   DC Machine Nomenclature and Units of Measurement

Emf generated in armature, E volts; terminal voltage, V volts; armature current, I amperes; line current, I_l amperes; shunt field current, I_f amperes; series field current, I_s amperes; armature resistance between brushes, R ohms; shunt field resistance including rheostat, R_f ohms; series field resistance including shunt, R_s ohms; number of poles, p; shunt field turns per pole, N_f; series field turns per pole, N_s; number of armature paths between terminals, m; number of armature conductors, Z; magnetic flux per pole, Φ webers; armature speed, n revolutions per minute; armature torque, T newton meters.

5.6.4   DC Machine Characteristics

Electromotive force (E) generated in the armature of a dc machine

$$E = \frac{p\Phi Zn}{60m} \text{ volts.} \tag{5.138}$$

Shunt field current (I_{fd}) equivalent to the demagnetizing magnetomotive force of the armature per pole when the armature current is I amperes and the brushes are shifted through an angle of θ space degrees from the neutral plane to improve commutation.

$$I_{fd} = \frac{ZI\theta}{360N_f m} \text{ amperes.} \tag{5.139}$$

Shunt field current (I_{fs}) equivalent to the magnetomotive force of the series turns per pole.

$$I_{fs} = \frac{N_s}{N_f} I_s \text{ amperes.} \tag{5.140}$$

Net field current (I_{fn}) at any load.

$$I_{fn} = I_f - I_{fd} \pm I_{fs} \text{ amperes.} \tag{5.141}$$

NOTE. The sign before I_{fs} is $+$ for a cumulative and $-$ for a differential compound machine.

Terminal voltage (V) of a shunt machine when the armature current is I amperes and the generated emf is E volts.

$$V = E \pm IR \text{ volts.} \tag{5.142}$$

NOTE. The sign before IR is $+$ for a motor and $-$ for a generator. In a series or long-shunt compound machine, $V = E \pm I(R + R_s)$ volts, and in a short-shunt compound machine, $V = E \pm IR \pm I_s R_s$ volts.

Armature speed (n) when the generated emf is E volts.

$$n = \frac{60Em}{p\Phi Z} \text{ revolutions/minute.} \tag{5.143}$$

Armature torque (T) of a dynamo when the armature current is I amperes.

$$T = \frac{0.159Z\Phi Ip}{m} \text{ newton meters.} \tag{5.144}$$

Rotational losses (P_r) of a machine that, operated as a shunt motor at no load with a voltage between brushes of V volts, takes an armature current of I amperes.

$$P_r = VI - I^2R \text{ watts.} \tag{5.145}$$

NOTE. To determine the rotational losses corresponding to a definite load, the machine operated as a shunt motor at no load must be run at the same speed and with the same generated emf as when running at the definite load.

Copper losses (P_k) at any load.

Shunt field:	$P_f = I_f^2 R_f = VI_f$ watts.		(5.146)
Series field:	$P_s = I_s^2 R_s$ watts.		(5.147)
Armature:	$P_a = I^2 R$ watts.		(5.148)

Power input (P_i) to a generator at any load.

$$
\begin{aligned}
P_i &= EI + P_r \text{ watts} \\
&= P_o + P_k + P_r \text{ watts} \\
&= 0.1420nT \text{ watts} \\
&= 1.903nT \times 10^{-4} \text{ horsepower.}
\end{aligned}
\tag{5.149}
$$

Power output (P_o) of a generator at any load.

$$
P_o = VI_1 \text{ watts.}
\tag{5.150}
$$

Power input (P_i) of a motor at any load.

$$
P_i = VI_1 \text{ watts.}
\tag{5.151}
$$

Power output (P_o) of a motor at any load.

$$
\begin{aligned}
P_o &= EI - P_r \text{ watts} \\
&= P_i - P_k - P_r \text{ watts} \\
&= 0.1420nT \text{ watts} \\
&= 1.903nT \times 10^{-4} \text{ horsepower.}
\end{aligned}
\tag{5.152}
$$

Efficiency (η) of a machine at any load.

$$
\eta = \frac{P_o}{P_i} = \frac{P_o}{P_o + P_k + P_r} = \frac{P_i - P_k - P_r}{P_i}.
\tag{5.153}
$$

5.7   ALTERNATING-CURRENT SYNCHRONOUS MACHINES

5.7.1   Conventions

All equations in this subsection apply to synchronous generators, motors, or condensers unless indicated otherwise. Sinusoidal voltages and currents are assumed, and their magnitudes expressed by effective values. Three-phase machines with wye-connected windings are assumed with all machine impedances, voltages, and currents being phase values for a wye connection. **Terminal voltage** of a wye-connected machine is $(3)^{1/2}$ times phase voltage.

5.7.2   AC Synchronous Machine Characteristics

Frequency (f) of the voltage generated in a synchronous machine having p poles, the speed of the rotating magnetic field being n revolutions per minute.

$$
f = \frac{pn}{120} \text{ hertz.}
\tag{5.154}
$$

Synchronous internal voltage or excitation voltage (E_i) generated in the armature of a synchronous machine.

$$E_i = 4.44fN\Phi \cos\frac{\beta}{2}\left(\frac{\sin\frac{m\alpha}{2}}{m\sin\frac{\alpha}{2}}\right) \text{ volts.} \tag{5.155}$$

NOTE. N is the number of armature series turns per phase or one-half the number of series conductors on the armature divided by the number of phases, Φ is the main field flux per pole in webers, β is the pitch deficiency or the difference in electrical degrees between the pole pitch (180°) and the coil pitch, m is the number of slots per pole per phase, and α is the angle between adjacent slot centers in electrical degrees. Electrical degrees equal space degrees multiplied by $p/2$. This equation assumes that the arrangement of the winding before each pole is the same and hence that there are an integral number of slots per pole per phase. If ϕ_r, the resultant air-gap flux per pole corresponding to the mmf R in Equation (5.158), is used instead of ϕ, the voltage given by Equation (5.155) is the air-gap voltage E_a.

Field magnetomotive force (F) in a cylindrical-rotor machine.

$$F = \frac{4}{\pi}N_fI_f\left(\frac{\sin\frac{n_f\alpha_f}{2}}{n_f\sin\frac{\alpha_f}{2}}\right) \text{ ampere-turns/pole.} \tag{5.156}$$

NOTE. F is the maximum value of the fundamental field mmf. N_t is the number of field turns per pole carrying the current I_f amperes per turn. n_t is the number of rotor slots per pole, and α_f is the electrical angle between adjacent slots in a belt. Equation (5.156) assumes that the field winding is a regular, distributed, full-pitch winding.

Armature magnetomotive force (A) in a cylindrical-rotor machine.

$$A = 0.90KN_aI \text{ ampere-turns/pole.} \tag{5.157}$$

NOTE. K equals $\cos(\beta/2)\{\sin(m\alpha/2)/[m\sin(\alpha/2)]\}$ as explained in Equation (5.155). N_a is the number of armature series turns per pole. I is the armature current. A may be obtained in equivalent field amperes by dividing by N_f.

Resultant magnetomotive force (R) in a machine with a field magnetomotive force F ampere-turns per pole and an armature magnetomotive force A ampere-turns per pole.

$$\mathbf{R} = \mathbf{F} + \mathbf{A} \text{ ampere-turns/pole.} \tag{5.158}$$

NOTE. The addition must be made vectorially. See Equations (5.155), (5.161), and (5.162).

Characteristic curves of a synchronous machine are shown in Fig. 5.16. Data are plotted as follows: OCC (open-circuit characteristic), terminal voltage at no load

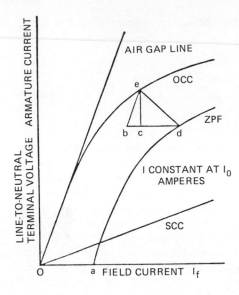

Fig. 5.16 No load and short circuit characteristics of a synchronous machine.

versus field current; the air-gap line is drawn tangent to the lower part of OCC; SCC (short-circuit characteristic), armature current with the armature terminals short-circuited versus field current; ZPF (zero-power-factor characteristic), terminal voltage versus field current with the armature supplying a constant current I_0 to a ZPF lagging load at its terminals, I_0 usually being taken equal to rated armature current.

Potier triangle *cde* (Fig. 5.16). Procedure in obtaining Potier triangle: Choose a point *d* well up on the curved part of ZPF; draw *db* parallel to and equal to *ao*; draw *be* parallel to the air-gap line; draw *ec* perpendicular to *db*; *cde* is then the Potier triangle.

Potier reactance (X_p) from Potier triangle (see Fig. 5.16).

$$X_p = \frac{ec \text{ in volts}}{I_0 \text{ in amperes}} \text{ ohms.} \tag{5.159}$$

NOTE. I_0 is the constant armature current for which ZPF is drawn. X_p is very nearly equal to, and is frequently used for, the armature leakage reactance X_a.

Armature magnetomotive force (A) corresponding to the armature current $I = I_0$ from Potier triangle (see Fig. 5.16).

$$A = cd \text{ equivalent field amperes.} \tag{5.160}$$

NOTE.   To obtain A in ampere-turns per pole as in Equation (5.157), substitute *cd* for I_f in Equation (5.156). A for any other value of I may be obtained by direct proportion.

General-method vector diagram for a cylindrical-rotor machine with armature current I, terminal voltage V, and phase angle θ between I and V. Figure 5.17 is drawn for generator operation with a lagging power-factor load. All voltage vectors are voltage rises. For motor operation, I and the voltage drop vector $-V$ are θ degrees out of phase.

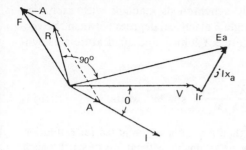

Fig. 5.17   Voltage and current vector diagram of a cylindrical-rotor synchronous machine.

Air-gap voltage E_a is given by

$$\mathbf{E}_a = \mathbf{V} + \mathbf{I}(r + jX_a), \tag{5.161}$$

in which the voltages $\mathbf{E}_a$ and $\mathbf{V}$ are rises and r is the armature resistance.

A is obtained from Equation (5.160) or (5.157).

R is obtained in equivalent field amperes by entering OCC (Fig. 5.16) with E_a volts and reading the corresponding field current.

Field current (I_f) required in a cylindrical-rotor machine with armature current I, terminal voltage V, and phase angle θ between I and V.

$$I_f = \text{magnitude of } (\mathbf{R} - \mathbf{A}) \text{ amperes.} \tag{5.162}$$

NOTE. R and A must be in equivalent field amperes in Equation (5.162); if they are in ampere-turns per pole, use Equation (5.156) to convert ampere-turns per pole into equivalent field amperes. See Fig. 5.17. R and A obtained as in Equation (5.161).

Excitation voltage (E_i) under conditions in Equation (5.162).

Enter OCC (Fig. 5.16) with I_f from Equation (5.162) and read the corresponding voltage.

For **salient-pole machines** the methods of Equation (5.156) to (5.162) will give approximately correct results. In Equation (5.156), F should be taken as $N_f I_f$, and in Equation (5.157) the factor 0.75 should be used instead of 0.9. For more accurate results the Blondel two-reaction method should be used.

Total power output (P_o) of a cylindrical-rotor synchronous machine with excitation voltage E_i volts, terminal voltage V volts, angle δ electrical degrees between the vectors $\mathbf{E}_i$ and $\mathbf{V}$, and synchronous reactance X_s ohms.

$$P_o = \frac{3VE_i}{X_s} \sin \delta \text{ watts.} \tag{5.163}$$

NOTE.   Losses are neglected. X_s should be properly adjusted for saturation. The maximum power output is $(3VE_i)/X_s$ watts; it may not be attainable without loss of synchronism. $(3VE_i)/X_s$ watts gives the breakdown or pull-out power for a cylindrical-rotor motor connected directly to a power system of capacity large compared to that of the motor.

Total power output (P_o) of a salient-pole synchronous machine with excitation voltage E_d volts, terminal voltage V volts, angle δ electrical degrees between the vectors $\mathbf{E}_d$ and $\mathbf{V}$, direct-axis synchronous reactance X_d ohms, and quadrature-axis synchronous reactance X_q ohms.

$$P_o = 3 \left[\frac{VE_d}{X_d} \sin \delta + \frac{V^2(X_d - X_q)}{2X_dX_q} \sin 2\delta \right] \text{watts.} \qquad (5.164)$$

NOTE. Losses are neglected. The reactances should be properly adjusted for saturation. The maximum value of P_o, as δ varies, may not be attainable without loss of synchronism. This maximum value gives the breakdown or pull-out power for a salient-pole motor connected directly to a power system of capacity large compared to that of the motor.

Efficiency (η) of a synchronous machine when the output is P_o watts.

$$\eta = \frac{P_o}{P_o + P_a + P_c + P_s + P_{fw} + P_f}. \qquad (5.165)$$

NOTE. All powers are total three-phase powers. P_a, the armature copper loss, equals three times the armature current squared times the ohmic resistance per phase. P_c is the open-circuit core loss (hysteresis and eddy-current losses). To determine P_c, enter a curve of open-circuit core loss versus open-circuit voltage with a voltage equal to the vector sum of the terminal voltage plus the armature resistance drop, and read the corresponding value of P_c. P_s is the stray-load loss (skin effect and eddy-current losses in the armature conductors plus local core losses due to armature leakage flux) and may be found from a curve of stray-load loss versus armature current. P_{fw} is the friction and windage loss, and P_f is the field copper loss.

5.8   ALTERNATING-CURRENT INDUCTION MACHINES

5.8.1   Conventions

Three-phase wye-wound machines are assumed throughout and unless indicated otherwise each formula applies to a generator or a motor. All rotor resistances and reactances are referred to the stator. All impedances, voltages, and currents are phase values for a wye connection. All values of power P are total three-phase powers.

5.8.2   AC Induction Machine Characteristics

Equivalent effective resistance (R_1) of an induction machine.

$$R_1 = \frac{P_i}{3I_i^2} \text{ohms.} \qquad (5.166)$$

NOTE. P_i is the power input on blocked-rotor and I_i is the stator blocked-rotor current.

Equivalent impedance (Z_1) of an induction machine.

$$Z_1 = \frac{V_1}{I_1} \text{ohms.} \qquad (5.167)$$

NOTE. V_1 is the stator terminal voltage during blocked-rotor and I_1 is the stator blocked-rotor current.

Equivalent reactance (X_1) of an induction machine.

$$X_1 = \sqrt{Z_1^2 - R_1^2} \text{ ohms.} \tag{5.168}$$

NOTE. Z_1 and R_1 are determined as in Equations (5.167) and (5.168).

Rotor resistance (r_2) of an induction machine referred to the stator.

$$r_2 = T_1^2\, r_2' \text{ ohms.} \tag{5.169}$$

NOTE. r_2' is the actual rotor resistance and T_1 is the ratio of transformation from stator to rotor or the ratio of the emf's induced in the stator and rotor, respectively, during blocked-rotor.

Rotor leakage reactance (x_2) of an induction machine referred to the stator.

$$x_2 = T_1^2 x_2' \text{ ohms.} \tag{5.170}$$

NOTE.   Read note to Equation (5.169), substituting x_2' for r_2' and reactance for resistance.

Equivalent effective resistance (R_1) of an induction machine of r_{1e} ohms effective stator resistance and r_{2e} ohms effective rotor resistance referred to the stator.

$$R_1 = r_{1e} + r_{2e} \text{ ohms.} \tag{5.171}$$

Equivalent reactance (X_1) of an induction machine of x_1 ohms stator leakage reactance and x_2 ohms rotor leakage reactance referred to the stator.

$$X_1 = x_1 + x_2 \text{ ohms.} \tag{5.172}$$

Synchronous speed (n_1) of an induction machine having p poles, the frequency of the impressed voltage being f hertz.

$$n_1 = \frac{120f}{p} \text{ revolutions/minute.} \tag{5.173}$$

Slip (s) of an induction machine of synchronous speed (n_1) revolutions per minute when the rotor speed is n_2 revolutions per minute.

$$s = 1 - \frac{n_2}{n_1}. \tag{5.174}$$

Equivalent circuit of an induction motor (Fig. 5.18), r_{1e} and r_{2e} as in Equation (5.171). x_1 and x_2 as in Equation (5.172). Z_n, the exciting impedance, and its components r_n and x_n may be determined from a no-load run. The determination is similar to that for R_1, Z_1, and X_1 in Equations (5.166), (5.167), and (5.168) except that

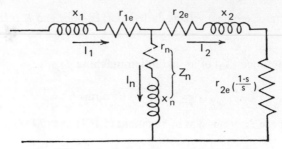

Fig. 5.18   Induction machine equivalent circuit.

the no-load voltage and current and the no-load power input minus friction and windage losses must be used. $r_{2e}\,[(1-s)/s]$ is the resistance equivalent for the rotational power, which is the shaft power output plus friction and windage and a small core loss.

Induced stator emf (E_n) of an induction machine at no load.

$$E_n = V_1 - I_n \sqrt{r_{1e}^2 + x_1^2} \text{ volts.} \tag{5.175}$$

NOTE. V_1 is the stator terminal voltage, I_n the no-load line current, r_{1e} and x_1 as in Equations (5.171) and (5.172).

Rotor current (I_2) referred to stator of an induction machine at slip (s).

$$I_2 = \frac{E_n}{\sqrt{\left(r_{1e} + \dfrac{r_{2o}}{s}\right)^2 + (X_1)^2}} \text{ amperes.} \tag{5.176}$$

NOTE.   For a wound-rotor induction motor with an effective external resistance R_e ohms per phase referred to the stator, substitute $(r_{2e} + R_e)$ for r_{2o}. To find the starting rotor current of an induction motor, make s equal one.

Stator current (I_1) of an induction machine at slip (s).

$$I_1 = \sqrt{I_2^2 + I_n^2 + 2I_2I_n \sin \alpha} \text{ amperes.} \tag{5.177}$$

NOTE. I_2 is determined by Equation (5.176), I_n is the no-load current, and $\alpha = \sin^{-1}$ (p.f. at no load) $+ \tan^{-1}(sx_2/r_{2o})$. For a wound-rotor motor, substitute $(r_{2e} + R_e)$ for r_{2o} in the expression for α. The starting stator current is given by making s equal one, the starting rotor current being determined as indicated in Equation (5.176).

Power output (P_o) of an induction machine at slip (s).

$$P_o = 3I_2^2 r_{2o} \left(\frac{1-s}{s}\right) - P_{fw} \text{ watts.} \tag{5.178}$$

NOTE.   For a wound-rotor motor, r_{2o} should be increased as indicated in Equation (5.176). When the slip is negative, P_o is negative and gives the power input to an induction generator. P_{fw} is the friction and windage loss.

Power input (P_i) to an induction machine at slip (s).

$$P_i = 3 \frac{I_2^2 r_{2o}}{s} + 3I_1^2 r_{1e} + P_n - 3I_n^2 r_{1e} \text{ watts.} \qquad (5.179)$$

NOTE. P_n is the power in watts taken at no load. For a wound-rotor motor, r_{2o} should be increased as indicated in Equation (5.176). When the slip is negative, P_i is negative and gives the power output of an induction generator.

Output torque (T) of an induction motor at slip (s).

$$T = 0.239 \frac{PI_2^2 r_{2o}}{fs} - T_{fw} \text{ newton meters.} \qquad (5.180)$$

NOTE. Read note on r_{2o} for a wound-rotor motor and s under starting conditions following Equation (5.176). T_{fw} is the friction and windage torque. If P_{fw} is known in watts, T_{fw} equals $\{0.0796 p P_{fw}/[f(1 - s)]\}$ newton meters.

Slip (s) of an induction motor at any stated load.

$$s = \frac{r_{2o} \left(\dfrac{3E_n^2}{P_o} - 2r_{1e} - 2r_{2o} \right)}{\left(\dfrac{3E_n^2}{P_o} - 2r_{1e} - 2r_{2o} \right)^2 - Z_1^2} \text{ approximately.} \qquad (5.181)$$

NOTE. Read note on r_{2o} for a wound-rotor motor following Equation (5.176).

Slip (s) of an induction generator at any stated load.

$$s = \frac{r_{2o} \dfrac{3E_n^2}{P_o}}{\left(\dfrac{3E_n^2}{P_o} \right) - Z_1^2 + r_{1e} \left(\dfrac{3E_n^2}{P_o} \right)} \text{ approximately.} \qquad (5.182)$$

Efficiency (η) of an induction machine.

$$\eta = \frac{P_o}{P_i}. \qquad (5.183)$$

5.8.3   Circle Diagram of an Induction Machine

Operation of a polyphase induction machine with equivalent circuit as shown in Fig. 5.18 can be illustrated by a two-quadrant circle (**Heyland**) diagram showing terminal voltages, currents, and phase angles for all conditions of operation as both a motor and generator (Fig. 5.19). **Voltage vectors V_M and V_G** represent terminal voltage as a motor and generator, respectively. Lines between point O and points on the circle are current vectors, the real power component of which is vertical and the reactive

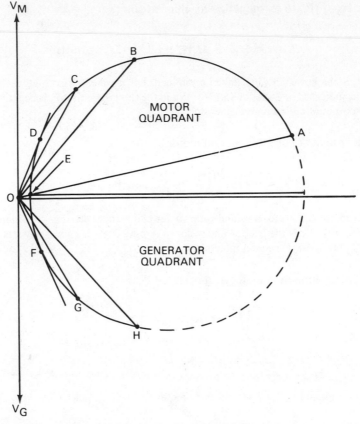

Fig. 5.19 Circle diagram of an induction macine.

power component of which is horizontal. In the motor quadrant, both **real** and **reactive power** go from the power source into the machine; in the generator quadrant, real power goes out of the machine while reactive power goes into it. **Slip** is positive in the motor quadrant and negative in the generator quadrant. The **current vectors** and their **phase angles** going counterclockwise around the circle from point A are:

Vector	Phase Angle	Operating Condition
OA	$V_M OA$	Motor, blocked-rotor
OB	$V_M OB$	Motor, maximum torque (pull-out torque)
OC	$V_M OC$	Motor, rated full load
OD	$V_M OD$	Motor, maximum power factor
OE	$V_M OE$	Motor, no load
OF	$V_G OF$	Generator, maximum power factor
OG	$V_G OG$	Generator, rated full load
OH	$V_G OH$	Generator, maximum torque (pull-away torque)

In the generating mode, if the electrical load is increased beyond rated load to the point of maximum torque, the prime mover will increase speed in a runaway condition. Speed regulation or protection against overspeed of the prime mover is required on an induction generator to prevent runaway.

5.9 TRANSFORMERS

5.9.1 Transformer Characteristics

Electromotive force (E) induced in N turns linked by a flux, $\phi = \Phi_m \sin 2\pi ft$ webers.

$$E = 4.44 Nf\Phi_m \text{ volts.} \tag{5.184}$$

Core loss (P_c) of a transformer at any load.

$$P_c = P_h + P_e \text{ watts.} \tag{5.185}$$

NOTE. P_h is the hysteresis loss and P_e the eddy-current loss in the magnetic circuit of the transformer in Equations (5.186) and (5.187).

Hysteresis loss (P_h) in a medium in which a variable magnetic flux of maximum density B_m teslas changes from positive to negative to positive maximum f times per second.

$$P_h = \eta f B_m^{1.6} \times 10^{-3} \text{ watts/centimeter}^3. \tag{5.186}$$

NOTE. This is an empirical equation and in some cases the exponent of B_m may differ appreciably from 1.6. The hysteresis loop must be symmetrical with no re-entrant loops. The hysteresis coefficient (η) varies in different materials as follows: cast iron, 0.012; cast steel, 0.005; hipernik (50 N), 0.00015; low-carbon sheet steel, 0.003; permalloy (78 N), 0.0001; pure Norway iron, 0.002; silicon sheet steel, 0.00046 to 0.001. Equation (5.186) does not apply to the hysteresis loss in iron rotated in a magnetic field. In the latter case at low flux densities the loss may be twice as much as that due to an alternating flux, but declines in value as the flux density increases. For soft iron the loss by either process will be about the same at 1.50 teslas, and at 2.00 teslas the loss due to rotation is practically zero.

Eddy-current loss (P_e) in thin laminations placed in a sinusoidally varying magnetic flux.

$$P_e = \frac{1.64(tfB_m)^2}{\rho \times 10^8} \text{ watts/centimeter}^3. \tag{5.187}$$

NOTE. t is the thickness of the laminations in centimeters, f is the frequency of flux variation in hertz, B_m is the maximum flux density in teslas, and ρ is the resistivity of the laminations in ohms per cubic centimeter.

Ratio of transformation (T_1) from primary to secondary of a transformer wound with two coils of N_1 (primary) and N_2 (secondary) turns, respectively.

$$T_1 = \frac{N_1}{N_2}. \tag{5.188}$$

NOTE. T_1 equals the ratio E_1/E_2 of the emf's induced, respectively, in the primary and secondary coils and equals approximately the ratio V_1/V_2 of terminal voltages of the primary and secondary coils or the ratio I_2/I_1 of the secondary and primary currents. The ratio of transformation (T_2) from secondary to primary equals $1/T_1$.

Magnetizing current (I_m) in a coil of N turns wound on a magnetic circuit of uniform maximum permeability (μ), l centimeters in mean length, A square centimeters in mean section, and conducting a flux, $\phi = \Phi_m \sin 2\pi f t$ webers.

$$I_m = \frac{10\Phi_m l \times 10^{-8}}{4\pi N \mu A \sqrt{2}} \text{ amperes (approx.).} \tag{5.189}$$

Core-loss current (I_c) in a coil containing an induced emf of E volts and wound on a magnetic circuit in which the core loss is P_c watts.

$$I_c = \frac{P_c}{E} \text{ amperes.} \tag{5.190}$$

No-load current (I_n) taken by a transformer which requires a magnetizing current of I_m amperes and a core-loss current of I_c amperes.

$$I_n = \sqrt{I_m^2 + I_c^2} \text{ amperes.} \tag{5.191}$$

Equivalent resistance (R_1) and **equivalent reactance** (X_1) between the primary terminals of a transformer which has a primary resistance of r_1 ohms, a primary leakage reactance of x_1 ohms, a secondary resistance of r_2 ohms, a secondary leakage reactance of x_2 ohms, and primary to secondary ratio of transformation of T_1.

$$R_1 = r_1 + T_1^2 r_2 \text{ ohms.} \tag{5.192}$$
$$X_1 = x_1 + T_1^2 x_2 \text{ ohms.} \tag{5.193}$$

NOTE. The equivalent resistance and reactance, respectively, between the secondary terminals is given by $R_2 = r_2 + T_2^2 r_1$ ohms and $X_2 = x_2 + T_2^2 x_1$ ohms. The equivalent impedance in each case equals $(R^2 + X^2)^{1/2}$ ohms and $Z_1 = T_1^2 Z_2$ ohms.

Equivalent resistance (R_1) between the primary terminals of a transformer which, with short-circuited secondary, absorbs P_i watts with a primary current of I_1 amperes.

$$R_1 = \frac{P_i}{I_1^2} \text{ ohms.} \tag{5.194}$$

Equivalent impedance (Z_1) between the primary terminals of a transformer which, with secondary short-circuited and with V_1 volts between the primary terminals, takes a primary current of I_1 amperes.

$$Z_1 = \frac{V_1}{I_1} \text{ ohms.} \tag{5.195}$$

Equivalent reactance (X_1) between the primary terminals of a transformer of equivalent resistance (R_1) ohms and equivalent impedance (Z_1) ohms between the primary terminals.

$$X_1 = \sqrt{Z_1^2 - R_1^2} \text{ ohms.} \tag{5.196}$$

Primary voltage (V_1) of a transformer of ratio of transformation (T_1), equivalent resistance and reactance, respectively, between secondary terminals (R_2) and (X_2) ohms, secondary terminal voltage (V_2) volts, secondary current (I_2) amperes, and power factor of the load on the secondary $(\cos \theta_2)$.

$$V_1 = T_1 \sqrt{(V_2 \cos \theta_2 + I_2 R_2)^2 + (V_2 \sin \theta_2 \pm I_2 X_2)^2} \text{ volts.} \tag{5.197}$$

NOTE. The sign before $I_2 X_2$ is $+$ for zero or lagging current phase and $-$ for leading current phase.

Voltage regulation (v.r.) of a transformer at any load, V_1, V_2, and T_1 as in Equation (5.197).

$$v.r. = \frac{V_1 - T_1 V_2}{T_1 V_2}. \tag{5.198}$$

Efficiency (η) of a transformer at any load.

$$\eta = \frac{I_2 V_2 \cos \theta_2}{I_2 V_2 \cos \theta_2 + I_2^2 R_2 + P_c}. \tag{5.199}$$

5.10   AC POWER TRANSMISSION

5.10.1   Conventions

Three-phase power networks are assumed throughout. All apparatus is assumed to be wye-connected, and all impedances and admittances are for one phase or line on this basis. All currents are line currents (phase currents for a wye-connection). Unless otherwise stated, all voltages except in Equations (5.200) and (5.201) are line-to-neutral voltages (phase voltages for a wye-connection). In Equations (5.200) and (5.201) line-to-line voltages are used.

5.10.2 Per Unit Quantities

Ohm's law and **Kirchhoff's laws** apply in circuits whether V, I, and Z are expressed in their absolute units or in per unit (percent/100) of a base value. The base may be chosen arbitrarily, but it must be consistent throughout any set of computations. For electrical equipment it is customary to use the equipment ratings as the base, that is, $V_{Rated} = 1.0$ per unit voltage, $I_{Rated} = 1.0$ per unit current, and rated kVA (kilovolt-amperes) $= 1.0$ per unit kVA. In a power network containing transformers, the same voltage, current, and kVA bases consistent with the transformer ratios must be used throughout.

Impedance and **admittance** are expressed in per unit of base quantities V_{Rated}/I_{Rated} and I_{Rated}/V_{Rated}, respectively. For changing from one base quantity to another, per unit impedances are directly proportional to the kVA base and inversely proportional to the square of the voltage base. Per unit admittances are inversely proportional to the kVA base and directly proportional to the square of the voltage base.

Per unit impedance (Z_{pu}) in terms of the impedance Z in ohms in one phase of a three-phase system, expressed on a three-phase base of kVA kilovolt-amperes and a line-to-line base voltage of V_b volts,

$$Z_{pu} = \frac{Z \times kva}{V_b^2} \times 10^3 \text{ per unit.} \qquad (5.200)$$

Per unit admittance (Y_{pu}) in terms of the admittance Y in siemens in one phase of a three-phase system, expressed on a three-phase base of kva kilovolt-amperes and a line-to-line voltage base of V_b volts,

$$Y_{pu} = \frac{YV_b^2}{kva} \times 10^{-3}. \qquad (5.201)$$

In a single-phase power system, with base kVA equal to rated kVA and base voltage equal to rated kV, the following relationships apply:

Base current in amps $=$ base kVA/base voltage.

$$\textbf{Base impedance in ohms} = \frac{\text{base voltage} \times 1000}{\text{base current}}$$
$$= \frac{\text{base voltage}^2 \times 1000}{\text{base kVA}}.$$

Per unit voltage $=$ actual voltage/base voltage.

Per unit current $=$ actual current/base current.

$$\textbf{Per unit impedance} = \text{actual impedance/base impedance}$$
$$= \frac{\text{actual impedance} \times \text{base kVA}}{\text{base voltage}^2 \times 1000}.$$

Per unit kVA $=$ actual kVA/base kVA.

In a three-phase system, with base KVA equal to total three-phase KVA, and base voltage the phase-to-neutral voltage in KV:

$$\text{Base current in amps} = \frac{\text{base kVA}}{3 \times \text{base voltage}}.$$

$$\text{Base impedance in ohms} = \frac{\text{base voltage} \times 1000}{\text{base current}}$$

$$= \frac{3 \times \text{base voltage}^2 \times 1000}{\text{base kVA}}.$$

Per unit voltage = actual voltage/base voltage.

Per unit current = actual current/base current.

Per unit impedance = actual impedance/base impedance

$$= \frac{\text{actual impedance} \times \text{base kVA}}{3 \times \text{base voltage}^2 \times 1000}.$$

Per unit kVA = actual kVA/base kVA.

To convert from one kVA base, kVA_1, to another, kVA_2:

Per unit impedance on base $kVA_2 = \dfrac{kVA_2}{kVA_1} \times$ per unit impedance on base kVA_1.

Per unit current on base $kVA_2 \quad = \dfrac{kVA_1}{kVA_2} \times$ per unit current on base kVA_1.

Per unit kVA on base $kVA_2 \quad = \dfrac{kVA_1}{kVA_2} \times$ per unit kVA on base kVA_1.

To convert from one kV base, kV_1, to another, kV_2:

Per unit impedance on base $kV_2 = \left(\dfrac{kV_1}{kV_2}\right)^2 \times$ per unit impedance on base kV_1.

Per unit current on base $kV_2 \quad = \dfrac{kV_2}{kV_1} \times$ per unit current on base kV_1.

Per unit voltage on base $kV_2 \quad = \dfrac{kV_1}{kV_2} \times$ per unit voltage on base kV_1.

5.10.3 Transmission Line Characteristics

Voltage (E_s) at the sending end of a transmission line of R ohms resistance per conductor and X ohms reactance per conductor conducting a current of I amperes, the voltage at the receiving end being E_r volts and the phase angle between the receiving-end voltage and the line current being θ_r.

$$E_s = \sqrt{(E_r \cos \theta_r + IR)^2 + (E_r \sin \theta_r \pm IX)^2} \text{ volts.} \qquad (5.202)$$

NOTE. See vector diagram, Fig. 5.20. When I lags or is in phase with E_r, the sign before IX is $+$, and when I leads E_r, the sign beofre IX is $-$. Equation (5.202) neglects capacitance. For transmission lines longer than 40 miles, the capacitance should be included. For cables longer than about 2 miles the capacitance should be included. See Equations (5.207), (5.216), and (5.218).

For application to single-phase lines, use $2R$ and $2X$ in place of R and X.

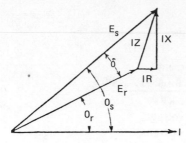

Fig. 5.20 Vector diagram of transmission line voltages and line current.

Phase angle (θ_s) between the sending-end voltage and the line current under the conditions stated in Equation 5.202.

$$\theta_s = \tan^{-1} \frac{E_r \sin \theta_r \pm IX}{E_s \cos \theta_s + IR}.$$ (5.203)

NOTE. The power factor at the sending end is $\cos \theta_s$.

Power input (P_s) to the sending end of a transmission line with δ degrees angular displacement between sending- and receiving-end voltages, the line having an impedance Z ohms and an impedance angle $\theta = \tan^{-1}(X/R)$.

$$P_s = \frac{E_s^2}{Z} \cos \theta - \frac{E_s E_r}{Z} \cos(\delta + \theta) \text{ watts.}$$ (5.204)

NOTE. δ is positive if E_s leads E_r. Capacitance is neglected. P_s will be total three-phase power if line-to-line voltages are used, and power per phase if line-to-neutral voltages are used.

The maximum value is $[(E_s E_r)/Z] - (E_s^2/Z) \cos \theta$; this maximum value may not be attainable without instability in the system.

Power output (P_r) at the receiving end of a transmission line under the conditions in Equation (5.204).

$$P_r = \frac{E_s E_r}{Z} \cos(\delta - \theta) - \frac{E_r^2}{Z} \cos \theta \text{ watts.}$$ (5.205)

NOTE. Read first paragraph of note to Equation (5.204), substituting P_r for P_s. The maximum value of P_r is $[(E_s E_r)/Z] - (E_r^2/Z) \cos \theta$; this maximum value may not be attainable without instability in the system.

Efficiency (η) of a transmission line under the conditions stated in Equation (5.202).

$$\eta = \frac{E_r I \cos \theta_r}{E_r I \cos \theta_r + I^2 R} = \frac{E_r \cos \theta_r}{E_s \cos \theta_s} = \frac{P_r}{P_s}.$$ (5.206)

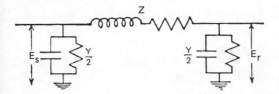

Fig. 5.21 Nominal pi equivalent circuit for a transmission line.

Nominal π (Fig. 5.21) and **nominal T** (Fig. 5.22) equivalent circuits for a transmission line of length l miles having an impedance z ohms per mile composed of resistance r ohms per mile and reactance x ohms per mile, and an admittance y siemens per mile composed of a leakage conductance g siemens per mile and a capacitive susceptance of b siemens per mile.

$$Z = zl \text{ ohms} \tag{5.207}$$
$$= (r + jx)l \text{ vector ohms.}$$
$$Y = yl \text{ siemens} \tag{5.208}$$
$$= (g + jb) \text{ vector siemens.}$$

NOTE. The admittance Y is almost always purely capacitive, leakage usually being negligible. Nominal π and T circuits are usually used for transmission lines of lengths between about 40 and 100 miles. Below 40 miles the capacitance may be neglected; above 100 miles the equivalent π or T should be used [see Equations (5.216) and (5.218)]. For cables, the nominal π and T circuits are usually used for lengths between about 2 and 5 miles. Below 2 miles the capacitance may be neglected; above 5 miles the equivalent π or T should be used.

The long transmission line. Nomenclature: length of line, l miles. Resistance, inductive reactance, capacitance, and leakage conductance, respectively, r ohms per mile, x ohms per mile, c farads per mile, and g siemens per mile; all per conductor. Impedance, $\mathbf{z}$ vector ohms per mile. Admittance, $\mathbf{y}$ vector siemens per mile. In the following equations, $\mathbf{z}$ and $\mathbf{y}$ are vectors equal, respectively, to $r + jx$ and $g + j\omega c$. Sending-end voltage and current, respectively, E_s vector volts and I_s vector amperes. Receiving-end voltage and current, respectively, E_r vector volts and I_r vector amperes. In the following equations, E_s, I_s, E_r, and I_r are vector quantities with a common, arbitrary reference axis.

Propagation constant (α) of a transmission line.

$$\alpha = \sqrt{\mathbf{zy}} = \sqrt{(r + jx)(g + j\omega c)} \text{ hyps/mile.} \tag{5.209}$$

NOTE. The real part of α is called the attenuation constant. The imaginary part is called the wavelength constant, phase constant, or velocity constant.

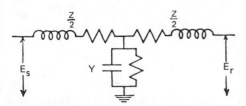

Fig. 5.22 Nominal T equivalent circuit for a transmission line.

Surge impedance or characteristic impedance (Z_0) of a transmission line.

$$Z_0 = \sqrt{\frac{z}{y}} = \sqrt{\frac{r + jx}{g + j\omega c}} \text{ vector ohms.} \qquad (5.210)$$

Hyperbolic angle (θ) of a transmission line.

$$\theta = \alpha l \text{ numerics.} \qquad (5.211)$$

Current (i) and **voltage (e)** at a point on the line distant x miles from the receiving end in terms of receiving-end voltage and current.

$$i = \mathbf{I}_r \cosh \alpha x + \frac{\mathbf{E}_r}{Z_0} \sinh \alpha x \text{ vector amperes.} \qquad (5.212)$$

$$e = \mathbf{E}_r \cosh \alpha x + \mathbf{I}_r Z_0 \sinh \alpha x \text{ vector volts.} \qquad (5.213)$$

NOTE.   To obtain $\mathbf{I}_s$ and $\mathbf{E}_s$, substitute θ for αx.

Current (i) and **voltage (e)** at a point on the line distant x miles from the receiving end in terms of sending-end voltage and current.

$$i = \mathbf{I}_s \cosh (l - x)\alpha - \frac{\mathbf{E}_s}{Z_0} \sinh (l - x)\alpha \text{ vector amperes.} \qquad (5.214)$$

$$e = \mathbf{E}_s \cosh (l - x)\alpha - \mathbf{I}_s Z_0 \sinh (l - x)\alpha \text{ vector volts.} \qquad (5.215)$$

NOTE.   To obtain $\mathbf{I}_r$ and $\mathbf{E}_r$, substitute θ for $(l - x)\alpha$.

Equivalent π circuit for a long transmission line. See Fig. 5.23.

$$Z' = zl \frac{\sinh \theta}{\theta} \text{ vector ohms.} \qquad (5.216)$$

$$Y' = yl \frac{\tanh \dfrac{\theta}{2}}{\dfrac{\theta}{2}} \text{ vector siemens.} \qquad (5.217)$$

NOTE.   Equivalent π and T circuits will represent exactly the performance of smooth transmission lines at their terminals under steady-state conditions. For transmission lines of lengths below about 100 miles, the nominal π and T circuits [see Equations (5.207) and (5.208)] may be used unless very precise results are desired.

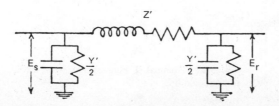

Fig. 5.23   Equivalent pi circuit for a long transmission line.

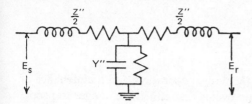

Fig. 5.24 Equivalent T circuit for a long transmission line.

Equivalent T circuit for a long transmission line. See Fig. 5.24.

$$Z'' = zl \frac{\tanh \dfrac{\theta}{2}}{\dfrac{\theta}{2}} \text{ vector ohms.} \qquad (5.218)$$

$$Y'' = yl \frac{\sinh \theta}{\theta} \qquad (5.219)$$

NOTE. See note to Equations (5.216) and (5.217).

General circuit constants (A, B, C, and D) are often used to express the steady-state performance of a network consisting of any combination of constant imped-ances to which power is supplied at one point and received at another. See Fig. 5.25 and note to Equations (5.220) and (5.221).

Sending-end voltage (E_s) and **current** (I_s) in terms of the receiving-end voltage E_r volts and current I_r amperes for the network of Fig. 5.25.

$$\mathbf{E}_s = A\mathbf{E}_r + B\mathbf{I}_r \text{ vector volts.} \qquad (5.220)$$
$$\mathbf{I}_s = C\mathbf{E}_r + D\mathbf{I}_r \text{ vector amperes.} \qquad (5.221)$$

NOTE. All quantities are vector quantities. These are the defining equations for general circuit constants. A and B are found for a given network by obtaining from circuit equations an expression for $\mathbf{E}_s$ in terms of $\mathbf{E}_r$ and $\mathbf{I}_r$; A is then the coefficient of $\mathbf{E}_r$ in this expression, and B is the coefficient of $\mathbf{I}_r$. C and D are found similarly from the expression for $\mathbf{I}_s$ in terms of $\mathbf{E}_r$ and $\mathbf{I}_r$. Thus, for a network consisting of a series impedance Z, $\mathbf{E}_s = \mathbf{E}_r + Z\mathbf{I}_r$, and $\mathbf{I}_s = \mathbf{I}_r$; hence for this network, $A = 1$, $B = Z$, $C = 0$, and $D = 1$.

Receiving-end voltage (E_r) and **current** (I_r) in terms of the sending-end voltage E_s volts and current I_s amperes for the network of Fig. 5.25.

$$\mathbf{E}_r = D\mathbf{E}_s - B\mathbf{I}_s \text{ vector volts.} \qquad (5.222)$$
$$\mathbf{I}_r = -C\mathbf{E}_s + A\mathbf{I}_s \text{ vector amperes.} \qquad (5.223)$$

NOTE. All quantities are vector quantities.

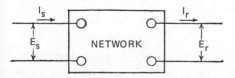

Fig. 5.25 Voltages and currents at sending and receiving ends of a network.

5.11 ELECTRICAL PROTECTION

5.11.1 Protection Principles

Good engineering practice dictates that in the design, operation, and maintenance of electrical equipment, natural and man-made events that can cause danger to people and damage to property must be planned for and protected against. Protection is the process of sensing an abnormal condition, applying logic to determine how and when to take corrective action, and taking the corrective action at the optimum time.

5.11.2 Abnormal Conditions

Abnormal electrical conditions are those during which measurable quantities being monitored, such as voltage, current, temperature, real power, reactive power, and phase angle, go outside of assigned safe limits. Common causes of abnormal conditions are overloads, switching operations, insulation breakdowns, and accidental short circuits or open circuits. Lightning is a frequent cause on high-voltage power systems. It is not always possible to accurately categorize an abnormal condition in terms of whether it is a cause or an effect. An initiating event itself is sometimes an effect of an obscure cause and it may result in a sequence of other events which might be termed either causes or effects.

5.11.3 Speed of Response for Corrective Action

Some abnormal conditions such as high-current short circuits and high transient voltages require corrective action as soon as possible after the initiating event, in an elapsed time of the order of milliseconds or even microseconds. Electromechanical protection devices such as circuit breakers and relays are considered **instantaneous** if they operate in 1 to 3 cycles (1 cycle = 16.7 milliseconds for 60-hertz power). Solid-state protection devices can operate at microseconds with precise timing. Time delays can be introduced in both types either electronically or by auxiliary time delay mechanical devices.

5.11.4 Protection Coordination

Electrical protection is applied to both individual components and to systems. Coordination is a method of protection for a system in which the failure of any individual component will isolate that component only from the system and allow the rest of the system to continue to operate normally. In practice it requires that the **disconnect of power** at devices farthest from the power source will always take place before disconnect of power closer to the source, and where a protective device is fed from more than one source it will be isolated from all its sources before any disconnects occur closer to the source.

5.11.5 Protection Devices

Devices used for protection of electric power equipment were assigned **device numbers** based on their definition and function by IEEE, and these numbers are incor-

Table 5.10 IEEE/ANSI Standard C37.2-1979 Protection Devices with Assigned Device Numbers

Device Name or Function	No.	Application[a]	Device Name or Function	No.	Application[a]
Accelerating	18	C	Jogging	66	C
Alarm	74	I	Level	63, 71	I, P
Annunciator	30	I,	Line	89	C
Atmospheric condition	45	I, P	Lockout	30, 86	P
Auxiliary motor, M-G	88	A	Manual transfer	43	C
Balance, current	61	P	Master contactor	4	C, P
Balance, voltage	60	P	Master element	1	C
Bearing	38	P	Mechanical condition	39	P
Blocking	68	P	Motor, M-G, auxiliary	88	A
Brush operating	35	C	Notching	66	C
Carrier	85	P	Operating mechanism	84	C
Checking	3	C, I	Out-of-step	78	P
Circuit breaker, ac	52	C, P	Overcurrent, ac	50, 51	P
Circuit breaker, anode	7	P	Overcurrent, dc	76	P
Circuit breaker, dc	54, 72	C, P	Overcurrent, directional	67	P
Circuit breaker, field	41	C, P	Overspeed	12	P
Circuit breaker,			Overvoltage	59	P
running	42	C	Permissive control	69	C, P
Contactor	4	C, P	Phase angle	78	P
Control power	8	C, P	Phase reversal, balance	46	P
Current balance	61	P	Phase sequence voltage	47	P
Current, directional	67	P	Pilot wire	85	P
Current, instantaneous	50	P	Polarity	36	P
Current, phase rev., bal.	46	P	Position	33	C
Current, time	51	P	Position changing	75	A
Current, under	37	P	Power, control	8	C, P
Decelerating	18	C	Power, directional	32, 92	P
Differential	87	P	Power factor	55	P
Directional current	67	P	Power, under	37	P
Directional power	32, 92	P	Pressure	63	P
Directional voltage	91, 92	P	Pulse transmitter	77	C, I
Discharge	17	P	Reclose, ac	79	P
Distance	21	P	Reclose, dc	82	P
Equalizer	22	C	Rectifier	35	C
Excitation	31, 53	C	Rectifier misfire	53	P
Field application	56	C	Regulate	90	C
Field breaker	41	C, P	Resistance, rheostat	70	C
Field change	93	C	Resistor contactor	73	C
Field excitation	40	P	Reverse	9	C
Flow	63, 80	C, P	Reverse phase, current	46	P
Frequency	81	P	Sealing, mechanical	39	P
Governor	65	C	Select	43, 83	C
Ground	64	P	Sequence change	10	C
Grounding	57	P	Sequence, incomplete	48	C, P
Incomplete sequence	48	C, P	Sequence, motor		
Interlocking	3	C, I	operated	34	C
Isolating	29	A	Sequence, unit starting	44	C

313

Table 5.10 (*Continued*)

Device Name or Function	No.	Application[a]	Device Name or Function	No.	Application[a]
Short circuit	57	P	Transition	19	C
Shunting	17	P	Trip, trip-free	94	P
Slip ring shorting	35	A	Unbalance, current	46	P
Speed	12, 14	P	Unbalance, voltage	47	P
Speed matcher	15	C	Undercurrent	37	P
Starting circuit breaker	6	C	Underpower	37	P
Start-run contactor	19	C	Underspeed	14	P
Stop	5	P	Undervoltage	27	P
Synchronizing, sync. check	13, 25	C, P	Unit sequence	10	C
Temperature	23	C	Valve, elec. operated	20	A
Temperature	26, 49	P	Vibration	39	P
Time delay	2, 62	C	Voltage balance	60	P
Transfer	43, 83	C	Voltage, directional	91, 92	P
			Voltage, under	27	P

[a]A = auxiliary equipment; C = control; I = indication; P = protection.

porated into ANSI Standard C37.2-1979 for power switchgear assemblies. Table 5.10 is an alphabetical index of these standard devices by name or function. In addition to the standard device number, it indicates whether the device is used primarily for **protection, indication, control,** or as **auxiliary equipment.** A numerical list of these device numbers is included in Section 9. These numbers are used in schematic diagrams, connection diagrams, specifications, and instruction literature of electric power equipment, with appropriate suffix letters when necessary to clarify functions they perform.

5.11.6 Voltage and Current Transformers

Protection devices that monitor voltage, current, or other quantities derived from voltage and current, are designed for 120 rated volts and 5 rated amperes. In power circuits with higher voltages and currents, transformers are connected between the power circuit and the device input circuits as shown in Fig. 5.26.

The **current transformer** is connected in series with the power line and transforms its nominal value of current to 5 amperes. The **voltage transformer** is connected across the power line and transforms its nominal voltage to 120 volts. To keep the secondary circuits safe, the secondary circuits of both voltage and current transformers must be grounded as shown. The **polarity markers** (black circles) and arrows indicate relative instantaneous directions of current in the transformer windings. They are important for correct operation of devices having more than one voltage or current input. They are always used in a manner that keeps direction of current flow in the secondary-connected device the same as if the primary circuit were itself connected to the device at that location. Data on standard current transformers are given in Tables 5.11, 5.12, and 5.13.

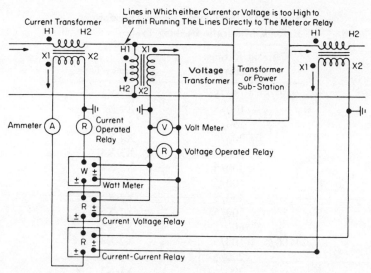

Fig. 5.26 Voltage and current (instrument) transformer connections. Used with permission of Westinghouse Electric Corporation.

Table 5.11 Standard Current Transformer Ratings, Single and Double Ratio

Single Ratio (amperes)	Double Ratio with Series, Parallel Primary Windings (amperes)	Double Ratio with Taps in Secondary Winding (amperes)
10/5	25 × 50/5	25/50/5
15/5	50 × 100/5	50/100/5
25/5	100 × 200/5	100/200/5
40/5	200 × 400/5	200/400/5
50/5	400 × 800/5	300/600/5
75/5	600 × 1,200/5	400/800/5
100/5	1,000 × 2,000/5	600/1,200/5
200/5	2,000 × 4,000/5	1,000/2,000/5
300/5		1,500/3,000/5
400/5		2,000/4,000/5
600/5		
800/5		
1,200/5		
1,500/5		
2,000/5		
3,000/5		
4,000/5		
5,000/5		
6,000/5		
8,000/5		
12,000/5		

© 1975 IEEE.

Table 5.12 Standard Current Transformer Ratings, Multiratio Bushing Type

Current Ratings (amperes)		Secondary Taps
600/5	50/5	X2-X3
	100/5	X1-X2
	150/5	X1-X3
	200/5	X4-X5
	250/5	X3-X4
	300/5	X2-X4
	400/5	X1-X4
	450/5	X3-X5
	500/5	X2-X5
	600/5	X1-X5
1200/5	100/5	X2-X3
	200/5	X1-X2
	300/5	X1-X3
	400/5	X4-X5
	500/5	X3-X4
	600/5	X2-X4
	800/5	X1-X4
	900/5	X3-X5
	1000/5	X2-X5
	1200/5	X1-X5
2000/5	300/5	X3-X4
	400/5	X1-X2
	500/5	X4-X5
	800/5	X2-X3
	1100/5	X2-X4
	1200/5	X1-X3
	1500/5	X1-X4
	1600/5	X2-X5
	2000/5	X1-X5
3000/5	1500/5	X2-X3
	2000/5	X2-X4
	3000/5	X1-X4
4000/5	2000/5	X1-X2
	3000/5	X1-X3
	4000/5	X1-X4
5000/5	3000/5	X1-X2
	4000/5	X1-X3
	5000/5	X1-X4

© 1975 IEEE.

The term **instrument transformer** is applied to these transformers in practice, whether they are used with instruments or with protective devices (relaying). The term **burden** is used with instrument transformers to differentiate this type of load from a primary circuit power load.

Current transformers used for relaying have an accuracy class designation consisting of the letter C or T followed by a number 10, 20, 50, 100, 200, 400, or 800.

Table 5.13 Standard Burden for Current Transformers

Standard Burden Designation	Characteristics		Characteristics for 60 Hz and 5 A Secondary Current		
	Resistance (ohms)	Inductance (mH)	Impedance (ohms)	Apparent Power[a] (VA)	Power Factor
B-0.1	0.09	0.116	0.1	2.5	0.9[b]
B-0.2	0.18	0.232	0.2	5.0	0.9[b]
B-0.5	0.45	0.580	0.5	12.5	0.9[b]
B-1	0.5	2.3	1.0	25	0.5[c]
B-2	1.0	4.6	2.0	50	0.5[c]
B-4	2.0	9.2	4.0	100	0.5[c]
B-8	4.0	18.4	8.0	200	0.5[c]

[a]At 5 A; note that VA = I^2Z, or 2 Ω at 5 A = 2×5^2 = 50 VA.
[b]Usually considered metering burdens, but data sheets may give metering accuracies at B-1.0 and B-2.0.
[c]Usually considered relaying burdens.
© 1975 IEEE.

C means the percent **ratio error** can be calculated; T means it has been determined by test. The number is the secondary terminal voltage the transformer will deliver to a standard burden as listed in Table 5.13 at 20 times normal secondary current without exceeding a 10% ratio error. Also, the ratio error should not exceed 10% at any current from 1 to 20 times rated current at any lesser burden. For example, a transformer rated C100 means the ratio error can be calculated and will not exceed 10% at any current from 5 to 100 amperes with a connected burden of 1 ohm.

Figure 5.27 shows typical excitation current/voltage characteristics for current transformers with ratios 50/5 to 600/5. In some applications where the ratio is low and burden is high, such as in detection of small ground fault currents, the ratio error is high, and it is necessary to add exciting current to burden current in order to determine the primary current at which a relay will operate. For example, a sensitive overcurrent relay with a minimum operating current of 0.25 ampere has a burden impedance of about 7.0 ohms. When used with the 50/5 ratio transformer of Fig. 5.27 its secondary resistance plus lead resistance adds about 0.1 ohm. Secondary voltage required to provide 0.25 ampere burden current is 7.1 × 0.25 = 1.8 volts. From Fig. 5.27, exciting current at 1.8 volts is 0.38 ampere and total secondary current is 0.63 ampere. Total primary current is 10 × 0.63 = 6.3 amperes to operate the relay, not 2.5 amperes which would be the value based on the turns ratio if there were no ratio error.

Voltage transformers, also called **potential transformers** (although the term voltage transformer is now preferred), have primary windings rated at nominal system voltages per Table 5.14 and secondary windings normally rated 120 volts to match the voltage rating of most protective devices. Standard ratings of voltage transformers are given in Table 5.15. **Accuracy classifications** of voltage transformers are stated by a number from 0.3 to 1.2, representing the percent ratio correction to obtain a true ratio. These accuracies are intended mainly for metering or indication purposes and are high enough that any standard voltage transformer is adequate for

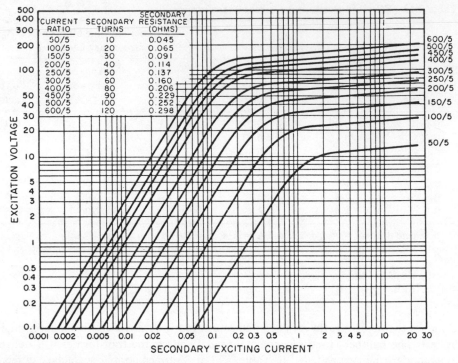

Fig. 5.27 Typical secondary excitation characteristics of current transformers. © 1975 IEEE.

protective relaying purposes as long as it is applied within its thermal and voltage limits.

5.11.7 Fuses

The simplest and oldest of protection devices, fuses, are now standardized in low, medium, and high voltage ratings (Table 5.16). They are used either for primary protection or for backup protection coordinated with circuit breakers. Low-voltage fuses are made in plug and cartridge housings up to 120 volts and in cartridges up to 600 volts. High-voltage power fuses have insulating tube cartridge housings. At voltages up to 34.5 kV they are made with replaceable fusible elements, above 34.5 kV the entire assembly is replaced after operation.

   Current-limiting fuses have fast-melting characteristics which cause interruption of fault current before it reaches its peak value on the first cycle of a short circuit. If a power circuit has total available fault current higher than the interrupting capability of its circuit breaker, current-limiting fuses used on the source side of the breaker will limit fault current and interrupt it before the breaker operates, thus preventing possible damage to the breaker.[1]

5.11.8 Synchronous Generator Protection

A typical generator protection scheme is shown in Fig. 5.28. The devices shown dashed are optional for small or low-voltage machines. Additional devices used for

Table 5.14 Voltage Classes and Nominal System Voltages

Voltage Class	Nominal System Voltage (volts)[a]			Maximum System Voltage	Reference
	Two-Wire	Three-Wire	Four-Wire		
Low voltage					
Single-phase systems	(120)	120/240		127 or 127/254	IEEE Std 100-1972 (ANSI C42.100-1972) defines low-voltage system as "less than 750 V." It is now being proposed as "maximum rms ac voltage of 1000 V."
Three-phase systems			208Y/120	220	
		(240)	240/120	254	
		(480)	480Y/277	508	
		(600)		635	
Medium voltage		(2,400)		2,540	These voltages are listed in ANSI C84.1-1970 but are not identified by voltage class.
		4,160		4,400	
		(4,800)		5,080	
		(6,900)		7,260	
			12,470Y/7,200	13,200	
		13,800	13,200Y/7,620	13,970	
		(23,000)	(13,800Y/7,970)	14,520	Note that additional voltages in this class are listed in ANSI C84.1-1970 but are not included because they are primarily oriented to electric-utility system practice.
				24,340	
			24,940Y/14,400	26,400	
		(34,500)	34,500Y/19,920	36,510	
		(46,000)		48,300	
		(69,000)		72,500	
High voltage		115,000		121,000	ANSI C84.1-1970 identifies these as "higher voltage three-phase systems" (as well as 46 kV and 69 kV).
		138,000		145,000	
		(161,000)		169,000	
		230,000		242,000	ANSI C92.2-1974
Extra high voltage		345,000		362,000	
		500,000		550,000	
		735,000–765,000		800,000	

[a]System voltages shown without parentheses are preferred.
© 1975 IEEE.

319

Table 5.15   Standard Voltage Transformer Ratings

Primary (volts)	Secondary (volts)	Ratio
120	120	1/1
240	120	2/1
480	120	4/1
600	120	5/1
2,400	120	20/1
4,200	120	35/1
4,800	120	40/1
7,200	120	60/1
8,400	120	70/1
12,000	120	100/1
14,400	120	120/1
24,000	120	200/1
36,000	120	300/1
48,000	120	400/1
72,000	120	600/1
96,000	120	800/1
120,000	120	1,000/1
144,000	120	1,200/1
168,000	120	1,400/1
204,000	120	1,700/1
240,000	120	2,000/1
300,000	120	2,500/1
360,000	120	3,000/1

© 1975 IEEE.

Table 5.16   Standard Fuse Ratings

Low Voltage

0–600 A			601–6000 A
250 V or Less (amperes)	300 V or Less (amperes)	600 V or Less (amperes)	600 V or Less (amperes)
0–30	0–15	0–30	601–800
31–60	16–20	31–60	801–1200
61–100	21–30	61–100	1201–1600
101–200	31–60	101–200	1601–2000
201–400		201–400	2001–2500
401–600		401–600	2501–3000
			3001–4000
			4001–5000
			5001–6000

Table 5.16 (*Continued*)

High Voltage

Nominal Rating (kV)	Maximum Continuous Current (amperes)	Maximum Three-Phase Symmetrical Interrupting Rating (MVA)
Boric-Acid Power Fuses, Refillable		
2.4	200, 400, 720	155
4.16	200, 400, 720	270
7.2	200, 400, 720	325
14.4	200, 400, 720	620
23	200, 300	750
34.5	200, 300	1000
Boric-Acid Power Fuses, Nonrefillable		
34.5	100, 200, 300	2000
46	100, 200, 300	2500
69	100, 200, 300	2000
115	100, 250	2000
138	100, 250	2000
Outdoor Expulsion-Type Power Fuses		
7.2	100, 200, 300, 400	162
14.4	100, 200, 300, 400	406
23	100, 200, 300, 400	785
34.5	100, 200, 300, 400	1174
46	100, 200, 300, 400	1988
69	100, 200, 300, 400	2350
115	100, 200,	3110
138	100, 200,	2980
161	100, 200,	3480
Current-Limiting Power Fuses		
2.4	100, 200, 450	155–210
2.4/4.16Y	450	360
4.8	100, 200, 300, 400	310
7.2	100, 200	620
14.4	50, 100, 175, 200	780–2950
23	50, 100	750–1740
34.5	40, 80	750–2600

© 1975 IEEE.

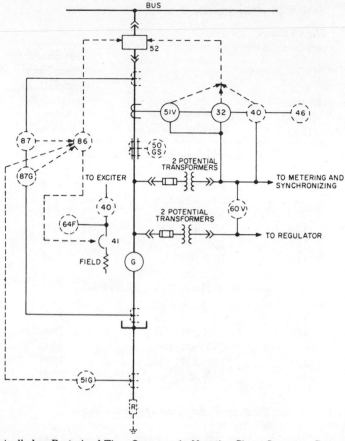

Fig. 5.28 Synchronous generator protection. © 1975 IEEE.

3 Voltage-Controlled or Restrained Time Over-current Relays (Device 51V)

1 Time Overcurrent Relay (Device 51G) (use if generator neutral is grounded)

1 Instantaneous Overcurrent Relay (Device 50GS) (use if generator neutral is not grounded)

1 Power Directional Relay (Device 32) (may be omitted if protective function is included with steam turbine)

1 Stator Impedance or Loss of Field Current Relay (Device 40)

1 Negative Phase Sequence Current Relay (Device 46)

1 Field Circuit Ground Detector (Device 64F)

1 Potential Transformer Failure Relay (Device 60V)

1 Lockout Relay (Device 86) (hand reset)

3 Fixed or Variable Percent Differential Relays (Device 87)

1 Current-Polarized Directional Relay (Device 87G)

protection of large power station generators are discussed in Chapter 6 of *Applied Protective Relaying* (see Bibliography).

5.11.9 Power Bus Protection

Power buses connected through circuit breakers to sources or as feeders to loads require protection against short circuits by the fastest possible means of sensing and breaker opening. Moreover, the protective relays must differentiate between a short

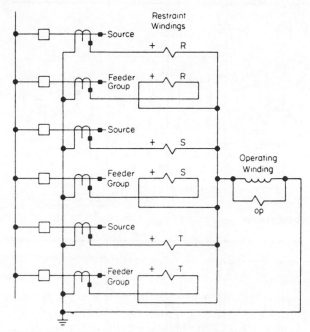

Fig. 5.29 Power bus protection with iron-core current transformers and differential current relay. Reproduced by permission of Relay-Instrument Division, Westinghouse Electric Corp.

circuit on the bus itself and one on the source or feeder, and operate the breaker only on the former. To accomplish this, special types of **differential relays** (device 87) with multiple **restraint windings** connected (Fig. 5.29) to current transformers on each source or feeder group are used. Under normal operation, the restraint winding currents are balanced and no current flows through the relay operating winding. A fault on the bus disturbs the balance and causes the relay to operate; a fault on the source or load side of any current transformer does not do so.

An alternative differential protection system using a sensitive **voltage relay** and **linear couplers** is shown in Fig. 5.30. Linear couplers are air-core bushing-type transformers wound on nonmagnetic toroidal cores, designed to produce 5 secondary volts per 1000 amperes of primary current. Under normal operation or external fault conditions, the secondary voltages in all linear couplers add up to zero; for a fault on the bus, their sum is a net voltage which operates the relay. Advantages of this system are that the linear couplers are not responsive to dc transients or saturation effects of iron, which sometimes cause false operation of differential relays using iron-core current transformers.

5.11.10   Transformer Protection

Primary **fuses** and a secondary **circuit breaker** operated by phase and ground (neutral) overcurrent relays (Fig. 5.31) form a satisfactory protection system for most small power transformers. On large transformers, the additional protection cost of a primary breaker, differential relay system (87T), and pressure switch (63) (Fig. 5.32) is recommended because of the very high outage and repair costs of an internal

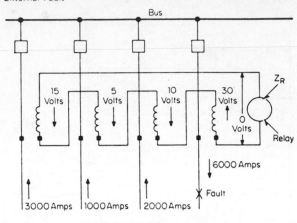

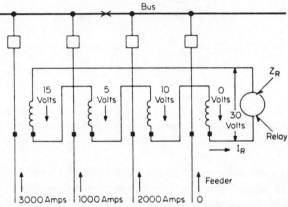

Fig. 5.30 Bus protection with air-core linear coupler sensors and sensitive voltage relay. Reproduced by permission of Relay-Instrument Division, Westinghouse Electric Corp.

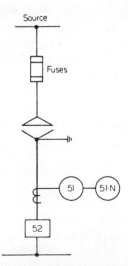

Fig. 5.31 Transformer protection with primary fuses and secondary breaker. Reproduced by permission of Relay-Instrument Division, Westinghouse Electric Corp. Note: Refer to Numerical List of Device Numbers, Sec. 9, for identification of numbered devices in this and subsequent figures.

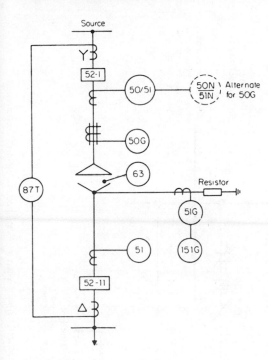

Fig. 5.32 Transformer protection with primary and secondary breakers and differential relay. Reproduced by permission of Relay-Instrument Division, Westinghouse Electric Corporation.

fault. Transformers are often paralleled in practice. A tie circuit breaker is used to isolate parallel secondary circuits for a fault on either side. Protection diagrams for these arrangements are shown in Chapter 8 of *Applied Protective Relaying* (see Bibliography).

5.11.11 Motor Protection

Ac motors vary in ratings from fractional horsepower to tens of thousands of horsepower and in speeds from 3600 revolutions per minute down to 180 or less. Physical size of a motor is directly proportional to horsepower rating and inversely proportional to speed. Standards for motors separate them into three size categories: fractional horsepower, medium (integral horsepower to approximately 1500 h.p. at 3600 rpm), and large (over 1500 h.p.).

Fractional h.p. motors can be adequately protected by a **temperature sensitive thermostatic switch** imbedded in the end windings and connected in series with the power source, if they operate from single-phase power.

Three-phase fractional h.p. and medium ac motors in most applications are protected by the combination of components called **motor control** shown in Fig. 5.33. Where control of many motors is desired from a single location, units of **motor control** consisting of these components are grouped in separate enclosure cubicals and the groups are called motor control centers. Function of the circuit breaker is to protect against short circuit currents in the motor or in the power leads connecting to the motor. The three-pole **magnetic contactor M** applies and disconnects power to the motor. Its auxiliary contacts operate one or more indicator lights to show oper-

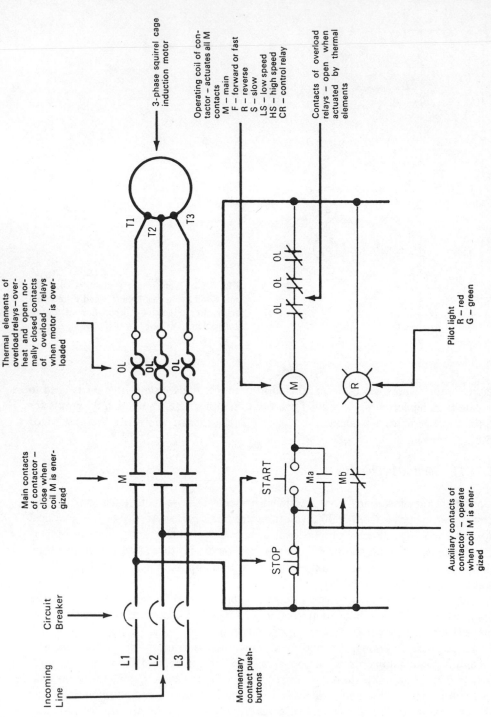

3-phase squirrel cage induction motor

Operating coil of contactor – actuates all M contacts
M – main
F – forward or fast
R – reverse
S – slow
LS – low speed
HS – high speed
CR – control relay

Contacts of overload relays – open when actuated by thermal elements

Thermal elements of overload relays – overheat and open normally closed contacts of overload relays when motor is overloaded

Pilot light
R – red
G – green

Main contacts of contactor – close when coil M is energized

Circuit Breaker

Incoming Line

L1
L2
L3

T1
T2
T3

OL OL OL

OL
OL
OL

M

M

R

START

STOP

Ma

Mb

Momentary contact push-buttons

Auxiliary contacts of contactor – operate when coil M is energized

Fig. 5.33 Protection and control for medium ac motors. Reproduced by permission of Westinghouse Electric Corporation.

326

ating conditions. The **overload relays** open the contactor circuit at preset conditions of overcurrent and time.

Where reliability of motors is important enough to warrant more complete and more sophisticated protection than provided by motor control, **relay protection** for operating circuit breakers is used as shown in Fig. 5.34, up to about 1500 h.p. This scheme provides **undervoltage** and **phase-sequence voltage protection** and sensitive **ground fault protection** in addition to **phase overcurrent** and **fault protection.**

For large motors above 1500 h.p. the complete complement of relay protection of Fig. 5.35 is typical. **Differential current protection** (device 87) for very fast **internal fault breaker opening** and **resistance temperature detectors (RTD)** in the stator windings for accurate alarm and protection on overloads are used in addition to the protective devices on smaller motors.

Protection of a large motor during a long acceleration period may be required in addition to protection while running. The time required for a motor to accelerate from zero to full speed is proportional to total inertia of the motor rotor, coupling, and load, and inversely proportional to average torque during the starting period, as indicated by Equation (5.135). Motor standards define a maximum value of total inertia for a given speed and horsepower rating that can safely be accelerated by a motor of standard electrical design without injurious temperature rise. In present practice, large motors are frequently required to accelerate loads with higher total inertia than these maximum values. To increase the physical size of a motor for the sole purpose of starting a high-inertia load may be uneconomical from the standpoints of both higher first cost and lower efficiency at its rated load. Therefore, in applications of motors driving high-inertia loads, special attention should be given to **starting protection.**

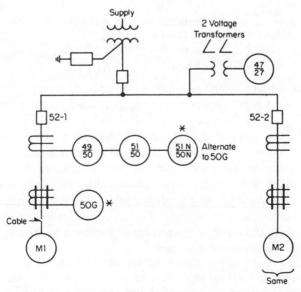

Fig. 5.34 Relay protection for motors with ratings up to 1500 horsepower. Reproduced by permission of Relay-Instrument Division, Westinghouse Electric Corp.

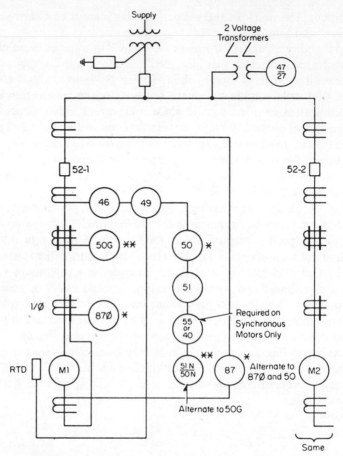

Fig. 5.35 Relay protection for large motors above 1500 horsepower. Reproduced by permission of Relay-Instrument Division, Westinghouse Electric Corp.

If the starting period is equal to or greater than the maximum permissible locked-rotor time for safe temperature rise, starting protection requires, in addition to an overcurrent relay set for maximum permissible locked-rotor time, a means of disabling (blocking) this relay at a time during the starting period when the rotor is known to be in motion. Two methods of detecting rotor motion during start and blocking the overcurrent relay as a result of this motion are in current use: (1) a **mechanical speed switch** coupled to the motor shaft with contacts set to open at a predetermined speed, and (2) an **impedance relay** responsive to the combination of motor voltage, current, and phase angle. The impedance relay has closed contacts at zero and low speeds that open at a set point represented by higher impedance and lower phase angle as the speed increases.

Voltage at the motor terminals during the starting period is another important consideration in motor starting protection because it affects the starting time. Large induction motors have **inrush currents** of 400 to 700% which decrease slowly until the rotor is almost up to full speed. At any time during the starting period, torque is proportional to the square of the voltage. If the voltage remains low during a large part of the period, the starting time increases as the square of the voltage drop. For example, at 70% voltage, the average torque is only about half of normal and the starting time is approximately doubled.

5.11.12 Transient Protection

Since abnormal short-time transient voltages and currents can occur in all high- and low-voltage circuits even under normal operating conditions, whenever the possibility exists for their causing malfunction or damage, protection measures must be taken to prevent or at least mitigate these effects.

In high-voltage power circuits the two most common causes of **transient surges** are **lightning** and **switching** operations. **Lightning arresters** are used between high-voltage terminals and ground on outdoor electrical equipment. These are nonlinear resistors made of zinc oxide blocks that act as insulators at the rated line voltage but pass very high currents during voltage surges. To reduce the magnitude of voltage surges during switching operations in high-voltage circuits, resistors are inserted across the breaker during the switching operation to reduce the rate of change of current during the operation. Detailed descriptions of up-to-date devices and practices for surge protection of high voltage equipment can be found in *Surge Protection of Power Systems* (see Bibliography).

From the standpoint of protection of equipment from malfunction, low-voltage devices must be protected against transients caused either within their own circuits or by **electromagnetic** or **electrostatic coupling** to high-voltage circuits. One or more of the following protective measures are applied in relaying and control circuits to guard against malfunction and damage due to transients: (1) **physical separation of coupled circuits** to reduce mutual inductance and mutual capacitance, (2) **series inductance** to slow the rate of change of current surges, (3) **parallel capacitance** or **resistance–capacitance** combinations to slow the rate of rise of voltage surges, (4) **parallel zener (clamping) diodes** to limit magnitude of voltage surges, (5) insertion of **resistance** across breaker or relay contacts during operation, (6) **shielding of control leads** to reduce capacitive coupling to other circuits, (7) **twisting of control leads** to reduce common mode coupling to other circuits, (8) use of **surge protection package circuits,** also called filters or buffers, in the input lead circuits of all solid-state

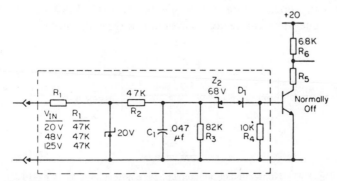

Fig. 5.36 Typical filter for transient protection of solid-state relays. Reproduced by permission of Relay-Instrument Division, Westinghouse Electric Corp.

relays. A typical such filter is shown in Fig. 5.36. For a detailed discussion of surge protection in power system relaying refer to Chapter 4 in *Applied Protective Relaying* (see Bibliography).

5.12 EMERGENCY POWER SYSTEMS

5.12.1 Reliability of Electric Utilities

Every means available to electric utilities is used in the design and operation of their distribution, transmission, and generation systems to give the highest possible reliability of service. **Parallel** or **looped feeders** provide alternate power paths from sources to consumers wherever possible. Low-voltage distribution systems are interconnected into **networks** to provide multiple power paths to commercial and industrial buildings in downtown and other areas where these buildings are concentrated. Protective devices restore power almost immediately in many cases where a fault occurs in a distribution system and can be isolated from the rest of the system. However, the needs of some consumers are so critical that the loss of power for even a very short time interval cannot be tolerated. For example, a hospital cannot have loss of power for more than a few seconds without possibly endangering someone's life. Industries such as those manufacturing semiconductors and photo processing require constant temperatures maintained by electric heat and electric control, where even a short power failure can be extremely costly. Communication services, particularly telephone systems, are another example. And one of the most critical applications where continuous power is essential is computers, where power interruptions as short as a few cycles (1 cycle = 0.0167 second) can cause serious problems.

5.12.2 Standby Power Sources

If power loss of a few seconds or a few minutes at most can be tolerated, an **alternate source,** normally not in use until the prime (utility) source fails and operating completely independent of the prime source, is used. The most common type of standby power source is an engine-driven generator set. On loss of voltage at the load bus a normally closed transfer switch disconnects the load from the utility source and initiates startup of the generator set. When rated voltage exists at the generator terminals a normally open switch connects it to the load. Most diesel and gasoline engine-driven generators can restore power in less than a minute.

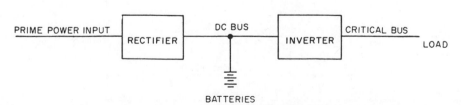

Fig. 5.37 Nonredundant uninterruptible power supply. © 1975 IEEE.

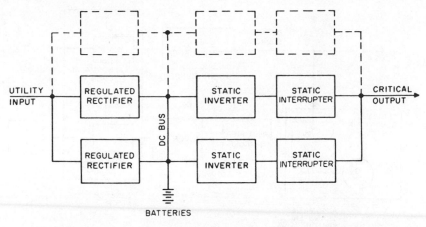

Fig. 5.38 Redundant uninterruptible power supply. © 1975 IEEE.

5.12.3 Uninterruptible Power Supplies (UPS)

To meet the operating needs of computers and critical manufacturing processes that cannot tolerate interruptions of even a few cycles, power supplies that operate in parallel with electric utility sources and supply their load without interruption when the utility source fails have become widely used during the past two decades. Most of these have batteries operating an inverter connected to the critical load at all times, the batteries being charged by a rectifier from the prime power source as shown in Fig. 5.37. Where utmost reliability is required, two or more redundant systems of rectifiers and inverters are connected in parallel with static selector switches on their load side as shown in Fig. 5.38. A UPS system with static selector switches connecting to either the utility prime source or to a rectifier, battery, and inverter source is shown in Fig. 5.39. Figure 5.40 goes one step further in capability for long-time emergency operation by providing alternate power to charge the batteries from an engine-driven generator. Reference 11 describes the preceding UPS systems and other emergency power systems in more detail.

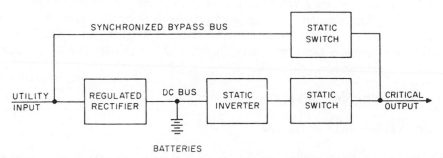

Fig. 5.39 Uninterruptible power supply with static switch bypass to prime source. © 1975 IEEE.

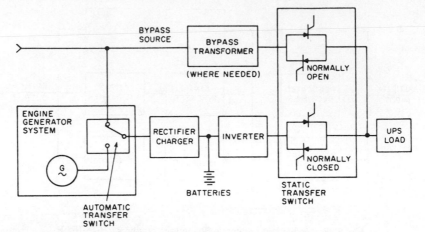

Fig. 5.40 Uninterruptible power supply with static switch bypass to prime source and automatic transfer switch to standby source.

5.13 ELECTRONICS

5.13.1 Definition and Scope

Electronics is the specialized field within electrical engineering that deals with electron devices and their utilization. An **electron device** is one in which current flows principally by the movement of electrons through a **vacuum, gas,** or **semiconductor.** In **semiconductors,** current flow may occur either by electron movement in the negative-to-positive direction, or by the apparent movement of electron sites, called holes, in the opposite direction.

5.13.2 Electronic Applications

The term "electronic" describes circuits, systems, and items of equipment utilizing electron devices, for example, electronic alarm and electronic voltage regulator.

5.13.3 Electronics Engineering

Branches of electrical engineering that deal primarily with electronics are radio, television, data communication, data and word processing, solid-state control and indication, and audio systems. Nearly every other engineering discipline becomes involved in electronics technology today because of the proliferation of electronic devices and techniques in other fields.

5.14 ELECTRONIC DEVICES

5.14.1 Vacuum and Gas Devices

Vacuum tubes contain a thermionic emitter of electrons, the **filament** or **cathode,** and an electron-collecting element called the **anode.** A tube with only these **two elements** forms a **diode** capable of rectification and other diode functions. **Triodes, tet-**

rodes, pentodes, and other multigrid configurations contain one or more grids between cathode and anode to control electron flow, thus making them capable of amplification, switching, and generation of ac voltages and currents by oscillation. Tubes containing gases, with or without thermionic emitters, in which the gas ionizes and allows current flow from anode to cathode are used as **rectifiers, voltage regulators,** and **light generators.**

Prior to the invention of the transistor in 1947, vacuum and gas tubes were used in nearly all electronic applications. In present technology they have been largely superseded by **semiconductor devices** which in most cases perform the same functions better and at less cost. Two exceptions are high-power vacuum tubes and gas-tube lighting devices. Up-to-date reference literature on these is available, and also on small vacuum tubes because of the large number of them still in use.[12,13]

5.14.2 Semiconductor Diode

When **impurities** are added to the semiconductor materials **silicon** and **germanium** in a process called **doping,** they acquire either an **excess** of **electrons** and become **N-type material,** or they acquire a **deficiency** of **electrons** and become **P-type material,** depending on the type of impurity added. A **diode** is formed by a surface or a point contact junction between P-type and N-type materials. When voltage is applied across it, positive to P-type and negative to N-type, current flows freely across the junction, labeled I_F in Fig. 5.41. When voltage is applied in the reverse direction, only a small leakage current flows until a breakdown voltage $-B_{VR}$ causes an avalanche current $-I_R$, as shown in Fig. 5.42.

Diodes are marked or labeled so as to indicate the direction of current flow through them (Fig. 5.43). This is opposite to the direction of electron flow, and is in the same direction as the flow of electron sites or holes.

The most common applications of diodes are their use as ac power rectifiers (see Figs. 5.44 through 5.48. Many other diode applications have resulted from unique properties of some types, as listed in Table 5.17.

5.14.3 Junction Transistor

A thin section of P-type material sandwiched between two thicker sections of N-type material forms an **NPN junction transistor** (Fig. 5.49a). The collector, base, and emitter are labeled C, B, and E, respectively. Figure 5.49c shows a PNP transistor

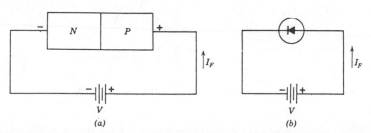

Fig. 5.41 Forward current in a diode. (a) Physical layout, (b) circuit symbol. From *Fundamentals of Electronics*, E. N. Lurch. © 1981 by John Wiley & Sons.

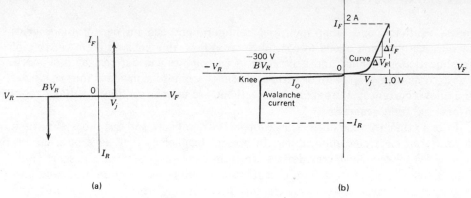

Fig. 5.42 Forward and reverse voltage-current characteristics. (*a*) Ideal diode, (*b*) actual diode. From *Fundamentals of Electronics*, E. N. Lurch. © 1981 by John Wiley & Sons.

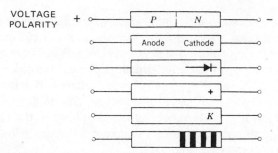

Fig. 5.43 Diode markings and voltage polarity for forward current flow. From *Fundamentals of Electronics*, E. N. Lurch. © 1981 by John Wiley & Sons.

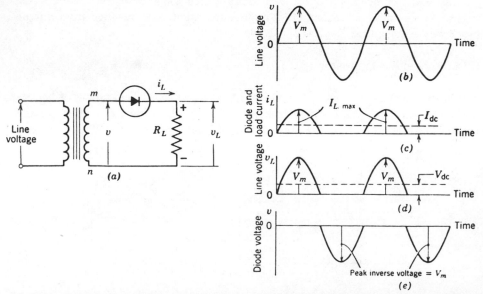

Fig. 5.44 The half-wave rectifier. (*a*) Circuit, (*b*) input voltage, (*c*) diode and load current, (*d*) load voltage, (*e*) diode voltage. From *Fundamentals of Electronics*, E. N. Lurch. © 1981 by John Wiley & Sons.

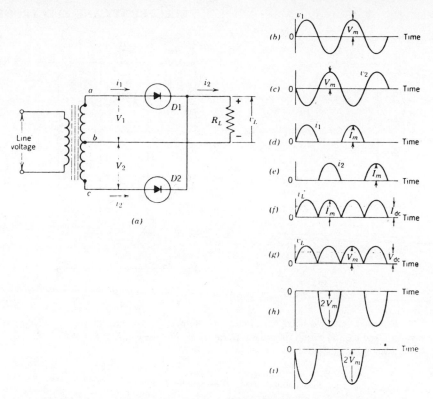

Fig. 5.45 The full-wave rectifier. (*a*) Circuit, (*b*) voltage applied to D1, (*c*) voltage applied to D2, (*d*) current in D1, (*e*) current in D2, (*f*) load current, (*g*) load voltage, (*h*) voltage across D1, (*i*) voltage across D2. From *Fundamentals of Electronics*, E. N. Lurch. © 1981 by John Wiley & Sons.

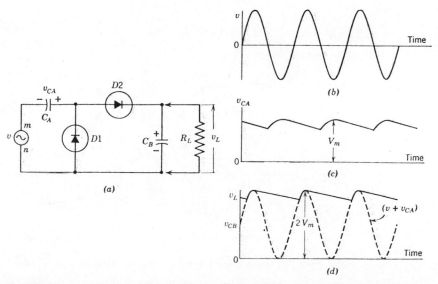

Fig. 5.46 Half-wave voltage doubler. (*a*) Circuit, (*b*) line voltage, (*c*) waveform across C_A, (*d*) waveform across C_B. From *Fundamentals of Electronics*, E. N. Lurch. © 1981 by John Wiley & Sons.

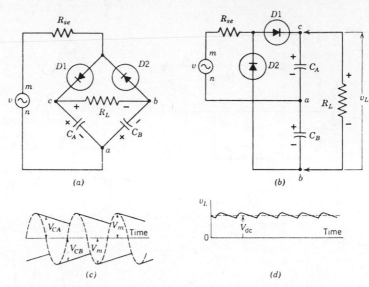

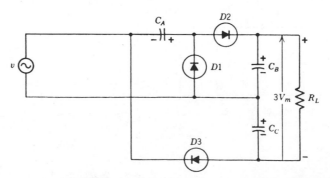

Fig. 5.47 Full-wave voltage doubler. (a) Circuit, (b) alternate form of circuit layout, (c) voltage wave-form across C_A and C_B, (d) output voltage waveform. From *Fundamentals of Electronics*, E. N. Lurch. © 1981 by John Wiley & Sons.

with an N section between two P sections, and the polarity of V_{BB} and V_{CC} reversed. Because junction transistors operate both by electron flow from negative to positive and by hole flow from positive to negative, they are also called **bipolar junction transistors (BJTs)**.

Junction transistors have collector current I_C larger than base current I_B, and emitter current:

$$I_E = I_B + I_C. \tag{5.224}$$

Fig. 5.48 Voltage tripler. From *Fundamentals of Electronics*, E. N. Lurch. © 1981 by John Wiley & Sons.

Table 5.17   Semiconductor Diode Applications

Application	Diode Name or Type	Diode Properties
Amplifier, radio frequency	Gunn (Tunnel)	Negative resistance at UHF and microwave frequencies
Capacitance, variable	Varactor	C variable with voltage
Frequency multiplier	PIN	Charge storage, variable R
Light source	Light-emitting (LED)	Light emission
Light to electric power	Photo-voltaic	Dc voltage and current proportional to light
Light sensor	Photo	Resistance variable with light
Noise source	Noise	Random noise generation
Oscillator	Gunn	Negative resistance at UHF and microwave frequencies
Oscillator frequency control	Varactor	C variable with voltage
Rectifier, ac power or signal	Silicon	High or lower power, usable at high temperatures
Rectifier, signal	Germanium	Low forward voltage drop
Switching, high speed	PIN	Low forward voltage drop, variable resistance
Switching, UHF and microwave	Hot carrier (HCD)	Low capacitance, high efficiency
Transient suppression	Zener	Unidirectional voltage limiting
Voltage reference	Zener	Constant voltage with variable current
Voltage regulator	Zener	Constant voltage with variable current

Dc beta (βdc) is defined at any operating point by

$$\beta_{dc} = \frac{I_C}{I_B}. \tag{5.225}$$

This ratio is equal to current gain when the transistor is used as a dc amplifier. **Dc alpha (α_{dc})** is defined by

$$\alpha_{dc} = \frac{I_C}{I_E}. \tag{5.226}$$

When transistors are used as ac amplifiers, three circuits are used, with different gain, input resistance, and phase shift characteristics: common-emitter, common-collector, and common-base.

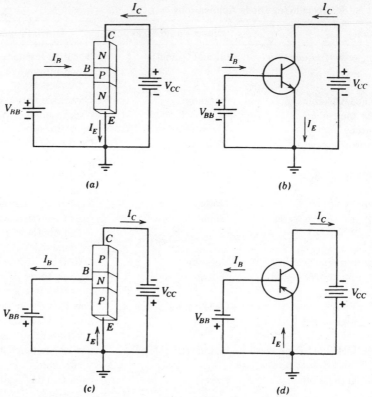

Fig. 5.49 Transistor bias connections. (*a*) and (*b*) NPN transistor, (*c*) and (*d*) PNP transistor. From *Fundamentals of Electronics*, E. N. Lurch. © 1981 by John Wiley & Sons.

Ac current gain (βac) is defined as the change in collector current caused by a given change in base current at the operating point:

$$\beta_{ac} = \frac{\Delta I_C}{\Delta I_B}. \tag{5.227}$$

Ac alpha (αac) is defined as the ratio of change in collector current to change in emitter current:

$$\alpha_{ac} = \frac{\Delta I_C}{\Delta I_E}. \tag{5.228}$$

Circuits and waveforms of a **common-emitter amplifier, common-collector amplifier,** and **common-base amplifier** are shown in Figs. 5.50, 5.51, and 5.52, respectively. Table 5.18 compares typical values of gain, input resistance, and phase shift of these amplifier circuits.

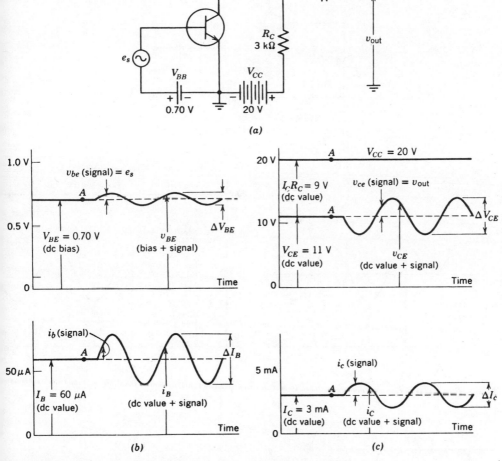

Fig. 5.50 Common-emitter amplifier. (*a*) Circuit, (*b*) base waveforms, (*c*) collector waveforms. From *Fundamentals of Electronics*, E. N. Lurch. © 1981 by John Wiley & Sons.

Table 5.18   Comparison of Basic Amplifier Circuits

	Common-Emitter	Common-Collector (Emitter Follower)	Common-Base
Current gain, A_i	50	51	$0.98 \approx 1$
Voltage gain, A_v	60	1	60
	Out of phase	In phase	In phase
Power gain, A_p	3000	51	$58.8 \approx 60$
Input resistance, r_{in}	250 Ω	155,500 Ω	49 Ω
Phase shift	180°	0°	0°

Source: Fundamentals of Electronics, E. N. Lurch, © 1981 by John Wiley & Sons. Reproduced with permission.

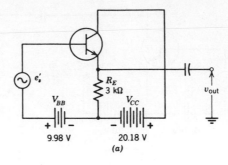

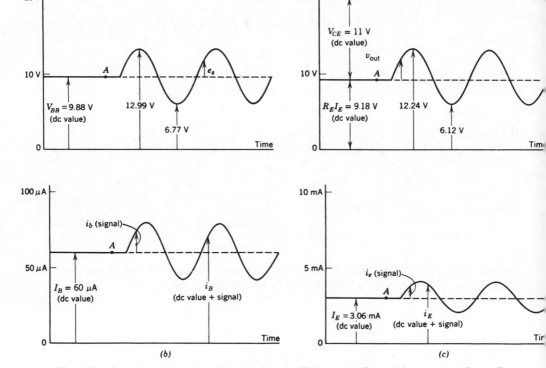

Fig. 5.51 Common-collector amplifier. (*a*) Circuit, (*b*) input waveforms, (*c*) output waveforms. From *Fundamentals of Electronics,* E. N. Lurch. © 1981 by John Wiley & Sons.

5.14.4 Junction Field-Effect Transistor (JFET)

The field-effect transistor has a high input resistance in the order of megohms and operates as an amplifier with negligible base current, thus overcoming one of the most important limitations of the junction transistor, and operating in a manner closely approximating that of a vacuum tube. Construction of an N-channel JFET is shown in Fig. 5.53; a P-channel unit has P-type material channels, an N-type gate, and the polarities of V_{GG} and V_{DD} reversed.

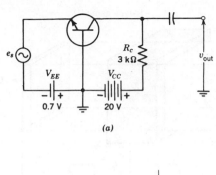

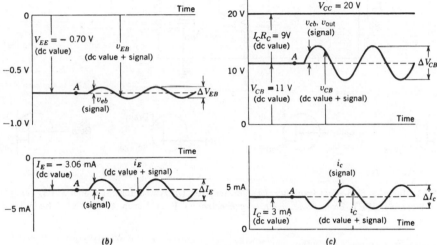

Fig. 5.52 Common-base amplifier. (*a*) Circuit, (*b*) emitter waveforms, (*c*) collector waveforms. From *Fundamentals of Electronics*, E. N. Lurch. © 1981 by John Wiley & Sons.

5.14.5 Metal Oxide Semiconductor Field-Effect Transistor (MOSFET)

A different type of field-effect transistor, in which the gate is insulated from the channel by a very thin layer of glass (SiO_2), is shown in Fig. 5.54. This type is also called **insulated gate field-effect transistor (IGFET).** The advantage of the insulated gate is an order of magnitude increase in input resistance, from around 10 megohms to 100 megohms. A disadvantage of this type is that because the gate is so sensitive to static or stray voltage, the insulating glass layer can easily be punctured and the transistor destroyed. The most recently designed MOSFETs have internal protective diodes between gate and source (Fig. 5.54*d*). Those without this feature must have all terminals short-circuited by grounding rings during handling, and the rings removed after the unit is wired into its circuit.

5.14.6 Unijunction Transistor

This is a three-terminal semiconductor device consisting of a P-material bar imbedded into an N-material block, with ohmic metallic contacts called Base 1 and Base

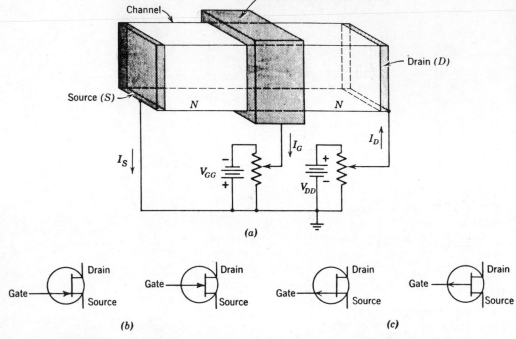

(a)

(b)   (c)

Fig. 5.53 Junction field-effect transistor. (*a*) Construction, (*b*) symbol for N-channel JFET, (*c*) symbol for P-channel JFET. From *Fundamentals of Electronics*, E. N. Lurch. © 1981 by John Wiley & Sons.

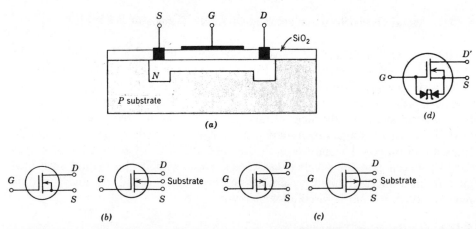

(a)   (d)

(b)   (c)

Fig. 5.54 Metal oxide silicon field-effect transistor. (*a*) cross-sectional view, (*b*) symbols for N-channel MOSFET, (*c*) symbols for P-channel MOSFET, (*d*) MOSFET with internal Zener diode protection. From *Fundamentals of Electronics*, E. N. Lurch. © 1981 by John Wiley & Sons.

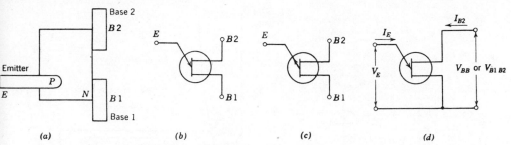

Fig. 5.55 Unijunction transistor. (a) Construction, (b) symbol for UJT with N-type base, (c) symbol for UJT with P-type base, (d) nomenclature. From *Fundamentals of Electronics,* E. N. Lurch. © 1981 by John Wiley & Sons.

2 welded to the N block without creating P–N junctions (Fig. 5.55a). When voltage is applied across the two bases, the emitter current voltage characteristic exhibits negative resistance over a portion of its range, that is, an increase in emitter current causes a decrease in emitter voltage. The negative resistance characteristic can be used in circuits of oscillators, timing circuits, triggering circuits, and voltage- and current-sensing applications. Its most common use is in relaxation oscillators.

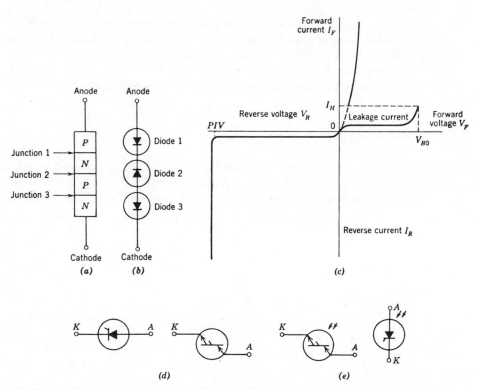

Fig. 5.56 Fundamental thyristor (NPNP transistor). (a) Construction, (b) diode model, (c) breakdown characteristics, (d) symbols, (e) symbols for light-activated switch (LAS). From *Fundamentals of Electronics,* E. N. Lurch. © 1981 by John Wiley & Sons.

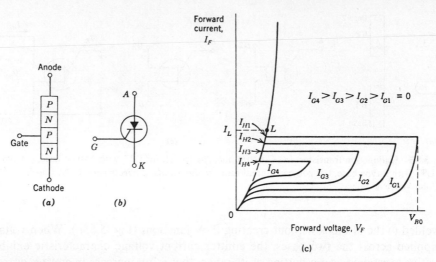

Fig. 5.57 Silicon-controlled rectifier (SCR). (*a*) Construction, (*b*) symbol, (*c*) characteristics. From *Fundamentals of Electronics*, E. N. Lurch. © 1981 by John Wiley & Sons.

5.14.7 Thyristor

The term thyristor was first used to describe a basic four-layer, three-junction diode that had two stable states of operation. It was also called **Shockley diode** and **reverse blocking diode thyristor.** Figure 5.56 shows its construction and characteristics. It is used in high-power applications comparable to those for the unijunction transistor. A light-activated switch (LAS) is a thyristor triggered by incident light instead of by voltage change.

5.14.8 Silicon-controlled Rectifier (SCR)

A thyristor with a gate like that shown in Fig. 5.57 is called a silicon-controlled rectifier, although in practice the term thyristor is becoming commonly used for either. SCRs have wide application in the control of ac power, power switching, dc to ac inverters, dc to dc converters, and var control in ac power transmission.

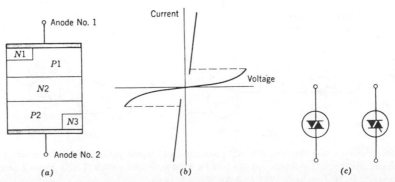

Fig. 5.58 Bidirectional thyristors. (*a*) Layer structure, (*b*) characteristics, (*c*) symbols. From *Fundamentals of Electronics*, E. N. Lurch. © 1981 by John Wiley & Sons.

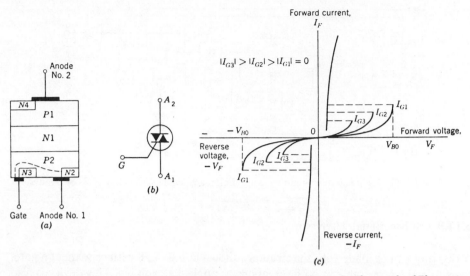

Anode
No. 2

N4	
	P1
	N1
	P2
N3	N2

Gate   Anode No. 1
(a)

A_2

G

A_1

(b)

Forward current,
I_F

$|I_{G3}| > |I_{G2}| > |I_{G1}| = 0$

$-V_{H0}$

Reverse
voltage,
$-V_F$

I_{G2} I_{G3}

I_{G1}

0

I_{G3}

I_{G2}

I_{G1}

Forward voltage,

V_{B0} V_F

Reverse current,
$-I_F$

(c)

Fig. 5.59 Triac. (*a*) Cross section, (*b*) symbol, (*c*) characteristics. From *Fundamentals of Electronics,*
E. N. Lurch. © 1981 by John Wiley & Sons.

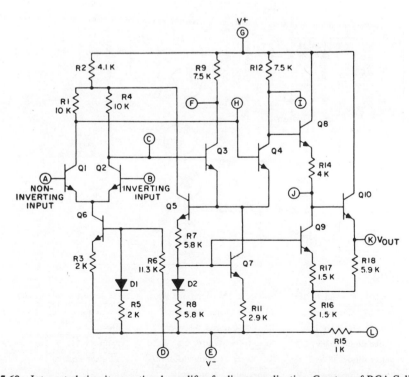

Fig. 5.60 Integrated circuit operational amplifier for linear application. Courtesy of RCA Solid State.

345

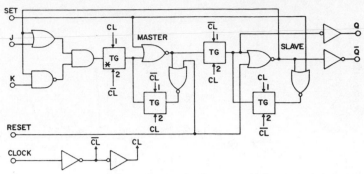

Fig. 5.61 Type J-K flip-flop integrated circuit block for digital applications. Courtesy of RCA Solid State.

5.14.9 Bidirectional Thyristor

A thyristor can be made with the structure shown in Fig. 5.58 with or without a gate, so as to make it operable in both directions of applied ac voltage. This device is also called a **bidirectional dipole thyristor,** or diac.

5.14.10 Triac

A triac is a form of SCR that can be triggered into conduction with polarity of either gate voltage or anode voltage (Fig. 5.59). These characteristics make it very useful

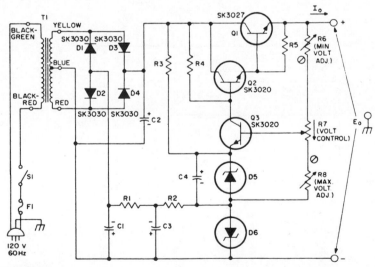

Parts List

C_1, $C_3 = 500$ μF, electrolytic, 25 V	R_1, $R_5 = 220$ ohms, 0.5 watt, 10%	$R_7 =$ potentiometer, 5000 ohms, linear taper
$C_2 = 5000$ μF, electrolytic, 25 V	$R_2 = 470$ ohms, 0.5 watt, 10%	$S_1 =$ toggle switch, 120 V, 1 ampere, single-pole, single-throw
$C_4 = 100$ μF, electrolytic, 10 V	$R_3 = 6800$ ohms, 0.5 watt, 10%	$T_1 =$ power transformer; primary 117 V; secondary 16 V, 1.5 ampere; Stancor TP-4 or equiv.
$D_5 =$ zener diode, 6.8 V, 1 watt	$R_4 = 10000$ ohms, 0.5 watt, 10%	
$D_6 =$ zener diode, 12 V, 1 watt	R_6, $R_8 =$ trimmer potentiometer, 5000 ohms, Mallory MTC-1 or equiv.	
$F_1 =$ fuse, 1 ampere, 120 V,		

Fig. 5.62 Full-range variable-voltage regulated dc power supply. Courtesy of RCA Solid State.

as a transient surge protector in ac power circuits, and in the control of ac power to a load without rectification.

5.14.11   Integrated Circuits

Combinations of diodes, transistors, and the passive electric circuit components (resistance, inductance, and capacitance) can be formed and connected together by deposition processes to form integrated circuits (ICs) on a silicon chip less than a square centimeter in size. Examples of ICs for linear and digital electronic applications are shown in Figs. 5.60 and 5.61. Standard ICs for digital applications are shown in Table 5.19. Large-scale integration (LSI) is the process by which IC blocks or modules with individual functions are combined on a larger silicon chip to form complete integrated assemblies of circuits, such as computer memories.

5.15   PRACTICAL ELECTRONIC CIRCUITS

Examples of circuits frequently used in practice are given in Figs. 5.62 through 5.69. Others can be found in Reference 15.

Table 5.19   Standardized Complementary Metal Oxide Semiconductor (CMOS) Integrated Circuits for Digital Applications

Gates

CD4000A	Dual 3-Input NOR Gate Plus Inverter
CD4001A	Quad 2-Input NOR Gate
CD4002A	Dual 4-Input NOR Gate
CD4011A	Quad 2-Input NAND Gate
CD4012A	Dual 4-Input NAND Gate
CD4019A	Quad AND-OR Select Gate
CD4023A	Triple 3-Input NAND Gate
CD4025A	Triple 3-Input NOR Gate
CD4030A	Quad Exclusive-OR Gate
CD4037A	Triple AND-OR Bi-Phase Pairs
CD4048A	Expandable 8-Input Gate

Flip-Flops

CD4013A	Dual D with Set/Reset Capability
CD4027A	Dual J-K with Set/Reset Capability
CD4047A	Monostable/Astable Multivibrator

Latches

CD4042A	Quad Clocked D Latch
CD4043A	NOR R/S Latch (3 Output States)
CD4044A	NAND R/S Latch (3 Output States)

Arithmetic Devices

CD4008A	Four-Bit Adder, Parallel Carry-Out
CD4032A	Triple Serial Adder, Internal Carry (Neg. Logic)
CD4038A	Triple Serial Adder, Internal Carry (Pos. Logic)
CD4057A	LSI 4-Bit Arithmetic Logic Unit

Table 5.19 *(Continued)*

Buffers

CD4009A	Hex Inverting Type
CD4010A	Hex Non-Inverting Type
CD4041A	Quad Inverting and Non-Inverting Type
CD4049A	Hex Buffer/Converter (Inverting)
CD4050A	Hex Buffer/Converter (Non-Inverting)

Complementary Pairs

CD4007A	Dual Complementary Pair Plus Inverter

Multiplexers and Decoders

CD4016A	Quad Bilateral Switch
CD4028A	BCD-to-Decimal Decoder
CD4066A	Quad Bilateral Switch

Counters

CD4017A	Decade Counter/Divider Plus 10 Decoded Decimal Outputs
CD4018A	Presettable Divide-by-N Counter
CD4020A	14-Stage Ripple Counter
CD4022A	Divide-by-8 Counter/Divider, 8 Decoded Outputs
CD4024A	7-Stage Ripple Counter
CD4026A	Decade Counter/Divider, 7-Segment Display Output
CD4029A	Presettable Up/Down Counter, Binary or BCD-Decade
CD4033A	Decade Counter/Divider, 7-Segment Display Output
CD4040A	12-Stage Binary/Ripple Counter
CD4045A	21-Stage Ripple Counter

Shift Registers

CD4006A	18-Stage Static Shift Register
CD4014A	8-Stage Synch Shift Register, Parallel-In/Serial-Out
CD4015A	Dual 4-Stage Shift Register, Serial-In/Parallel-Out
CD4021A	8-Stage Asynchronous Shift Register, Parallel-In/Serial-Out
CD4031A	64-Stage Static Shift Register
CD4034A	Parallel-In/Parallel-Out Shift Register (3 Output States)
CD4035A	4-Bit Parallel-In/Parallel-Out Shift Register, J-$\overline{\text{K}}$ In, True-Comp. Out

Phase-Locked Loop

CD4046A	Micropower Phase-Locked Loop

Memories

CD4036A	4-Word by 8-Bit RAM (Binary Addressing)
CD4039A	4-Word by 8-Bit RAM (Word Addressing)
CD4061A	256-Word by 1-Bit State RAM

Drivers

CD4054A	4-Line Liquid-Crystal-Display Driver
CD4055A	BCD to 7-Segment Decoder/Driver
CD4056A	BCD to 7-Segment Decoder/Driver

Courtesy of RCA Solid State.

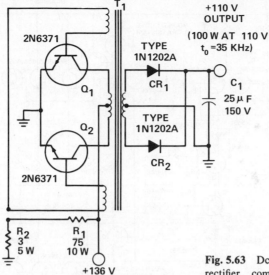

Fig. 5.63 Dc to ac power inverter and rectifier combination (dc–dc converter). Courtesy of RCA Solid State.

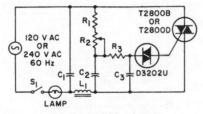

Parts List

120-Volt, 60-Hz Operation

C_1, C_2 = 0.1 μF, 200 V
C_3 = 0.1 μF, 100 V
L_1 = 100 μH
R_1 = 1000 ohms, 0.5 watt
R_2 = light control, poten-

tiometer, 0.1 megohm, 0.5 watt

240-Volt, 60-Hz Operation

C_1 = 0.1 μF, 400 V
C_2 = 0.05 μF, 400 V

C_3 = 0.1 μF, 100 V
L_1 = 100 μH
R_1 = 7500 ohms, 2 watts
R_2 = light control, potentiometer, 0.2 megohm, 1 watt
R_3 = 7500 ohms, 2 watts

Fig. 5.64 Triac light dimmer. Courtesy of RCA Solid State.

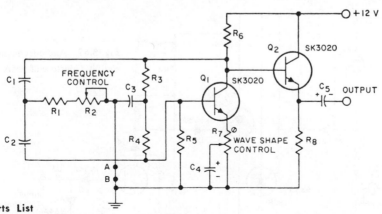

Parts List

C_1, C_2 = see chart for value, mica or paper
C_3 = twice the value of C_1, mica or paper
C_4 = 1 μF, electrolytic, 12 V
C_5 = 300 μF for frequencies below 2000 Hz or 5 μF for frequencies

above 2000 Hz, electrolytic, 6 V
C_6 = 20 μF, electrolytic, 6 V
R_1 = 2700 ohms, 0.5 watt
R_2 = Frequency control, potentiometer, 5000 ohms, 0.5 watt

R_3, R_4 = 51000 ohms, 0.5 watt
R_5 = 22000 ohms, 0.5 watt
R_6 = 4700 ohms, 0.5 watt
R_7 = Wave-shape control, potentiometer, 250 ohms, 0.5 watt
R_8 = 820 ohms, 0.5 watt

Fig. 5.65 Audio frequency oscillator. Courtesy of RCA Solid State.

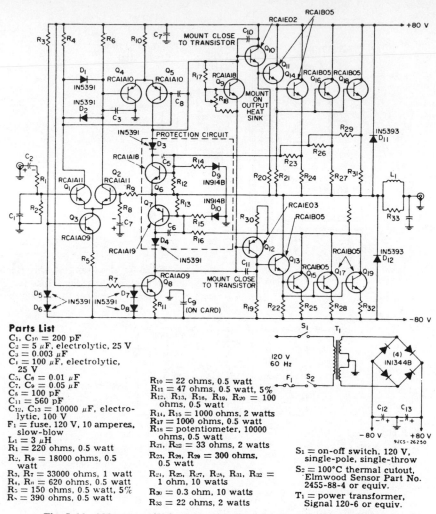

Parts List

C₁, C₁₀ = 200 pF
C₂ = 5 μF, electrolytic, 25 V
C₃ = 0.003 μF
C₄ = 100 μF, electrolytic, 25 V
C₅, C₆ = 0.01 μF
C₇, C₉ = 0.05 μF
C₈ = 100 pF
C₁₁ = 560 pF
C₁₂, C₁₃ = 10000 μF, electrolytic, 100 V
F₁ = fuse, 120 V, 10 amperes, slow-blow
L₁ = 3 μH
R₁ = 220 ohms, 0.5 watt
R₂, R₉ = 18000 ohms, 0.5 watt
R₃, R₇ = 33000 ohms, 1 watt
R₄, R₈ = 620 ohms, 0.5 watt
R₅ = 150 ohms, 0.5 watt, 5%
R₆ = 390 ohms, 0.5 watt

R₁₀ = 22 ohms, 0.5 watt
R₁₁ = 47 ohms, 0.5 watt, 5%
R₁₂, R₁₃, R₁₆, R₁₉, R₂₀ = 100 ohms, 0.5 watt
R₁₄, R₁₅ = 1000 ohms, 2 watts
R₁₇ = 1000 ohms, 0.5 watt
R₁₈ = potentiometer, 10000 ohms, 0.5 watt
R₂₁, R₂₂ = 33 ohms, 2 watts
R₂₃, R₂₆, R₂₉ = 300 ohms, 0.5 watt
R₂₄, R₂₅, R₂₇, R₂₈, R₃₁, R₃₂ = 1 ohm, 10 watts
R₃₀ = 0.3 ohm, 10 watts
R₃₃ = 22 ohms, 2 watts

S₁ = on-off switch, 120 V, single-pole, single-throw
S₂ = 100°C thermal cutout, Elmwood Sensor Part No. 2455-88-4 or equiv.
T₁ = power transformer, Signal 120-6 or equiv.

Fig. 5.66 200-watt audio frequency amplifier. Courtesy of RCA Solid State.

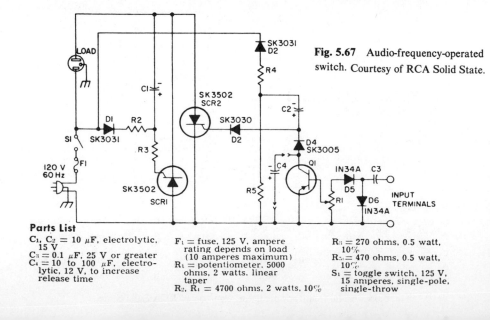

Fig. 5.67 Audio-frequency-operated switch. Courtesy of RCA Solid State.

Parts List

C₁, C₂ = 10 μF, electrolytic, 15 V
C₃ = 0.1 μF, 25 V or greater
C₄ = 10 to 100 μF, electrolytic, 12 V, to increase release time

F₁ = fuse, 125 V, ampere rating depends on load (10 amperes maximum)
R₁ = potentiometer, 5000 ohms, 2 watts, linear taper
R₂, R₄ = 4700 ohms, 2 watts, 10%

R₃ = 270 ohms, 0.5 watt, 10%
R₅ = 470 ohms, 0.5 watt, 10%
S₁ = toggle switch, 125 V, 15 amperes, single-pole, single-throw

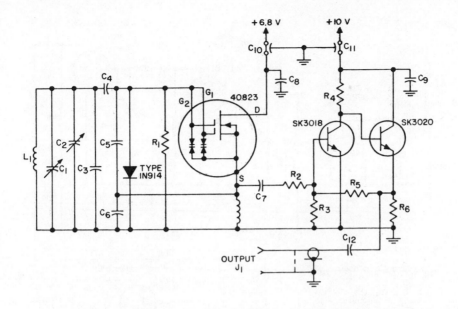

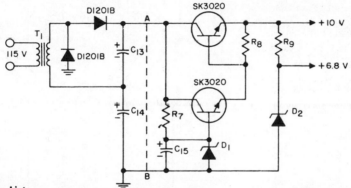

Parts List

C_1 = Double-bearing variable capacitor, Millen 23100 or 23050 (or equiv.) depending upon frequency range (see Tuned-Circuit Data)

C_2 = Air-type trimmer capacitor, 25 pF maximum, Hammarlund APC-25 or equiv.

C_3, C_4, C_5, C_6 = silver-mica capacitors (see Tuned-Circuit Data for values)

C_7 = 2200 pF, silver mica

C_8 = 0.05 pF, ceramic disc, 50 V.

C_9 = 0.1 pF, ceramic disc, 50 V.

C_{10}, C_{11} = 1500 pF, feed-through

C_{12} = 0.025 μF, ceramic disc, 50 V.

C_{13} = 500 μF, electrolytic, 12 V.

C_{14} = 500 μF, electrolytic, 12 V.

C_{15} = 50 μF, electrolytic, 12 V.

D_1 = Zener diode, 12-volt, 1-watt

D_2 = Zener diode, 6.8 volt, 1-watt

J_1 = Coaxial connector

L_1 = Variable inductor (see Tuned-Circuit Data for details)

L_2 = Miniature rf choke,

R_1 = 22000 ohms, 0.5 watt

R_2 = 12000 to 47000 ohms, 0.5 watt; select value for 2-volt peak output level at input to transmitter

R_3 = 12000 ohms, 0.5 watt

R_4 = 820 ohms, 0.5 watt

R_5 = 47000 ohms, 0.5 watt

R_6 = 240 ohms, 0.5 watt

R_7 = 2200 ohms, 0.5 watt

R_4 = 220 ohms, 0.5 watt

R_9 = 180 ohms, 0.5 watt

T_1 = 6.3-volt, 1.2-ampere filament transformer

Fig. 5.68 Stable tunable radio frequency oscillator. Courtesy of RCA Solid State.

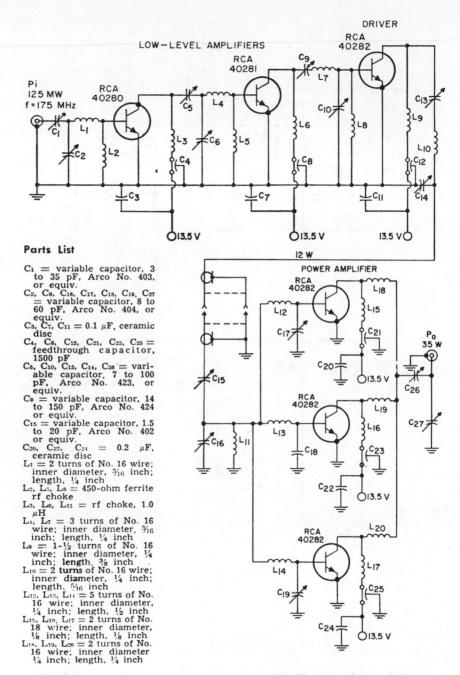

Parts List

C_1 = variable capacitor, 3 to 35 pF, Arco No. 403, or equiv.

C_2, C_6, C_{16}, C_{17}, C_{18}, C_{19}, C_{27} = variable capacitor, 8 to 60 pF, Arco No. 404, or equiv.

C_3, C_7, C_{11} = 0.1 μF, ceramic disc

C_4, C_8, C_{12}, C_{21}, C_{23}, C_{25} = feedthrough capacitor, 1500 pF

C_5, C_{10}, C_{13}, C_{14}, C_{26} = variable capacitor, 7 to 100 pF, Arco No. 423, or equiv.

C_9 = variable capacitor, 14 to 150 pF, Arco No. 424 or equiv.

C_{15} = variable capacitor, 1.5 to 20 pF, Arco No. 402 or equiv.

C_{20}, C_{22}, C_{24} = 0.2 μF, ceramic disc

L_1 = 2 turns of No. 16 wire; inner diameter, $\frac{3}{16}$ inch; length, $\frac{1}{4}$ inch

L_2, L_5, L_8 = 450-ohm ferrite rf choke

L_3, L_6, L_{11} = rf choke, 1.0 μH

L_4, L_7 = 3 turns of No. 16 wire; inner diameter, $\frac{3}{16}$ inch; length, $\frac{1}{4}$ inch

L_9 = 1-$\frac{1}{2}$ turns of No. 16 wire; inner diameter, $\frac{1}{4}$ inch; length, $\frac{3}{8}$ inch

L_{10} = 2 turns of No. 16 wire; inner diameter, $\frac{1}{4}$ inch; length, $\frac{5}{16}$ inch

L_{12}, L_{13}, L_{14} = 5 turns of No. 16 wire; inner diameter, $\frac{1}{4}$ inch; length, $\frac{1}{2}$ inch

L_{15}, L_{16}, L_{17} = 2 turns of No. 18 wire; inner diameter, $\frac{1}{8}$ inch; length, $\frac{1}{8}$ inch

L_{18}, L_{19}, L_{20} = 2 turns of No. 16 wire; inner diameter $\frac{1}{4}$ inch; length, $\frac{1}{4}$ inch

Fig. 5.69 175 Mhz 35-watt radio frequency amplifier. Courtesy of RCA Solid State.

5.16 NATIONAL ELECTRICAL CODE

The National Electrical Code (NEC) adopted by the National Fire Protection Association and published by them (1981) as Volume 6 of the National Fire Codes has been widely mandated as the electrical standard for licensing bodies in the United States. Volume 6 includes the National Electrical Code, an electrical code for one- and two-family dwellings, and electrical requirements for employee workplaces.

The NEC includes sections on Wiring Design and Circuit Protection, Wiring Methods and Materials, General Electrical Equipment, Special Occupancies (types of facilities), Special Equipment (e.g., Signs, Escalators, Electroplating), Special Conditions, Communications Systems, and Tables and Examples. It should be utilized where electrical design work deals with these subjects. It is revised and reissued approximately triannually.

REFERENCES

1. *IEEE Standards Catalog and Standards Listing,* Institute of Electrical and Electronic Engineers, New York, 1982.
2. *ANSI Catalog of American National Standards,* American National Standards Institute, New York, 1982.
3. *NEMA Standards Publications,* National Electrical Manufacturers Association, Washington, DC, 1982.
4. *Quick Reference to IEEE Standards (QRIS),* Institute of Electrical and Electronic Engineers, New York, 1980.
5. *IEEE Standard Dictionary of Electrical and Electronics Terms,* ANSI/IEEE Standard 100-1977, Institute of Electrical and Electronic Engineers, New York, 1977.
6. *Metric Practice,* ANSI Standard Z210.1-1976, IEEE Std 268-1976, American Society for Testing and Materials (ASTM) Std E380-76, 1976.
7. *IEEE Standard Letter Symbols for Units of Measurement,* ANSI/IEEE Std 260-1978, Institute of Electrical and Electronic Engineers, New York, 1973.
8. *Letter Symbols for Quantities Used in Electrical Science and Electrical Engineering,* ANSI/IEEE Std 280-1968, Institute of Electrical and Electronic Engineers, New York, 1968.
9. *Electrical and Electronics Graphic Symbols and Reference Designations,* Std 76-ANSI/IEEE Y32E, Institute of Electrical and Electronic Engineers, New York, 1976.
10. *Recommended Practice for Protection and Coordination of Industrial and Commercial Power Systems,* IEEE Standard 242-1975, Chapter 5, Institute of Electrical and Electronic Engineers, New York, 1975.
11. *IEEE Recommended Practice for Emergency and Standby Power Systems for Industrial and Commercial Application,* IEEE Standard 446-1980, Chapter 4, Institute of Electrical and Electronic Engineers, New York, 1980.
12. *Receiving Tube Manual,* RC-30, RCA Corporation, Camden, N.J., 1975.
13. *Transmitting Tubes,* Technical Manual TT, RCA Corporation, Camden, N.J.
14. *Lighting Handbook,* Westinghouse Electric Corporation, Bloomfield, N.J., 1978.
15. *Solid-state Devices Manual,* SC-16, RCA Solid-State Div., Somerville, N.J., 1975.

BIBLIOGRAPHY

Applied Protective Relaying, Westinghouse Electric Corp., Coral Springs, Florida, 1979.

Baumeister, T. and Avallone, E. A., *Marks' Standard Handbook for Mechanical Engineers,* 8th edition, McGraw-Hill, New York, 1978.

Bonebreak, R. L., *Practical Techniques of Electronic Circuit Design,* John Wiley & Sons, New York, 1982.

Cahill, S. J., *Digital and Microprocessor Engineering*, John Wiley & Sons, New York, 1982.

Cowles, L. G., *Transistor Circuits and Applications*, 2nd edition, Prentice-Hall, Inc., Englewood Cliffs, N.J., 1974.

DeSa, A., *Principles of Electronic Instrumentation*, John Wiley & Sons, New York, 1981.

Distribution Systems Reference Book, Westinghouse Electric Corp., Pittsburgh, Pa., 1965.

Electrical Transmission and Distribution Reference Book, Westinghouse Electric Corp., Pittsburgh, Pa., 1964.

Eshbach, O. W. and Souders, M., *Handbook of Engineering Fundamentals*, 3rd edition, John Wiley & Sons, New York, 1975.

Faber, R. B. *Applied Electricity and Electronics for Technology*, 2nd edition, John Wiley & Sons, New York, 1982.

Fink, D. G. and Beaty, H. W., *Standard Handbook for Electrical Engineers*, 11th edition, McGraw-Hill, New York, 1978.

Ginsberg, G. L., *A User's Guide to Selecting Electronic Components*, John Wiley & Sons, New York, 1981.

Howes, M. J. and Morgan, D. V., *Large Scale Integration Devices, Circuits and Systems*, John Wiley & Sons, New York, 1981.

Krauss, H. L., Bostian, C. W., and Raab, F. H., *Solid State Radio Engineering*, John Wiley & Sons, New York, 1980.

Krutz, R. L., *Microprocessors and Logic Design*, John Wiley & Sons, New York, 1980.

Lurch, E. N., *Fundamentals of Electronics*, 3rd edition, John Wiley & Sons, New York, 1981.

Sen, P. C., *Thyristor DC Drives*, John Wiley & Sons, New York, 1981.

Sessions, K. W., *IC Schematic Sourcemaster*, John Wiley & Sons, New York, 1978.

Seymour, J., *Electronic Devices and Components*, John Wiley & Sons, New York, 1981.

Skilling, H. H., *Electrical Engineering Circuits*, 2nd edition, John Wiley & Sons, New York, 1965.

Sonde, B. S., *Introduction to System Design Using Integrated Circuits*, John Wiley & Sons, New York, 1980.

Surge Protection of Power Systems, Westinghouse Electric Corp., Pittsburgh, Pa., 1971.

Sze, S. M., *Physics of Semiconductor Devices*, John Wiley & Sons, New York, 1981.

Wildi, T., *Electrical Power Technology*, John Wiley & Sons, New York, 1981.

SECTION 6
CONTROLS

6.1 CONTROL THEORY

Control theory deals with methods for changing an output (or process) based on preestablished performance values (**set points**). A control loop can be generalized as shown in Fig. 6.1. The typical response of a controlled system will vary depending on the type of control exercised.

Automatic control can be broadly categorized into the following types:

On–off
Proportional (P)
Integral (I)
Proportional plus reset (PI)
Proportional plus integral plus derivative or error rate (PID)

6.1.1 On–Off Control

An **on–off controller** acts by applying full corrective action whenever an error is detected. It has only two states, one of which corrects positive deviations of the system, the other negative. The control signal applied is thus either 100% or zero.

A system controlled in this way is always in a transient state, and oscillation between limits will always occur. The controller compensates for disturbances, applying its full signal when its set point is reached. The system performance will be dependent on the rate of system deviation (in the off condition) and the rate of system recovery after control signal initiation (the on condition). Figure 6.2 shows a typical time response of a system under on–off control. A common example is the furnace control in a home heating system.

6.1.2 Proportional Control (P)

Proportional or throttling control (Fig. 6.3) is the simplest linear control, in which output of the controller is a linear algebraic function of its input. A change in the measured variable produces a proportionate change in the control signal.

Thus: Control output $= K \times E$ (6.1)

where: K = proportional constant
E = error

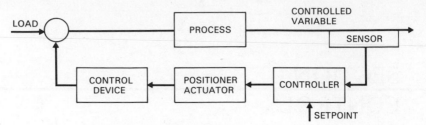

Fig. 6.1   Block diagram of a generalized control loop.

Since the difference between the measured variable and its set point controls the magnitude of the correction signal, once correction is under way, this signal is reduced, leading to progressively smaller correction signals; the difference between the measured variable and the set point is called **proportional offset.**

$$\text{Proportional offset} = SP - MV \qquad (6.2)$$

where: MV = set point
         MV = measured variable

The ratio of control signal change to measured variable change is called the **gain of the controller.** The **proportional band** of the controller is defined as 100 times the percent input change divided by the percent output change. Thus a gain of 0.1 is equivalent to 1000% proportional band, a gain of 1.0 is 100% proportional band, and so on. Low gain is equivalent to a wide proportional band, high gain to a narrow proportional band (Fig. 6.4).

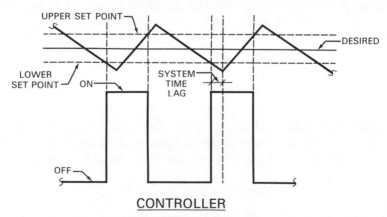

Fig. 6.2   On–off control.

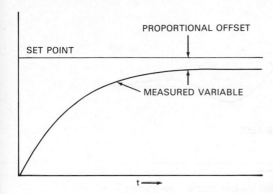

Fig. 6.3 Proportional control.

6.1.3 Integral Control (I) (Also Called Reset Control)

The effect of proportional offset can be eliminated by using a control system in which the measured error signal is continuously integrated with respect to time.

Recent-model, commercially available controllers with integral action, although not perfect integrators, can closely approach a true integral for the periods of time under concern. The gain at very low frequencies is limited and commonly called the **reset gain** of the controller. Too low a value for this figure is undesirable because it can lead to **static offset.** Proportional control can be fast acting. However, it does give static offset. Integral control, on the other hand, minimizes static offset but is sluggish because of its attenuation of controller output at higher frequencies. A mode of control that can be used to give the best features of both is proportional plus integral.

6.1.4 Proportional Plus Integral Control (PI)

PI control is probably the most commonly used mode of control. Its principal disadvantage is that extended open loop operation can lead to comparatively large overshoot in the integral part of the action through the effect known as "reset windup." All modern controllers are designed to minimize this effect.

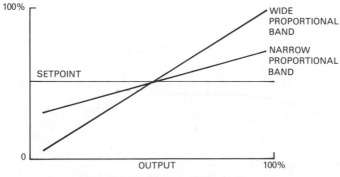

Fig. 6.4 Output for proportional bands.

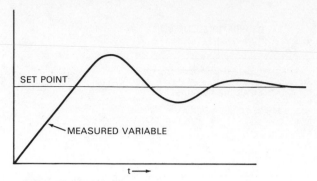

Fig. 6.5 Proportional plus integral control, "pure action."

PI control can be expressed:

$$\text{Controller output} = K \times E + K_I \times E \int dt \qquad (6.3)$$

where K_I = constant of integration (adjustable).

A typical time response of proportional plus integral (PI) controller is shown in Fig. 6.5, which also indicates how the pure-action controllers would behave under the same stimulus; Fig. 6.6 shows their actual response.

6.1.5 Proportional Plus Integral Plus Derivative or Error Rate Control (PID)

A third linear control mode is occasionally used in which the output of the controller is proportional to the rate of change of the error signal.
Thus, if an error begins to develop, a controller with "rate action" can anticipate and begin to take action before the error becomes large.

If the action were perfect, the gain would increase steadily with the frequency, and the phase lead would always be 90°. This sort of control action would be highly undesirable, since high-frequency noise would be excessively amplified. Derivative controllers must, therefore, be modified to limit their action at high frequencies.

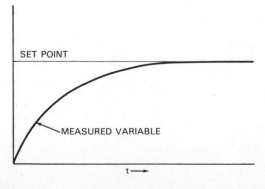

Fig. 6.6 Proportional plus integral control, "actual."

A pure derivative controller would not be a controller at all, since there is no unique relationship between error and control action. Derivative action is therefore used only to augment proportional or proportional-plus-integral controllers to improve their high-frequency response.

$$\text{Controller output} = K\,E + K_I E \int dt + K_d \frac{de}{dt} \qquad (6.4)$$

where K_d = rate constant (adjustable).

6.2 PROCESS CONTROL

Control systems can be modeled mathematically and the required characteristics of the control elements thus determined. The process, however, can be laborious and subject to inaccuracies in predicting performance and characteristics of the process. Thus, for **process control**, the **sensing elements, transmitters,** and **final control elements** are selected based on the process requirements. Off-the-shelf controllers are used, and the system is "tuned" during startup of the process to provide the mode or combinations of modes of control needed. With the recent advent of microprocessors and digital control, this has been greatly facilitated. Most commercially available controllers have all the control modes previously described either switch selectable, as modules, or by program selection.

Some typical applications of control modes to industrial processes are:

P	Blending to meet a desired composition
I	Speed control of centrifugal machines
	Liquid flow when a fast actuation is needed
PI	Majority of industrial processes
	Gas pressure
	Liquid flow with slow actuation
	Concentration controllers
PID	Temperature control where sensing lag is
	significant compared to process lag

6.2.1 Control Systems

Plant control systems can be visualized as a pyramidal hierarchy of control (Fig. 6.7) truncated depending on the degree of centralized control desired.

Hierarchy of Control

Item	Level
Local/hand	Component
Sub-automation (local loop)	Sub-system
Automation	System
Integrated plant control	Integrated plant

All of the levels of controls are compatible with and typically form part of some degree of an integrated plant control system. The degree of central control can be varied widely depending on process or economic requirements.

Figure 6.8 and the table which follows indicate broadly the degree of current (1983) process control capability readily achievable both as **local** and **centralized control.**

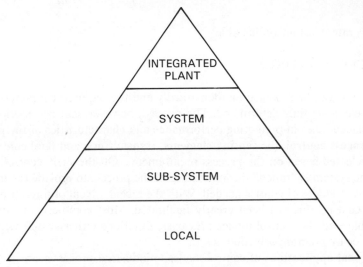

Fig. 6.7 Pyramidal hierarchy.

Control technology is subject to high rates of change (and obsolescence), as witnessed by the recent developments in computer chip capability, Cathode Ray Tube (CRT), and communication technology. Further, as processes have become more complex, instrumentation and control requirements have increased.

Microprocessor-based control systems offer an attractive solution to many control problems because it is possible to **analog** (represent or duplicate) any of the P, PI, PD, or PID control modes by software programming and thus modify the control mode with minimum physical changes. They have the added advantage of low cost, thus permitting the incorporation of redundant microprocessors to improve reliability. With microprocessors it is possible to analog all elements in a single controller whose mode can be modified by software changes only; in some instances, several controllers can be housed in one case.

6.2.2 Degree of Automatic Control

The **degree of automatic control** applied depends on the process requirements, which may range from batch mode to continuous applications. Considerations include the length of time for batch processes, the economic consequences of outage time, and accuracy and degree of control required. Other significant factors include the response time required, the control limits and the effect of variation within that for

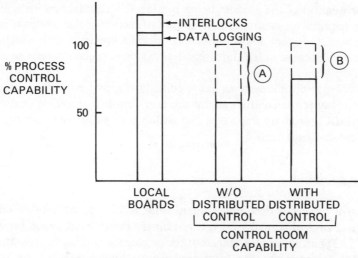

LIMITATIONS: (A) SPACE; INFORMATION HANDLING CAPABILITY
(B) ECONOMICS

Fig. 6.8 Local and centralized control.

Without Distributed Control	With Distributed Control
Advantages	*Advantages*
Lower capital cost	Lower installation cost
	Few operators, more control and data in central control room
Less sophisticated personnel requirements	Adaptive control strategies easier to implement
	Process control changes implemented largely by software changes (interactive systems)
Disadvantages	*Disadvantages*
Controls/data breadth limited by human capability	Higher capital cost
More operating personnel required	Information handling in control room greater and possibly more complex
Control changes cumbersome	More sophisticated maintenance required
	Some applications may require backup computer
	Increased operator training and technical support requirements

the value (or quality) of the product being produced. Other factors include system lag, the time required for control to take effect (a function of the combined response times of the control system and the process), and the precision with which control can be exercised. Because of lag time, sensitive systems frequently require anticipatory control.

Today (1983), electronic systems are specified twice as frequently as pneumatic for large control systems. Control systems are also classified as **analog** or **direct digital control** (**DDC**) types. An important and widely used subset of direct digital control is distributed digital control.

6.2.3   Analog Control

Analog control utilizes a continuous signal with respect to both value and time, whereas **digital control** normally operates in binary digits or bits, each bit equaling either 0 or 1. Typically, analog transmitters produce a continuous signal directly proportional to the variable, thus for digital control applications, analog to digital (A/D) converters are normally provided. Similarly, digital to analog converters (D/A) are used to adapt digital signals to analog process control units.

6.2.4   Analog Application Considerations

These systems typically utilize 3 to 15 psig (48 to 103 kPa) air or low current (4 to 20 or 10 to 50 mA dc) as the variable signal. Analog control is simple and relatively easy to understand because of the limited number of process variables involved.

Similarly, maintenance of the hardware circuitry is relatively simple. Analog controllers performing only a single function can achieve control very close to the set point. Analog instrumentation may be used to implement advanced control (i.e., control in terms of the final objective, such as energy optimization) rather than merely controlling flows, pressures, temperatures, and so on. The adjustability of the control modes are limited, however, and typically are as follows:

Proportional Band: 1 to 1000% (Gain: 100 to 0.1)
Integral: 0.02 to 100 minute (50 to 0.01 minutes/repeat)
Derivative: 0 to 10 minutes

The limitations are determined by controller pneumatic bellows sizing or electronic controller capacitor capability. Signals can also be provided at 1 to 5 V dc, ±10 V dc, and on occasion ±10, 20, and 50 mV dc.

Calculation accuracy for each element is typically 0.25 to 0.5%; thus, a multielement calculation may have an overall accuracy of 2 to 4%, which is usually sufficient, but may not be adequate for specialized applications. Analog systems tend to require more components, thus lowering overall reliability; and modifications to equipment are awkward once the equipment is installed. Providing compensation for transmitter failure is difficult and adds considerable complications to systems.

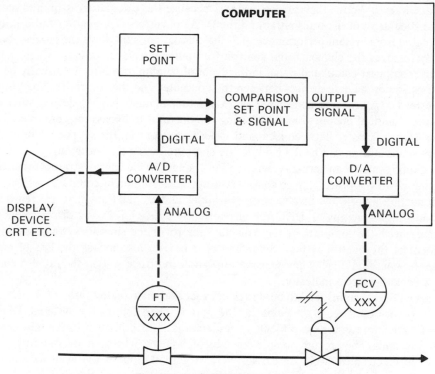

Fig. 6.9 Typical DDC control loop.

6.2.5 Direct Digital Control (DDC)[7]

In **direct digital control (DDC)** a central computer performs the entire control function; the computer contains the control function or **algorithm** for all controls in the system. Field data in the form of analog inputs and outputs are transmitted to and from this central computer via "data highways." Because of the large number of data manipulations required, fast computers are necessary. Figure 6.9 shows direct digital control of a process flow. The computer receives an analog signal from the flow transmitter, converts this signal into a numerical value, compares this value with a set point, calculates the proper output signal, converts the result into an analog value, and sends a signal corresponding to this value to the control device. Indication of the value of the flow is performed by a computer-operated display device, such as a cathode ray tube.

6.2.6 DDC Application Considerations[7]

The control modes and their settings are part of the microprocessor controller software, thus the combinations that may be used, proportional band, reset and rate times are relatively unlimited. A digital computer system can take into account a significantly larger number of variables than an equivalent analog system. Calculation accuracy is better than analog systems because the calculation is limited only by the accuracies of the transmitter and the D/A converters. Transmitter failure can be isolated from system performance such that it does not shut down the process, but merely removes the element from control. Computer interface stations, which hold the last computer-calculated signal in the event of computer failure, are usually provided as analog backup devices between the computer and the controller. The pulse converter (D/A) may be considered as a computer manual (C/M) device, since it includes a manual operation mode allowing the operator to bypass the computer. To allow smooth transfer between local and supervisory control, the set point output of the interface station can be fed back to the computer as an analog input.

A digital system can normally make online changes in the control strategy without shutting down the process, since control strategy is a function of computer software. Optimization algorithms involve many variables and calculations that are feasible only on a digital computer. DDC systems normally utilize CRTs as a display device. Interface with the computer by these displays improves and simplifies operator management of the control system. Since loss of computer also means the loss of any computer-operated display device, each important interface station should also contain a backup process indicator.

Large DDC systems tend to be slower in execution than analog systems. Thus the last loop variations from set point can be greater than for analog systems. DDC systems are more complex, difficult to understand, and difficult to maintain than analog systems. They require maintenance by technicians with an understanding primarily of electronics. Modularized software largely eliminates the need for reprogramming and permits relatively easy modifications after initial installation.

6.2.7 Distributed Digital Control[7]

With the recent development of economical, dependable microprocessors and miniaturized computers, it is increasingly practical to utilize distributed digital control for systems having from only 1 loop to those with 1000 or more loop applications. **Distributed digital control** distributes the control intelligence to devices at or near the point of control, and only exception and special requests are transmitted to and from the central control computer. As a result, slower **data highways** are satisfactory because of the lower quantity of data being handled. Distributed control "highways" can interface with pneumatic, hydraulic, or electronic local control systems of the analog type, eliminating the need for all controls to be the direct digital type. For distributed digital control systems, large central processors are usually needed only for data reduction and evaluation but not for control.

Distributed digital control permits substantial savings in cable quantities, raceways, and so on, because it allows the process controllers to be placed close to the process; permits ready modification of control mode, loop performance, and display systems over the life of the plant; and broadly simplifies the problem of central control. Care must be given in the initial planning to space allocation, ambient environmental conditions, and data highway type and routings. Changes, loop additions, and other modifications can be readily accomplished providing flexibility in application.

Both distributed and direct digital control can benefit from the use of **data highways.** All communication in these control systems is by a **single coaxial cable** (rather than multipair cables). The system can be readily expanded but has length limitations on the order of 3000 feet (900 m) between the computer and the farthest component. Usual points of connection with the data highway are **multiplexers** or **data terminals.** Some manufacturers offer prefabricated cables to simplify installation. In addition to the savings in wiring, its space requirements, and so forth, data highways are relatively immune to electrical noise.

6.2.8 Functional Distribution Concepts

With small modularized microprocessor-based systems it is possible to add input/ output modules, CRTs, and so on, and thus expand the capability readily. These modularized units are then interconnected through the data highways to provide an integrated control system.

The near-term trends appear to be in the direction of heuristic (learning) type control, more sophisticated networking, and improved diagnostic capability. Problems remain in the areas of equipment component standardization, commonality to permit different vendors' products to be used in a common system, and the availability of extensive, complete system documentation, communication checks, and, to a lesser extent, system security (reliability).

6.3 CONTROL SYSTEMS EQUIPMENT

Control systems equipment includes **sensors, transmitters, indicators, recorders, controllers, process computers,** and **final control elements.** Frequently, functions are grouped in a single unit, and with the advent of microprocessors, this trend is accelerating.

6.3.1 Sensors

Sensors are available in two basic types: those in **direct** contact with the fluid for normal applications and the **indirect** (noncontact) type for applications that involve abrasive or corrosive materials, materials that tend to build up, or where leakage or contamination are undesirable. A frequently used system injects a **purge liquid** (or gas) and reads differentials between it and the measured variable, thus avoiding contact between the fluid and the measuring system. Typical **commercial sensor accuracy** is $\pm$ 0.5% of range with **precision sensors** available with accuracies of $\pm 0.2\%$ (or better) of the range.

Pressure sensors for analog systems tend to be of the bourdon tube type or in some cases diaphragm (bellows) type suitable for pressures up to the 50,000 psi (35 $\times$ 10^4 kPa) range. Electronic control systems use capacitance, inductance, and resonant wire type primary sensing elements, due to their improved accuracy and direct conversion from measured variable to the process signal. Differential pressure sensors normally use coupled diaphragms.

Temperature sensors commonly used are bimetallic or filled fluid systems utilizing liquid, vapor, or gas. These can handle temperatures from -400 to $+1400°F$ (-240 to $760°C$). For high temperatures to $+2700°F$ ($1500°C$), **thermocouples** are commonly used, whereas **resistance temperature detectors (RTDs)** are frequently used for midrange temperatures (-300 to $+1200°F$; -185 to $650°C$). For electronic control systems, thermocouples and RTDs are used for virtually all temperature ranges.

Level can be determined by displacers or floats (either fixed or operating through a linkage or by stationary probes that electrically ground out when liquid levels reach them), differential pressure cells, or capacitance devices. Pressure cells are also frequently used with bubblers (measuring backpressure) used for fluids that plug or coat normal sensors.

Density and thickness can be determined by absorption of radiation using a low-level source located on one side of a pipe (or barrier) and a sensor on the other. Where piping is carrying the fluid to be measured, rather careful calibration is needed to properly compensate for pipe wall thickness absorption, which is usually greater than changes in fluid density. Thickness can also be measured indirectly, although easily, by reference to a specific datum such as roll setting or by direct measuring devices such as calipers.

Flowmeters are available as turbine (bladed), target (paddle), vortex shedding, magnetic flow tube, or orifice types.

Other more specialized sensors are available for conductivity, relative humidity, sound intensity, vibration, various types of gas analysis (including gas chromatograph and infrared), pH, and similar variables.

6.3.2 Transmitters

Transmitters are commercially available either in combination with sensors or as separate units. Depending on the control mode employed, they can be obtained with 3- to 15-psi (48- to 103-kPa) pneumatic, 4- to 20- or 10- to 50-mA dc, outputs. Other versions also available yield 20- to 100-kPa signals. Transmitter signals are essentially proportional to the change in the measured variable. **Converters** are also available to convert pneumatic to electronic signals [pressure to current (P/I) and vice versa (I/P)], thus permitting adaptability and interchange of sensors.

6.3.3 Indicators

Indicators are sometimes included as an integral part of the sensor (local indicators), but remote indicator dials, vertical scales, counters, and so forth, are available for mounting at other locations. They can accept mechanical, pneumatic, or electronic inputs and are available in a wide variety of types and sizes.

6.3.4 Recorders

Recorders are usually of the **circular** or **strip-chart type,** almost all today of the electronic type (both chart and computerized). Able to accept a variety of inputs, they are usually panel mounted and can include controllers.

6.3.5 Controllers

Controllers are available in designs similar to recorders but are frequently intended for panel mounting; thus they are often of the vertical scale type. To facilitate reading and reduce reading error, matching scale pointers are frequently utilized.

They are compatible with a wide variety of inputs and can be obtained with any combination of control modes desired, although, more recently, PID controllers with switchable mode selection are most widely used.

Ratio controllers, autoselector systems, cascade controllers, and computing devices are also available with functions that can be tailored to specific control requirements.

6.3.6 Scanners

Scanners are used to monitor multiple sensors and alarm when out-of-range conditions are detected. Frequently they are high-speed type, some models scanning thousands of points per second. Although their functions are somewhat similar, scanners are considered distinct from **multipoint recorders,** which may have alarms. Some examples of scanners include fire eye, smoke detectors, and RTD or thermocouple scanning systems for bearing and electrical winding protection.

6.3.7 Trend Indicators

Trend indicators may be continuous or noncontinuous operating instruments. The noncontinuous type is normally wired to start at a preset level of the variable or upon command. Their purpose is to establish or display a trend of the variable and thus provide operators with process direction changes in advance of reaching process control limits. Frequently they include some internal reduction (mathematical) capability. They are typically of the strip chart variety in board-mounted cases up to 4 inches wide.

6.3.8 Annunciators

Annunciators are available in a variety of types producing audible, visual, or other types of signals. For control rooms, the **drop window** type is frequently used. When actuated, it provides a flashing light behind a translucent engraved window together with an audible signal. This continues until the operator silences the audible signal; the light at that time remains on, but no longer flashes. Subsequently, the light is extinguished when the alarm signal is terminated.

Normally the annunciator windows are arranged in groups above the uppermost operating controls; large control rooms may have several hundred windows. The windows are engraved with the name of the problem (e.g., high bearing temperature, low water level) and are available in a variety of colors, although translucent white or cream is most common. When annunciators are specified, it is considered good practice to provide from 10 to 25% spares (unassigned) for future needs.

6.3.9 Data Loggers

Data loggers are normally of the printing type. Frequently they operate noncontinuously, provide a record of data or events, and are triggered to function as event recorders. Normally electronic, they are capable of operating and/or recording data at high speed. They can accept input from a wide variety of sensors.

6.3.10 Accessories

A family of accessories is available for instrument systems including alarms, ratio controllers, and computing devices. Alarms are available with most instruments and usually consist of electrical contacts whose settings are adjustable to operate at selected process signal levels.

6.3.11 Enclosures

Enclosures are available for most types of instruments for use in outdoor, corrosive, explosive, or other special environments. In some cases specially treated purge air systems are provided to pressurize instrument cases to avoid contamination.

Enclosures provide mechanical and electrical protection for the operator and equipment. Brief descriptions of the more common types of enclosures follow. NEMA standards are used to designate the type of enclosure used. (See NEMA Standards Publication No. 250-1979 for more comprehensive descriptions, definitions, and/or test criteria.)

NEMA Type 1—For Indoor Use

Suitable for most applications where unusual service conditions do not exist and where a measure of protection from accidental contact with enclosed equipment is required. Designed to meet tests for Rod Entry and Rust Resistance.

NEMA Type 3R—For Outdoor Use

Primarily intended for applications where falling rain, sleet, or external ice formation are present. Gasketed cover. Designed to meet tests for Rain, Rod Entry, External Icing, and Rust Resistance.

NEMA Type 4—For Indoor or Outdoor Use

Provides a measure of protection from splashing water, hose-directed water, and wind-blown dust or rain. Constructed of sheet steel or stainless steel with gasketed cover. Designed to meet tests for Hosedown, Dust, External Icing, and Rust Resistance.

NEMA Type 4X—Nonmetallic for Indoor or Outdoor Use

Corrosion resistant, glass polyester or similar construction. Excellent for applications where wind-blown dust, rain, hose-directed water, or splashing water are present. Designed to meet tests for Hosedown, Dust, External Icing, and Corrosion Resistance.

NEMA Type 7—For Hazardous Gas Locations

For use in Class 1, Group C or D indoor locations as defined in the National Electrical Code, NEMA Type 7 enclosures must withstand the pressure generated by explosion of internally trapped gases and be able to contain the explosion so that gases in the surrounding atmosphere are not ignited. Under normal operation, the surface temperature of the enclosure must be below the point where it could ignite explosive gases present in the surrounding atmosphere. Designed to meet Explosion, Temperature, and Hydrostatic design tests.

NEMA Type 9—For Hazardous Dust Locations

For use in Class II, Group E, F, or G indoor locations as defined in the National Electrical Code. Heat-generating devices within the enclosure are designed to maintain the surface temperature of the enclosure below a point where it could ignite the dust–air mixture in the surrounding atmosphere or cause discoloration of surface dust. These enclosures are designed to meet tests for Dust Penetration, Temperature, and Aging of Gaskets (if used).

NEMA Type 12—For Indoor Use

Provides a degree of protection from dripping liquids (noncorrosive), falling dirt, and dust. Designed to meet tests for Drip, Dust, and Rust Resistance.

NEMA Type 13—For Indoor Use

Provides dust protection and protection against water, oil, or noncorrosive coolant spray. Designed to meet tests for Oil Exclusion and Rust Resistance.

NOTE. Enclosures are not designed to protect equipment against condensation, corrosion, icing, or contamination that can occur inside the enclosure or enter through a conduit or unsealed opening. Provisions to safeguard against such conditions must be made by the user.

6.4 APPLICATION CONSIDERATIONS

6.4.1 Reliability and Redundancy

Reliability is critical to many process control applications. It is usually not practical to provide duplicate *control* devices, and further, these devices have proven to be relatively trouble-free and in themselves highly dependable. **Sensors** and **transducers** tend to be the least reliable components in the loop, mainly due to environment. To achieve reliability, critical applications frequently have these components duplicated to provide reliability through **redundancy.** Special requirements for reliability are frequently achieved by hard-wired control circuits (rather than using electronic controls running through computers), the use of backup computers where the computer is used in an interactive control mode, and in some cases by redundant control systems where it is absolutely critical that one of the systems functions.

6.4.2 Fail-Safe Requirements

Fail-safe control systems assure that failure of a device or a control element does not adversely affect the operation of a system nor cause a hazardous condition. These can be broadly described as **stored energy systems,** those depending on **natural forces** or those in which the process is **self-compensating.**

Stored energy systems in many cases use springs, pneumatic fluid under pressure (such as bottled high-pressure gases), or in some cases explosive (chemical energy) actuators such as explosive valves for critical services. Other stored energy systems may use battery packs or chemical reactions to cause a control action. Examples of these are emergency lighting systems and the common soda–acid fire extinguisher.

Natural force systems frequently use gravity; for example, those systems in which a weight may be available but prevented from acting by a fusible link, as in fire doors.

Self-compensating systems are those in which the system tends to be self-regulating. An example of this is the float on a level control valve that uses the buoyant force of the controlled liquid acting on the float to provide the force necessary to operate the valve itself.

6.4.3 Requirements for Local Control

It is a matter of good practice to provide for **local control** wherever practical. In some cases the governing codes, frequently the National Electrical Code, will require that certain control devices be located within sight of the process or within sight of the operator so that in the event of a malfunction the energy source can be rapidly disconnected. The degree to which local control is incorporated depends not only on code requirements but also on the funding available for allocation to centralized process control. It may also depend on convenience when facilities are spread over a wide area, thus rendering local control inconvenient. Where processes are controlled from a remote location, it is good safety practice and often a legal requirement to provide sufficient warning signs, guards, and similar devices to avoid hazard to personnel or equipment in the area.

6.4.4 Interrelationships

Typically, basic plant service systems such as compressed air and service water are automated and controlled separately from other processes. To minimize operational difficulty, **control interconnections** between service systems and process systems should be minimized. Peripheral and secondary process systems are automated next, and, finally, the main process systems are themselves automated. This separation minimizes the potential for major difficulties because the building-block approach proceeds from those systems that are less complex and have the least impact on the process to those that are more sophisticated and have the greatest impact.

6.4.5 Interlocks

Although **interlocks** are used to protect downstream equipment and components, their control typically cascades upstream. Examples of this are conveyors, hydraulic controls, and level controls. These interlocks may be mechanical, electrical, hydraulic, or electronic.

6.5 CONTROL VALVES[2,3,5]

Control valve bodies are often at least one size smaller than the line in which they are installed, and it is not uncommon for them to be two sizes smaller. It is not good practice to install control valves with bodies smaller than two sizes less than the connecting lines unless a stress analysis is performed to ensure that the valve body is not overstressed due to thermal transients or dynamic forces.

To fit an accepted definition of a control valve, the valve body must be linked with an **actuator** mechanism which receives, and acts upon, a signal from the controlling system. One of the more common types of actuator uses a **spring/diaphragm** combination. Control valves are frequently provided with **shut-off (block) valves** on each side to permit them to be isolated for maintenance, and, if continuity of service is important, a bypass globe pattern valve is also provided. Many control valves employ a **characterized plug or cage;** that is, a plug or cage that is specially contoured to provide some specific relationship between valve stem (and hence plug) position and valve flow area. To avoid **wire drawing** (seat damage) due to throttling, seats, and in some cases plugs, are hardened or hard surfaced.

6.5.1 Valve Characterization

The **flow characteristic** of a control valve is defined as the relationship that exists between the **flow rate** through the valve and the **plug travel** from zero to 100%. The flow rate through the valve is also a function of the pressure drop across the valve. Since the pressure drop in an actual installation is usually not constant and ordinarily is different for each application, the recognized way to compare one valve with another is under similar conditions of constant pressure drop. When the pressure drop is held constant, the relationship of flow rate to plug travel is known as the **inherent flow characteristic.** There is also an **installed flow characteristic,** which is obtained under the varying pressure drop condition of an actual installation.

Although there are numerous inherent flow characteristics, only four of the more common ones are discussed here. They are: **quick-opening, linear, equal-percentage,** and **modified parabolic** and are shown graphically in Fig. 6.10.

The **quick-opening flow characteristic** provides maximum change in flow rate at low plug travels, with a fairly linear relationship. Additional increases in plug travel give sharply reduced changes in flow rate, and when the plug nears the wide-open position, the change in flow rate approaches zero. This characteristic is used primarily for on–off service where the flow must be established quickly when the valve begins to open. A typical example is in relief valve applications. In many applications it is possible to use a quick-opening characteristic where a linear characteristic would normally be specified, because the quick-opening characteristic is essentially linear up to about 70% of the maximum flow rate.

The **ideal linear flow characteristic** curve indicates a flow rate directly proportional to the plug travel. This proportional relationship produces a characteristic with a constant slope so, with constant pressure drop, the valve gain is the same at all flow rates. A linear characteristic is commonly specified for liquid level control and for certain flow control applications requiring constant gain.

With the **ideal equal-percentage flow characteristic,** equal increments of plug travel produce flow rate changes, which are equal percentages of the existing flow. The change in flow rate is always proportional to the flow rate that exists just before the change in plug position is made. When the plug is near its seat and the flow rate is small, the change in flow rate will be small; with a large flow rate, the change in flow rate will be large. These are generally used on pressure-control applications, and on other applications where a large percentage of the pressure drop is normally absorbed by the system itself with only a relatively small percentage available at the

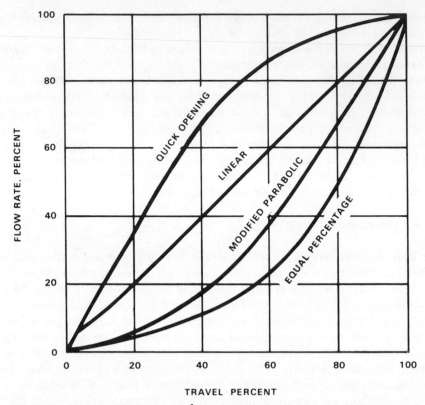

Fig. 6.10   Flow characteristic curves.[3] Courtesy of Fisher Controls International, Inc.

control valve. Equal-percentage valve characteristics should also be considered where highly varying pressure drop conditions can be expected.

The **modified parabolic flow characteristic** curve falls between the linear and equal-percentage characteristics. It provides close throttling action at low valve-plug travel and approximately linear characteristics for upper portions of plug travel. Linear or equal-percentage characteristics can be substituted with little change in performance.

Two coefficients are used for sizing and comparing control valve characteristics: C_v, valve sizing coefficient for liquids; and C_g, valve sizing coefficient for gases.

6.5.2   Liquid Flow[3]

$$Q = C_v \sqrt{\frac{\Delta P}{G}} \qquad (6.5)$$

where: Q = flow (gpm)
   ΔP = pressure differential across valve (psi)
   G = specific gravity of fluid (water = 1.0)
   C_v = valve flow coefficient (includes effects of flow area, discharge coefficient, etc.)

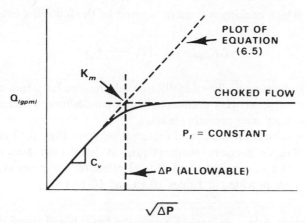

Fig. 6.11 Flow curve[3] showing C_v and K_m. Courtesy of Fisher Controls International, Inc.

C_v equals the number of U.S. gallons of water that will flow through the valve in 1 minute when the water temperature is 60°F (15.6°C) and the pressure differential across the valve is 1 psi (6.89 kPa). Thus C_v provides an index for comparing the liquid flow capacities of different types of valves under a standard set of conditions.

Control valves are available in **high** and **low (pressure) recovery** designs. **High recovery** is achieved by streamlining the internal design of the valve (e.g., high recovery—a V-notch ball valve; low recovery—a globe pattern valve).

Choked flow is a condition where the flow of a fluid through a control valve remains constant when the pressure differential across the valve increases. Figures 6.11 and 6.12 show this condition. Although the exact mechanisms of the choking process are currently unknown, the presence of a certain amount of vapor in the liquid is associated with this phenomenon.

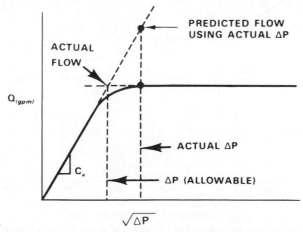

Fig. 6.12 Relationship between actual ΔP and ΔP allowable.[3] Courtesy of Fisher Controls International, Inc.

The point at which choking will occur is given by the following equation:

$$\Delta P_{\text{allow}} = K_m(P_i - r_c P_v) \tag{6.6}$$

where: ΔP_{allow} = maximum allowable differential pressure for sizing purposes (psi)
$\quad\quad K_m$ = valve recovery coefficient from manufacturer's literature
$\quad\quad P_i$ = body inlet pressure (psia)
$\quad\quad r_c$ = critical pressure ratio (determined from Figs. 6.13 and 6.14)
$\quad\quad P_v$ = vapor pressure of the liquid at body inlet temperature (psia)
$\quad\quad\quad$ (Vapor pressures and critical pressures for many common liquids
$\quad\quad\quad$ are provided in Tables 10.13 and 10.14.)

After calculating ΔP_{allow}, substitute it into the basic liquid sizing equation $Q = C_v (\Delta P/G)^{1/2}$ to determine either Q or C_v. If the actual ΔP is less than ΔP_{allow}, then the actual ΔP should be used in the sizing equation.

Use the curve in Fig. 6.13 for water. Enter on the abscissa at the water vapor pressure at the valve inlet. Proceed vertically to intersect the curve. Move horizontally to the left to read the critical pressure ratio r_c on the ordinate.

Use the curve in Fig. 6.14 for liquids other than water. Determine the vapor pressure/critical pressure ratio by dividing the liquid vapor pressure at the valve inlet by the critical pressure of the liquid. Enter on the abscissa at the ratio just calculated and proceed vertically to intersect the curve. Move horizontally to the left and read the critical pressure ratio r_c on the ordinate.

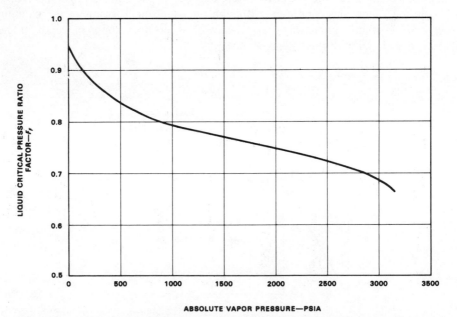

Fig. 6.13 Critical pressure ratios for water.[2] Courtesy of Fisher Controls International, Inc.

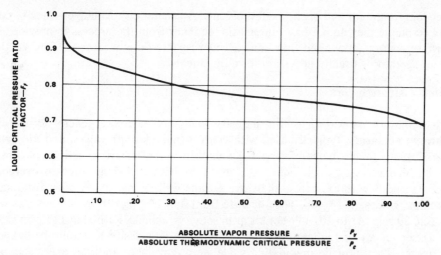

Fig. 6.14 Critical pressure ratios for liquids other than water.[2] Courtesy of Fisher Controls International, Inc.

6.5.3 Gaseous Flow[3]

Sizing of control valves for gases is more complex due to problems of critical flow and gas compressibility. One prominent manufacturer has developed the following empirical formula to reflect valve performance over its entire pressure range and for either high- or low-recovery valves. Because of the use of several coefficients this formula must be used in conjunction with performance data from the manufacturer. This requirement is similar to but varies among the manufacturers since valve performance is sensitive to valve body design (C_g) as well as the gas characteristics for the specific application.

$$Q = 1.06 \sqrt{d_i P_i}\ C_g \sin \left(\frac{3417}{C_1} \sqrt{\frac{\Delta P}{P_1}} \right)_{\text{Deg.}} \qquad (6.7)$$

where: $Q_{\text{lb/hr}}$ = gas, steam, or vapor flow (lb/hr)
  d_i = inlet gas density (lb/ft^3)
  C_1 = ratio gas to liquid flow coefficients, range 16–37 (High-recovery valves have low C_1 values, low-recovery valves have high C_1 values.)

6.6 CONTROL DEVICES

6.6.1 Manual Actuators

Manual actuators are the most commonly found actuators in control processes. Typical of them are handwheels on valves, adjustable weirs, and adjustable orifices. Major advantages to the manual devices are low cost, dependability, and relative freedom from wear and operational problems. A general disadvantage is lack of control response; they do not have automatic feedback from the process to provide any form of control action, and they have a slow response time because the intervention of an operator is usually necessary for adjustments.

6.6.2   Air Actuators

Air actuators are widely used in industry today and take two general forms: **diaphragm actuators,** typically used with control (modulating) valves, and **air cylinders,** usually found on on–off service. Diaphragm actuators are sometimes of considerable diameter (e.g., 24 inches; 0.6 m) to provide sufficient force to overcome unbalanced flow forces across the valve seat as well as internal friction from stem packing, guides, and so on. They operate from 3 to 15 psig (48 to 103 kPa) as well as 6 to 30 psig (41 to 207 kPa) of air or in some air cylinders utilizing 125-psig (870-kPa) service air or higher pressures where there is a desire to minimize cylinder diameter. They provide modulating valve stem movement and can operate in any position, although a diaphragm or cylinder arrangement above the valve body is preferred in order to avoid moisture accumulation within the diaphragm or cylinder. Major advantages of air actuators are low cost and flexibility in application, since they can be used with a wide range of air pressures and in some cases can operate directly from a low-pressure control transmitter. Response time may be delayed due to compressibility of the actuating fluid.

For systems utilizing **instrument air systems,** air is normally provided by **special compressors** which are of the **nonlubricated** variety, typically carbon or teflon ring. Oil contamination must be avoided in these systems as it may cause plugging of the fine orifices and flapper valves found within instruments. To avoid dew point changes from expansion of air both at the compressor and at point of use, **air dryers** are normally installed immediately downstream of the instrument air compressors. The air dryers are usually of the regenerative type and provide air having a dew point of approximately $-40°F$ ($-40°C$) at standard conditions. Air actuators have been widely used for many years, are well proven, and are available in standard designs and sizes for a variety of services and applications.

Diaphragm type actuators provide linear motion, limited to the amount of diaphragm deflection permitted, frequently to 4 inches (100 mm). Wear over a period of years in some installations can become a problem because the diaphragms occasionally tend to harden and crack. Piston seal wear can be a problem with air cylinder actuators, and "boots" are sometimes installed to keep contaminants off the piston rod surface.

A typical diaphragm actuated control valve is shown in Fig. 6.15.

6.6.3   Hydraulic Actuators

Hydraulic actuators are used for systems that require rapid response or large forces. They can range from low-pressure systems to those that operate on 3000 to 5000 psig ($21–35 \times 10^3$ kPa) pressure and that generally require a local, dedicated source of high-pressure fluid such as a special pump unit. Usually the working fluid is **oil,** the control mode being exercised by **spool-type valves** operating through various

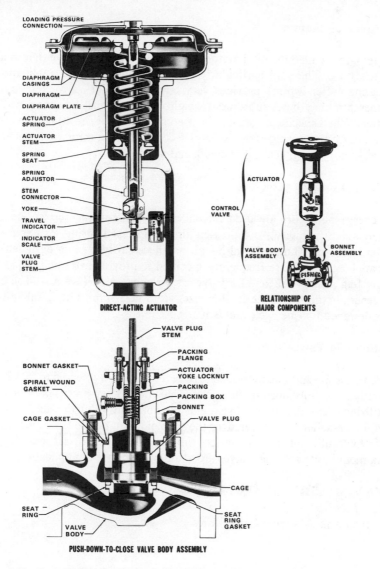

LOADING PRESSURE CONNECTION
DIAPHRAGM CASINGS
DIAPHRAGM
DIAPHRAGM PLATE
ACTUATOR SPRING
ACTUATOR STEM
SPRING SEAT
SPRING ADJUSTOR
STEM CONNECTOR
YOKE
TRAVEL INDICATOR
INDICATOR SCALE
VALVE PLUG STEM

DIRECT-ACTING ACTUATOR

ACTUATOR
CONTROL VALVE
VALVE BODY ASSEMBLY
BONNET ASSEMBLY

RELATIONSHIP OF MAJOR COMPONENTS

VALVE PLUG STEM
PACKING FLANGE
BONNET GASKET
ACTUATOR YOKE LOCKNUT
SPIRAL WOUND GASKET
PACKING
PACKING BOX
BONNET
CAGE GASKET
VALVE PLUG
CAGE
SEAT RING
SEAT RING GASKET
VALVE BODY

PUSH-DOWN-TO-CLOSE VALVE BODY ASSEMBLY

Fig. 6.15 Typical control valve assembly. Courtesy of Fisher Controls International, Inc.

drilled and ported blocks having very close clearances and usually O-ring seals to minimize bypass leakage and thus avoid valve drift.

Filtration of the fluid to the 1-micron range designed to remove particulate material, is an essential and usually integral part of the high-pressure hydraulic package. The hydraulic actuators have the advantage of very rapid response rates, they can be small and light in weight, and are fairly easy to install and tune up. The cost for providing a dedicated source of hydraulic fluid and drainage lines back to a sump are the principal disadvantages. Some types of hydraulic fluids also present a fire hazard and special precautions such as extinguishing systems may be necessary.

6.6.4   Motor Actuators

Motor actuators called motor operators (electric) are suitable for either modulating or on–off service. They are particularly useful for high-pressure applications since their gearing (often worm) provides very large forces for valve stem movement. These operators may be of substantial weight and this must be considered in their application. When specified, they can be used on valves for throttling service. Motor operator closure times range from several seconds to several minutes, depending on valve size, motor sizing, gearing selected, and so forth.

6.6.5   Stored Energy Actuators

Stored energy actuators are used for on–off or single-stroke service, generally of either the weight (gravity), compressed air, compressed spring, or explosive type. These actuators are "tailored" (although not necessarily custom designed) for the specific application and are frequently specified to provide extremely rapid opening or closure [e.g., $\frac{1}{10}$ to $\frac{2}{10}$ second for valves as large as 36 in. (0.9 m)]. The explosive type is generally limited to the smaller sizes (to 3 in.; 76 mm) and, unlike the other types, is designed for one cycle and is not reuseable.

6.6.6   Solenoid Valves

Solenoid valves are suitable for on–off service and utilize a solenoid to move the valve disc to either the fully open or fully closed position. They are usually of the globe type, utilizing a circular disc and seat and are normally of moderate size—2 inches (51 mm) and smaller. For larger sizes the solenoid may be used as a pilot to operate an air or hydraulic actuator. Response time is very rapid; $\frac{1}{10}$ second is typical. In addition, modulating solenoid valves are available for special applications.

6.7   MAN/MACHINE INTERFACE[1,4,6,8]

6.7.1   Human vs. Machine Capabilities

Limitations of Humans

  Accuracy—Susceptible to constant and variable errors.
  Speed—Time required for decision and movement.
  Force—Depends on body member in use and fatigue.
  Computing—Slow, inaccurate (limited to single integration and differentiations).
  Decision making—Optimum strategy not always used; preservation exists.
  Information input rate—Susceptible to overloading; stress and boredom affect
    performance.

Advantages of Humans

  Detection—Can detect a wide range of signals, strong to weak.
  Perception—Can see through complex situations, have constancy, and can detect
    signals through noise.

Flexibility—Can shift rapidly in attention, can revert to alternate modes of operation.

Judgment—Have inductive reasoning, incidental intelligence, "hunches."

Reliability—Satisfactory performance under adverse conditions is possible; can perform when parts are out of order; performance good when highly motivated.

Limitations of Machines

Maintenance is required.
Monitoring is required.
Decision making is limited.

Advantages of Machines

Speed is possible.
Accuracy is possible.
Short-term memory.
Simultaneous activities are possible.
Complex problems can be handled.
Good for repetitive tasks.
Do not tire.

A human being is more desirable than an automatic regulator only when the eye and brain take an essential part in the regulation process (e.g., in flying an aircraft in difficult situations or driving a car in traffic). However, regulation of temperature, of sound volume, or physical quantities can be performed better by an automatic device.

Humans have two faculties than cannot be matched by technical transmission systems:

The human eye is extremely efficient, especially in visual acuity.
The connection between eye and brain makes possible a logical discrimination of things that are seen.

6.8 PANEL DESIGN

Panel design must consider the characteristics of the human body and the functions to be exercised.

6.8.1 General

HEIGHT. For optimum results, locate panel devices below eye level but not lower than a sight line 30° below horizontal. Easy eye movements are from horizontal to 30° below.

DISTANCE.   Optimum reading radius is 18 to 22 inches (0.4 to 0.6 m). Standard
    displays are designed to be read at a 28-inch (0.7-m) radius from the eye but
    distance can be greatly increased if the display is designed accordingly and
    reach is not a factor. Minimum acceptable reading distance is 13 inches (0.3
    m). Normal reading distance for standard cathode ray tubes is 14 to 18 inches
    (0.35 to 0.46 m).

COMPACTNESS.   Avoid extended displays to minimize scanning.

PRIORITY.   Locate important and frequently used instruments nearest the normal
    line of sight, which varies for the seated or standing posture.

STANDARDIZATION.   Primary instruments should be located in standard arrange-
    ment from one situation to another to prevent error in reading.

ASSOCIATED MANUAL CONTROLS.   Associated knobs and switches should be
    located below the instrument or to the right to prevent visual interference due
    to hand movements.

6.8.2   Panels

Panels are frequently laid out to indicate the process being controlled. In the
arrangement called **mimic panels,** main process flow paths and equipment are indicated in schematic form, and instruments, status lights, and so on, are located
approximately matching the process locations of the sensors, control elements, and
so on. **Semimimic panels** are a more generalized layout concept and are usually substantially cheaper to produce. For processes subject to change, several manufacturers
offer a modular grid mimic panel where grid elements can be changed to suit process
changes.

   More recently, the expanded use of CRTs to display process diagrams, including
in some cases flow rates and temperatures, has reduced the interest in mimic panels.
The programming costs and complications must be considered when making the
decision to use CRTs for this purpose.

   CRT Consoles.   Several manufacturers have begun offering CRT consoles as an
alternate (small systems) or supplement (large systems) to conventional control
panels. The CRTs partially or completely replace the panels, switches, annunciators,
and so forth, and utilize one or more color CRTs with an associated typewriter-style
keyboard. The CRTs display either the manufacturer's standard or customized for-
mats and graphics. The displays are in color and can incorporate system graphics,
performance and status data in bar graph form, and flashing (alarm) indications.
They can also permit selection of level of detail displayed as well as specific parameters of interest. Through the associated keyboard, set points can be changed, motors
started, valves positioned, and so on. In some cases, **interactive graphics** utilizing
light pens are provided. These permit the operator to actuate components displayed
on the CRT by touching the pen to the display and depressing an actuation device.
They can provide overview data, trend data, and permanent records via line printers.

   The CRT consoles, although providing major space savings and improving operator span of control, are more costly than conventional instrumentation and may
require more programming and operator training.

6.8.3 Instruments

Kinds	Reading Error
Counters	0.4%
Open reading dials	0.5%
Round dials	10.9%
Semicircular dials	16.6%
Horizontal dials	27.5%
Vertical dials	35.5%

Preference.   Direct-reading counters and circular dials are best; all linear scales lead to errors.

6.8.4   Round Dials

Standard dial face diameters of 2¾ to 3 inches (70 to 76 mm) are recommended. A fixed dial with moveable pointer is superior to a moving face and fixed pointer, which is not readable during rapid motion.
All dials should increase from zero in a clockwise direction.
Locate zero value at 9 to 12 o'clock positions. Locate zero value at 12 o'clock position if values go from plus to minus.
Dials with the fewest markings can be read the most rapidly.
Scales numbered by intervals of 1, 10, 100, and so on and subdivided by ten graduation intervals are superior to other acceptable scales. Avoid irregular scales except logarithmic and special scales.
Except for locations higher than the head, titles should be placed above instruments and manual controls to prevent visual interference.

6.8.5   Pictorial Indicators

Used to assist in interpretation of special relationships [e.g., to symbolize aircraft attitude (i.e., orientation in space), pitch, and roll].

6.8.6   Signal Lights

Alerting with lights.
Signal light colors:
   Red—critical, malfunction
   Amber—cautionary
   White—general status
   Green—safe
   Yellow—caution

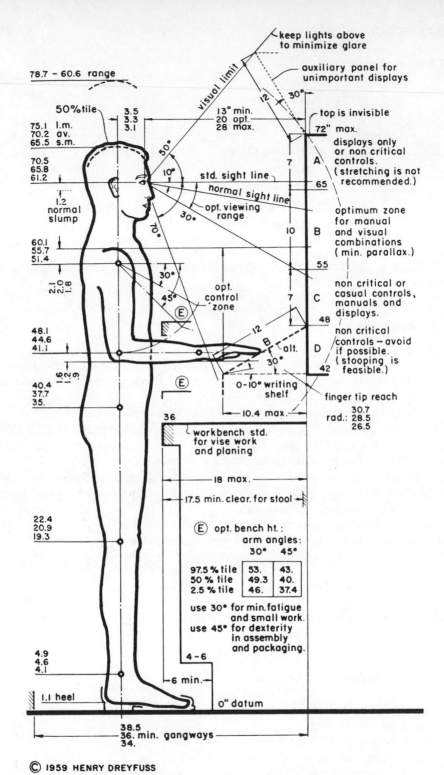

Fig. 6.16 Anthropometric data, adult male standing at control board.[1] © 1959, 1960, 1967, by Henry Dreyfuss & Associates. Reprinted by permission of The Whitney Library of Design.[1]

Warning lights for critical functions must be within 30° of the normal line of sight. A central warning light may be used to direct attention to another panel area.

Keep number and colors to a minimum.

Locate on or near the associated control to facilitate appropriate action.

Normal operating signals should be arranged in patterns or superimposed on diagrams compatible with the equipment they symbolize.

6.8.7 Cathode Ray Tubes

Scope Size (diagonal).   Five inches (125 mm) min.; 19 inches (480 mm) typ.; 25 inches (623 mm) max.

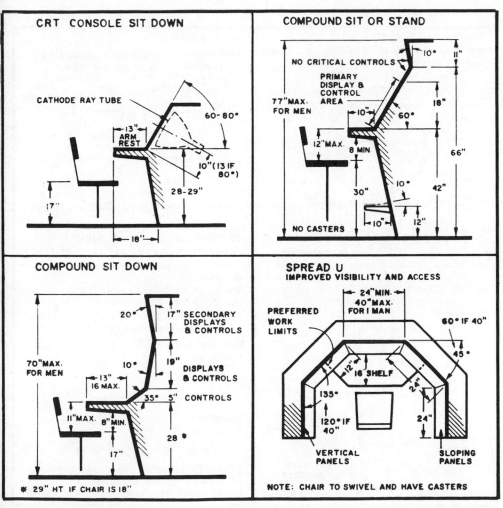

© 1966 HENRY DREYFUSS

Fig. 6.17 Preferred console dimensions.[1]

Scope Position. Viewing distance, 12 to 18 inches (300 to 450 mm). For 25-inch (635-mm) screens, distances to 6 feet (1.8 m) are satisfactory. Screen should be perpendicular to normal line of sight or not greater than 30° off.

6.8.8   Audible Signals

Buzzer. Use in quiet locations, commands attention without causing undue alarm (e.g., individual operator alert).

Bells. Penetrating (low-frequency noise). Abrupt onset demands attention and fast response (e.g., fire alarm).

Chimes. Used for nonurgent actions. Chimes do not cause undue alarm.

Tones. Can be used over electric intercom system.

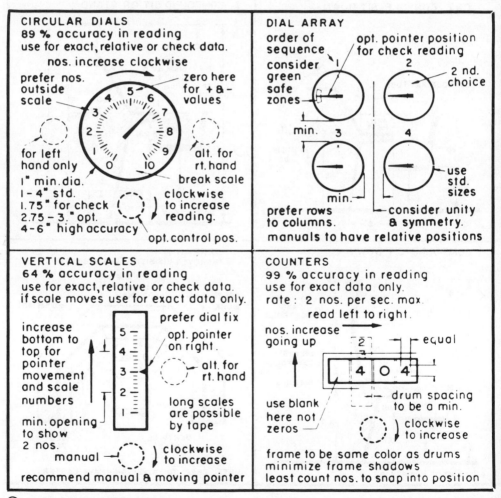

© 1960 HENRY DREYFUSS

Fig. 6.18   Basic display data.[1]

Recommended Design Practice

Warning signals should be at least 10 dB above noise level.

Signals should not exceed 110 dB unless ear protectors are used. Preferred frequency, 200 to 5000 cps (sensitive range of hearing).

6.8.9 Anthropometric Data

Standard anthropometric data together with information on displays and panel design is shown in Fig. 6.16.

6.8.10 Consoles

Selected dimensional data on consoles is shown in Fig. 6.17.

6.8.11 Basic Display Data

Orientation and other data on basic displays are shown in Fig. 6.18.

REFERENCES

1. Henry Dreyfuss, *The Measure of Man,* Whitney Library of Design, c/o Watson-Guptill Publications, New York, 1967.
2. *Control Valve Handbook,* 2nd edition, Fisher Controls International, Inc., Marshalltown, Iowa, 1977.
3. *Technical Monographs* 29, 30, and 31, Fisher Controls International, Inc., Marshalltown, Iowa.
4. John Hammond, *Understanding Human Engineering: An Introduction to Ergonomics,* David and Charles Inc., North Pomfret, Vt., 1979.
5. J. W. Hutchinson, *ISA Handbook of Control Valves,* 2nd edition, Instrument Society of America, Pittsburgh, Pa., 1976.
6. Ernest J. McCormick, *Human Factors Engineering,* 3rd edition, McGraw-Hill, New York, 1970.
7. Clifford Warren, *A Comparison of Analog and Digital Control Techniques,* Fisher Controls International, Inc., Marshalltown, Iowa, 1979.
8. Wesley E. Woodson and Donald W. Conover, *Human Engineering Guide for Equipment Designers,* University of California Press, Berkeley, Ca., 1964.

SECTION 7
ECONOMICS/STATISTICS

7.1 INTEREST

Interest is the time value of money—often referred to as the cost of money.

7.1.1 Simple Interest

Simple interest assumes that the debt (principal) is repaid at the end of the period with interest earned over the period. Thus:

$$I = Pni, \qquad\qquad (7.1)$$

where: I = interest
$\qquad$ P = principal

and

$$T = P(I + ni), \qquad\qquad (7.2)$$

where: n = time (usually number of years)
$\qquad$ i = interest rate per unit time (usually per year)
$\qquad$ T = total amount due

7.1.2 Compound Interest

Compound interest assumes the periodic interest payments are added to the principal and subsequent interest is earned on the resulting larger principal:

$$P \text{ at end of period} = P(1 + i)^n. \qquad\qquad (7.3)$$
$$I \text{ earned during period} = P(1 + i)^{n-1}i. \qquad\qquad (7.4)$$
$$T \text{ due at end of period} = P(1 + i)^n. \qquad\qquad (7.5)$$

As a matter of convenience, interest rates compounded daily are often based on a 360-day year to permit simplified conversion to annual rates, rather than using a 365-day year. Also, extensive tables are available for interest and principal values (factors).

A useful rule of thumb for quickly estimating the effect of compound interest is the **rule of 72.** If 72 is divided by the interest rate, the answer is a very close approximation of the number of periods required to double the principal. Thus at 6% annual interest, compounded, the principal doubles in 12 years, at 8%, 9 years, and so on.

7.1.3   Present Worth

Present worth is the value of a principal to be paid (or redeemed) at a future date, the principal bearing compound interest during the intervening time:

$$P_w = \frac{1}{(1 + i)^n},$$
(7.6)

where P_w = present worth of the principal. (Tables of P_w are found in Section 7.6)

7.1.4   Discounted Value

The discounted value is the present value of a principal to be redeemed or paid at a future date. The **discount rate** (d) is the discount on the principal of one over one unit of time. Thus:

where d = discount rate and i the effective interest rate is:

$$i = \frac{d}{1 - d}.$$
(7.7)

7.1.5   Annuities

An **annuity** is a series of equal periodic transactions that earn interest. The calculation of annuities can involve either deposits or withdrawals:

Amount of Annuity

$$A_n = \frac{[(1 + i)^n - 1]}{i},$$
(7.8)

where A_n = Amount of the Annuity

(Tables of A_n are found in Section 7.6.)

Present Worth of an Annuity

$$P_w = \frac{1 - (1 + i)^{-n}}{i},$$
(7.9)

(Tables of P_w are found in Section 7.6.)

7.2   DEPRECIATION

Depreciation represents a charge reflecting the deterioration in a capital asset. In business this permits accumulation of funds to replace assets as they wear out. Another form of deterioration is **obsolescence,** which represents a limited economic life of an asset due to technological change. Both of these can be provided for by setting aside funds usually called "reserves for depreciation" to permit replacement of the asset.

Allowable periods of write-off for depreciation permitted by the Internal Revenue Code do not necessarily reflect the actual economic life of the asset, but may be structured to encourage investment or plant modernization as a matter of governmental policy.

Depreciation allowances may be calculated in several ways. (Note that none of these formulas make allowance for the effect of inflation; i.e., the new asset costs more than the asset being replaced. To some extent sinking funds that earn interest can compensate for the effect of inflation, however.)

7.2.1 Straight Line

The simplest, most widely used formula assumes uniform depreciation over the economic life of the asset:

$$d = \frac{C - C_s}{L},$$

(7.10)

where: d = depreciation per year
$\quad\quad C$ = original cost
$\quad\quad C_s$ = salvage value
$\quad\quad L$ = life (years)

7.2.2 Sinking Fund

The **sinking fund** method assumes that a sinking fund receiving periodic payments and earning interest is established. Thus:

$$d = (C - C_S) \frac{1}{S_L}.$$

(7.11)

7.2.3 Accelerated Depreciation

Accelerated depreciation assumes the asset declines more rapidly in value during its early years. There are several types commonly used; for example, double declining and sum of the digits.

Double declining depreciation utilizes deductions of twice the straight line amount applied in each year to the remaining balance, thus accelerating the amount of depreciation deducted during the early life of the asset. For an asset with a 10-year life, the depreciation deducted is 20% the first year, 16% (20% of the remaining 80%) the second year, and so on.

The **sum of the digits** method utilizes the life of the asset as the numerator and the sum of the digits of the life of the asset as the denominator. Thus for a 10-year life, the sum of the digits 1 through 10 is 55, and the depreciation taken the first year is 10/55, the second year, 9/55, and so on.

$$d = \frac{n'}{n\frac{(n+1)}{2}},$$
(7.12)

where: d = depreciation taken in year n'
$\qquad$ n' = particular year
$\qquad$ n = life of asset remaining (number of years)

7.3 STATISTICS

7.3.1 Measures of Central Tendency

The two most useful measures of central tendency are the **arithmetic mean** and the **median.**
$\qquad$ The **arithmetic mean:**

$$\bar{x} = \Sigma\frac{x}{n},$$
(7.13)

where: $\bar{x}$ = arithmetic mean
$\qquad$ x = individual observed values
$\qquad$ n = number of values observed

The **median** is the individual observed value that has an equal number of observed values above and below it; that is, it is the middle observed value in a series. In small or skewed statistical populations, it may vary widely from the arithmetic mean, whereas in large nonskewed populations it will closely approach the arithmetic mean. **Population** is the term applied to the entire group from which characteristics will be determined.

7.3.2 Measures of Dispersion

Observations will tend to cluster around the arithmetic mean with the degree of their dispersion defined as the **standard deviation:**

$$s = \sqrt{\frac{\Sigma(x - \bar{x})^2}{n - 1}}.$$
(7.14)

where: s = standard deviation

For normal (nonskewed) distributions, a predictable proportion of the values will fall within a range of standard deviations:

Range	Percent of Values Included
$-1s$ to $+1s$	68.26
$-1.96s$ to $+1.96s$	95.0
$-2s$ to $+2s$	95.46
$-3s$ to $+3s$	99.73

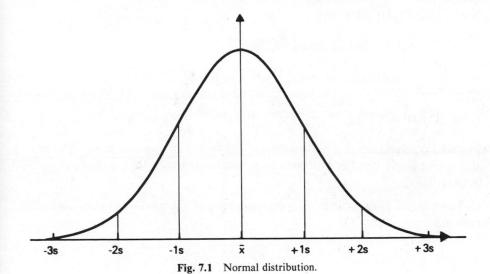

Fig. 7.1 Normal distribution.

The **coefficient of variation** is a useful measure to compare dispersion of data from different sources or bases. This can be particularly useful when establishing correlations:

$$V = \frac{s}{\bar{x}} . \tag{7.15}$$

where: V = coefficient of variation

7.3.3 Sampling

Sampling provides a method by which the characteristics of a population can be predicted by examination of only a small portion of the population. To achieve this it is necessary that the sample be representative (usually achieved by random sampling). Where all items in the population are examined, the sample is 100%. This is sometimes also referred to as exhaustive sampling.

Sampling frequently uses a single sample whose size varies depending on the size of the population (or lot) and the **confidence level** desired. The confidence level desired will have the larger effect on sample size required. Predetermined acceptance levels are compared to the sample characteristics to establish acceptability or rejectability of the lot.

Where a **single sample** indicates noncompliance with acceptable criteria, **double (or multiple) sampling** is frequently employed. This requires larger progressive samples with a reduced proportion of nonacceptable items permitted.

After determining the **acceptable quality level (AQL)** (the level of defects acceptable) together with the **confidence level (CL)** of achieving or bettering that level, standard published tables of acceptance sampling plans can be referred to for lot sizes, sample sizes, acceptance and reject quantities for each, and so on. Among the more widely used of these are:

MIL-STD-105 (Military Standard 105)
Dodge–Romig Sampling Plans
DOD Interim Handbooks H106, H107, and H108

7.3.4 Probability

Probability establishes the mathematical likelihood of an event occurring. The probability of a single event is expressed by a number between 1.0 (certainty) and 0.0 (impossibility).

The probability of a series of events occurring is the product of their individual probabilities:

$$P_s = P_a \cdot P_b \cdot P_c \cdots P_n, \tag{7.16}$$

where: P_s = Probability of the series of events
P_a, P_b, P_c, P_n = Probability of the individual events.

The probability of a series of events can be greatly improved by identifying and improving those items having low probability. Alternatively, the probability or performance of a system of components can be improved by establishing parallel circuits:

$$P_s = 1 - (1 - P_a)(1 - P_b) \cdots (1 - P_n), \tag{7.17}$$

where P_s = probability of the system.

Thus for a **series system** (Fig. 7.2) with a low-probability (reliability) component, for example:

$$P_a = 0.98$$
$$P_b = 0.88$$
$$P_c = 0.65$$
$$P_d = 0.98,$$

$$P_s = (.98)(.88)(.65)(.98) = .549. \tag{7.18}$$

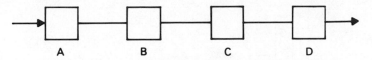

Fig. 7.2 Series system.

Duplicating the low-reliability component in **parallel** (Fig. 7.3) yields:

$$P_s = (.98)(.88)[1 - (1 - .65) (1 - .65)](.98) =$$
$$\cdot \cdot [1 - (.35)(.35)] \cdot \cdot$$
$$[1 - .123],$$

$$(.98)(.88)[.88](.98) = .744 \tag{7.19}$$

7.3.5 Frequency Diagram

A **frequency diagram** is a powerful diagnostic tool to establish causes of events. Where events are grouped by cause (or type) and the number of occurrences of each are plotted as a simple bar graph, the diagram will clearly establish those items that occur most frequently and that deserve first attention.

As is apparent in Fig. 7.4, Event Types E, B, and H are the major repetitive ones and should be given first attention. Of the 80 events, 46 are of only three types while the remaining 34 comprise 10 different event types. The **Pareto principle,** which describes this, states that only a few causes account for the majority of the events. When testing data, a rule of thumb often used is "20% of the causes generate 80% of the events"—**the 80-20 rule.**

7.4 ECONOMIC CHOICE

7.4.1 Minimum-Cost Point

The **minimum-cost point** is time independent. When only immediate expenditures are relevant to the choice among alternatives, the time value of money is not involved. The problem then is to obtain the design with the minimum first cost. These problems

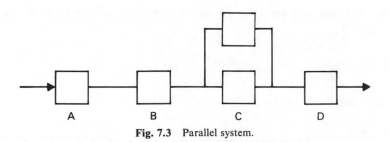

Fig. 7.3 Parallel system.

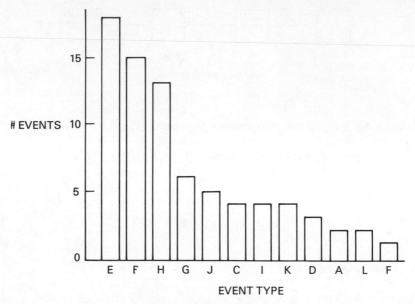

Fig. 7.4   Frequency diagram.

are solved graphically or analytically by expressing the costs as functions of the design variable, adding the separate costs, and locating the point of minimum total cost.

Economic Balance.   When increase in annual capital charges results in decreased annual operating charges, both expressed in terms of a common design variable, the solution for minimum total annual cost is called **economic balance.** Although graphical or tabular constructions are generally applicable, many of these problems may be solved analytically. For example, if x is the common design variable; a, b, c, and d are constants related to the specific problem; and

$$\text{Annual capital cost} = ax + c, \tag{7.20}$$

$$\text{Annual operating cost} = \frac{b}{x} + d, \tag{7.21}$$

$$\text{Total annual cost} = ax + \frac{b}{x} + (c + d), \tag{7.22}$$

differentiating total cost and equating to zero gives the optimum solution as

$$x = \sqrt{\frac{b}{a}}. \tag{7.23}$$

When there are two common design variables x and y,

$$\text{Total annual cost} = ax + \frac{b}{xy} + cy + d, \tag{7.24}$$

from which a plot of total annual cost versus x at various values of y gives a cost curve with the minimum point as the optimum.

7.4.2 Measures of Profitability

In the choice among alternative investments, the objective is to make the best possible use of a limited resource, capital. Decision criteria are thus required, among which the principal ones are described here using the convention of **continuous interest** r with periodic receipts and disbursements. To convert these equations to periodic interest, substitute $(1 + i)^n$ for e^{rn}.

Periodic and continuous interest should not be intermixed in economic comparisons. All results should be converted to one form of interest before comparing profitabilities.

Minimum Acceptable Return

In the array of prospective investments, some more profitable than others, a minimum standard of comparison is needed. **Minimum acceptable return p** may be looked upon as the interest rate of the marginal investment that is generally available to the investor. It is commonly taken as the average rate of earnings on the total assets of the company, often called the **pool rate.** Rates of return are usually computed after income tax.

Internal Rate Method

This well-known method, recently referred to as **discounted cash flow** and **interest rate of return,** serves adequately for the great majority of investment and budgetary decisions. It may be applied to incremental cash flow from an incremental investment in a project as well as to an array of unrelated projects.

The **internal rate method** postulates that the algebraic sum of the compound amounts of all cash flows for a project is zero at some internal rate of return found by trial-and-error solution of Equation (7.26).

$$\sum_{n=1}^{n=L} R_n e^{r(L-n)} - \sum_{n=0}^{n=L} D_n e^{r(L-n)} = 0 \tag{7.25}$$

or

$$\sum_{n=1}^{n=L} R_n e^{-rn} - \sum_{n=0}^{n=L} D_n e^{-rn} = 0. \tag{7.26}$$

The decision criterion is the internal rate r, larger r being preferred.

This method is not entirely satisfactory because of two inherent defects: (1) it is based on the questionable assumption that the receipts from a project will be reinvested in an equally profitable investment, and (2) the solution for r may be indeterminate (imaginary or multiple roots) when there is more than one reversal in the direction of annual net cash flow.

Proportional Gain Method

This method, attributed to Bernoulli, avoids trial-and-error solutions and is suitable for choosing between mutually exclusive alternates or for ranking an array of investment opportunities. This formulation postulates that the net receipts are accumulated in one account and the net investments in another account, both at interest rate p. When the project is terminated at time L, the relative gain G is the ratio of the two accounts or the ratio of their present worths:

$$\frac{\displaystyle\sum_{n=1}^{n=L} R_n e^{p(L-n)}}{\displaystyle\sum_{n=0}^{n=L} D_n e^{p(L-n)}} = \frac{\displaystyle\sum_{n=1}^{n=L} R_n e^{-pn}}{\displaystyle\sum_{n=0}^{n=L} D_n e^{-pn}} = G = e^{kL}. \qquad (7.27)$$

The proper decision criterion is G, not k as erroneously used by some authors. Since this method is biased in favor of long-term investments, it is generally reliable only for comparing investments with nearly equal lives as originally proposed by Bernoulli.

Present Worth Method

This method, also referred to as **venture worth** and **incremental present worth,** is restricted to comparison of projects that have identical lives L or that cover the same total time span. If it can be assumed that the costs and returns of replacements will repeat those of the original asset, multiple cycles of a short-term project may be compared with a single long-term project covering the same time span (e.g., three 5-year lives considered equivalent to one 15-year life).

With this method, present worth is the decision criterion, larger present worth being preferred. The present worth P of each project is computed, with r equal to the minimum acceptable return, as the algebraic sum of the present worths of the annual net cash flows, with salvage value taken as a receipt at the end of the project life.

$$P = \sum_{n=1}^{n=L} R_n e^{-rn} - \sum_{n=0}^{n=L} D_n e^{-rn} \qquad (7.28)$$

This method is suitable for projects that have no positive receipts, a situation in which the other two methods are indeterminate.

7.5   QUALITY ASSURANCE/RELIABILITY

Quality assurance practices are based on organization and control of activities, whereas **reliability** deals with predictability of results. Quality assurance requires organization, planning of activities, control of those activities, cause determination and loop closing of deviations, and independent reviews of effectiveness. The United States government in 10 Code of Federal Regulations, Part 50, Appendix B, set forth 18 criteria to which nuclear projects must conform.[1] These are typical of the philosophy and scope of formal quality assurance programs applied throughout industry, although all elements may not be appropriate for a particular activity or component. In summary form these elements are:

I.   An organization with sufficient independence shall be established to control and verify the performance of functions affecting quality.

II. A documented program defining scope and responsibilities shall be established.

III. Measures to assure proper control of design activities including verification or testing shall be applied.

IV. Documents which control procurements shall include the applicable Quality Assurance program requirements.

V.   Activities affecting quality shall be documented, accomplished in accordance with these documents (procedures) and include acceptance criteria.

VI. The review, issuance and change of documents affecting quality is controlled.

VII. Conformance to procurement documents of purchased materials, equipment and services shall be established, including both control and documentation.

VIII. Identification and control of material, parts and components shall be established.

IX. Welding, heat treating, nondestructive testing and other special processes are to be controlled.

X.   A program for the inspections of activities affecting quality shall be established.

XI. A controlled testing program shall be established and its results shall be documented.

XII. Measuring and test equipment shall be controlled and calibrated.

XIII. Measures to control handling, shipping and storage to prevent damage or deterioration shall be established.

XIV. Measures shall be established to control and indicate inspection, test and operating status.

XV. Nonconforming materials shall be identified and controlled to prevent their installation or inadvertent use.

XVI. Conditions adverse to quality shall be identified, corrected and for significant items reported to management.

XVII. Sufficient records shall be maintained to furnish evidence of activities affecting quality.

XVIII. A system of planned, periodic audits shall be established to verify compliance with and the effectiveness of the Quality Assurance program.

To avoid excessive cost and waste, it is of great importance that the requirements for quality assurance and reliability be selectively applied to only those components and/ or characteristics necessary.

7.6 TABLES

Table 7.1   Amount at Compound Interest $(1 + i)^n$

n	1¼%	2%	2½%	3%	3½%	4%	4½%	5%
1	1.01500	1.02000	1.02500	1.03000	1.03500	1.04000	1.04500	1.05000
2	1.03023	1.04040	1.05062	1.06090	1.07122	1.08160	1.09203	1.10250
3	1.04568	1.06121	1.07689	1.09273	1.10872	1.12486	1.14117	1.15763
4	1.06136	1.08243	1.10381	1.12551	1.14752	1.16986	1.19252	1.21551
5	1.07728	1.10408	1.13141	1.15927	1.18769	1.21665	1.24618	1.27628
6	1.09344	1.12616	1.15969	1.19405	1.22926	1.26532	1.30226	1.34010
7	1.10984	1.14869	1.18869	1.22987	1.27228	1.31593	1.36086	1.40710
8	1.12649	1.17166	1.21840	1.26677	1.31681	1.36857	1.42210	1.47746
9	1.14339	1.19509	1.24886	1.30477	1.36290	1.42331	1.48610	1.55133
10	1.16054	1.21899	1.28008	1.34392	1.41060	1.48024	1.55297	1.62889
11	1.17795	1.24337	1.31209	1.38423	1.45997	1.53945	1.62285	1.71034
12	1.19362	1.26824	1.34489	1.42576	1.51107	1.60103	1.69588	1.79586
13	1.21355	1.29361	1.37851	1.46853	1.56396	1.66507	1.77220	1.88565
14	1.23176	1.31948	1.41297	1.51259	1.61869	1.73168	1.85194	1.97993
15	1.25023	1.34587	1.44830	1.55797	1.67535	1.80094	1.93528	2.07893
16	1.26899	1.37279	1.48451	1.60471	1.73399	1.87298	2.02237	2.18287
17	1.28802	1.40024	1.52162	1.65285	1.79468	1.94790	2.11338	2.29202
18	1.30734	1.42825	1.55966	1.70243	1.85749	2.02582	2.20848	2.40662
19	1.32695	1.45681	1.59865	1.75351	1.92250	2.10685	2.30786	2.52695
20	1.34685	1.48595	1.63862	1.80611	1.98979	2.19112	2.41171	2.65330
21	1.36706	1.51567	1.67958	1.86029	2.05943	2.27877	2.52024	2.78596
22	1.38756	1.54598	1.72157	1.91610	2.13151	2.36992	2.63365	2.92526
23	1.40838	1.57690	1.76461	1.97359	2.20611	2.46472	2.75217	3.07152
24	1.42950	1.60844	1.80873	2.03279	2.28333	2.56330	2.87601	3.22510
25	1.45095	1.64061	1.85394	2.09378	2.36324	2.66584	3.00543	3.38635

n	5½%	6%	7%	8%	10%	12%	15%	20%
1	1.055	1.060	1.070	1.080	1.100	1.120	1.150	1.200
2	1.113	1.124	1.145	1.166	1.210	1.254	1.322	1.440
3	1.174	1.191	1.225	1.260	1.331	1.405	1.521	1.728
4	1.239	1.262	1.311	1.360	1.464	1.574	1.749	2.074
5	1.307	1.338	1.403	1.469	1.611	1.762	2.011	2.488
6	1.379	1.419	1.501	1.587	1.772	1.974	2.313	2.986
7	1.455	1.504	1.606	1.714	1.949	2.211	2.660	3.583
8	1.535	1.594	1.718	1.851	2.144	2.476	3.059	4.300
9	1.619	1.689	1.838	1.999	2.358	2.773	3.518	5.160
10	1.708	1.791	1.967	2.159	2.594	3.106	4.046	6.192
11	1.802	1.898	2.105	2.332	2.853	3.479	4.652	7.430
12	1.901	2.012	2.252	2.518	3.138	3.896	5.350	8.916
13	2.006	2.133	2.410	2.720	3.452	4.363	6.153	10.699
14	2.116	2.261	2.579	2.937	3.797	4.887	7.076	12.839
15	2.232	2.397	2.759	3.172	4.177	5.474	8.137	15.407
16	2.355	2.540	2.952	3.426	4.595	6.130	9.358	18.488
17	2.485	2.693	3.159	3.700	5.054	6.866	10.761	22.186
18	2.621	2.854	3.380	3.996	5.560	7.690	12.375	26.623
19	2.766	3.026	3.617	4.316	6.116	8.613	14.232	31.948
20	2.918	3.207	3.870	4.661	6.727	9.646	16.367	38.338
21	3.078	3.400	4.141	5.034	7.400	10.804	18.821	46.005
22	3.248	3.604	4.430	5.437	8.140	12.100	21.645	55.206
23	3.426	3.820	4.741	5.871	8.954	13.552	24.891	66.247
24	3.615	4.049	5.072	6.341	9.850	15.179	28.625	79.497
25	3.813	4.292	5.427	6.848	10.835	17.000	32.919	95.396

Table 7.2   Present Worth $(1 + i)^{-n}$

n	2%	3%	4%	5%	6%	7%	8%	10%
1	0.98039	0.97087	0.96154	0.95238	0.94340	0.9346	0.9259	0.9091
2	0.96117	0.94260	0.92456	0.90703	0.89000	0.8734	0.8573	0.8264
3	0.94232	0.91514	0.88900	0.86384	0.83962	0.8163	0.7938	0.7513
4	0.92385	0.88849	0.85480	0.82270	0.79209	0.7629	0.7350	0.6830
5	0.90573	0.86261	0.82193	0.78353	0.74726	0.7130	0.6806	0.6209
6	0.88797	0.83748	0.79031	0.74622	0.70496	0.6663	0.6302	0.5645
7	0.87056	0.81309	0.75992	0.71068	0.66506	0.6227	0.5835	0.5132
8	0.85349	0.78941	0.73069	0.67684	0.62741	0.5820	0.5403	0.4665
9	0.83676	0.76642	0.70259	0.64461	0.59190	0.5439	0.5002	0.4241
10	0.82035	0.74409	0.67556	0.61391	0.55839	0.5083	0.4632	0.3855
11	0.80426	0.72242	0.64958	0.58468	0.52679	0.4751	0.4289	0.3505
12	0.78849	0.70138	0.62460	0.55684	0.49697	0.4440	0.3971	0.3186
13	0.77303	0.68095	0.60057	0.53032	0.46884	0.4150	0.3677	0.2897
14	0.75788	0.66112	0.57748	0.50507	0.44230	0.3878	0.3405	0.2633
15	0.74301	0.64186	0.55526	0.48102	0.41727	0.3624	0.3152	0.2394
16	0.72845	0.62317	0.53391	0.45811	0.39365	0.3387	0.2919	0.2176
17	0.71416	0.60502	0.51337	0.43630	0.37136	0.3166	0.2703	0.1978
18	0.70016	0.58739	0.49363	0.41552	0.35034	0.2959	0.2502	0.1799
19	0.68643	0.57029	0.47464	0.39573	0.33051	0.2765	0.2317	0.1635
20	0.67297	0.55368	0.45639	0.37689	0.31180	0.2584	0.2145	0.1486
21	0.65978	0.53755	0.43883	0.35894	0.29416	0.2415	0.1987	0.1351
22	0.64684	0.52189	0.42196	0.34185	0.27751	0.2257	0.1839	0.1228
23	0.63416	0.50669	0.40573	0.32557	0.26180	0.2109	0.1703	0.1117
24	0.62172	0.49193	0.39012	0.31007	0.24698	0.1971	0.1577	0.1015
25	0.60953	0.47761	0.37512	0.29530	0.23300	0.1842	0.1460	0.0923

n	12%	15%	20%	25%	30%	35%	40%	45%
1	0.8929	0.8696	0.8333	0.8000	0.7692	0.7407	0.7143	0.6897
2	0.7972	0.7561	0.6944	0.6400	0.5917	0.5487	0.5102	0.4756
3	0.7118	0.6575	0.5787	0.5120	0.4552	0.4064	0.3644	0.3280
4	0.6355	0.5718	0.4823	0.4096	0.3501	0.3011	0.2603	0.2262
5	0.5674	0.4972	0.4019	0.3277	0.2693	0.2230	0.1859	0.1560
6	0.5066	0.4323	0.3349	0.2621	0.2072	0.1652	0.1328	0.1076
7	0.4523	0.3759	0.2791	0.2097	0.1594	0.1224	0.0949	0.0742
8	0.4039	0.3269	0.2326	0.1678	0.1226	0.0906	0.0678	0.0512
9	0.3606	0.2843	0.1938	0.1342	0.0943	0.0671	0.0484	0.0353
10	0.3220	0.2472	0.1615	0.1074	0.0725	0.0497	0.0346	0.0243
11	0.2875	0.2149	0.1346	0.0859	0.0558	0.0368	0.0247	0.0168
12	0.2567	0.1869	0.1122	0.0687	0.0429	0.0273	0.0176	0.0116
13	0.2292	0.1625	0.0935	0.0550	0.0330	0.0202	0.0126	0.0080
14	0.2046	0.1413	0.0779	0.0440	0.0254	0.0150	0.0090	0.0055
15	0.1827	0.1229	0.0649	0.0352	0.0195	0.0111	0.0064	0.0038
16	0.1631	0.1069	0.0541	0.0281	0.0150	0.0082	0.0046	0.0026
17	0.1456	0.0929	0.0451	0.0225	0.0116	0.0061	0.0033	0.0018
18	0.1300	0.0808	0.0376	0.0180	0.0089	0.0045	0.0023	0.0012
19	0.1161	0.0703	0.0313	0.0144	0.0068	0.0033	0.0017	0.0009
20	0.1037	0.0611	0.0261	0.0115	0.0053	0.0025	0.0012	0.0006
21	0.0926	0.0531	0.0217	0.0092	0.0040	0.0018	0.0009	0.0004
22	0.0826	0.0462	0.0181	0.0074	0.0031	0.0014	0.0006	0.0003
23	0.0738	0.0402	0.0151	0.0059	0.0024	0.0010	0.0004	0.0002
24	0.0659	0.0349	0.0126	0.0047	0.0018	0.0007	0.0003	0.0001
25	0.0588	0.0304	0.0105	0.0038	0.0014	0.0006	0.0002	0.0001

Table 7.3   Amount of Annuity $[(1 + i)^n - 1]/i$

n	1½%	2%	2½%	3%	3½%	4%	4½%	5%
1	1.00000	1.00000	1.00000	1.00000	1.00000	1.00000	1.00000	1.00000
2	2.01500	2.02000	2.02500	2.03000	2.03500	2.04000	2.04500	2.05000
3	3.04522	3.06040	3.07562	3.09090	3.10623	3.12160	3.13702	3.15250
4	4.09090	4.12161	4.15252	4.18363	4.21494	4.24646	4.27819	4.31013
5	5.15227	5.20404	5.25633	5.30914	5.36247	5.41632	5.47071	5.52563
6	6.22955	6.30812	6.38774	6.46841	6.55015	6.63298	6.71689	6.80191
7	7.32299	7.43428	7.54743	7.66246	7.77941	7.89829	8.01915	8.14201
8	8.43284	8.58297	8.73612	8.89234	9.05169	9.21423	9.38001	9.54911
9	9.55933	9.75463	9.95452	10.1591	10.3685	10.5828	10.8021	11.0266
10	10.70272	10.94972	11.2034	11.4639	11.7314	12.0061	12.2882	12.5779
11	11.86326	12.16872	12.4835	12.8078	13.1420	13.4864	13.8412	14.2068
12	13.04121	13.41209	13.7956	14.1920	14.6020	15.0258	15.4640	15.9171
13	14.23683	14.68033	15.1404	15.6178	16.1130	16.6268	17.1599	17.7130
14	15.45038	15.97394	16.5190	17.0863	17.6770	18.2919	18.9321	19.5986
15	16.68214	17.29342	17.9319	18.5989	19.2957	20.0236	20.7841	21.5786
16	17.93237	18.63929	19.3802	20.1569	20.9710	21.8245	22.7193	23.6575
17	19.20136	20.01207	20.8647	21.7616	22.7050	23.6975	24.7417	25.8404
18	20.48938	21.41231	22.3863	23.4144	24.4997	25.6454	26.8551	28.1324
19	21.79672	22.84056	23.9460	25.1169	26.3572	27.6712	29.0636	30.5390
20	23.12367	24.29737	25.5447	26.8704	28.2797	29.7781	31.3714	33.0660
21	24.47052	25.78332	27.1833	28.6765	30.2695	31.9692	33.7831	35.7193
22	25.83758	27.29899	28.8629	30.5368	32.3289	34.2480	36.3034	38.5052
23	27.22514	28.84496	30.5844	32.4529	34.4604	36.6179	38.9370	41.4305
24	28.63352	30.42186	32.3490	34.4265	36.6665	39.0826	41.6892	44.5020
25	30.06302	32.03030	34.1578	36.4593	38.9499	41.6459	44.5652	47.7271

n	5½%	6%	7%	8%	10%	12%	15%	20%
1	1.00000	1.00000	1.000	1.000	1.000	1.000	1.000	1.000
2	2.05500	2.06000	2.070	2.080	2.100	2.120	2.150	2.200
3	3.16803	3.18360	3.215	3.246	3.310	3.374	3.472	3.640
4	4.34227	4.37462	4.440	4.506	4.641	4.779	4.993	5.368
5	5.58109	5.63709	5.751	5.867	6.105	6.353	6.742	7.442
6	6.88805	6.97532	7.153	7.336	7.716	8.115	8.754	9.930
7	8.26689	8.39384	8.654	8.923	9.487	10.089	11.067	12.916
8	9.72157	9.89747	10.260	10.637	11.436	12.300	13.727	16.499
9	11.2563	11.4913	11.978	12.488	13.579	14.776	16.786	20.799
10	12.8754	13.1808	13.816	14.487	15.937	17.549	20.304	25.959
11	14.5835	14.9716	15.784	16.645	18.531	20.655	24.349	32.150
12	16.3856	16.8699	17.888	18.977	21.384	24.133	29.002	39.580
13	18.2868	18.8821	20.141	21.495	24.523	28.029	34.352	48.497
14	20.2926	21.0151	22.550	24.215	27.975	32.393	40.505	59.196
15	22.4087	23.2760	25.129	27.152	31.772	37.280	47.580	72.035
16	24.6411	25.6725	27.888	30.324	35.950	42.753	55.717	87.442
17	26.9964	28.2129	30.840	33.750	40.545	48.884	65.075	105.931
18	29.4812	30.9057	33.999	37.450	45.599	55.750	75.836	128.117
19	32.1027	33.7600	37.379	41.446	51.159	63.440	88.212	154.740
20	34.8683	36.7856	40.995	45.762	57.275	72.052	102.443	186.688
21	37.7861	39.9927	44.865	50.423	64.002	81.699	118.810	225.025
22	40.8643	43.3923	49.006	55.457	71.403	92.502	137.631	271.031
23	44.1118	46.9958	53.436	60.893	79.543	104.603	159.276	326.237
24	47.5380	50.8156	58.177	66.765	88.497	118.155	184.167	392.484
25	51.1526	54.8645	63.249	73.106	98.347	133.334	212.793	471.981

Table 7.4 Present Worth of Annuity $[1 - (1 + i)^{-n}]/i$

n	2%	3%	4%	5%	6%	7%	8%	10%
1	0.98039	0.97087	0.96154	0.95238	0.94340	0.935	0.926	0.909
2	1.94156	1.91347	1.88609	1.85941	1.83339	1.808	1.783	1.736
3	2.88388	2.82861	2.77509	2.72325	2.67301	2.624	2.577	2.487
4	3.80773	3.71710	3.62990	3.54595	3.46511	3.387	3.312	3.170
5	4.71346	4.57971	4.45182	4.32948	4.21236	4.100	3.993	3.791
6	5.60143	5.41719	5.24214	5.07569	4.91732	4.767	4.623	4.355
7	6.47199	6.23028	6.00205	5.78637	5.58238	5.389	5.206	4.868
8	7.32548	7.01969	6.73274	6.46321	6.20979	5.971	5.747	5.335
9	8.16224	7.78611	7.43533	7.10782	6.80169	6.515	6.247	5.759
10	8.98258	8.53020	8.11090	7.72173	7.36009	7.024	6.710	6.144
11	9.78685	9.25262	8.76048	8.30641	7.88687	7.499	7.139	6.495
12	10.57534	9.95400	9.38507	8.86325	8.38384	7.943	7.536	6.814
13	11.34837	10.6350	9.98565	9.39357	8.85268	8.358	7.904	7.103
14	12.10625	11.2961	10.5631	9.89864	9.29498	8.745	8.244	7.367
15	12.84926	11.9379	11.1184	10.3797	9.71225	9.108	8.559	7.606
16	13.57771	12.5611	11.6523	10.8378	10.1059	9.447	8.851	7.824
17	14.29187	13.1661	12.1657	11.2741	10.4773	9.763	9.122	8.022
18	14.99203	13.7535	12.6593	11.6896	10.8276	10.059	9.372	8.201
19	15.67846	14.3238	13.1339	12.0853	11.1581	10.336	9.604	8.365
20	16.35143	14.8775	13.5903	12.4622	11.4699	10.594	9.818	8.514
21	17.01121	15.4150	14.0292	12.8212	11.7641	10.836	10.017	8.649
22	17.65805	15.9369	14.4511	13.1630	12.0416	11.061	10.201	8.772
23	18.29220	16.4436	14.8568	13.4886	12.3034	11.272	10.371	8.883
24	18.91393	16.9355	15.2470	13.7986	12.5504	11.469	10.529	8.985
25	19.52346	17.4131	15.6221	14.0939	12.7834	11.654	10.675	9.077

n	12%	15%	20%	25%	30%	35%	40%	45%
1	0.893	0.870	0.833	0.800	0.769	0.741	0.714	0.690
2	1.690	1.626	1.528	1.440	1.361	1.289	1.224	1.165
3	2.402	2.283	2.106	1.952	1.816	1.696	1.589	1.493
4	3.037	2.855	2.589	2.362	2.166	1.997	1.849	1.720
5	3.605	3.352	2.991	2.689	2.436	2.220	2.035	1.876
6	4.111	3.784	3.326	2.951	2.643	2.385	2.168	1.983
7	4.564	4.160	3.605	3.161	2.802	2.507	2.263	2.057
8	4.968	4.487	3.837	3.329	2.925	2.598	2.331	2.109
9	5.328	4.772	4.031	3.463	3.019	2.665	2.379	2.144
10	5.650	5.019	4.192	3.571	3.092	2.715	2.414	2.168
11	5.938	5.234	4.327	3.656	3.147	2.752	2.438	2.185
12	6.194	5.421	4.439	3.725	3.190	2.779	2.456	2.196
13	6.424	5.583	4.533	3.780	3.223	2.799	2.469	2.204
14	6.628	5.724	4.611	3.824	3.249	2.814	2.478	2.210
15	6.811	5.847	4.675	3.859	3.268	2.825	2.484	2.214
16	6.974	5.945	4.730	3.887	3.283	2.834	2.489	2.216
17	7.120	6.047	4.775	3.910	3.295	2.840	2.492	2.218
18	7.250	6.128	4.812	3.928	3.304	2.844	2.494	2.219
19	7.366	6.198	4.844	3.942	3.311	2.848	2.496	2.220
20	7.469	6.259	4.870	3.954	3.316	2.850	2.497	2.221
21	7.562	6.312	4.891	3.963	3.320	2.852	2.498	2.221
22	7.645	6.359	4.909	3.970	3.323	2.853	2.498	2.222
23	7.718	6.399	4.925	3.976	3.325	2.854	2.499	2.222
24	7.784	6.434	4.937	3.981	3.327	2.855	2.499	2.222
25	7.843	6.464	4.948	3.985	3.329	2.856	2.499	2.222

Table 7.5 U.S. Economic Indicators[a]

Year	G.N.P. ($ MM)	Inflation Rate (%)	Consumer Price Index (1967 = 100)	Prime Rate (%)	T-Bill Rate (90-day)	Dow-Jones Ind. Avg. (Yr-End)	Unemployment (%)	Finished Goods ($) (1967 = 100)
1950	286	4.9	72.1	2.1	1.2	235.41	5.1[b]	79.0
1955	396[b]	4.6	80.2	3.2	1.7	488.40	5.3[b]	85.5
1960	506	1.3	88.7	4.8	4.0[c]	615.89	5.8	93.7
1965	748[b]	3.5	94.5	4.6	4.2[c]	969.26	4.3	95.7
1970	990	6.0	119[b]	7.9	7.4[c]	838.92	5.0	110.4
1972	1190	5.2	124[b]	5.3	5.9[c]	1020.02	5.6	128.5[b]
1974	1430	12.2	147.7	10.8	7.8[c]	616.24	5.9	147.5
1976	1720	5.4	170.5	6.9	5.0	974.92	7.7	170.6
1978	2160	9.1	195.4	9.6	7.2	820.23	6.0	195.9
1980	2630	12.3	246.8	15.3	11.5	891.41	7.4	247.0

[a]Year-End Data or Average (if more indicative).
[b]Approximate.
[c]Two-year rate.

REFERENCES

1. 10 Code of Federal Regulations, Part 50, Appendix B.
2. CPI Detailed Report, U.S. Department of Labor, Bureau of Labor Statistics, August 1981.
3. Economic Indicators, Joint Economic Committee by the Council of Economic Advisors, U.S. Government Printing Office, Washington, D.C., 1981.
4. Employment and Earnings, U.S. Department of Labor, Bureau of Labor Statistics, January 1981.

BIBLIOGRAPHY

Grant and Ireson, *Principles of Engineering Economy,* 7th edition, Ronald Press, New York, 1982.

Juran, Gryna, and Bingham, *Quality Control Handbook,* 3rd edition, McGraw-Hill, New York, 1974.

Marguglio, *Quality Systems in the Nuclear Industry,* American Society for Testing Materials, 1977.

White, Agee, and Case, *Principles of Engineering Economic Analysis,* John Wiley & Sons, New York, 1977.

SECTION 8
ENERGY SOURCES

8.1 GENERAL

Energy sources can be classified broadly depending on fuel types. A typical classification identifies them as solid, liquid, gaseous, nuclear, geothermal, solar, wind, and advanced (still largely experimental) forms such as fuel cells and magnetohydrodynamics. This ranking is relatively widely used and broadly follows the common applications of energy sources around the world. It is possible also to categorize fuels and energy sources as to whether they are renewable or nonrenewable; thus wind and solar would be considered renewable sources, whereas coal, lignite, geothermal in some cases, petroleum, and natural and liquified petroleum gases would be considered nonrenewable.

Historical data on consumption of energy in the United States is shown in Fig. 8.1. There are numerous projections as to the kind of fuel that will be in use in the various nations around the world and within the United States in particular. Broadly, there is a projected increase in fuel costs, as has been the case in the United States in the last 6 or 8 years, although the forecast annual increases are smaller. The use of smaller automobiles (a major consumer of liquid fuels) and the transition from petroleum to coal, with its larger reserves available in the United States, will be accelerated. This will tend to shift the use of fuels away from those that are imported to those that are indigenous. Further, the higher costs will tend to shift the use of fuels away from those that are imported to those that are indigenous. Further, the higher costs will tend to provide increased economic incentives for improvements in the utilization of fuels through use of small automobiles, improved insulation in homes, more widespread use of heat-recovery units in industrial processes, and improved efficiency in heat utilization equipment. Concurrently, there will be a movement toward increased use of the renewable sources of energy; windpower and solar are the two principal sources given increased emphasis in the near term. These sources are currently undergoing fairly extensive development and testing in the United States to determine their applicability for broader applications.

8.2 HEATING VALUE AND COMBUSTION

Higher Heating Value for a Solid Fuel. In the case of a solid fuel that contains C percent* of fixed and volatile carbon, H percent of hydrogen, O percent of oxygen, and S percent of sulfur, the heating value (Q) per pound of fuel-as-fired is

$$Q = 14,540C + 62,030\left(H - \frac{O}{8}\right) + 4,050S \text{ Btu.} \qquad (8.1)$$

*Percentages by weight are expressed as a decimal fraction.

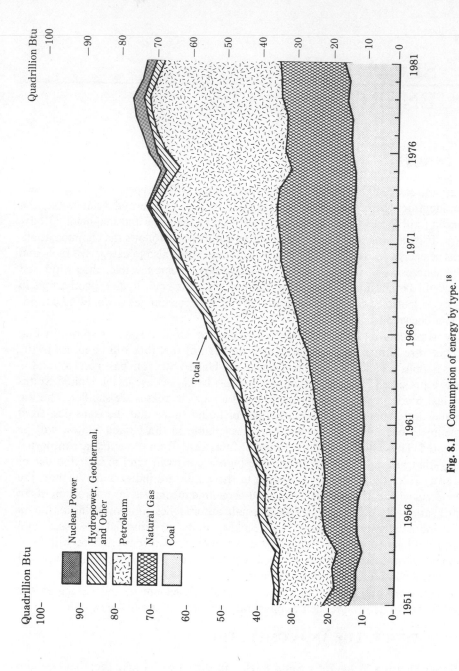

Fig. 8.1 Consumption of energy by type.[18]

The higher heating value (**HHV**) of a fuel is closely produced in a bomb-type calorimeter when the condensation and cooling of the hydrogen products (water) of combustion occurs. This adds approximately 1050 Btu/pound (710 kcal/kg) of H_2O formed (heat of condensation) to the lower heating value (**LHV**) of the fuel. For processes where vapor is not condensed and cooled, LHV is more indicative. Because of the difficulties of measuring LHV, the prevailing practice in the United States is to use HHV, whereas in Europe LHV is frequently used.

Heating Value for Liquid Fuel. In the case of a liquid fuel that contains C percent of carbon and H percent of hydrogen, the heating value Q per pound of fuel is

$$Q = 13,500C + 60,890H \text{ Btu.} \tag{8.2}$$

If the Baumé reading is known, the heating value Q per pound of fuel is

$$Q = 18,650 + 40 \text{ (Baumé reading} - 10) \text{ Btu.} \tag{8.3}$$

Weight of Dry Flue Gases per Pound of Carbon. The flue gas analysis indicates the percentage CO_2, CO, O_2, and N_2 by volume. The weight (G_1) of dry flue gas per pound of carbon is given by

$$
\begin{aligned}
G_1 &= \frac{11 \text{ CO}_2 + 8 \text{ O}_2 + 7 \text{ (CO} + \text{N}_2)}{3 \text{ (CO}_2 + \text{CO)}} \text{ pounds} \\
&= \frac{4 \text{ CO}_2 + \text{O}_2 + 700}{3 \text{ (CO}_2 + \text{CO)}} \text{ pounds.}
\end{aligned}
\tag{8.4}
$$

Weight of Dry Flue Gases per Pound of Coal-as-Fired. The percentage by weight of carbon in the coal (C_c) and in the ash (C_a) can be determined from the coal and ash analyses. If the pounds of coal-as-fired (M_c), the pounds of ash (M_a), and the gas analysis are known, then the weight (G_2) of dry flue gas per pound of coal-as-fired is given by

$$G_2 = \frac{M_c C_c - M_a C_a}{M_c} \left[\frac{4 \text{ CO}_2 + \text{O}_2 + 700}{3 \text{ (CO}_2 + \text{CO)}} \right] \text{ pounds.} \tag{8.5}$$

Actual Weight of Dry Air per Pound of Coal-as-Fired. If air is assumed to be 77% nitrogen (N_2) by weight, the weight (G_3) of dry air per pound of coal-as-fired is given by

$$G_3 = \frac{M_c C_c - M_a C_a}{M_c} \left(\frac{3.032 \text{ N}_2}{\text{CO}_2 + \text{CO}} \right) \text{ pounds.} \tag{8.6}$$

Theoretical Weight of Dry Air per Pound of Coal-as-Fired. In the case of a coal-as-fired that contains C percent by weight of fixed and volatile carbon, H percent of hydrogen, O percent of oxygen, and S percent of sulfur, the theoretical weight (G_4) of dry air per pound of coal-as-fired is given by

$$G_4 = 11.57C + 34.8\left(H - \frac{O}{8}\right) + 4.35S \text{ pounds.} \qquad (8.7)$$

Percentage of Excess Air.   Since G_3 gives the actual weight and G_4 the theoretical weight of air required for combustion, it follows that

$$\text{Excess air} = \left(\frac{G_3 - G_4}{G_4}\right) 100 \text{ percent.} \qquad (8.8)$$

Some common chemical reactions of combustion are shown in Table 8.1, which follows.

8.3   SOLID FUELS

8.3.1   Introduction

Solid fuels are widely distributed and have formed the basis for most energy development until the 1930s. For large stationary energy applications, they are still considered a primary candidate.

8.3.2   Characteristics of Common Solid Fuels

The more common solid fuels, in descending order from coal, are shown in Table 8.2.

8.3.3   Coal

Coal is an extremely widely used fuel not only in the United States but throughout the rest of the world. It is the most generally disbursed, commercially valuable fuel and it is available in a variety of types, sizes, and heating values. The coals are typ-

Table 8.1   Common Chemical Reactions of Combustion[17]

Combustible	Reaction
Carbon (to CO)	$2C + O_2 = 2CO$
Carbon (to CO_2)	$C + O_2 = CO_2$
Carbon monoxide	$2CO + O_2 = 2CO_2$
Hydrogen	$2H_2 + O_2 = 2H_2O$
Sulfur (to SO_2)	$S + O_2 = SO_2$
Sulfur (to SO_3)	$2S + 3O_2 = 2SO_3$
Methane	$CH_4 + 2O_2 = CO_2 + 2H_2O$
Acetylene	$2C_2H_2 + 5O_2 = 4CO_2 + 2H_2O$
Ethylene	$C_2H_4 + 3O_2 = 2CO_2 + 2H_2O$
Ethane	$2C_2H_6 + 7O_2 = 4CO_2 + 6H_2O$
Hydrogen sulfide	$2H_2S + 3O_2 = 2SO_2 + 2H_2O$

Table 8.2 Typical Characteristics of Solid Fuels

Type	HHV (Btu/lb)	LHV (Btu/lb)	Fixed Carbon (%)	Moisture (%)	Volatile Matter (%)	Ash (%)
Anthracite coal	14,600	12,900	92–98	5	2–8	11
Bituminous coal	15,200	13,800	69–78	3–5	22–31	7
Subbituminous coal	9,500	9,000	52	17	39	8
Lignite	7,500	6,900	40	35	45	12
Peat	10,000[a]	3,500	23	70	67	2–70
Wood	8,800	8,300[c]	20	45	78	2
Bagasse	8,300	3,000	45	50	—	3
Hogged fuel	8,800	8,300	20	50	78	2
Municipal garbage	9,500[b]	—	—	35–50	—	8–12
Manure	7,400	—	—	45–65	—	16

[a]Moisture and ash free.
[b]Varies widely.
[c]Can be as low as 4300 Btu with 50% moisture.

ically ranked as anthracite, bituminous and lignites, from the very hard to the very soft. Characteristics of coals vary widely and each deposit should be properly sampled to provide firm data for design.

The coals are solid materials found principally underground in seams as narrow as a few inches to very thick beds of coal up to 50 feet (15 m) thick that can be handled with large-scale excavating equipment. Thick seams are frequently found on or near the surfaces in the United States, Germany, and elsewhere, where they are mined with very large excavating equipment such as drag lines and mining wheels. The coal must be handled as a solid and conveyed either as a solid or (with newer coal technology currently being developed) as particles of solid in suspension in a liquid.

Care must be taken in stockpiling the coal at the point of use, because spontaneous combustion may begin if the coal pile is not well sealed. Such a fire typically extends into the interior of the coal pile itself where it is most difficult to extinguish. Thus, for any significant application of coal, it is essential that the coal pile be properly sealed (airtight) by compacting and rolling to avoid supporting combustion.

The burning of coal involves combustion of the material either as a solid on a grate, which permits air to pass through the burning fuel, or as a finely divided suspension in an air stream (as a pulverized material). The use of pulverizing equipment is typically limited to rather large installations, such as central power plants or large industrial boilers, and is not widely found in small installations. Small installations typically use a grate with a series of slots through which combustion air is introduced sufficient for the rapid combustion of the coal. In the case of pulverized coal, some air is introduced with the coal while other air is introduced around the coal and provides the added oxygen needed for combustion. In the burning of pulverized coal, particularly where the coal tends to be low in volatiles or of the lower rankings, it is frequently necessary to burn fuel oil as a stabilizing agent. This assures that the combustion is stable and avoids puffing or other types of intermittent combustion.

Furnaces have been known to experience severe flashbacks and blowbacks where a flame has been lost and fuel had continued to be fed into the furnace. With their low volatility, this is not so much a problem with the coals and is more usually a concern with gaseous or liquid fuels.

When burning noncleaned coals as pulverized coal, the coal may very often contain significant amounts of sulfur, perhaps as much as 2 to 3%. Current technology is developing methods to introduce desulfurizing agents into the coal at the time of burning on the grate. One method is the so-called **fluidized bed** method wherein the chemical reaction between the sulfur and the desulfurization agent (typically dolomite—$Ca,MgCO_3$) takes place on the grate surface itself. This requires the introduction of large amounts of air and a rather open grate structure, but has the distinct advantage that virtually the entire desulfurization takes place as a part of the combustion process rather than farther back in the gas stream flow path where the gases are cooler. This contrasts with other sulfur removal systems when the gas stream is scrubbed (i.e., desulfurized), usually involving both a cooling and chemical combination and thus creating a significant efficiency penalty for the cycle.

Coal contains a substantial amount of ash or nonburnable material. The ash can be broadly divided into the categories of fly ash and bottom ash. Fly ash is the very light ash that finds its way out of the furnace combustion zone and is carried by the air stream up through the furnace and on out toward the stack. Bottom ash is a heavy, more granular type of material that may stick to the lower walls of the furnace or fall through the grate structure itself and be discharged from the actual combustion zone in that fashion. An important part of the concern for handling of both the fly ash and the bottom ash is their temperature; that is, they are at roughly the combustion temperature of the furnace and must be cooled, removed from the furnace or the gas stream, and disposed of. Furnaces in the larger sizes are frequently built with a wet bottom in which the bottom ash is discharged into a sealed wet tank integral with the bottom of the furnace where the ash is typically sluiced away by a water stream. Alternatively, a dry-bottom furnace may be employed where the ash is dumped into a pit below the furnace, mixed with water and sluiced or pumped away as a slurry. Fly ash, being very light, is carried with the gas stream and will be discharged up the stack unless fly ash removal equipment is installed. In many large installations today, such as power utility boilers, the cost of fly ash equipment may cost in the tens of millions of dollars. The removal of fly ash is frequently handled by an electrostatic system. The electrostatic generator places a charge on a series of wires or plates, which selectively attract and hold the fly ash. The electrostatic system then bypasses one or more of the sections of the fly ash collector and the electric charge is shut off. The plates or wires are then rapped or shaken, the fly ash then dumping into a hopper system beneath. Alternatively, many furnaces employ a passive system of (filter) bag collectors; the bag collectors have proven to be simple to operate and require no power consumption. Although the bag systems have the disadvantage of being unable to operate at very high temperatures, they have proven to be extremely useful for many installations. The bags act as a filtering agent to remove the fly ash from the flue gas stream, and again, periodically, the gas stream is diverted and the bags are shaken or moved to break loose the accumulated ash from the bag surface. The ash is collected in a series of hoppers from which it is disposed.

Of particular concern with the design of furnaces burning coal is the ash fusion temperature. The bottom ash has a tendency to fuse on the walls of the furnace and can agglomerate into extremely large nodules of several hundred pounds if the ash fusion temperature is substantially exceeded. Thus, the ash fusion temperature for the coals to be burned must be established rather carefully to assure that furnace design does not create a situation in which large masses of ash accumulate on the lower walls of the furnaces. Where this does happen, it may be necessary to shut down the furnace, enter it, and break up the agglomerate with jackhammers, pinch bars, or other mechanical means.

One of the more significant areas of potential savings in a furnace is the use of air preheaters. Typically, for large furnaces, the combustion air, which will be introduced with the fuel, flows through an air preheater where its temperature is increased. The air preheater is an air heat exchanger, exchanging the heat from the discharge gas stream with the incoming cold air, thus improving cycle efficiency.

8.3.4 Lignite and Peat

Both **lignite** and **peat** are solid fuels having a high moisture content and are low-ranked coals in terms of the coal formation geological process. They are the predecessors of the bituminous and the anthracite coals. Because of their high moisture content, large amounts of combustion air are needed to carry off the steam resulting from the water given off during combustion; and, as a consequence, fuels generally have a reduced heating value since their net heating value or lower heating value includes the energy required to drive the moisture out of the fuel itself.

Because of the weight of water in the fuel and its relatively low heating value, they are not considered economic to transport over long distances. Where these deposits occur close to the point of use, they can be economically exploited; for example, in Germany where the so-called brown coal region burns a very low-grade coal similar to a lignite or peat, much local industry has been based on this source of fuel.

8.3.5 Wood, Hogged Fuel, and Bagasse

These **solid fuels** are in the wood family with **bagasse** being the residual from sugarcane processing (stalks, etc.). In more recent years, wood and its by-product, **hog fuel,** which is a waste product of wood (chips, branches, slashings, etc.), have been more valuable as building and furniture products in themselves than as a fuel. In some areas of the world, where they exist in surplus, they are used for fuels. Again, the economics of transportation apply and tend to limit their usable geographic area, although it typically is wider than the peat/lignite area previously described. Wood, hogged fuel, and bagasses have a high volume/weight ratio since they are in the form of chips or random sizes, and rail cars that carry these normally have high extended sides to permit loading large volumes of this material in a car.

Normally, the combustion air requirements for these fuels are not excessive since large amounts of surface area are available when the fuel is burned and there is a relatively free flow of air through the burning material itself. When sawdust is burned, however, particularly if fresh and fairly damp, it tends to compact, requiring more combustion air and occasional agitation. Usually these materials are burned on a grate or in a stoker furnace. If the sawdust or a finely divided wood material is burned in suspension, it is important that there be sufficient residence time within the furnace to dry the material prior to combustion.

The moisture in these woody or lignocellulose-type fuels varies widely depending on how recently the fuel was cut, its form, matter of transport, and so forth. Of the several fuels, hogged fuel is more predictable, because it is typically produced in a chipper, which yields flakes or chips of a relatively uniform size. Bagasse tends to have a high content of dirt, silica, ash, and other noncombustibles and extra provisions must be provided to handle the large quantities of ash that occur with its combustion.

8.3.6 Municipal Garbage

Municipal garbage is a unique and highly variable fuel. Typically, it contains substantial amounts of inert or metallic materials. These range from small metallic materials, such as nails and metal fasteners, to objects as large as entire bed springs. Other typical materials in the as-received condition include glass bottles, organic materials, plastics, aluminum containers, and utensils.

Normally, in handling municipal garbage there is a system for removal of the metals from the incoming material flow; magnets are used to remove the ferrous materials and gravity separation is used for dropping out materials such as aluminum or other nonferrous or glass materials. Hammer mills are usually used to reduce the size of the materials prior to charging into the furnace.

Fairly high quantities of combustion air are required because of typically high amounts of moisture in the fuel. With the widely varying mix of materials present in the municipal garbage, grate-type furnaces producing extended residence time are normally employed to burn this material. Because of transportion costs, availability of this fuel is limited to the garbage produced in the immediate area. Its composition will vary over the years as increased conservation measures reduce the amounts of recoverable materials in the garbage stream such as aluminum, steel, and plastics. The municipal garbage installations to date are able to operate on a self-sustaining basis as regards the energy available versus the energy used and in fact have a slight net excess operating energy. It is relatively small, however, and has been used more widely primarily as a disposal method rather than a system for power generation.

8.4 LIQUID FUELS

8.4.1 Introduction

Liquid fuels principally used are hydrocarbons of petroleum base and are usually ranked as shown in Table 8.3.

8.4.2 Characteristics of Liquid Fuels

Liquid fuels can be conveniently ranked as shown in Table 8.3. (See also Table 10.14, Physical Constants of Hydrocarbons.)

Liquid hydrocarbon fuels rank in decreasing volatility as follows:[11]

VOLATILE PRODUCTS.   Liquified gases and natural gasoline.

LIGHT OILS.   Gasolines; jet and tractor fuels; kerosene.

Table 8.3 Typical Characteristics of Liquid Fuels[a]

Type	H. H. V. (Btu/gal)	Lb/Gal at 60°F	Specific Gravity	Pour Point (°F)	Viscosity (Centistokes at 100°F)	API Gravity (60°F)	Sulfur (Max. %)	Sediment and Water (%)	Recommended Pumping Temp. (°F)
#1 Fuel oil (light distillate)	137,000	6.87	0.825	< 0	1.6[b]	40	0.1	Trace	Ambient
#2 Fuel oil (distillate)	141,000	7.21	0.865	< 0	2.7	32	0.5	Trace	Ambient
#4 Fuel oil (very light residual)	146,000	7.73	0.928	10	15.0	21	1.0	0.5 (max.)	15 (min.)
#5 Fuel oil (light residual)	148,000	7.94	0.953	30	50.0	17	2.0 (max.)	1.0 (max.)	35 (min.)
#6 Fuel oil (residual)	150,000	8.21	0.986	65	360.0	12	2.8 (max.)	2.0 (max.)	100 (min.)
Gasoline (90 octane)	127,000	6.25	0.750	< 0	[b]	—	—	—	Ambient
Methanol (106 octane)	66,700	6.60	0.790	< 0	[b]	—	Nil	Nil	Ambient

[a]See also Table 10.14, Physical Constants of Hydrocarbons.
[b]Viscosity not usually a consideration.

413

HEAVY DISTILLATES.   Burner oil, furnace distillates, diesel fuel, and gas oil.
RESIDUES.   Fuel oil, asphalt, coke.

These reflect the results of the initial separation of crude petroleum into several "cuts" of average boiling point. The "lighter" fractions (low boiling point) become gasoline directly or through intermediate processing. "Heavier" fractions become gasoline, distillates, or residual fuel oil.

  Gasoline is the major objective of the refinery and is the end product of a variety of alternative processes. Light fractions from the crude are "hydrotreated" to remove contaminants and "re-formed" to raise octane rating. Heavier fractions are "cracked," to break the high-molecular-weight hydrocarbon into lower-weight components either by purely thermal treatment or by catalytic cracking or hydrocracking in the presence of hydrogen.

  Other products, of lesser production importance, are lubricating oils, grease or wax, petroleum coke, and residue. Feedstock for conversion to petrochemicals is a potential refinery product as well. All of the crude petroleum feedstock may be converted to useful forms.

  Depending on geographic source, **crude petroleum** will reflect a predominance of one or more of the following series, that is, the "base."*

  Paraffin series, saturated chain compounds.
  Olefin series, unsaturated chain compounds.
  Naphthene series, saturated ring compounds.
  Aromatic series, unsaturated ring compounds.

Fuel and lubrication products[13] are produced by splitting the crude into fractions, mainly by distillation. Each fraction is a mixture whose average molecular weight is related to its boiling range, the "heavier" fractions are those having higher boiling points. For most purposes, these stocks can be defined in terms of simple physical properties.

8.4.3   Tests and Measures[11,13]

API GRAVITY.   A measure of specific gravity, according to:

$$\text{Degrees API} = (141.5/\sigma) - 131.5, \tag{8.9}$$

  where σ = specific gravity, referred to water at 60°F.
BAUMÉ GRAVITY.   A measure of specific gravity, according to:

$$\text{Degrees Baumé} = (140/\sigma) - 130 \tag{8.10}$$

  [For liquids heavier than water = $145 - (145/\sigma)$].

---

*Naphthene base is sometimes termed asphalt base.

SAYBOLT UNIVERSAL VISCOSITY. A measure of the kinematic viscosity of a liquid; specifically the time, in seconds for 60 cc liquid to drain through a standardized orifice at a specified temperature.

FLASH POINT. The minimum temperature of a liquid at which vapor issuing from liquid exposed to air will momentarily ignite.

FIRE POINT. The minimum temperature, as above, at which the vapor from the liquid will remain ignited.

POUR POINT. The maximum oil temperature plus 5°F (2.8°C) at which no discernible movement is apparent in a jar of oil within 5 seconds after being tipped 90°.

REID VAPOR PRESSURE. A standardized measure of the volatility of liquid; approximately equal to the true vapor pressure of the test liquid.

FRACTIONATION. The process of distillation by which raw petroleum liquid is separated into mixtures ("cuts") of hydrocarbons of relatively narrow boiling ranges.

ASTM DISTILLATION. An ASTM nonrectifying test in which a liquid sample is distilled at uniform rate, yielding percent vaporized as a function of liquid temperature, the 10, 50, and 95% points being of particular interest (ASTM D-86). Analogous results can be obtained by the Hempel, Saybolt, Engler, and flash vaporization tests.

CETANE NUMBER. A measure of ignition quality of a liquid fuel; specifically, the volume percent of cetane in a mixture with alpha methyl naphthalene showing the same ignition characteristics as the fuel.

OCTANE NUMBER. A measurement of the knock characteristics in a spark-inputed engine, the percentage by volume of iso-octane which must be mixed with normal heptane to match the knock intensity of the fuel under test. The "Motor" and "Research" octane numbers are measured at different engine speeds.

8.4.4 Viscosity Conversion

Viscosity standards have been developed using a variety of instruments. Their conversion is shown in Fig. 8.2.

8.4.5 Liquified Petroleum Gases[19]

Liquified petroleum gases can be classified as:

COMMERCIAL PROPANE. A hydrocarbon product for use where high volatility is required.

COMMERCIAL BUTANE. A hydrocarbon product for use where low volatility is required.

COMMERCIAL PB MIXTURES. Mixture of propane and butane for use where intermediate volatility is required.

SPECIAL-DUTY PROPANE. A high-quality product composed chiefly of propane, which exhibits superior antiknock characteristics when used as an internal combustion engine fuel.

$$\mu = \nu\rho' = \nu S$$

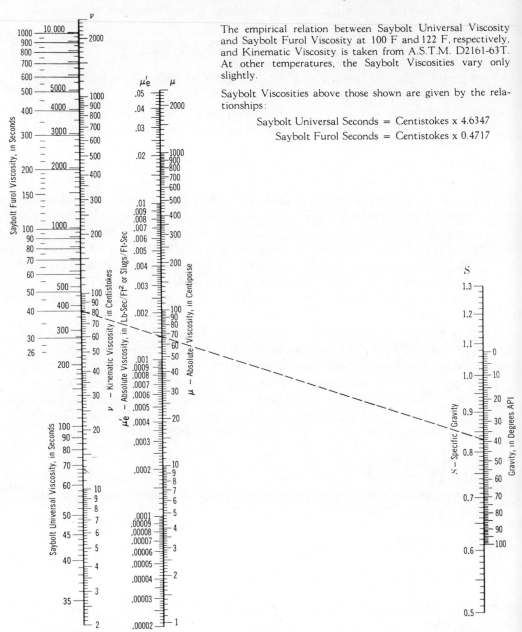

The empirical relation between Saybolt Universal Viscosity and Saybolt Furol Viscosity at 100 F and 122 F, respectively, and Kinematic Viscosity is taken from A.S.T.M. D2161-63T. At other temperatures, the Saybolt Viscosities vary only slightly.

Saybolt Viscosities above those shown are given by the relationships:

Saybolt Universal Seconds = Centistokes x 4.6347

Saybolt Furol Seconds = Centistokes x 0.4717

Fig. 8.2 Equivalents of kinematic, Saybolt universal, Saybolt furol, and absolute viscosity. From Technical Paper No. 410, Courtesy of Crane Co.[4]

ASTM Standard Specification D1835-76 establishes detail requirements as shown in Table 8.4.

8.4.6   Transportation Fuels[15,16,19,20]

Automotive gasolines are defined by the SAE as the fuels for internal combustion engines in which ignition is spark induced and which are used primarily for passenger car and truck service. They are blends of petroleum hydrocarbons, with additives to suppress bad properties or promote good ones. Their properties are a function of geographic location in the United States. The *SAE Handbook* provides recommendations of volatility class for all states in the United States. ASTM Standard Specification D439-79 provides detailed requirements.

Flash point	$-45°F$ ($-43°C$) (approx.)
Autoignition temperature	$500°F$ ($260°C$) (approx.)
Flammability limits, volume in air	1.4 to 7.6%
Vapor pressure at $70°F$ ($21°C$)	4 to 8 psi (28 to 55 kPa)
Concentration in saturated air at $68°F$ ($20°C$), vol.	25 to 50% (SAE)

Gasoline flash point is so low that a flammable mixture is always presumed to exist over the liquid in the open. In covered tanks the saturated vapor is likely to be too rich to ignite, but explosive conditions may exist at extremely low temperatures or during transfer operations.

Diesel fuel oils[16] are evaluated by Standard Specification ASTM D-975 for Grades 1D, 2D, and 4D. Ignition quality (ASTM D-613) is determined by the cetane numbers, or from API gravity and midboiling point. Heating value (ASTM D-240) can be either measured in the calorimeter or estimated from the aniline point (ASTM D-611) and the API gravity of the fuel.

Jet fuels[16] are kerosene or a kerosene/gasoline mixture having low Reid vapor pressure, 0.2 to 3 psi (1.4 to 2.1 kPa) and very low freezing points, -58 to $-80°F$ (-50 to $-62°C$). Although similar to motor gasolines, quality control for this fuel must be more rigorous for reasons of safety. Tetraethyllead is used to control premature ignition or knocking, which is usually inaudible and can damage or destroy an engine. Substitution of motor gasoline for aviation use introduces additional hazards of vapor lock and carburetor icing.

8.4.7   Fuel Oils[16]

Fuel oils are categorized as either distillates or residual oils. ASTM Standard Specification D-396 specifies ranges of properties conducive to good burner performance. The oil grades are:

GRADE NO. 1.   A light distillate of high volatility for burners of the vaporizing type.

GRADE NO. 2.   A distillate heavier than Grade 1 oil, for use in burners of the atomizing type to meet domestic or medium commercial duty.

Table 8.4 Detail Requirements for Liquified Petroleum Gases

Characteristic	Product Designation			
	Commercial Propane	Commercial Butane	Commercial PB Mixtures	Special-Duty Propane[a]
Vapor pressure at 100°F (37.8°C), max., psig	208	70	[b]	208
kPa	1430	485		1430
Volatile residue:				
Evaporated temperature, 95%, max., °F	−37	36	36	−37
°C	−38.3	2.2	2.2	−38.3
or				
Butane and heavier, max., vol. %	2.5	—	—	2.5
Pentane and heavier, max., vol. %	—	2.0	2.0	—
Propylene content, max., vol. %	—	—	—	5.0
Residual matter:				
Residue on evaporation 100 ml, max., ml	0.05	0.05	0.05	0.05
Oil stain observation	pass[c]	pass[c]	pass[c]	pass[c]
Relative density (specific gravity) at 60/60°F (15.6/15.6°C)	[d]	[d]	[d]	—
Corrosion, copper, strip, max.	No. 1	No. 1	No. 1	No. 1
Sulfur, grains/100 ft³ max. at 60°F and 14.92	15	15	15	10
psia mg/m³ (15.6°C and 101 kPa)	343	343	343	229
Hydrogen sulfide content	—	—	—	pass[e]
Moisture content	pass	—	—	pass
Free water content	—	none[f]	none[f]	—

[a]Equivalent to Propane HD-5 of GPA Publication 2140.

[b]The permissible vapor pressures of products classified as PB mixtures must not exceed 20 psig (1380 kPa) and additionally must not exceed that calculated from the following relationship between the observed vapor pressure and the observed specific gravity:

Vapor pressure, max. = 1167 − 1880 (sp. gr. 60/60°F) or 1167 − 1880 (density at 15°C).

A specific mixture shall be designated by the vapor pressure at 100°F in pounds per square inch gauge. To comply with the designation, the vapor pressure of the mixture shall be within +0 to −10 psi of the vapor pressure specified.

[c]An acceptable product shall not yield a persistent oil ring when 0.3 ml of solvent residue mixture is added to a filter paper, in 0.1-ml increments and examined in daylight after 2 min as described in Method D 2158.

[d]Although not a specific requirement, the specific gravity must be determined for other purposes and should be reported. Additionally, the specific gravity of PB mixture is needed to establish the permissible maximum vapor pressure (see Footnote b).

[e]An acceptable product shall not show a distinct coloration.

[f]The presence or absence of water shall be determined by visual inspection of the samples on which the gravity is determined.

418

GRADES NO. 4 AND NO. 4 (LIGHT).   Light residual oil or heavy distillate capable
   of atomization at relatively low storage temperatures.
GRADES NO. 5 AND NO. 5 (LIGHT).   Residual fuels of viscosity intermediate
   between those of Grades 4 and 6, occasionally requiring preheating in cold
   climates.
GRADE NO. 6.   Sometimes referred to as Bunker C, a high-viscosity residual oil
   used for commercial and industrial heating, requiring preheating at the storage
   tanks to aid pumping, and preheating at the burners to support atomization.

   Petroleum products are widely available around the world and vary from the light
crude oils to residual fuels or the extremely heavy petroleum cokes, the residues from
the refining processes. Typically, the lighter grades of products such as the gasolines,
No. 2 oils, and so forth, are used for transportation purposes and chemical feedstocks.
These grades have low flash points and ventilation requirements become important
because combustion may occur if insufficient care is given during the transportation,
processing, handling, and storage of them. In most regions, lightning protection must
be provided as well as explosion-proof electrical systems, and so on, to avoid initiating
combustion explosions.
   Heavier fuels are used in stationary engines or large energy users such as central
station power boilers, refinery boilers, and other major installations that can make
an investment in the heating and other equipment necessary to handle the high-viscosity fuels together with their contaminants such as sulfur and vanadium. The heavier fuels require significant heating to reduce the viscosity and may require systems
such as steam-traced piping and equipment to maintain the viscosity sufficiently low
to permit ready pumping and firing.
   The heavy fuels in some cases have a substantial amount of vanadium (as much
as 2%), which creates problems of erosion in large boilers or gas turbines using these
fuels. For these cases, water washing and centrifuging is sometimes used to remove
the vanadium, or chemical agents such as magnesium and zinc may be added to the
fuel to reduce the effect of vanadium. The lighter fuels, such as gasoline and
naphthas, are liquid at normal temperatures and despite their volatility, their handling and transport are relatively easy. Typically, filtration of these lighter fuels will
be necessary because of end-use equipment. In the case of diesel engine unit injectors,
filtration systems removing very small particles in the range of 1 micron are
necessary.
   A typical system for serving a burner with Bunker C grade oil is illustrated in
Fig. 8.3. Provision for heating the fuel and recirculating it with positive displacement
pumps to the burners is shown. The heating medium—steam—can also be used to
smother flames at the holding tank if necessary. The lines are flushed with light oil
at system shutdown to prevent freezeup.

8.4.8   Methanol and Ethanol[3,12]

Methanol is a versatile liquid fuel produced by the distillation of biomass materials,
such as wood, peat, coal, and farm products, or by chemical conversion from natural
gas. It is relatively low in cost, less flammable than gasoline, and is extremely clean
burning, yielding no particulates or sulfur dioxides. Because of its relatively low com-

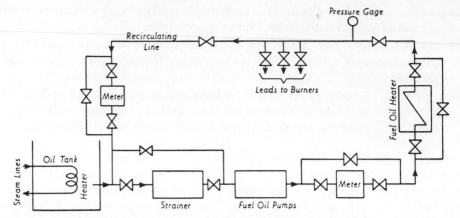

Fig. 8.3 Schematic diagram of fuel-oil system. From *Power Plant Theory and Design* by Philip J. Potter. Copyright © 1959 by The Ronald Press Co. Reprinted by permission.[14]

bustion temperature, fewer nitrogen oxides are produced than with petroleum-based fuels. Methanol dissolves in water and also has important industrial uses as a solvent. It has a very high octane rating (from 106 to 120) and up to 10% can be added to unleaded gasoline to improve its performance. It can be burned at a nominal 100% concentration as a fuel, and small amounts of water (1 to 3%) may be added to assist combustion.

Methanol has been used as an automotive fuel since the 1900s, particularly for high-performance engines where cost is not a significant factor. At the present time (1983), it's being produced and sold commercially at a cost roughly three-fourths that of gasoline. Large-scale production plants coming on line are expected to further

Table 8.5   Methanol and Gasoline Characteristics[8]

Characteristic	Methanol	Gasoline
Specific gravity, 60°F (15.6°C)	0.796	0.72
Boiling point	148°F (64°C)	100–400 °F (38–204°C)
Freezing point	−143.7°F (−97.6°C)	−40°F (−40°C)
Viscosity, 77°F (25°C)	0.5 cps	0.5 cps
Vapor pressure, 100°F (38°C)	0.32 kg/cm²	0.6–0.84 kg/cm²
Heat of vaporization at 68°F (20°C)	506 Btu/lb (342 kcal/kg)	150 Btu/lb (101 kcal/kg)
Heat of combustion	8640 Btu/lb (5830 kcal/kg)	18,650 Btu/lb (12,590 kcal/kg)
Flashpoint, closed cup	52°F (11°C)	−40°F (−40°C)
Ignition temperature	878°F (470°C)	800–950°F (427–510°C)
Flammable limits	6–37%	1–8%
Flame speed	1.6 fps (0.48 m/s)	1.1 fps (0.34 m/s)
Octane number	110–112	90–100

reduce its cost. Recently it has also become the feedstock in a developing process for synthesizing gasoline.

Internal combustion engines can utilize methanol at compression ratios of 12 to 16 or higher to 1, with relatively high combustion efficiencies (roughly twice those utilized for gasoline). To a large extent, this compensates for its lower heating value per unit volume. Diesel engines can use regular fuel for starting and warmup and then switch over to pure methanol so that of the total fuel used approximately 90% is methanol and 10% diesel fuel. For combustion turbines or for boilers, only a minor change of burner tips or combustors is necessary to adapt to this fuel with significantly higher combined cycle efficiency.

Ethanol can be produced by fermentation of any carbohydrate, such as saccharin (sugarcane, sugar beets, molasses, and other juices), starch (grains and potatoes), or cellulose (wood, bagasse, and straw). The thermal properties are similar to those of methanol, although chemically it is quite different and much less stable than methanol. Its production cost today (1983) is substantially higher than methanol.

8.4.9 Oil Sands[2]

The **oil sands** of Canada and the United States are a large-scale source of "bitumen," a crude hydrocarbon exhibiting the properties of an asphalt or tar but capable of being upgraded to fuels. The oil sands contain bitumen, which acts as a matrix holding together particles of sand. When treated with hot water, the agglomerate releases the hydrocarbon, which can be separated as a heavy viscous liquid (8 to 12°API), suitable as a refinery feedstock. The raw material is readily mined as a solid using bucket wheel excavators or draglines from surface deposits.

The hydrocarbon is easily displaced from the mineral by the strong wetting action of water, aided if necessary by surfactants. Commercial ventures use the "hot water" process to effect this treatment.

A family of methods for development of synthetic oils and extraction of oils from nonconventional sources is under way. The processes are still largely either experimental or developmental at the pilot plant stage and have varying degrees of potential for large-scale economic development. Typically these processes for shale oil, coal liquification, and coal-derived oils require extensive investment in processes and facilities and their development is dependent on and heavily influenced by world market crude oil pricing, with which they directly compete.

8.4.10 Shale Oil

Shale oil is widely found in the United States and elsewhere. In the United States, significant deposits are located in the Rocky Mountains [estimated at 2.2 trillion barrels (bbl)] in thick deposits and in a belt including the Western Appalachian Mountains, Indiana, Illinois, and Michigan (estimated at 400 billion bbl).

The shale contains a form of bitumen that, when heated to 900°F (480°C), yields 30 to 40 gallons of kerogen/ton (0.12 to 0.17 gal/kg) for Western shale and 15 to 18 gallons/ton (0.06 to 1.075 gal/kg) for Eastern shale. **Kerogen** is a heavy synthetic oil whose release is accompanied by high-quality synthetic gas. Eastern shales tend to produce coke when heated; thus, some current processes use hydrogen to reduce this tendency. Other processes utilize the coke directly as a process fuel, yielding oil as well as significant amounts of waste heat.

Current extraction methods utilize mining, crushing, and retorting the shale in large "mine mouth" type plants. These have the disadvantages of the mining and transportation costs as well as the disposal of spent shales. Various *in-situ* processes are under development to liberate the oil by burning a fraction of it underground in place, the resultant heat liberating the kerogen, which flows to sumps where the liquid is pumped to the surface. This process avoids most of the surface mining problems, but presents other technical problems such as methodology and control of underground shale fracturing, maintenance and control of combustion air, rate of advance, and temperature of flame front.

8.4.11   Coal Liquifaction

A variety of processes are available for the production of liquid fuels from coal. Unless the resultant liquids are burned directly as a fuel, these generally require further refining to produce other hydrocarbons of commercial value. At the present time (1983), with the price of crude oil in the $25 to $32/bbl range, none of the processes are sufficiently (economically) viable to be a significant source of fuel. It appears crude oil prices in excess of $35/bbl or low interest rates (e.g., 5 to 7%) are necessary to stimulate the large investment necessary for full-scale production facilities.

8.5   GASEOUS FUELS

8.5.1   Introduction

Gaseous fuels are considered by many to be the ideal fuel since they are extremely easy to handle, can be readily controlled, are economically competitive with other fuels, and are clean burning, leaving little or no ash or residue. Their disadvantage lies in their relatively low heating value per unit volume, which requires either a continuous piped supply or a compressed-gas storage system, rendering them somewhat unattractive for portable uses.

8.5.2   Characteristics of Gaseous Fuels

Table 8.6 lists the characteristics of the more commonly used gaseous fuels. See also Table 10.14, Physical Constants of Hydrocarbons.

8.5.3   Natural Gas

Natural gas is found compressed in porous rock, shale formations, or overlying oil deposits at pressures up to several thousand psi (20×10^3 kPa) or more. As the gas is removed from the field, the pressure drops until the gas in the field can no longer be withdrawn. In many areas, however, where the gas field is large and rock porosity

Table 8.6 Typical Characteristics of Gaseous Fuels

Type	HHV (Btu/cf, dry)	Specific Gravity (Air = 1.0)	Nitrogen (%)
Natural gas	1100	0.62	0.5
Blast furnace gas	110	0.97	58
Producer gas	160	0.84	53
Propane (C_3H_8)	2520	1.56	—
Butane (C_4H_{10})	3390	2.01	—
Hydrogen	324	0.07	—

sufficient, pressure reduction is not a problem and production pressure can be considered a constant over substantial periods of time.

Natural gas consists principally of methane with smaller quantities of other hydrocarbons, particularly ethane, although carbon dioxide and nitrogen are usually present in small amounts. Appreciable amounts of hydrogen sulfide are also sometimes present; these are usually removed at the field before transmission.

Oxygen is present only where there has been atmospheric air infiltration, although in some fields nitrogen (in the range of 5 to 30%) or carbon dioxide (to 6.0%) may be found. In some fields in Texas as much as 1% helium is also found.

Natural gas characteristics are determined by underground conditions. Where the field contains oil deposits, the gas may contain heavy saturated hydrocarbons, which are liquid at ordinary temperatures and pressures. This "wet" gas is dried by stripping out the liquids at the wellhead. Where sulfur is present in the oil, the gas will contain hydrogen sulfide. "Dry" gas contains less than 0.1 gallon of gasoline vapor per 1000 feet^3 (0.013 liter/10^3 cl) and is produced by wells generally remote from oil-bearing areas.

Natural gases are also classified as "sweet" or "sour." The "sour" gases contain a high percentage, frequently to 7% or more, of hydrogen sulfide as well as some mercaptans.

The higher heating value of natural gas is usually around 1000 Btu/feet^3 (8.9 kcal/liter) and it can be computed by adding together the heat contributed by the volumetric percentage of the various component gases. Natural gas normally is odorless and methyl mercaptan, a distinctive aromatic, is added to make leak detection easier.

8.5.4 Biomass[1]

The main process by which energy may be obtained from **biomass** includes direct combustion, pyrolysis, hydrogasification, anaerobic digestion, alcoholic fermentation, and biophotolysis. Each technology has advantages and these depend on the biomass source and the type of energy needed.

Among the advantages of utilizing biomass materials as an energy source are: (1) biomass provides an effective low-sulfur fuel; (2) it can provide an inexpensive source of energy (e.g., fuel-wood, dry animal manure, methane gas), and (3) in some cases, processing biomass materials for fuel reduces the environmental impact for these materials (e.g., biomass from sewage and processing wastes).

The major difficulties in utilizing biomass materials from solar energy conversion are: (1) the relatively small percentage (less than 0.1%) of radiant energy converted into biomass by plants; (2) the relatively sparse and low concentration of biomass per unit area of land and water (causing high labor costs for collection and transport); (3) the scarcity of additional land suitable for growing plants; and (4) the high moisture content (50 to 95%) of biomass that makes collection and transport expensive and energy conversion relatively inefficient. All of these factors make biomass energy costly in terms of energy and other activities expended in the conversion process and reduce the net energy yield.

In general, the use of wastes such as livestock manure and municipal sewage for biomass generation would help reduce environmental problems associated with the management of these wastes. In addition, the residual materials remaining after processing can be used as fertilizers.

Some environment benefits occur in the conversion of urban and industrial wastes for heat and electricity. These organic wastes generally have a low sulfur content, their combustion under carefully managed conditions would cause minimal air pollution problems and reduce the need for landfills, and the residue ash can (dependent on chemical makeup) be used as a fertilizer for crops and lawns. Urban and industrial wastes can also be converted into liquid and solid fuels by pyrolysis. The pyrolysis of these materials requires the majority of the combustion energy of the original wastes and, therefore, is less attractive as an energy source. Production of oil, char, and gas gives more freedom to the energy needs with the fuels produced.

The use of surplus sugar crops for the production of ethanol is feasible in areas where availability of fertile land is not a constraint and when plentiful solar radiation and manpower are available. The lack of enough land for sugarcane raising and the comparatively lower solar radiation are among the causes for the low potential contribution of sugar crops to provide energy in the United States.

The potential of biomass energy conversion from various biological materials can be categorized as shown in Table 8.7.

Table 8.7   Production of Biofuels from various Biomass Sources[a]

Biomass Substrate	Conversion Technology	Biofuels Produced
Livestock manure	Anaerobic digestion	CH_4
Urban refuse	Pyrolysis	Fuel oil
		Gas
		Char
Urban refuse	Incineration	Heat/electricity
Food processing waste	Anaerobic digestion	CH_4
Sugar crops	Ethanol fermentation	C_2H_5OH
Forest biomass	Incineration	Heat/electricity
Municipal sewage	Anaerobic digestion	CH_4

[a]*Source:* From Auer, *Advances in Energy Systems and Technology, Vol 1,* © *1978, Academic Press. Used with permission.*

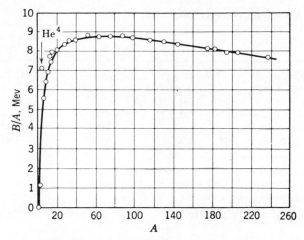

Fig. 8.4 Average binding energy per nucleon.

8.6 NUCLEAR ENERGY

8.6.1 Equivalence of Energy and Mass

In nuclear physics it is the composite of mass and energy that is conserved. From *Einstein's law*, when velocity is small relative to c,

$$E = mc^2. \tag{8.11}$$

For example, complete conversion of 1 gram of matter into energy yields 9×10^{20} ergs or 2.5×10^7 kilowatt hours of energy. In the practical development of nuclear power, the yield of energy is very small, only about 0.1% of the mass of U-235 being converted to energy in the fission process, for example.

Mass Defect. When a compound nucleus has less mass than the sum of the masses of its particles or nucleons, it is said to have a mass defect. The mass defect is equivalent to the energy radiated when the particles combined or to the energy required to separate the nucleus into its particles, which is called the **binding energy.** The binding energy per nucleon increases up to mass number of about 50 and then decreases gradually with increasing mass number (see Fig. 8.4).

8.6.2 Nuclear Reactions

The conventional notation for a nuclear reaction is

$$(_zC^A)_1 + (_zC^A)_2 \rightarrow (_zC^A)_3 + (_zC^A)_4 + Q \tag{8.12}$$

in which z is the number of protons, A is the mass number, C is the chemical symbol for the atom, electron, or nucleon, and Q is the energy released.

Table 8.8   Some Nuclear Constants

	amu	grams $\times 10^{-24}$	Rest energy, Mev
Unit mass, m	1	1.65990	931.16
Electron, $_1e^0$ or β^-	0.00054862	0.00091091	0.51083
Proton, p or $_1p^1$	1.007595	1.67247	938.17
Neutron, n or $_0n^1$	1.008983	1.67472	939.43
Hydrogen atom, $_1H^1$	1.00812	1.67338	938.68
Alpha particle, α or $_2He^4$	4.00280	6.64424	3727.07

Charge on electron $(1.60206 \pm .00007) \times 10^{-19}$ coulomb
Radius of electron $(2.81784 \pm .00010) \times 10^{-13}$ cm
Radius of nucleus $(1.5 \pm .15)\sqrt{A} \times 10^{-13}$ cm

The conservation equations are,

$$\sum Z_i = 0 \quad \text{and} \quad \sum A_i = 0. \tag{8.13}$$
$$\text{Initial mass} - \text{final mass} = Q. \tag{8.14}$$

An example is the **fission reaction:**

$$_{92}U^{235} + {_0}n^1 \rightarrow {_{92}}U^{236} \rightarrow {_{38}}Sr^{94} + {_{54}}Xe^{140} + 2{_0}n^1 + Q. \tag{8.15}$$

The strontium and xenon products are highly radioactive and decay further to other products. The final result is a spectrum of products.

Table 8.8 gives frequently encountered constants.

Modes of Radioactive Decay

Negative beta (electron) emission:

$$_0n^1 \rightarrow {_{-1}}e^0 + {_1}H^1 + \text{neutrino}. \tag{8.16}$$

$$_{38}Sr^{94} \xrightarrow{\beta^-} {_{39}}Y^{94} \xrightarrow{\beta^-} {_{40}}Zr^{94}. \tag{8.17}$$

$$_{54}Xe^{140} \rightarrow 4({_{-1}}e^0) + {_{58}}Ce^{140}. \tag{8.18}$$

Positive beta (positron) emission:

$$_7N^{13} \rightarrow {_{+1}}e^0 + {_6}C^{13} + \text{neutrino}. \tag{8.19}$$

$$_{+1}e^0 + {_{-1}}e^0 \rightarrow 2 \text{ gammas of 0.51 MeV each}. \tag{8.20}$$

Alpha emission:

$$_{94}Pu^{239} \rightarrow {_2}He^4 + {_{92}}U^{235}. \tag{8.21}$$

Neutron emission:

$$_{53}I^{137} \xrightarrow{\beta^-} {}_{54}Xe^{137} \rightarrow {}_0n^1 + {}_{54}Xe^{136}. \tag{8.22}$$

Orbital electron (K) capture:

$$_{29}Cu^{64} + {}_{-1}e^0 \rightarrow {}_{28}Ni^{64}. \tag{8.23}$$

Gamma emission by:

1. Ejection of a gamma photon from an excited nucleus
2. Isomeric transition of a nucleus from one energy level to another
3. Annihilation of an electron following positive beta emission

Nuclei with an excess of neutrons are usually electron emitters. Among nuclei with a deficiency of neutrons, the heavy ones usually decay by alpha emission, and the light ones by positron emission or orbital electron capture. Gamma emission often accompanies other types of decay.

8.6.3 Decay with Time

From statistical considerations, the rate of decay (alpha emission) is proportional to the number N of radioactive nuclei present,

$$\frac{dN}{dt} = -\lambda N, \tag{8.24}$$

where the proportionality constant λ is called the **disintegration constant** or the **radioactive decay constant.** Integration of Equation (8.24) gives

$$N = N_0 e^{-\lambda t}, \tag{8.25}$$

where N_0 is the initial number of radioactive nuclei. The **half-life** $t_{1/2}$, or the time required for one-half of the original atoms to decay, by substitution in Equation (8.25), is

$$t_{1/2} = \frac{0.693}{\lambda}. \tag{8.26}$$

When there is more than one radioisotope in the decay chain,

$$P \xrightarrow{\lambda_1} Q \xrightarrow{\lambda_2} R \xrightarrow{\lambda_3} \tag{8.27}$$

for the second member or daughter Q

$$\frac{N_Q}{N_{P_1}} = \frac{\lambda_1}{\lambda_2 - \lambda_1} e^{-\lambda_1 t} + \frac{\lambda_1}{\lambda_1 - \lambda_2} e^{-\lambda_2 t}. \tag{8.28}$$

If the half-life of the parent P is longer than the half-life of the daughter $Q(\lambda_2 > \lambda_1)$, after a lapse of time $e^{-\lambda_2 t}$ becomes negligible and

$$\frac{N_Q}{N_P} = \frac{\lambda_1}{\lambda_2 - \lambda_1} e^{-\lambda_1 t}. \tag{8.29}$$

With naturally occuring radioisotopes, the half-life of the parent is very long compared to that of the daughter, so that Equation (8.29) reduces to

$$\frac{N_Q}{N_P} = \frac{\lambda_1}{\lambda_2} e^{-\lambda_1 t}. \tag{8.30}$$

8.6.4 Characteristics of Selected Radioisotopes

Half-life and emitted particle energy for selected radioisotopes are shown in Table 8.9.

8.6.5 Biological Effects

Relative biological effectiveness (RBE) is the ratio of gamma radiation to another type of radiation that produces the same biological effect, both expressed in rads. The **roentgen equivalent man (rem)** is the unit of absorbed radiation dose for biological effects. It is defined as

$$\text{rem} = \text{RBE} \times (\text{number of rad}). \tag{8.31}$$

The maximum permissible external radiation dose for adult workers in the radiation industries is 1.25 R per quarter. A single chest X-ray is about (0.2 rem) and continuous exposure of man to natural background radiation at sea level and 50° latitude is about one-thirtieth of this rate (0.0033 rem per week).

The ingestion of radioactive materials in air, water, or food presents hazards that depend not only on the concentration and absorption but also on the rate of elimination. For example, Sr-90 with a long half-life is absorbed in bone structure and persists for years.

8.6.6 Shielding

Since alpha and beta particles have relatively short range, shielding is provided mainly for neutron and gamma radiation. Shielding for neutrons usually involves thermalization of the neutrons followed by their absorption in a material with high-absorption cross section. Neutron shielding also involves provision for photons from neutron–gamma reactions in the shield. Gamma shielding normally is provided by a material with a high absorption coefficient or by large thickness. Design considerations include geometrical arrangement and balance of thickness and effectiveness

Table 8.9   Characteristics of Selected Radioisotopes

Element	Mass Number	β (MeV)	γ (MeV)	Half-life
Aluminum	29	2.5	—	6.7 m
Antimony	122	1.36, 1.94	0.57	2.8 d
	124	0.74, 2.45	1.72	60 d
	125	0.3, 0.7	0.55	2.7 y
Arsenic	77	0.8	None	40 h
Beryllium	10	0.56	None	3×10^6 y
Bismuth	210	1.17	None	5 d
Bromine	82	0.465	0.547, 0.787, 1.35	34 h
Cadmium	115	0.6, 1.11	0.65	2.8 d
Calcium	45	0.2, 0.9	None	180 d
	49	2.3	0.8	2.5 h
Carbon	11	$0.97e^+$	—	20.5 m
	14	0.145	None	5100 y
Cerium	141	0.6	0.21	28 d
Chlorine	36	0.66	None	10^6 y
	38	1.1, 2.8, 5.0	1.65, 2.15	37 m
Cobalt	60	0.31	1.10, 1.30	5.3 y
Columbium	95	0.15	0.75	35 d
Copper	61	0.9, 1.23	None	3.4 h
Europium	154	0.9	1.4	5.4 y
	155	0.23	0.084	2 y
Fluorine	18	0.7	—	1.86 h
Gallium	72	0.8, 3.1	0.84, 2.25	14.1 h
Germanium	71	1.2	—	40 h
	77	1.9	—	12 h
Gold	198	0.98	0.12, 0.41	2.7 d
	199	1.01	0.45	3.3 d
Hafnium	181	0.8	0.5	46 d
Hydrogen	3	0.011, 0.015	None	12 y
Iodine	131	0.315, 0.600	0.367, 0.080, 0.284, 0.638	8.0 d
Iridium	192	0.67	$0.137 \rightarrow 0.651(12\gamma)$	74.7 d
	194	0.48, 2.18	0.38, 1.43	19 h
Iron	59	0.26, 0.46	1.1, 1.3	46.3 d
Krypton	85	1.0	0.17, 0.37	4.5 h
Lanthanum	140	1.32, 1.67, 2.26	$0.093 \rightarrow 2.5$	40 h
Magnesium	27	0.9, 1.80	0.64, 0.84, 1.02	9.58 m
Mercury	203	0.205	0.286	43.5 d
Molybdenum	99	0.445, 1.23	0.04, 0.741, 0.780	68.3 h
Neodymium	147	0.17, 0.78	0.035, 0.58	11 d
Nitrogen	13	$1.23e^+$	None	10.1 m
Osmium	191	0.142	0.039, 0.127	15.0 d
	193	1.15	1.58	32 h
Phosphorus	32	1.718	None	14.3 d
Platinum	197	0.65	None	18 h
Potassium	40	1.40	1.45, K	9.9×10^8 y
	42	3.5	None	12.4 h
Praeseodymium	142	0.636, 2.154	1.57	19.1 h
	143	0.932	None	13.8 d
Promethium	147	0.229	None	2.26 y
Rhenium	186	0.64, 0.95, 1.09	0.132, 0.275, 1.70	92.8 h
Rhodium	105	0.57	0.33	36.2 h

Table 8.9 *Continued*

Element	Mass Number	β (MeV)	γ (MeV)	Half-life
Rubidium	86	0.72, 1.80	1.08	19.5 d
Ruthenium	103	0.205, 0.670	0.494	42 d
Samarium	153	0.68, 0.80	0.070, 0.103, 0.61	47 h
Scandium	46	0.36, 1.49	0.89, 1.12	85 d
Silicon	31	1.5	None	2.59 h
Silver	110	0.087, 0.53	0.885, 0.935, 0.1389, 1.516	270 d
	111	1.06	None	7.5 d
Sodium	22	$0.58e^+$	1.3	2.6 y
	24	1.390	1.380, 2.758	15.0 h
Strontium	89	1.50	None	53 d
	90	0.54	None	19.9 y
Sulfur	35	0.167	None	87.1 d
Tantalum	182	0.52 m, 1.1	$0.05 \rightarrow 1.24(33\gamma)$	115 d
Technetium	97	IT	0.097	90 d $> 10^3$ y
	99	0.30	0.140	2.1×10^5 y
Tellurium	127	IT $\frown$ 0.8	0.089	90 d, 9.3 h
	129	IT 1.46	0.102, 0.3, 0.8	32 d, 72 m
	131	IT > 1.8	0.177	30 h, 25 m
Thallium	206	0.58	None	2.7 y
Titanium	51	0.36	1.0	72 d
Tungsten	185	0.428	None	73.2 d
	187	0.627, 1.318	0.086, 0.70	24.1 h
Yttrium	90	2.24	None	61 h

of materials against cost. Effectiveness depends both on the material and on the energy of radiation, as shown in Table 8.9.

The attenuation with distance x of the intensity I of gamma radiation from a plane source is given by

$$\frac{I_x}{I_0} = be^{-\mu x}, \tag{8.32}$$

and from a point source surrounded by a spherical shield by

$$J = \frac{bI_0 e^{-\mu x}}{4\pi x^2}, \tag{8.33}$$

where I_0 is the intensity at the source (MeV/cm²-sec), μ is the attenuation coefficient (see Table 8.10), b is the buildup factor for shields of appreciable thickness, and J is the current density (MeV/cm²-sec). The factor b, determined experimentally, is found to depend on μx and I_0, as well as on the shield material. Equations (8.32) and (8.33) can be applied to the attenuation of thermal neutrons where $\Sigma = \mu$. Attenuation of fast neutrons is a much more complex problem and requires specialized knowledge.

The energy of radiation attenuated by a shield is nearly all converted to heat

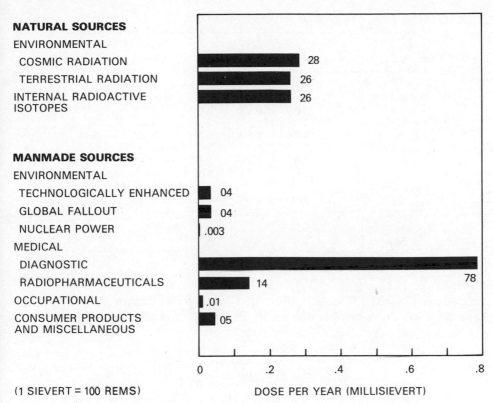

Fig. 8.5 Radiation exposure. From "The Biological Effects of Low-Level Ionizing Radiation," by Arthur C. Upton. Copyright © 1982 by Scientific American, Inc. All rights reserved.[15]

energy. With the assumption that all the energy is absorbed within one free path, the maximum temperature rise in a plane shield is given by

$$\Delta T = \frac{I_0}{K\mu}, \tag{8.34}$$

and in a spherical shield by

$$\Delta T = \frac{I_0}{4\pi K} \left(\frac{1}{r_1} - \frac{1}{r_2} \right), \tag{8.35}$$

where K is the thermal conductivity of the shield (in MeV/sec-cm-°C).

Shielding properties of some selected materials are shown in Table 8.10.

8.7 SOLAR ENERGY[6,9]

8.7.1 Solar Characteristics

The relationship of the earth and sun with respect to the incident solar energy is shown in Fig. 8.6. Variation of earth–sun distances during the year affect the amount

Table 8.10 Shielding Properties of Materials

	Water	Iron	Lead	Portland Concrete	Barite Concrete
Density, g/cm³	1.00	7.78	11.3	2.37	3.49
Thermal neutrons					
Σ, cm⁻¹	0.100	0.156	0.113	0.094	0.094
γ at 0.5 Mev					
μ, cm⁻¹	0.096	0.653	1.72	0.20	0.30
Build-up b					
At $\mu x = 1$	2.46	2.80	1.51		
At $\mu x = 10$	71.5	34.2	3.01		
γ at 6 Mev					
μ, cm⁻¹	0.027	0.237	0.503	0.065	0.109
Build-up b					
At $\mu x = 1$	1.46	1.30	1.14		
At $\mu x = 10$	5.18	7.10	4.20		

of radiation received (Fig. 8.7). Note that the radiation theoretically received is actually lower in the summer because the earth–sun distance is greater.

Solar radiation received can be estimated:

$$H_{av} = H_o'\left(a + b\,\frac{n}{N}\right) \tag{8.36}$$

where: H_{av} = average horizontal radiation for the period in question (e.g., month)
$\qquad$ H_o' = clear day horizontal radiation for the same period
$\qquad$ n = average daily hours of bright sunshine for same period
$\qquad$ N = maximum daily hours of bright sunshine for same period
$\qquad$ a, b = constants

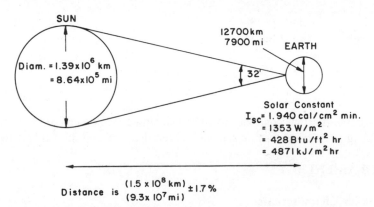

Fig. 8.6 Sun–earth relationships. From *Solar Thermal Energy Process* by Duffie and Beckman. Copyright © 1974 by John Wiley & Sons. Reprinted by permission.[6]

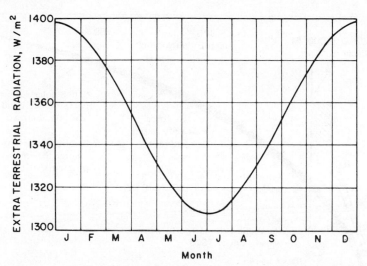

Fig. 8.7 Variation of the extraterrestrial solar radiation with time of the year. From *Solar Thermal Energy Process* by Duffie and Beckman. Copyright © 1974 by John Wiley & Sons. Reprinted by permission.

Values frequently used are $a = 0.35$ and $b = 0.61$. Values of H'_o for use in Equation (8.36) can be obtained from Fig. 8.8.

8.7.2 Solar Utilization Systems

Solar utilization systems can be defined as either concentration or collector systems, with direct conversion a further variation. The use of systems with reflectors or concentrators provide ways to increase the energy level to the point that steam can be

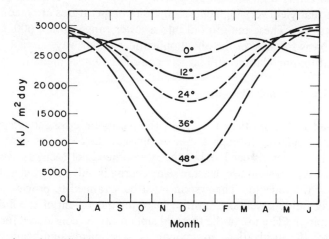

Fig. 8.8 Clear day solar radiation on a horizontal plane for various latitudes. From *Solar Thermal Energy Process* by Duffie and Beckman. Copyright © 1974 by John Wiley & Sons. Reprinted by permission.[6]

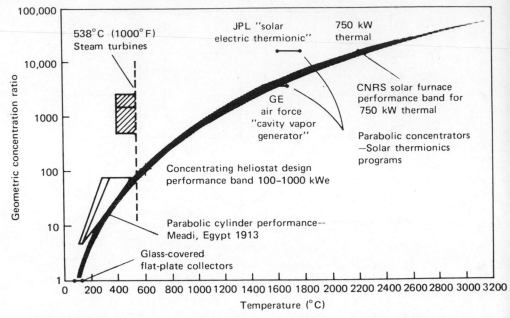

Fig. 8.9 Typical temperature performance of solar collectors. From *Principles of Solar Engineering* by F. Kreith and J. F. Kreider, copyright © 1978 by Hemisphere Publishing Corporation. Reprinted by permission.[9]

generated from solar rays and the steam used to generate electric power. This development appears most promising, although it does have the disadvantage of high first cost and the requirement for a large amount of land to gather the relatively low-energy incident sunlight. A potential problem with this is that little data is available on maintenance requirements for the solar collector arrays and the problems of maintaining a sufficiently high level of reflectivity of the mirror reflector. High-concentration ratios that yield very high operating fluid temperatures are readily achievable with some systems. These tend to fall into a lower range group, 600°C and below, which is the basis for most of the pilot applications under way (circa 1983) and very high temperature applications still largely considered experimental.

8.7.3 Self-Contained Collectors

Self-contained collectors typically are of the parabolic cross-section type with the working fluid circulating through a pipe or pipes at the focal line of the mirror system. A tracking system is needed, but the system need not focus as accurately as in a concentrator-type system because the solar energy is only reflected a short distance to the focus of the reflector. The system must be arranged to provide for insulation of the working fluid as well as a positive head for movement of the fluid (generally through pumping). (The use of the thermal siphon can be considered for those installations where the utilization equipment is positioned significantly above the collector.)

Flat-plate collectors are also suitable for large arrays and can be made relatively

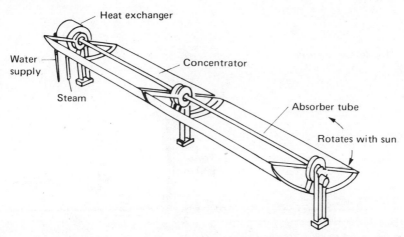

Fig. 8.10 Typical self-contained collector. From Principles of Solar Engineering by F. Kreith and J. F. Kreider, copyright © 1978 by Hemisphere Publishing Corporation. Reprinted by permission.[9]

efficient. The flat plate mirror systems have the advantage of low-cost fabrication and low maintenance, but require careful focusing (as with the single- and double-curvature devices) to achieve reasonable concentration ratios.

8.7.4 Mirror and Tracking Systems[9]

Mirror and tracking systems have high efficiency factors and can achieve temperatures above 3000°C when grouped in multiple arrays. Arrays can be furnished as either single- or double-curvature devices; the multiple-curvature devices provide higher concentration ratios.

   Tracking types include either intermittent tilt change types or continuously tracking reflectors, refractors, or receivers. If oriented east–west, they require an approximate ±30°/day motion; if north–south, a 15°/hr motion. Both must accommodate to a ±23°/yr declination excursion. Single-curvature devices may be of either type but double-curvature devices are usually of the continuously tracking, high-concentration ratio type.

   One company has developed a concentrating solar collector that can yield temperatures up to about 530°F (277°C). The system uses a parabolic trough reflector with special heat-transfer fluid flowing through tubes placed in transparent envelopes at the parabola's focus; air having been exhausted from the envelopes. The reflector tracks the sun; its movement around a single axis is controlled by a microcomputer that draws relevant information from a series of sensors.

   The reflectors and collectors are built in standard modules with an area of 135 feet^3 (12.5 m^2) each. There is an automatic flushing device and wiper to remove accumulating dust. As many modules as desired can be assembled, which are controlled by a single computer unit that tracks the sun, activates the cleaning equipment, operates the heat-transfer fluid pump, and otherwise manages the system for optimum productivity. The units have a design capability to collect an average of 2.7 kWh/day/m^2 of collector area in sunny locations.

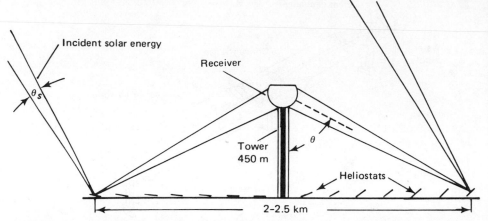

Fig. 8.11 Cross-sectional view of solar power tower system. From *Principles of Solar Engineering* by F. Kreith and J. F. Kreider, copyright © 1978 by Hemisphere Publishing Corporation. Reprinted by permission.[9]

8.7.5 Direct Conversion[9]

Thermal conversion utilizes the solar absorption phenomenon with its accompanying heating. Solar energy collectors working on this principle consist of a dark surface facing the sun, which transfers part of the energy it absorbs to a working fluid in contact with it. To reduce atmospheric heat losses, one or two sheets of glass are usually placed over the absorber surface to improve its efficiency. These types of thermal collectors suffer from heat losses due to radiation and convection, which increase rapidly as the temperature of the working fluid increases. Improvements such as the use of selective surfaces, insulation, evacuation of the collector to reduce heat losses, and special kinds of glass are used to increase the efficiency of these devices.

These simple thermal-conversion devices are called flat-plate collectors. They are available today for operation over a range of temperatures up to approximately 200°F (93°C). These collectors are suitable mainly for providing hot service water and space heating and may operate absorption-type air-conditioning systems.

The thermal utilization of solar energy for the purpose of generating low-temperature heat is at the present time (1983) technically feasible and economically viable for producing hot water and heating swimming pools. In some parts of the world thermal low-temperature utilization is also economically attractive for heating and cooling buildings. Other applications, such as the production of low-temperature steam are under development.

Production of higher working temperatures requires the use of focusing devices in connection with a basic absorber–receiver. Operating temperatures as high as 3700°C (6740°F) have been achieved in the Odeillo solar furnace in France, and the generation of steam to operate pumps for irrigation purposes has also proved technologically feasible. At the present time a number of focusing devices for the generation of steam to produce electric power are under construction in different regions of the world, and estimates suggest that the cost of solar power equipment in favorable locations may be reduced to the range of $1500/kw of capacity. To compare with costs for conventional power sources, the percent availability of sunshine must be considered (i.e., number of hours/day of sunshine, number of days with significant cloud cover, rain, etc.).

8.7.6 Photovoltaic Conversion[9]

Photovoltaic solar cells utilize energetic photons of the incident solar radiation directly to produce electricity. This technique is often referred to as **direct solar conversion.** Conversion efficiencies of thermal systems are limited by collector temperatures, whereas the conversion efficiency of photo cells is limited by other factors.

The conversion of solar radiation into electrical energy by means of solar cells has been developed as a part of satellite and space-travel technology. The theoretical efficiency of solar cells is about 24%, and, in practice, efficiencies as high as 15% have been achieved with silicon photovoltaic devices. The technology of photovoltaic conversion is well developed, but large-scale application is hampered by the high price of photo cells. A cost reduction of the order of 100 to 1 will be necessary before photovoltaic conversion can be economically viable.

Efforts are currently underway to develop new technologies for producing solar cells in the hope that such methods might lead to an appreciable lowering in price. Silicon cells, which have performed well for spacecraft application, are still extremely expensive. Efforts are directed toward producing thin monocrystalline foil-bands because such a method might lead to appreciable lowering of the production cost. Also, cell materials are being considered, but unfortunately the lower price is counterbalanced by lower efficiency of these polycrystalline cells. Cadmium sulfide–copper sulfide cells (one of the candidates) have an efficiency of only about 6 or 7%.

8.8 GEOTHERMAL ENERGY[5]

Geothermal energy utilizes the heat within the earth, either from steam venting through natural or man-made vents (wells) or by heating water injected into the earth through wells. In several locations, Larderello, Italy; Iceland; New Zealand; and California, U.S.A., geothermal energy has been sufficiently developed to yield significant amounts of energy.

Geothermal steam generally includes large amounts of both water and noncondensibles (gases) as well as hydrogen sulfides and carbonates. When condensed, the hydrogen sulfides convert to weak sulfuric acid requiring the widespread use of corrosion-resistant materials (typically Type 304 or 316 stainless steel for wetted portions). For power recovery applications utilizing turbines, and so on, steam treatment including water and suspended solids removal is necessary. For closed systems very large air ejector capacity is necessary to remove the noncondensibles and prevent "air binding." Chemical treatment may also be necessary to assure proper pH of throttle steam.

Wellhead steam conditions are typically near saturation, with perhaps 10 to 20°F (12 to 16°C) of superheat, flows of 50,000 pounds/hour (22,700 kg/hr) per well, 8 to 10 inches (200 to 250 mm) in diameter at pressures roughly one-half the saturation pressure of water corresponding to the well depth [e.g., 200 psig (1380 kPa) for an 800- to 1000-foot (240- to 300-m) deep well] are common.

Table 8.11 Typical Geothermal Data[5]

Place	Activity	Description
Italy Larderello	Power generation	420,000 kW, 27 units
Avg. well 37,500 lb/hr (17,000 kg/hr) at 30–100 psig (210–700 kPa), 285–430°F (140–204°C), 1–20% noncondensibles		
Iceland Reykjavik	Primarily space heating plus power generation	65% of total island usage for 214,000 people, 100% of Reykjavik, 30,000 kW
Avg. steam content, 1000 ppm dis. solids, 300–350°F (149–177°C), 1% noncondensibles		
New Zealand Wairakei, Kawerau, Rotorua.	Power generation plus lumber drying, space heating	200,000 kW
Avg. well 20–35,000 lb/hr (9–16,000 kg/hr) at 90–125 psig (628–873 kPa), 350 °F ± (177°C), 0.5–2.5% noncondensibles		
California The geysers	Power generation	2,000,000 kW, 22 units,
Avg. well 200,000 lb/hr (91,000 kg/hr) at 140 psig (978 kPa), 350°F ± (177°C), 0.3–1% noncondensibles		

Frequently, 2 to 3 pounds (0.9 to 1.4 kg) of water are produced per pound of steam flow. As with other natural processes, however, these vary widely with location, and each well and site must be separately evaluated. Disposal of spent geothermal fluids needs to be carefully considered, with disposition of solids, pH control, and odor emissions the most prevalent problems. Depletion of well capacity needs to be evaluated also since this can occur in some locations.

8.9 WIND ENERGY

Wind energy, although in use for centuries as distributed single-purpose installations, has not until recently been seriously considered for electric power generation or central power grid use. Previous uses included grain milling, water pumping, and some general manufacturing applications. Recently, with the development of inexpensive voltage regulators, small windmill output can be compatible with the voltages used by public utility systems, and units of the 500- to 3000-W size are being installed for both private and industrial applications. Larger units are available in sizes up to 2000 kW, or more, with design lives of 30 years. Installed in clusters at favorable locations, they are becoming more common and, although still largely experimental, have the potential for significant power generation.

Typical of the large machines being installed are several in Washington, USA, rated 2500 kW each at 17.5 rpm in a wind speed of 27 mph (43 km/hr). Located on 200-foot (61-m) towers, with two-bladed steel propellors with a diameter of 300 feet (91 m), the blades have variable pitch capability (over part of their length) and their speed is stepped up to 1800 rpm by a gear box integral with the tower-located generator. Designed to start with a 14-mph (4.27-km/hr) wind, they are designed to withstand up to 125-mph (200-km/hr) winds.

8.10   FUEL CELLS[7]

Fuel cells are direct energy conversion devices in which the conversion from chemical to electrical energy occurs without the intermediate steps of combustion, conversion to mechanical energy, and (by a generator) conversion to electrical energy. The fuel cell process is a simple step where the chemical energy from a methanol or a hydrocarbon fuel such as coal, oil, or gas and an oxidizing agent such as oxygen or air are converted directly to electrical energy. Since the conversion is direct, it is extremely efficient and fuel cells used in aerospace applications have achieved efficiencies greater than 75%.

Basically, the fuel cell consists of three principal elements: an anode, an electrolyte, and a cathode. Hydrogen-rich fuel is fed to the anode, where the hydrogen loses its electrons. On the other side of the cell, oxygen is fed to the cathode, which picks up electrons and develops a positive charge on the cathode. The flow of excess elec-

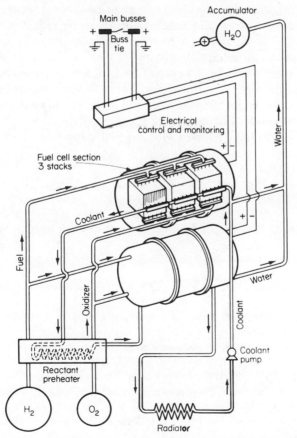

Fig. 8.12   Gemini fuel cell schematic diagram. From *Fuel Cells* by McDougall. © 1976. Reprinted by permission of Macmillan, London and Basingstoke.[10]

trons proceeds from the anode toward the cathode and creates the electric power, which is the useful output of the cell. The waste product of the excess hydrogen ions from the anode and the oxygen ions from the cathode pass into the electrolyte where they combine to form water, which is exhausted from the cell. Because of its elevated operating temperature, the water is exhausted as steam. The electron flow produces dc energy which for many applications must be converted to ac, requiring the use of a converter in the system. Typically, ac to dc converters have efficiencies on the order of 90%, so that the conversion is highly efficient and little energy is lost in this step.

The flexibility of fuel cells when arranged in systems is a major advantage because the fuel cells can be stacked and arranged in a modular fashion and connected in parallel to provide sufficient power. Potentially, there are no major restrictions as to size or capacity, only those limited by the mechanical strength of the structural array. Fuel cells operate efficiently at part load as well as full load and provide very rapid response to variations to electrical demand. For this reason they have a high potential for use in systems that have widely varying demands.

A major advantage of the fuel cells is the flexibility to accept a variety of fuels, including methanol and hydrocarbons such as light distillates, natural gas, and high-, medium- or low-Btu gases, although for some fuels it may be necessary to process the fuel to assure that it is clean enough to avoid contamination of the anode. Other advantages of the fuel cell plants include low water requirements, quiet operation, and limited emissions that are environmentally compatible. These help to make fuel cell plants an attractive concept for distributed power generation. It is possible to cool these plants by low-speed fans and, since combustion processes are not involved, the emissions from these plants are low temperature and do not yield the nitrous oxides, unburned hydrocarbons, and other typical products of combustion that we find with combustion engines or the particulate discharge and the sulfur compounds in the flue gases released with larger size fossil fuel burning units.

REFERENCES

1. Peter Auer, *Advances in Energy Systems and Technology*, Vol. 1, Academic Press, New York, 1978.

2. *Alternative Liquid Fuels*, Report 652-1981, Stanford Research Institute, Menlo Park, Calif., 1981.

3. D. M. Considine, *Chemical and Process Technology Encyclopedia*, McGraw-Hill, New York, 1974.

4. *Flow of Fluids Through Valves, Fittings and Pipe*, Technical Paper 410, Crane Co., New York, 1980.

5. Ronald DiPippo, *Geothermal Energy as a Source of Electricity*, U.S. Department of Energy, Washington, D.C., 1980.

6. John A. Duffie and Wm. A. Beckman, *Solar Energy Thermal Processes*, Wiley-Interscience, New York, 1974.

7. *Fuel Cells*, Publ. 0024-1978, U.S. Dept. of Energy, Washington, D.C., 1978.

8. R. E. Kirk and D. F. Othmer, *Encyclopedia of Chemical Technology*, 3rd edition, Wiley-Interscience, New York, 1981.

9. Frank Kreith and Jan F. Kreider, *Principles of Solar Engineering*, Hemisphere Publishing Corp., New York, 1978.

10. Angus McDougall, *Fuel Cells*, Halsted Press, New York, 1976.

11. W. L. Nelson, *Petroleum Refinery Engineering*, McGraw-Hill, New York, 3rd ed., 1949.

12. D. F. Othmer, *Methanol as a Fuel, Methanol Production and Use as a Low Cost Fuel;* and *Methanol—The Efficient Conversion of Valueless Fuels into Versatile Fuel and Chemical Feedstock,* Polytechnic Institute of New York, Brooklyn, New York, 1981 and 1982 (respectively).

13. J. H. Perry, *Chemical Engineers' Handbook,* McGraw-Hill, New York, 5th ed., 1973.

14. Phillip J. Potter, *Power Plant Theory and Design,* The Ronald Press, New York, 1959.

15. Arthur C. Upton, The Biological Effects of Low-Level Ionizing Radiation, *Scientific American,* February 1982, W. H. Freeman, San Francisco, Calif.

16. Ella Mae Shelton, *Aviation Turbine Fuels,* 1980, DOE/BETC/PPS-81/2; *Diesel Fuel Oils,* 1981, DOE/BETC/PPS-81/5; *Heating Oils,* 1981, DOE/BETC/PPS-81/4; and *Motor Gasolines,* Winter 1980–1981, DOE/BETC/PPS-81/3, Dept. of Energy, Bartlesville, Okla.

17. Joseph G. Singer, *Combustion—Fossil Power Systems,* Combustion Engineering Inc., Windsor, Conn., 1981.

18. *1981 Annual Report to Congress,* Vol. 2, *Energy Statistics,* U.S. Dept. of Energy, Washington, D.C., May 1982.

19. *1981 Annual Book of ASTM Standards,* American Society of Testing and Materials, Philadelphia, Penn.

20. *1981 SAE Handbook,* Society of Automotive Engineers, Warrendale, Penn.

SECTION 9
ENGINEERING PRACTICE

9.1 CALCULATIONS

9.1.1 Preparation and Checking

When preparing calculations it is normal practice that the basis for the design be clearly stated at the beginning of the calculation. Typically, the calculation includes statements for those items that are **given,** those items that are **to be found** or calculated, and the necessary **assumptions** that were used in the calculations. It is most important that all calculations be **signed** and **dated** by the person performing them. Calculations must be legible, sufficiently clear, and arranged in a logical sequence so that someone unfamiliar with them, performing an independent check, can readily follow them and verify their accuracy. Further, the calculations should be so arranged that someone of experience equal to the originator can readily check them and determine their accuracy. Thus they should call out codes, standards, references used, and so on, so that both the originator and the checker can return to them at a later date if necessary and reestablish both their applicability and their accuracy.

All calculations should be checked. The checking may include:

A rough rule-of-thumb check
An independent separate check using different but equivalent formulas*
An estimate by a highly experienced person*
A detailed, meticulous check of each of the steps in the calculation*

Calculations can have various levels of accuracy depending on their end use. Thus, a calculation for a precision machine part might have a very high level of accuracy, while a calculation used in a gross analysis of earth to be moved for a large construction project might be suitable at a level of rough approximation only. It is important to state the level of accuracy desired in the calculations so that the usage is clear and the purpose of the calculation is clearly understood.

Calculations are typically prepared on quadrille paper, usually of the vellum variety. This provides ease of reproduction by the ozalid as well as the photocopy processes. The quadrille ruling makes it convenient to prepare those sketches that are a part of the calculation.

---

*To avoid the natural phenomenon of overlooking one's own error.

9.1.2 Graph Papers

Graph papers are available in a wide variety of sizes and styles with rulings appropriate to almost every area of engineering specialization. For normal activities a useful selection of 8½ × 11-inch papers includes:

> Quadrille ruled, 5 or 10 lines per inch; preferred because curve plotting on 4 lines per inch (each fourth line accented) is less rapid, although 4 lines per inch is very useful for sketching
> Quadrille ruled, 4 lines per inch (for desk sketches)
> Isometric paper; 30° axis
> Scheduling/calendar paper; 1, 2, and 5 years × 100 divisions
> Semi-log paper; 1, 2, and 5 cycles × 10 divisions
> Log paper; 1, 2, and 5 cycles—both axes
> Polar coordinate paper; by degrees, 10° accented

Paper having extremely fine divisions (such as 1 mm) tends to be of limited use. Most papers are available printed either on bond or vellum. The catalogs of major suppliers of drafting papers and equipment cover the wide range of special papers available and should be consulted for specialized uses.

9.1.3 Computer Applications

With the explosion in capability of computer technology, three fairly distinct groupings of computer applications are now emerging. These categories, or "tiers," of information processing applications may be interconnected through suitable communications networks.

Central applications include engineering design and analysis, general purpose information processing and retrieval systems, large-document word processing, and so on. These large, complex systems:

> Require high-capacity computing facilities, large shared data bases and/or geographic sharing of information.
> Involve extensive analysis and programming by information processing specialists.

Distribution applications include computer-aided drafting (CAD), computer-aided manufacturing (CAM), interactive computer-aided engineering (CAE), clustered-word processing, and remote batch entry and output. These applications:

> Utilize shared, special-purpose computing facilities and data located in distributed and central data bases.
> Serve several related functional groups, usually performing a defined set of tasks.

Local applications include computer-aided drafting, personal computing, word processing, graphics and data preparation, entry and inquiry. These applications:

Provide convenient operation on a wide range of personal computers and intelligent workstations, frequently by users who are not specialists in information processing.

Access central or distributed information systems to obtain additional processing power or data, as needed, and, if established, to provide an extensive array of software tools that permits users to develop their own applications without the need to rely on programmers.

Serve specific, single tasks, usually performed by one group or person.

Long-range planning is critical in achieving application compatibility among the tiers and to permit sharing of data among applications. The importance of long-range planning is accentuated by rising costs of software development, the widening spectrum of alternatives, and the lengthening lead time for major developments.

Effective flow of data from one application to another is vital to productivity. Cross-application data flow encompasses engineering design, computer-aided drafting, material specification, and so on.

9.2 SPECIFICATIONS

Specifications fall into two distinct categories **performance** or **design**. **Performance specifications** state input and output parameters but do not normally restrict or establish the way in which these are met. They may also in a nondetailed way establish the features required. They have the advantage of permitting the supplier to offer his standard unit (or alternative units) most nearly matching the overall required performance and place no responsibility on the purchaser. They are most applicable to standard items for which some design, manufacturing, and operating experience is available.

Design specifications normally state not only input and output parameters but specific design requirements, materials of construction, and so on. They establish not only what must be achieved, but to some extent how to do it as well. As a consequence, standard designs rarely apply, alternates are limited, and the purchaser may bear some degree of responsibility if performance is not met. They are most appropriate for innovative designs or developmental work, particularly where new technologies are involved. Frequently, the number of sources of supply is limited and prequalification of bidders may be necessary to assure technical competence.

When preparing specifications, preciseness is essential. Terms such as "fitness for use," "good commercial quality," and so on, are not sufficient to define requirements for complex, costly, or high-impact items. For these cases, specific **acceptance criteria** should be stated, and, where codes are involved, they should be referenced or preferably included in the body of the specification.

The acceptance criteria should be limited to significant measurable requirements (to avoid overspecifying). Tolerancing, if omitted from drawings issued for use by others, can create major problems and should be considered a part of acceptance criteria where appropriate. Acceptance criteria can include acceptable quality levels (AQLs) and consumer's risk or confidence levels (CLs) to provide a measure of deviation allowed from 100% conformance. AQLs are typically stated as either percent or maximum number of defects permitted per hundred, per thousand, and so

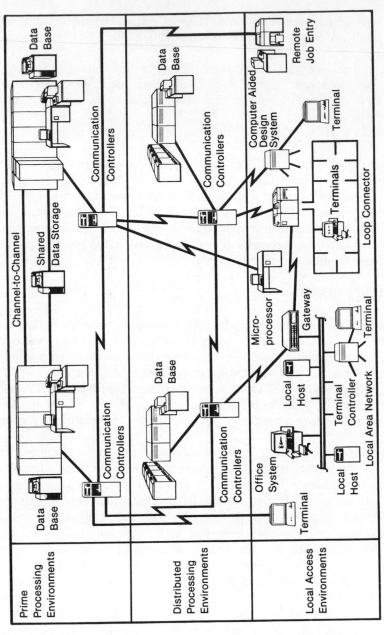

HORIZONTAL INTEGRATION

Fig. 9.1 Three-tier computer network architecture.

on. Consumer's risk or CLs are stated as decimals, frequently 0.05, and can range from 0.01 or less to 0.10.

MIL-STD-105D provides precalculated values of accept/reject quantities for various lot sizes for different AQLs and is widely used as an acceptance standard for inspection acceptances. It also contains tables of single- and multiple-sample data for use when more or less rigorous inspection may be appropriate.

Specifications will normally take precedence over drawings where conflicts exist, and contracts, if properly drawn, will state the precedence of the documents. There are many rules of interpretation that a court might apply in determining the intention of the parties and the precedence of documents. The best course is that the contract be clear and unambiguous on these matters. Such contract provisions will govern, notwithstanding that the parties may have selected an order of precedence different from that which is usual or customary.

A typical tier of documents in descending order of precedence is:

Equipment- or component-unique specifications
Basic design or data sheets
Process requirements
Project, equipment, or component drawings
Standard specifications
Standard drawings
Reference drawings
Industrial standards

9.3 GRAPHIC STANDARDS

9.3.1 Drawing Sizes

Standard sizes for drawings are 8½ × 11 inches (outside sheet dimensions) or multiples thereof:

8½ × 11 inches	11 × 17 inches
17 × 22 inches	22 × 34 inches
34 × 44 inches	

This arrangement of sizes permits folding into letter size (8½ × 11 inches) for convenient mailing or filing in standard office files.

Drawing sheets using preprinted borders are preferred to hand-drawn borders and are typically available from engineering and drafting supply firms in the sizes noted; in many cases sheets with preprinted title blocks bearing the company name, standard numbering systems, and so forth, are stocked.

9.3.2 Scales and Materials

Scale requirements vary widely depending on the type and purpose of the drawing. Scales of ³⁄₃₂ inch and ³⁄₁₆ inch equal one foot should be avoided where possible.

Since most drawings are reproduced, transparent drawing sheet material is typically used. Vellum has been the standard material used very satisfactorily for several decades. More recently, Mylar drawing sheets have come into wide use. Mylar has the advantage of permitting more repeated erasures without damaging the transparent characteristics of the sheets, permits cleaner erasures without "ghosts," and has greater resistance to tearing or creasing than vellum. The disadvantages of Mylar are its higher cost and the requirements to use special leads and erasers.

9.3.3   Drafting Practice

Some drafting practices that are economic and minimize error or confusion include:

Draw repetitive details only once and reference on other drawings.
Eliminate unnecessary views.
Make full use of industry standard details.
Use templates for drawing symbols and common shapes.
Cross-referencing must be accurate.
Independent checking is essential and should be indicated in the sign-off block.
Details of backgrounds should be omitted unless essential to a clear understanding and use of the drawing.

9.3.4   Revision Control and Check Prints

Absolute control of revisions is essential to any drafting operation. A convention of using letter designations for revisions to "in house" issues and number designations for revisions issued for external use is widely followed. Regardless of the convention adopted, total control using a formal register or control record is essential to avoid reuse of a revision with different dates and, more importantly, different information.

Check prints used for internal review or coordination need not in most cases be controlled as closely as formal revisions. One system having wide acceptance is to indicate a check mark and date in the revision block of the drawing, thus clearly identifying the purpose and timing of the print.

9.3.5   Cancelled Numbers

Cancelled drawing numbers should be withdrawn from use and not reused. The register should list the drawing number and a notation such as "Superseded by Drawing XXX," or "Cancelled." The same practice should be followed with specifications or other types of unique documentation.

9.4   PROCESS DIAGRAMS

Diagrams of several different types are used to depict and develop process designs. **Flow diagrams** may range from simple block diagrams to those carrying complete and detailed flow or material balance data. **Piping and instrumentation diagrams** permit line sizing, valve selection, and instrument and control function determination/inclusion. **Logic diagrams,** for more complex control circuits, indicate functions such as "and," "or," and "not" required for control purposes.

SYMBOL	FIRST LETTER MEASURED OR INITIATING VARIABLE	SENSING DEVICE PRIMARY ELEMENT	TRANSMITTER BLIND	TRANSMITTER INDICATING	DISPLAY DEVICE INDICATOR	RECORDER	ALARM (NOTE 5) LOW	ALARM HIGH	ALARM HIGH & LOW	CONTROL DEVICE (NOTE 3) CONTROLLER BLIND	CONTROLLER INDICATING	CONTROLLER RECORDING	CONTROL VALVE (NOTE 6)	SELF-ACTUATED VALVE	SWITCH (NOTE 3)
( )	TYPICAL SYMBOL	( )E	( )T	( )IT	( )I	( )R	( )L	( )H	( )HL	( )C	( )IC	( )RC	( )V	( )CV	( )S
A	ANALYSIS (NOTE 2)	AE	AT	AIT	AI	AR	AAL	AAH	AAHL	AC	AIC	ARC	AV		AS
B	BURNER, FIRE & FLAME	BE						BAH							
C	CONDUCTIVITY	CE	CT	CIT	CI	CR	CAL	CAH	CAHL				CV		CS
D	DENSITY	DE	DT	DIT	DI	DR	DAL	DAH	DAHL	DC	DIC	DRC	DV		DS
E	VOLTAGE (EMF)	EE	ET	EIT	EI	ER	EAL	EAH	EAHL	EC	EIC	ERC			ES
F	FLOW	FE	FT	FIT	FI	FR	FAL	FAH	FAHL	FC	FIC	FRC	FV	FCV	FS
G	GAGING	GE	GT	GIT	GI	GR	GAL	GAH	GAHL	GC	GIC	GRC			GS
H	HAND (MANUAL)									HC	HIC	HRC	HV	HCV	HS (NOTE 9)
I	CURRENT	IE	IT	IIT	II	IR	IAL	IAH	IAHL	IC	IIC	IRC			IS
J	POWER	JE	JT	JIT	JI	JR	JAL	JAH	JAHL	JC	JIC	JRC			JS
K	TIME	KE	KT	KIT	KI		KAL	KAH	KAHL	KC	KIC	KRC	KV		KS
L	LEVEL	LE	LT	LIT	LI	LR	LAL	LAH	LAHL	LC	LIC	LRC	LV	LCV	LS
M	MOISTURE	ME	MT	MIT	MI	MR	MAL	MAH	MAHL	MC	MIC	MRC	MV		MS
N	UNCLASSIFIED (NOTE 4)														
O	TORQUE	OE	OT	OIT	OI	OR	OAL	OAH	OAHL	OC	OIC	ORC	OV		OS
P	PRESSURE	PE	PT	PIT	PI	PR	PAL	PAH	PAHL	PC	PIC	PRC	PV	PSV (NOTE 7)	PS
PD	PRESSURE DIFFERENTIAL	PDE	PDT	PDIT	PDI	PDR	PDAL	PDAH	PDAHL	PDC	PDIC	PDRC	PDV	PDCV	PDS
Q	QUANTITY OR EVENT		QT	QIT	QI	QR	QAL	QAH	QAHL	QC	QIC	QRC	QV		QS
R	RADIOACTIVITY	RE	RT	RIT	RI	RR	RAL	RAH	RAHL	RC	RIC	RRC	RV		RS
S	SPEED OR FREQUENCY	SE	ST	SIT	SI	SR	SAL	SAH	SAHL	SC	SIC	SRC			SS
T	TEMPERATURE	TE	TT	TIT	TI	TR	TAL	TAH	TAHL	TC	TIC	TRC		TCV	TS
TV	TELEVISION	TVE	TVT	TVIT	TVI	TVR				TVC	TVIC	TVRC			TVS
U	MULTIVARIABLE				UI	UR	UAL	UAH	UAHL	UC	UIC	URC	UV		US
V	VISCOSITY	VE	VT	VIT	VI	VR	VAL	VAH	VAHL	VC	VIC	VRC	VV		VS
W	WEIGHT (NOTE 10)	WE	WT	WIT	WI	WR	WAL	WAH	WAHL	WC	WIC	WRC	WV		WS
X	UNCLASSIFIED (NOTE 4)														
Y	OBJECT OR MOTION SENSOR	YE	YT	YIT	YI	YR	YAL	YAH	YAHL	YC	YIC	YRC	YV		YS
Z	POSITION	ZE (NOTE 11)	ZT	ZIT	ZI	ZR	ZAL	ZAH	ZAHL	ZC	ZIC	ZRC			ZS

NOTE:

1. THE INSTRUMENT LEGEND IS BASED ON ISA STANDARD S5.1 1975.

2. THE LETTER "A" IS USED FOR ALL ANALYSIS VARIABLES. TERMS ARE PLACED OUTSIDE THE INSTRUMENT CIRCLE OF A LOOP TO DENOTE THE SPECIFIC VARIABLES. SOME EXAMPLES ARE:

CO — CARBON MONOXIDE
COMB — COMBUSTIBLES
H_2^D — DISSOLVED HYDROGEN
H_2^G — GASEOUS HYDROGEN
Na — SODIUM
M^e — METHANE
NOx — NITROGEN OXIDES

DO — DISSOLVED OXYGEN
O_2 — GASEOUS OXYGEN
pH — PERCENT HYDROGEN
Cl_2 — CHLORINE
SMOKE — SMOKE DENSITY
SO_2 — SULPHUR DIOXIDE
TRB — TURBIDITY
TSP — TOTAL SUSPENDED PARTICULATE

3. A DEVICE THAT CONNECTS, DISCONNECTS, MODIFIES OR TRANSFERS ONE OR MORE CIRCUITS MAY BE EITHER A SWITCH, A RELAY, OR AN ON-OFF CONTROLLER, DEPENDING ON THE APPLICATION.

• A SWITCH, IF IT IS ACTUATED BY HAND OR THE DEVICE IS USED FOR ALARM, PILOT LIGHT, SELECTION, INTERLOCK, OR SAFETY.
• A CONTROLLER IF THE DEVICE IS USED FOR NORMAL ON-OFF OPERATING CONTROL SUCH AS A SIMPLE HEATING THERMOSTAT.
• THE LETTERS H AND L ARE ADDED TO THE MEASURED VARIABLES FOR HIGH AND LOW RESPECTIVELY, LETTER AS FOR ALARMS (LSH, PSL, ETC.).
• A CONTROL OR SENSING DEVICE HAVING A DISPLAY FUNCTION SHOULD HAVE THE APPROPRIATE DISPLAY LETTERS ADDED AFTER THE MEASURED VARIABLE DESIGNATION, E.G. AIC DESIGNATES ANALYSIS INDICATING CONTROL STATION.

4. AN UNCLASSIFIED LETTER MAY BE USED FOR UNLISTED MEANINGS THAT WILL BE USED REPETITIVELY ON A PARTICULAR PROJECT. THE MEANINGS WILL BE DEFINED ONLY ONCE FOR THAT PROJECT AND HAVE ONE MEANING AS THE FIRST LETTER AND ANOTHER SINGLE MEANING AS THE SUCCEEDING LETTER.

5. HIGH-HIGH ALARMS HAVE ( ) AHH AND LOW-LOW ALARMS HAVE ( ) ALL IN INSTRUMENT CIRCLE E.G. LAHH — DESIGNATES HIGH-HIGH LEVEL ALARM LALL — DESIGNATES LOW-LOW LEVEL ALARM

6. VALVES:
• IF A DEVICE MANIPULATES A FLUID PROCESS STREAM AND IS NOT A MANUALLY ACTUATED ON-OFF BLOCK VALVE, IT SHALL BE DESIGNATED AS A CONTROL VALVE.
• A HAND CONTROL VALVE HCV IS A MANUALLY ACTUATED VALVE THAT EITHER MODULATES(THROTTLES) A PROCESS STREAM OR IS USED AS AN INSTRUMENT DEVICE.
  MOTORIZED VALVES ARE DESIGNATED THE SAME AS OTHER CONTROL VALVES, E.G. FV, PV, HCV, HV ETC.
  AN ON-OFF VALVE REMOTELY CONTROLLED BY A HAND SWITCH IS DESIGNATED AS A HAND VALVE HV.

7. THE DESIGNATION PSV APPLIES TO ALL VALVES INTENDED TO PROTECT AGAINST EMERGENCY PRESSURE CONDITIONS. RUPTURE DISCS SHALL BE DESIGNATED PSE.

8. USE OF MODIFYING TERMS HIGH, LOW, AND MIDDLE OR INTERMEDIATE SHALL CORRESPOND TO VALUES OF THE MEASURED VARIABLE, NOT OF THE SIGNAL, UNLESS OTHERWISE NOTED.

9. SPECIAL SWITCH CONTROLS SHALL BE DENOTED IN POSITION C OF SWITCH SYMBOLS.
JOG — JOGGLE CONTROLS
KEY — KEY-LOCKED SWITCHES

10. LOAD AND PRESSURE CELLS ARE USUALLY DENOTED "WE".

11. DENOTES STRAIN GAUGE.

Fig. 9.2 Selected piping and instrumentation diagram legend.

449

9.4.1   Piping and Instrumentation Diagrams

Piping and instrumentation diagrams (P&IDs) utilize a wide variety of symbols and alphabetic codes to describe functions and are relatively standardized. A selection of the more common standard symbols and a portion of a typical P&ID are shown in Figs. 9.2 through 9.4.

9.4.2   Logic Diagrams

Logic diagrams establish the requirements for signals (generally electrical) to cause resultant actions. They utilize standardized symbolism and frequently are used in conjunction with P&IDs to define the control logic required. Examples of some logic diagrams are shown in Figs. 9.5a through 9.5c.

9.5   ELECTRICAL DIAGRAMS

9.5.1   One-Line Diagrams

Power generation, transmission, and distribution systems are frequently presented in a one-line format sometimes called **single-line diagrams.** The term **"riser diagram"** is sometimes used for power distribution systems in commercial or industrial buildings. These diagrams are analogous to a mechanical flow or P&ID and, by use of symbols and single lines, show the generators, transformers, buses, switchgear, circuit breakers, disconnect switches, and so on, and their interconnections for all or a portion of an electrical power system. Equipment ratings are usually also given. Instrument transformers, meters, and protective relays are frequently shown also. Standard device numbers are used where appropriate. Fig. 9.6 shows a typical single line diagram

9.5.2   Schematic Diagrams

Schematic diagrams, sometimes called **elementary diagrams,** are prepared for power systems, or portions thereof, and for control and instrumentation circuits. These diagrams are in three-line or two-line format and show individual items of equipment, devices within equipment, their coils, contacts, windings, terminals, and so on, and each connection (wire, cable, or bus) between equipment or devices. Schematics can be interpreted to indicate the function of the circuits. They are used in the preparation of wiring diagrams for the internals of individual items of equipment and connection diagrams for field wiring to equipment. Fig. 9.7 shows a typical schematic diagram.

9.5.3   Wiring Diagrams

Wiring diagrams are based on the schematic diagrams and are prepared to permit shop wiring of individual items of equipment. They show each device terminal, terminal point for external connection, and the wires interconnecting these terminals.

VALVES

GATE　GLOBE　NEEDLE　CHECK　　PLUG　BUTTERFLY　ANGLE　　BALL　　3 WAY

STATUS

LC　　NO　　NC

OPEN　CLOSED　LOCKED　NORMALLY　NORMALLY
　　　　　　CLOSED　OPEN　　CLOSED

ACTUATORS

PNEUMATIC　ROTARY　PNEUMATIC　HYDRAULIC
DIAPHRAGM　MOTOR　CYLINDER　CYLINDER
　　　　　　　　　　SINGLE　DOUBLE
　　　　　　　　　　ACTING　ACTING

PROCESS PIPING AND FITTINGS

REDUCERS　　CAPS

CONCENTRIC　　SCREWED

MAIN LINE　SECONDARY　ECCENTRIC　WELDED　　　BLIND　LOOP　VENTS　SIGHT
　　　　　　　　　　　　　　　　　　　　　　FLANGE　SEAL　　　　DRAIN

SIGNALS AND LINES

CONNECTION TO　　PNEUMATIC OR　　ELECTRICAL　　　HYDRAULIC
PROCESS OR　　　AIR SUPPLY
MECHANICAL LINK

EQUIPMENT

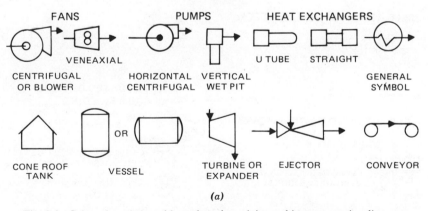

FANS　　　　　PUMPS　　　HEAT EXCHANGERS

VENEAXIAL　　　　　　U TUBE　STRAIGHT

CENTRIFUGAL　HORIZONTAL　VERTICAL　　　　　　　　GENERAL
OR BLOWER　CENTRIFUGAL　WET PIT　　　　　　　　SYMBOL

OR

CONE ROOF　　VESSEL　　TURBINE OR　EJECTOR　　CONVEYOR
TANK　　　　　　　　　EXPANDER

(a)

Fig. 9.3　Selected symbols and legend used on piping and instrumentation diagrams.

GENERAL INSTRUMENTS

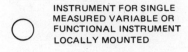 INSTRUMENT FOR SINGLE MEASURED VARIABLE OR FUNCTIONAL INSTRUMENT LOCALLY MOUNTED

 INSTRUMENT FOR MORE THAN ONE FUNCTION

TYPICAL CONNECTION ANY VARIABLE

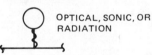 DIRECT CONNECTION

CAPILLARY FILLED SYSTEM WITH CHEMICAL SEAL

 IN LINE DEVICE

OPTICAL, SONIC, OR RADIATION

SELF-ACTUATED DEVICES—FLOW

 FLOW REGULATOR SELF-CONTAINED

SELF-ACTUATED DEVICES—LEVEL

 LEVEL REGULATOR WITH MECHANICAL LINKAGE

SELF-ACTUATED DEVICES— PRESSURE

 PRESSURE REDUCING REGULATOR, SELF-CONTAINED

 PRESSURE REDUCING REGULATOR WITH EXTERNAL PRESSURE TAP

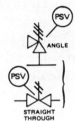 PRESSURE RELIEF OR SAFETY VALVE, SPRING OR WEIGHT LOADED, OR WITH INTEGRAL PILOT

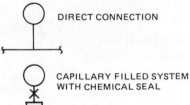 PRESSURE RELIEF OR SAFETY VALVE, ANGLE PATTERN, TRIPPED BY INTEGRAL SOLENOID

 RUPTURE DISK OR SAFETY HEAD FOR PRESSURE RELIEF

SELF-ACTUATED DEVICES— TEMPERATURE

 TEMPERATURE REGULATOR, FILLED-SYSTEM TYPE

 SELF ACTUATED TEMPERATURE REGULATOR

(b)

Fig. 9.3 (*Continued*)

452

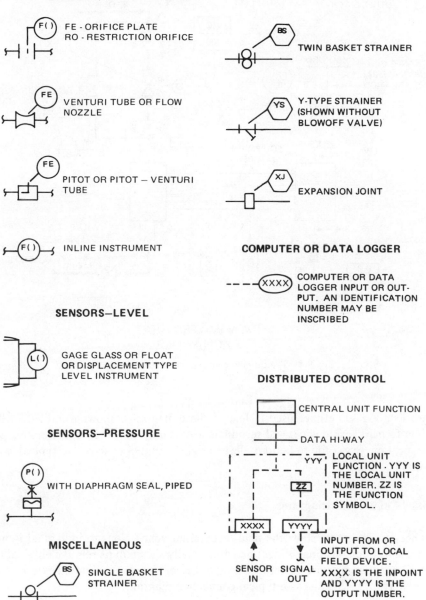

SENSORS—FLOW

FE - ORIFICE PLATE
RO - RESTRICTION ORIFICE

VENTURI TUBE OR FLOW NOZZLE

PITOT OR PITOT — VENTURI TUBE

INLINE INSTRUMENT

TWIN BASKET STRAINER

Y-TYPE STRAINER (SHOWN WITHOUT BLOWOFF VALVE)

EXPANSION JOINT

COMPUTER OR DATA LOGGER

COMPUTER OR DATA LOGGER INPUT OR OUTPUT. AN IDENTIFICATION NUMBER MAY BE INSCRIBED

SENSORS—LEVEL

GAGE GLASS OR FLOAT OR DISPLACEMENT TYPE LEVEL INSTRUMENT

DISTRIBUTED CONTROL

CENTRAL UNIT FUNCTION

DATA HI-WAY

LOCAL UNIT FUNCTION . YYY IS THE LOCAL UNIT NUMBER. ZZ IS THE FUNCTION SYMBOL.

SENSORS—PRESSURE

WITH DIAPHRAGM SEAL, PIPED

MISCELLANEOUS

SINGLE BASKET STRAINER

INPUT FROM OR OUTPUT TO LOCAL FIELD DEVICE. XXXX IS THE INPOINT AND YYYY IS THE OUTPUT NUMBER.

SENSOR IN   SIGNAL OUT

(c)

Fig. 9.3   (*Continued*)

453

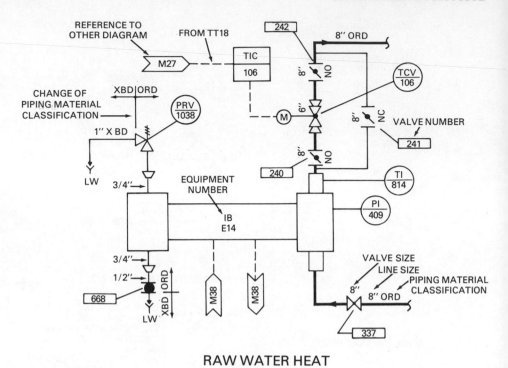

**RAW WATER HEAT
EXCHANGER**

Fig. 9.4 Typical piping and instrumentation diagram.

Wire lists are sometimes used in lieu of these diagrams, showing individual wires in line form. These devices and terminals are usually shown in their relative physical location (e.g., top-to-bottom, left-to-right). Fig. 9.8 shows a typical wiring diagram.

9.5.4   Connection Diagrams

Connection diagrams show the field or external connections to individual items of equipment. Each terminal is identified and usually shown in its relative physical location. Connection lists are sometimes used in lieu of showing individual external wires in line form. Fig. 9.9 shows a typical connection diagram.

9.5.5   Electrical Devices

Various **electrical devices,** including circuit breakers, relays, and switches have been given standardized device numbers for use on electrical single-line, schematic, and wiring diagrams. These devices are listed in ANSI C37.2-1979, and some of the more commonly used ones are shown in Table 9.1. A more complete alphabetical listing is given in Table 5.10. These device numbers are typically used with suffixes

Function	Symbol	Definition
MANUAL INPUT		Momentary hand switch input to logic
		Maintained hand switch input to logic
AND		Output exists only when all inputs are present.
OR		Output exists only when one or more inputs are present.
NOT		Output exists only when input is not present.
ON DELAY		Output exists only when input has been continuously present for a preset time and remains present.
OFF DELAY (TIMED MEMORY)		Output exists only when input is present and for a preset time after the input is not present.
MEMORY		Set output exists when set input is present and continues until the reset input is present. Reset output exists only when set output is not present.
COINCIDENCE MATRIX		Output exists only when at least A out of B inputs are present.
LOW BISTABLE		Digital output exists only when analog input is lower than set point.
HIGH BISTABLE		Digital output exists only when analog input is higher than set point.
ISOLATION		Output is electrically isolated from input.
TEST DEVICE		Test signal can be inserted manually in place of normal signal.
LIGHT		RED—Operating, flowing, or increasing GREEN—Not operating, not flowing, or decreasing AMBER—Automatic, standby, or intermediate WHITE—Manual or protective trip BLUE—
ANNUNCIATOR		Input to annunciator
COMPUTER		Input to computer
CONTINUATION		Logic continuation

1. Logic symbols represent system functions and do not necessarily duplicate circuit arrangement or devices. System control logic diagrams do not inherently imply energized, de-energized, or other circuit operation states.

2. Process equipment will change state when a change is initiated, and will remain in that state until a change to another state is initiated.

3. Process equipment will remain in, or return to, the original state after a loss and restoration of power, unless otherwise noted.

4. Inherent equipment interlocks such as circuit breaker trip free and reversing starter cross interlocks are not shown.

5. Some protection actions are shown also as start permissives. Trip free design prevents equipment operation when a protection action exists, even if a start permissive is not provided.

6. Final instrument set points are shown elsewhere. Set points shown on system control logic diagrams are approximate.

7. See electrical drawings for details of equipment electrical overcurrent, short circuit, and differential protection and space heaters.

8. The memory, reset, and start permissive logic associated with the operation of electrical protection devices is not shown. Electrical auxiliary system breakers are reset by operation of the control room switch to trip. Mechanical auxiliary system circuits are reset by operation of a switch at the switchgear or motor control center.

9. The test control switches at the switchgear which function only when a circuit breaker is in the test position are not shown.

10. All circuit controls, except interlocks with other equipment, function when a circuit breaker is in the test position to allow circuit testing.

11. The logic to show that valve and damper position lights are both on when the equipment is in an intermediate position is not shown.

12. Limit and torque switches to stop valve and damper motor actuators at the end of travel are not shown in the logic. The valve type and required actions will be noted on the diagram when available.

ABBREVIATIONS

C01	—	Unit control panel
C02	—	Auxiliary control panel
C03	—	Hot shutdown panel
L	—	Local to controlled equipment
MCC	—	Motor control center
SWGR	—	Switchgear

(a)

Fig. 9.5 a Logic diagrams legend.

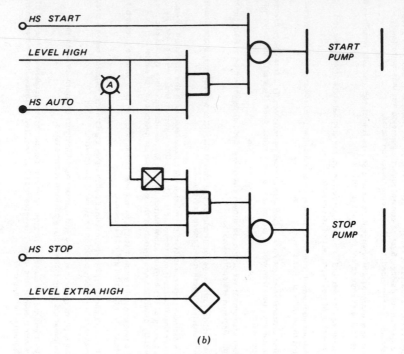

(b)

Fig. 9.5 b   Typical logic diagram.

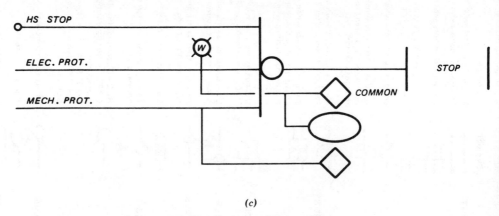

(c)

Fig. 9.5 c   Typical logic diagram.

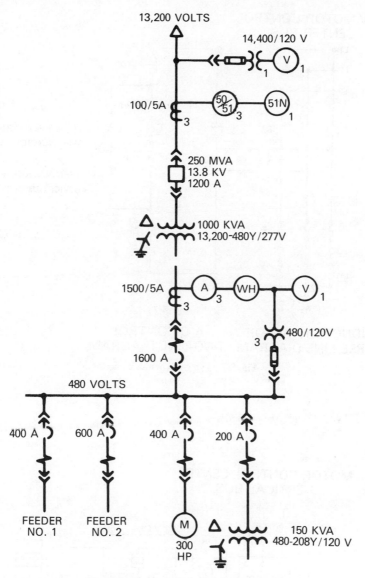

Fig. 9.6 Typical single-line diagram.

to indicate auxiliary devices, condition, location, and so forth, and may be given a unique identification number as well. Because of this, electrical drawings typically carry legends or notes to assist the user in understanding the specific meaning intended.

9.5.6 Selected ANSI Device Numbers

Table 9.1 shows some of the more commonly used ANSI device numbers.

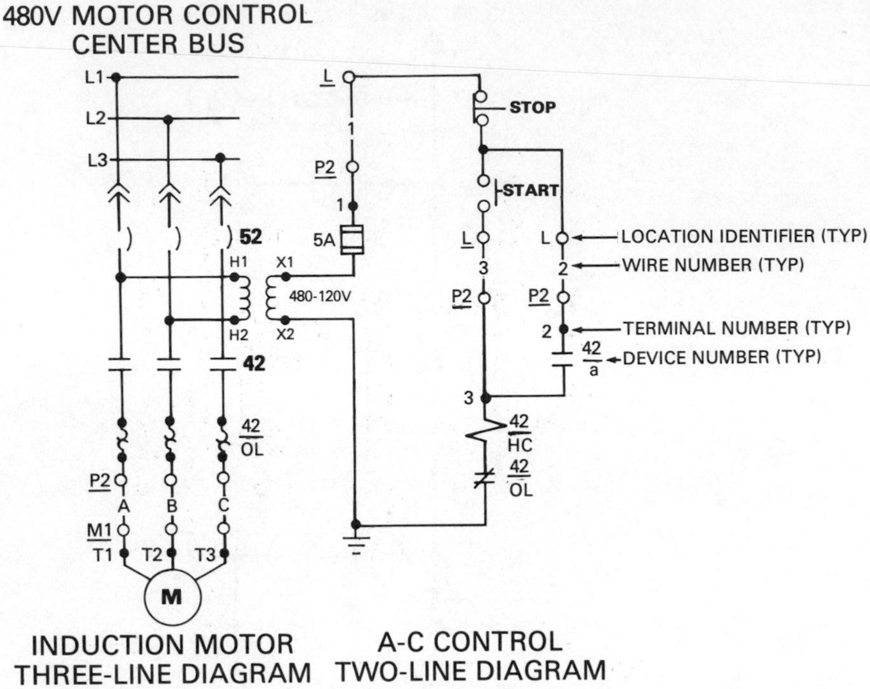

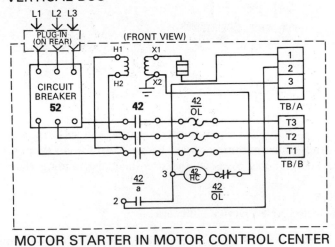

480V MOTOR CONTROL CENTER BUS

L1
L2
L3

52

H1 X1
480-120V
H2 X2

42

42
OL

P2
A   B   C

M1
T1 T2 T3

M

INDUCTION MOTOR THREE-LINE DIAGRAM

L
P2
1
5A

STOP

START

L L ← LOCATION IDENTIFIER (TYP)
3 2 ← WIRE NUMBER (TYP)
P2 P2
2 ← TERMINAL NUMBER (TYP)
42/a ← DEVICE NUMBER (TYP)

3
42/HC
42/OL

A-C CONTROL TWO-LINE DIAGRAM

Fig. 9.7 Typical schematic diagram.

MOTOR CONTROL CENTER VERTICAL BUS

L1 L2 L3

PLUG-IN (ON REAR)

(FRONT VIEW)

H1   X1
H2   X2

CIRCUIT BREAKER
52

42

42/OL

1
2
3
TB/A

T3
T2
T1
TB/B

42/a
3   42/HC   42/OL
2

MOTOR STARTER IN MOTOR CONTROL CENTER

Fig. 9.8 Typical wiring diagram.

458

Table 9.1 Selected ANSI Device Numbers

Device Number	Function
1.	Master element (initiating device)
2.	Time-delay starting or closing relay
3.	Checking or interlocking relay
20.	Electrically operated valve
21.	Distance relay
23.	Temperature control device
25.	Synchronizing or synchronism-check device
27.	Undervoltage relay
30.	Annunciator relay
32.	Directional power relay
33.	Position switch
37.	Undercurrent or underpower relay
40.	Field relay
41.	Field circuit breaker
42.	Running circuit breaker (motor starter contactors)
43.	Manual transfer or selective device
44.	Unit sequence starting relay
46.	Reverse-phase or phase balance current relay
47.	Phase sequence voltage relay
48.	Incomplete sequence relay
49.	Machine or transformer thermal relay
50.	Instantaneous overcurrent or rate-of-rise relay
51.	Ac time overcurrent relay
52.	Ac circuit breaker
53.	Exciter or dc generator relay
56.	Field application relay
59.	Overvoltage relay
62.	Time-delay stopping or opening relay
63.	Pressure switch
64.	Ground detector relay
67.	Ac directional overcurrent relay
69.	Permissive control
71.	Level switch
72.	Dc circuit breaker
74.	Alarm relay
76.	Dc overcurrent relay
78.	Phase-angle measuring or out-of-step protective relay
79.	Ac reclosing relay
80.	Flow switch
81.	Frequency relay
82.	Dc reclosing relay
83.	Automatic selective control or transfer relay
85.	Carrier or pilot-wire receiver relay
86.	Lockout relay
87.	Differential protective relay
90.	Regulating device
92.	Voltage and power directional relay

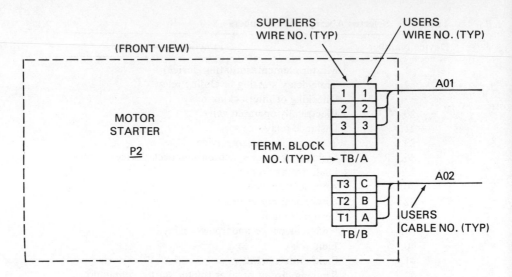

MOTOR STARTER IN MOTOR CONTROL CENTER

Fig. 9.9 Typical connection diagram.

9.5.7   Selected Suffix Letters

Suffix letters are used with device numbers for various purposes. In order to prevent possible conflict, any suffix letter used singly, or any combination of letters, denotes only one word or meaning in individual equipment. Suffix letters generally form part of the device function designation and thus are written directly behind the device number, such as 23X, 90V, or 52BT. Some selected suffix letters are shown in Tables 9.2 and 9.3.

Table 9.2   Separate Auxiliary Devices

Suffix	Device
C	Closing relay or contactor
CL	Auxiliary relay, closed (energized when main device is in closed position)
CS	Control Switch
D	"Down" position switch relay
L	Lowering relay
O	Opening relay or contactor
OP	Auxiliary relay, open (energized when main device is in open position)
PB	Push button
R	Raising relay
U	"Up" position switch relay
X	Auxiliary relay
Y	Auxiliary relay
Z	Auxiliary relay

Table 9.3   Condition or Electrical Quantity

Suffix	Denotes
A	Air, or amperes
C	Current
F	Frequency, or flow
L	Level, or liquid
P	Power, or pressure
PF	Power factor
S	Speed
T	Temperature
V	Voltage, volts, or vacuum
VAR	Reactive power
W	Water, or watts

Location of the main device in the circuit, or the type of circuit in which the device is used, or the type of circuit or apparatus with which it is associated, can be shown by the suffixes in Table 9.4.

In addition to the suffix letters, other letters are used to denote parts of the main device or other features, characteristics, or conditions. These are generally shown directly below the device number, such as $\frac{20}{2S}$ or $\frac{52}{a}$. A partial listing of these letters is shown in Table 9.5.

Table 9.4   Typical Location or Circuit Identification

Suffix	Denotes
A	Alarm or auxiliary power
AC	Alternating current
BK	Brake
C	Capacitor, or condenser, or compensator, or carrier current
CA	Cathode
DC	Direct current
E	Exciter
F	Feeder, or field, or filament
G	Generator, or ground
M	Motor, or metering
N	Network, or neutral
P	Pump
R	Reactor, or rectifier
S	Synchronizing
T	Transformer, or test, or thyratron
TM	Telemeter
U	Unit

Table 9.5 Typical Device Parts Identification[a]

Letter	Denotes
C	Coil, or condenser, or capacitor
CC	Closing coil
HC	Holding coil
LS	Limit switch
M	Operating motor
S	Solenoid
TC	Trip coil
a	Auxiliary switch, open when the main device is in the de-energized or nonoperated position
b	Auxiliary switch, closed when the main device is in the de-energized or nonoperated position
A	Accelerating or automatic
B	Blocking, or backup
C	Close, or cold
D	Decelerating, or detonate, or down
E	Emergency
F	Failure, or forward
H	Hot, or high
HR	Hand reset
HS	High speed
IT	Inverse time
L	Left, or local, or low, or lower, or leading
M	Manual
OFF	Off
ON	On
O	Open
P	Polarizing
R	Right, or raise, or reclosing, or receiving, or remote, or reverse
S	Sending, or swing
T	Test, or trip, or trailing
TDC	Time-delay closing
TDO	Time-delay opening
U	Up

[a]Typical examples of the use of device numbers, suffixes, and other identifiers are:

$$\frac{52G}{a} = \text{Auxiliary contact, normally open, on ac generator circuit breaker.}$$

86X = Auxiliary relay to lockout relay.

87G = Differential protective relay for generator.

$$\frac{74T}{TDC} = \text{Time-delay closing contact on temperature alarm relay.}$$

9.6 REPORT WRITING

Report writing is an essential skill of the engineer and must be performed with a view toward maximizing the amount of information for the reader or user of the report. Typically, reports are structured with sections titled:

Purpose
Scope
Conclusions and Recommendations
Discussion
Appendices

Further, a portion of the report (typically the introduction) would indicate who participated in the preparation of the report, when the report was prepared, and so forth. This provides a source for clarifications and also permits the reader to determine the validity of the report based on the credentials or reputation of the author(s) and contributors.

The **purpose** sets forth why the report was written, the factors that led to its writing, etc. The **scope** describes the extent of the report, limitations on its coverage, and so on. This serves to put the background of the report in clear perspective so that the reader will understand fully not only the purpose of the report but any limitations imposed.

Conclusions and recommendations may be one section or separate sections, but should be placed toward the front of the report, so that a busy person can easily locate this section and readily determine the findings of the report. Each conclusion must be supported by material elsewhere in the report. Conclusions are rarely useful without recommendations. Therefore, typically, recommendations will be included and are essential to a full use of the report.

Sections on **discussion** and **appendices** are self-evident and become a convenient way of dividing the material between that found in the course of the investigation (Discussion) and Appendices, which typically are secondary or reference material of limited interest to most users of the report, although of possible interest to some.

For extensive or lengthy reports a separate **executive summary** is often prepared which, although placing major emphasis on the conclusions and recommendations, provides an overview of the entire report.

Reports should be written in clear concise language with emphasis on short sentences and clarity of thought. Ponderous or obscure language should be avoided and technical or complex material should be put into lay terms to the maximum extent possible. Wherever possible, abstractions should be avoided and the report made as concrete and specific as possible.

Graphical material is preferred over tabular material, although tabular material can be placed in the discussion or appendices if necessary.

9.7 BOOK LIST/INFORMATION SOURCES

Although the specific requirements for references will vary depending on the type of engineering being performed, certain volumes are useful as a core library of ready reference data. The titles cited have proven particularly useful to the author.

Webster's New Collegiate Dictionary, G. & C. Merriam.
Roget's International Thesaurus, Crowell.
Harbrace College Handbook, Hodges and Whitten, Harcourt, Brace, Jovanovich.
World Almanac, Newspaper Enterprise Assoc., Inc.
The Wiley Engineer's Desk Reference (successor volume to the *Engineers' Manual*), Wiley.
Handbook of Engineering Fundamentals, Eshbach, Wiley.
The Way Things Work, Simon and Schuster.
Engineering Economy, Grant, Ireson, and Leavenworth, Ronald Press.
The Procedure Handbook of Arc Welding, The Lincoln Electric Co.
Materials and Processes in Manufacturing, DeGarmo, Macmillian.
Manual of Steel Construction, American Institute of Steel Construction.

Some sources of **engineering information** include:[3]

Science and Engineering Literature: A Guide to Reference Sources, Robert H. Molinowsky, Libraries Unlimited, Inc., 1980.

A master index listing reference sources by particular fields of engineering.

Guide to Basic Information Sources in Engineering, Ellis Mount, New York: Wiley, 1976, 196 pp., index. (A Halsted Press Book; Information Resources Series.)

A small book for the engineering student and researcher covering in four broad categories: Technical Literature—What It Is, Where to Find It; Books; Periodicals and Technical Reports; and Other Sources of Information. Under each category there are entries for bibliographies, dictionaries, encyclopedias, handbooks, guides to literature, histories, and so on.

Use of Engineering Literature, K. W. Mildren (Ed.), Woburn, Mass.: Butterworths, 1976, 621 pp., illus. index. (Information Sources for Research and Development.)

This handbook represents a survey of the fields of engineering, covering electronics, communications, control engineering, aeronautics and astronautics, chemical engineering, production engineering, and soil engineering. Each of the chapters includes information on classification and indexing, journals, conferences and theses, translations, reports, patents, standards, product information, and selected coverage of publications of government and international organizations.

Applied Science and Technology Index, New York: H. W. Wilson, 1913– (monthly except August) with quarterly and annual cumulations. (From 1913 to 1957 known as *Industrial Arts Index.*)

A subject index to about 225 English language periodicals, covering both the theoretical sciences and their engineering applications. It includes pure physics, chemistry and geology, mathematics, metallurgy, and computer science from the major scientific periodicals in their fields. It also provides substantial coverage of science applications, that is, the various subfields of engineering.

Bibliographic Guide to Technology, 1975, annual.

Twelve volumes include the library's 185,000-volume collection of books, journals, films, technical reports, and unpublished manuscripts in all fields of engineering. The supplements, now called *Bibliographic Guide to Technology,* appear on an annual basis, making this a prime source of current bibliography. These data extend also into the geoscience subfields of geology and geophysics.

Engineering Index, New York: Engineering Index, Inc., 1884– (monthly with annual cumulation).

An English-language abstracting service in engineering, including all of the subdivisions. It indexes more than 2000 journals and all pertinent reports, symposium papers, patents, books, and miscellaneous serials published in 20 or more languages. Indicative abstracts are arranged in classified subject order. Each entry provides a complete citation followed by an abstract. Engineering Index, Inc. also maintains a machine-readable data base with partial funding by the National Science Foundation. Its services, in addition to the printed *Engineering Index* include:

COMPENDEX, a computerized magnetic tape version of *Engineering Index,* issued monthly, well in advance of the printed versions.

ENGINEERING INDEX MICROFILM EDITION.

CARD-A-LERT, a weekly current awareness service issued on 3 × 5-inch cards.

SHE: Subject Headings for Engineering, New York: Engineering Index, Inc., 1972. 149 pp.

An alphabetical list of terms currently in use by EI technical editorial specialists as a controlled vocabulary for indexing transdisciplinary literature of engineering and related sciences. Includes many cross-references and scope notes.

Encyclopedia of Engineering Materials and Processes, H. R. Clauser (Ed.), New York: Reinhold, 1963, 787 pp., diagrams, tables, graphs.

Materials treated range from ABS plastics to zirconium; processes discussed include heat treating, welding, electroplating, and tanning. Excellent cross-references connect the 300 articles, written by more than 200 contributors.

9.8 EXPERT WITNESS

The engineer is frequently called upon to testify or give depositions as an **expert witness.** Testimony occurs in a courtroom with the usual complement of a judge, perhaps a jury, counsel, the defendant, plaintiff, and so on. **Depositions** are typically sworn statements with legal counsel present, taken by a legal stenographer at a location distant from the court. Depositions are given at a location generally of convenience to the deposer, that is, the person testifying. The deposition is given under oath and counsel for both parties are usually present. In every aspect the deposition is identical to testimony given in the court and constitutes admissible testimony.

Typically, the person testifying will initially be asked to establish his credentials. This takes the form of a series of questions that establish the educational background, experience levels, and other aspects of the training and experience of the expert—permitting the testimony to be accepted as unique and expert and outside the sphere of the layperson. In most cases, the testimony of the expert witness is not contested and the testimony of the witness is used to provide an improved understanding of an engineering matter for the benefit of the court. For example, a witness might testify, in the case of a failure of a concrete structure, on the rate at which concrete acquires strength during the curing process; present test data indicating a strength of a certain member for a vehicular component where a case of product liability was involved and so forth.

The testimony of the witness needs to be clear and concise and in lay terms as much as possible. Although the expert witness is engaged because of his ability to understand and evaluate the problem, the testimony will be used either by a judge who may not be familiar with the technological details, or a jury who are themselves not experts in the field. The engineer's testimony, therefore, needs to be in a language that the layperson can readily understand. The use of similes, parallels, and the drawing of comparisons are very useful ways in which the testimony can be made realistic and understandable to the jury. The use of tabular data should be minimized, but charts, visual aids, and models should be maximized, as these provide a very ready way in which the jury can understand the item being presented.

The testimony of the expert witness must be conducted with suitable decorum, and absolute integrity is essential in the presentation of material to the court. Since the layperson does not, in many cases, understand concepts such as design margins and statistical analysis, simplifications that may be necessary should be carefully chosen to avoid distorting the basic information. The maintenance of integrity is absolutely essential for the purpose of establishing and maintaining the credibility of the testimony. Typically, the expert witness will be counseled not to volunteer information, but merely to answer the questions asked of him or her by the attorneys. Normally, the testimony of the witness is carefully reviewed before the actual deposition is taken to assure the attorney understands fully the scope and implications of the material and thus can properly explore all aspects through his or her line of questioning.

9.9   REGISTRATION

Registration, although not a legal requirement for all engineers working for corporations, is nevertheless considered essential for professional growth. In the case of engineers in consulting practice (as distinguished from corporate practice), registration is virtually mandatory. In many jurisdictions the engineer in responsible charge of the work must be registered in that jurisdiction (normally state), and the drawings and sometimes calculations supporting them must be sealed (have the engineer's registration stamp affixed or drawings embossed).

Registration authorities are normally concerned with both technical competence and moral character of applicants. The technical competence is usually established by a two-part examination system covering (1) engineering theory, mathematics, physics, and so on; and (2) a second portion, typically a few years later, covering application of engineering knowledge to practical problems. Usually the law requires that during this period actual engineering work be performed under the supervision of a registered engineer. In addition, references active in the engineering field are required to attest to the moral character of the applicant.

In general, application for registration should be made as early in one's career as possible, while the theory remains fresh.

Registration supports the status as expert witness—an increasingly important aspect of professional engineering.

9.10 ETHICS

The engineering profession has, over the years, developed a code of ethics intended to assure integrity and competence in the conduct of the engineering profession. A copy of this standard and its detailed guidelines are shown in Fig. 9.10.

9.11 COMMERCIAL CONSIDERATIONS

A prime concern in operating an engineering practice is the basis on which the engineering services will be rendered. Broadly, the contractual bases available are lump sum, cost plus fee (either fixed or percentage), and unit priced.

9.11.1 Lump Sum

Lump sum contractual arrangements are preferred by clients because the client assumes no risks for overruns on the part of the engineer. Since the engineer assumes all risk, he therefore must take great pains to carefully describe the **scope** of his activities such that the lump sum offering can be properly defined and **limited.** With a lump sum arrangement, **scope changes** are the basis for "extras," which require negotiation and can be acrimonious. Nevertheless, where indicated, these scope changes should be pursued since their cost is directly a "bottom line" cost to the engineer. Particular care should be given to time delays, as these can have serious effects on lump sum work, usually causing cost overruns.

9.11.2 Cost Plus Fee

Cost-plus-fee arrangements are preferred by the engineer when scope cannot be well defined or subject to changes, when schedules are indeterminant, or when new technologies are being utilized for the development of the engineering work. In a cost-plus-fee arrangement the purchaser, that is, the client, assumes the risk for overruns but gains the benefit of underruns. Since it is a separate item, the fee represents a secure form of profit for the engineer. However, the fee may not be totally net profit since entertainment and other costs may be charged against it. Obviously, for those cases where some form of reimbursable contract is entered into, the client will typically expect to involve himself or herself to a much greater extent in those engineering decisions made as the work proceeds on a day-by-day basis. This has the potential disadvantage of causing a loss of much of the engineer's freedom of action since the

CODE OF ETHICS OF ENGINEERS

THE FUNDAMENTAL PRINCIPLES

Engineers uphold and advance the integrity, honor and dignity of the engineering profession by:

I.   using their knowledge and skill for the enhancement of human welfare;

II. being honest and impartial, and serving with fidelity the public, their employers and clients;

III. striving to increase the competence and prestige of the engineering profession; and

IV. supporting the professional and technical societies of their disciplines.

THE FUNDAMENTAL CANONS

1. Engineers shall hold paramount the safety, health and welfare of the public in the performance of their professional duties.

2. Engineers shall perform services only in the areas of their competence.

3. Engineers shall issue public statements only in an objective and truthful manner.

4. Engineers shall act in professional matters for each employer or client as faithful agents or trustees, and shall avoid conflicts of interest.

5. Engineers shall build their professional reputation on the merit of their services and shall not compete unfairly with others.

6. Engineers shall act in such a manner as to uphold and enhance the honor, integrity and dignity of the profession.

7. Engineers shall continue their professional development throughout their careers and shall provide opportunities for the professional development of those engineers under their supervision.

Approved by the Board of Directors, October 5, 1977

Fig. 9.10 Code of ethics.

Engineers' Council for Professional Development

SUGGESTED
GUIDELINES FOR USE WITH
THE FUNDAMENTAL CANONS OF ETHICS

Engineers shall hold paramount the safety, health and welfare of the public in the performance of their professional duties.

a. Engineers shall recognize that the lives, safety, health and welfare of the general public are dependent upon engineering judgments, decisions and practices incorporated into structures, machines, products, processes and devices.

b. Engineers shall not approve nor seal plans and/or specifications that are not of a design safe to the public health and welfare and in conformity with accepted engineering standards.

c. Should the Engineers' professional judgment be overruled under circumstances where the safety, health, and welfare of the public are endangered, the Engineers shall inform their clients or employers of the possible consequences and notify other proper authority of the situation, as may be appropriate.

(c.1) Engineers shall do whatever possible to provide published standards, test codes and quality control procedures that will enable the public to understand the degree of safety or life expectancy associated with the use of the design, products and systems for which they are responsible.

(c.2) Engineers will conduct reviews of the safety and reliability of the design, products or systems for which they are responsible before giving their approval to the plans for the design.

(c.3) Should Engineers observe conditions which they believe will endanger public safety or health, they shall inform the proper authority of the situation.

d. Should Engineers have knowledge or reason to believe that another person or firm may be in violation of any of the provisions of these Guidelines, they shall present such information to the proper authority in writing and shall cooperate with the proper authority in furnishing such further information or assistance as may be required.

(d.1) They shall advise proper authority if an adequate review of the safety and reliability of the products or systems has not been made or when the design imposes hazards to the public through its use.

(d.2) They shall withhold approval of products or systems when changes or modifications are made which would affect adversely its performance insofar as safety and reliability are concerned.

e. Engineers should seek opportunities to be of constructive service in civic affairs and work for the advancement of the safety, health and well-being of their communities.

f. Engineers should be commited to improving the environment to enhance the quality of life.

2. Engineers shall perform services only in areas of their competence.

a. Engineers shall undertake to perform engineering assignments only when qualified by education or experience in the specific technical field of engineering involved.

b. Engineers may accept an assignment requiring education or experience outside of their own fields of competence, but only to the extent that their services are restricted to those phases of the project in which they are qualified. All other phases of such project shall be performed by qualified associates, consultants, or employees.

c. Engineers shall not affix their signatures and/or seals to any engineering plan or document dealing with subject matter in which they lack competence by virtue of education or experience, nor to any such plan or document not prepared under their direct supervisory control.

3. Engineers shall issue public statements only in an objective and truthful manner.

a. Engineers shall endeavor to extend public knowledge, and to prevent misunderstandings of the achievements of engineering.

b. Engineers shall be completely objective and truthful in all professional reports, statements, or testimony. They shall include all relevant and pertinent information in such reports, statements, or testimony.

c. Engineers, when serving as expert or technical witnesses before any court, commission, or other tribunal, shall express an engineering opinion only when it is founded upon adequate knowledge of the facts in issue, upon a background of technical competence in the subject matter, and upon honest conviction of the accuracy and propriety of their testimony.

d. Engineers shall issue no statements, criticisms, nor arguments on engineering matters which are inspired or paid for by an interested party, or parties, unless they have prefaced their comments by explicitly identifying themselves, by disclosing the identities of the party or parties on whose behalf they are speaking, and by revealing the existence of any pecuniary interest they may have in the instant matters.

e. Engineers shall be dignified and modest in explaining their work and merit, and will avoid any act tending to promote their own interests at the expense of the integrity, honor and dignity of the profession.

4. Engineers shall act in professional matters for each employer or client as faithful agents or trustees, and

shall avoid conflicts of interest.

a. Engineers shall avoid all known conflicts of interest with their employers or clients and shall promptly inform their employers or clients of any business association, interests, or circumstances which could influence their judgment or the quality of their services.

b. Engineers shall not knowingly undertake any assignments which would knowingly create a potential conflict of interest between themselves and their clients or their employers.

c. Engineers shall not accept compensation, financial or otherwise, from more than one party for services on the same project, nor for services pertaining to the same project, unless the circumstances are fully disclosed to, and agreed to, by all interested parties.

d. Engineers shall not solicit nor accept financial or other valuable considerations, including free engineering designs, from material or equipment suppliers for specifying their products.

e. Engineers shall not solicit nor accept gratuities, directly or indirectly, from contractors, their agents, or other parties dealing with their clients or employers in connection with work for which they are responsible.

f. When in public service as members, advisors, or employees of a governmental body or department, Engineers shall not participate in considerations or actions with respect to services provided by them or their organization in private or product engineering practice.

g. Engineers shall not solicit nor accept an engineering contract from a governmental body on which a principal, officer or employee of their organization serves as a member.

h. When, as a result of their studies, Engineers believe a project will not be successful, they shall so advise their employer or client.

i. Engineers shall treat information coming to them in the course of their assignments as confidential, and shall not use such information as a means of making personal profit if such action is adverse to the interests of their clients, their employers, or the public.

(i.1) They will not disclose confidential information concerning the business affairs or technical processes of any present or former employer or client or bidder under evaluation, without his consent.

(i.2) They shall not reveal confidential information nor findings of any commission or board of which they are members.

(i.3) When they use designs supplied to them by clients, these designs shall not be duplicated by the Engineers for others without express permission.

(i.4) While in the employ of others, Engineers will not enter promotional efforts or negotiations for work or make arrangements for other employment as principals or to practice in connection with specific projects for which they have gained particular and specialized knowledge without the consent of all interested parties.

j. The Engineer shall act with fairness and justice to all parties when administering a construction (or other) contract.

k. Before undertaking work for others in which Engineers may make improvements, plans, designs, inventions, or other records which may justify copyrights or patents, they shall enter into a positive agreement regarding ownership.

l. Engineers shall admit and accept their own errors when proven wrong and refrain from distorting or altering the facts to justify their decisions.

m. Engineers shall not accept professional employment outside of their regular work or interest without the knowledge of their employers.

n. Engineers shall not attempt to attract an employee from another employer by false or misleading representations.

o. Engineers shall not review the work of other Engineers except with the knowledge of such Engineers, or unless the assignments/or contractual agreements for the work have been terminated.

(o.1) Engineers in governmental, industrial or educational employment are entitled to review and evaluate the work of other engineers when so required by their duties.

(o.2) Engineers in sales or industrial employment are entitled to make engineering comparisons of their products with products of other suppliers.

(o.3) Engineers in sales employment shall not offer nor give engineering consultation or designs or advice other than specifically applying to equipment, materials or systems being sold or offered for sale by them.

5. Engineers shall build their professional reputation on the merit of their services and shall not compete unfairly with others.

a. Engineers shall not pay nor offer to pay, either directly or indirectly, any commission, political contribution, or a gift, or other consideration in order to secure work, exclusive of securing salaried positions through employment agencies.

b. Engineers should negotiate contracts for professional services fairly and only on the basis of demonstrated competence and qualifications for the type of professional service required.

c. Engineers should negotiate a method and rate of compensation commensurate with the agreed upon scope of services. A meeting of the minds of the parties to the contract is essential to mutual confidence. The public interest requires that the cost of engineering services be fair and reasonable, but not the controlling consideration in selection of individuals or firms to provide these services.

(c.1) These principles shall be applied by Engineers

in obtaining the services of other professionals.

d. Engineers shall not attempt to supplant other Engineers in a particular employment after becoming aware that definite steps have been taken toward the others' employment or after they have been employed.

(d.1) They shall not solicit employment from clients who already have Engineers under contract for the same work.

(d.2) They shall not accept employment from clients who already have Engineers for the same work not yet completed or not yet paid for unless the performance or payment requirements in the contract are being litigated or the contracted Engineers' services have been terminated in writing by either party.

(d.3) In case of termination of litigation, the prospective Engineers before accepting the assignment shall advise the Engineers being terminated or involved in litigation.

e. Engineers shall not request, propose nor accept professional commissions on a contingent basis under circumstances under which their professional judgments may be compromised, or when a contingency provision is used as a device for promoting or securing a professional commission.

f. Engineers shall not falsify nor permit misrepresentation of their, or their associates', academic or professional qualifications. They shall not misrepresent nor exaggerate their degree of responsibility in or for the subject matter of prior assignments. Brochures or other presentations incident to the solicitation of employment shall not misrepresent pertinent facts concerning employers, employees, associates, joint ventures, or their past accomplishments with the intent and purpose of enhancing their qualifications and work.

g. Engineers may advertise professional services only as a means of identification and limited to the following:

(g.1) Professional cards and listings in recognized and dignified publications, provided they are consistent in size and are in a section of the publication regularly devoted to such professional cards and listings. The information displayed must be restricted to firm name, address, telephone number, appropriate symbol, names of principal participants and the fields of practice in which the firm is qualified.

(g.2) Signs on equipment, offices and at the site of projects for which they render services, limited to firm name, address, telephone number and type of services, as appropriate.

(g.3) Brochures, business cards, letterheads and other factual representations of experience, facilities, personnel and capacity to render service, providing the same are not misleading relative to the extent of participation in the projects cited and are not indiscriminately distributed.

(g.4) Listings in the classified section of telephone directories, limited to name, address, telephone number and specialties in which the firm is qualified without resorting to special or bold type.

h. Engineers may use display advertising in recognized dignified business and professional publications, providing it is factual, and relates only to engineering, is free from ostentation, contains no laudatory expressions or implication, is not misleading with respect to the Engineers' extent of participation in the services or projects described.

i. Engineers may prepare articles for the lay or technical press which are factual, dignified and free from ostentations or laudatory implications. Such articles shall not imply other than their direct participation in the work described unless credit is given to others for their share of the work.

j. Engineers may extend permission for their names to be used in commercial advertisements, such as may be published by manufacturers, contractors, material suppliers, etc., only by means of a modest dignified notation acknowledging their participation and the scope thereof in the project or product described. Such permission shall not include public endorsement of proprietary products.

k. Engineers may advertise for recruitment of personnel in appropriate publications or by special distribution. The information presented must be displayed in a dignified manner, restricted to firm name, address, telephone number, appropriate symbol, names of principal participants, the fields of practice in which the firm is qualified and factual descriptions of positions available, qualifications required and benefits available.

l. Engineers shall not enter competitions for designs for the purpose of obtaining commissions for specific projects, unless provision is made for reasonable compensation for all designs submitted.

m. Engineers shall not maliciously or falsely, directly or indirectly, injure the professional reputation, prospects, practice or employment of another engineer, nor shall they indiscriminately criticize another's work.

n. Engineers shall not undertake nor agree to perform any engineering service on a free basis, except professional services which are advisory in nature for civic, charitable, religious or non-profit organizations. When serving as members of such organizations, engineers are entitled to utilize their personal engineering knowledge in the service of these organizations.

o. Engineers shall not use equipment, supplies, laboratory nor office facilities of their employers to carry on outside private practice without consent.

p. In case of tax-free or tax-aided facilities, engineers should not use student services at less than rates of other employees of comparable competence, including fringe benefits.

6   Engineers shall act in such a manner as to uphold and enhance the honor, integrity and dignity of the profession.

a. Engineers shall not knowingly associate with nor permit the use of their names nor firm names in business ventures by any person or firm which they know, or have reason to believe, are engaging in business or professional practices of a fraudulent or dishonest nature.

b. Engineers shall not use association with non-engineers, corporations, nor partnerships as 'cloaks' for unethical acts.

7. Engineers shall continue their professional development throughout their careers, and shall provide opportunities for the professional development of those engineers under their supervision.

a. Engineers shall encourage their engineering employees to further their education.

b. Engineers should encourage their engineering employees to become registered at the earliest possible date.

c. Engineers should encourage engineering employees to attend and present papers at professional and technical society meetings.

d. Engineers should support the professional and technical societies of their disciplines.

e. Engineers shall give proper credit for engineering work to those to whom credit is due, and recognize the proprietary interests of others. Whenever possible, they shall name the person or persons who may be responsible for designs, inventions, writings or other accomplishments.

f. Engineers shall endeavor to extend the public knowledge of engineering, and shall not participate in the dissemination of untrue, unfair or exaggerated statements regarding engineering.

g. Engineers shall uphold the principle of appropriate and adequate compensation for those engaged in engineering work.

h. Engineers should assign professional engineers duties of a nature which will utilize their full training and experience insofar as possible, and delegate lesser functions to subprofessionals or to technicians.

i. Engineers shall provide prospective engineering employees with complete information on working conditions and their proposed status of employment, and after employment shall keep them informed of any changes.

Engineers' Council for Professional Development
345 East 47th Street
New York, N.Y.

client participation is so great. This involvement can also adversely affect schedule, since clients have difficulty understanding the effects of even small changes that may be requested. In the lump sum contract, for example, the engineer can ask for total freedom with no restraints or involvement by the client until the finished product is delivered, because the engineer is furnishing the entire package and assuming the financial risk if the lump sum work overruns its contractual amount. It is not possible to take this position with cost-plus-fee work.

The fee determination for either **fixed or percentage fee** can be arranged as necessary to suit the preferences of the owner. Typically, fixed fees are used for those cases where it is possible to reasonably establish the scope in a very broad way, and the fixed fee is not normally subject to renegotiation except in those cases where the scope changes have been significant, say on the order of 20% or more. The use of percentage fee is not usually favored because of the possibility of expanding the basic scope to automatically earn additional fee.

9.11.3   Unit Pricing

For some work where the scope and the extent of the engineering is either totally unpredictable or difficult to anticipate, unit pricing may instead be used. Typically, the unit pricing includes both **direct** and **indirect payroll costs, overhead (burden),** and **fee,** all normally structured on the basis of dollars per manhour. **Unit pricing** will usually establish different unit prices for different categories of work. Thus, the unit price in dollars per hour for a draftsman will differ from that for a checker, a design engineer, a supervising design engineer, and so on.

9.11.4   Cost Structure

Pricing for services typically includes (1) direct payroll (gross amounts paid employees); (2) an allowance for indirect payroll costs, which is a direct multiplier on payroll costs, ranging from 25 to 40% to include the costs for employer payroll taxes, vacation, sick leave, retirement, and so on; (3) an overhead or burden percentage on the order of 75 to 100% to cover a large variety of office expenses (e.g., office rent, furniture and fixtures, insurance, personnel recruiting, training, development and management, legal services, technical society participation, and sales costs including proposal preparation); and (4) an allowance for fee (profit) often arbitrarily set at 10 or 15%.

Typically, the overall multiplier of direct payroll (including indirect costs) will range from 2½ to 3½; this includes fee (profit) plus overhead costs as well. As a part of providing engineering services, there are frequently costs incurred for laboratory testing, transportation and travel, consultants' fees, and so on; all accrued for the benefit of the client. Typically, these costs are passed along to the client with no or a modest 10 to 15% mark-up only. Normal engineering practice does not include the addition of fees to these types of costs.

9.11.5   Fee Curves

The ASCE has published fee curves which, although intended for civil engineering, are widely used as guides in engineering consulting practices more generally. The fee curves represent total remuneration, as in the case of lump sum work, and not merely that portion of fee (profit) described earlier.

9.11.6   Contracts

For all engineering work, even of the most simple type, some form of contract is essential. The contract may range from a simple letter of agreement to a complex lengthy document, spelling out all of the details of the work, contingencies, payment schedules, unit prices, and so forth. There are some key points that should be covered in any contract, however abbreviated:

A brief description of the scope of the project
A description of the services to be provided
Basis for the service—reimbursability and fee arrangement
Overall schedule
Delays
Changes
Force majeure
Liability limitations
Termination
Payment

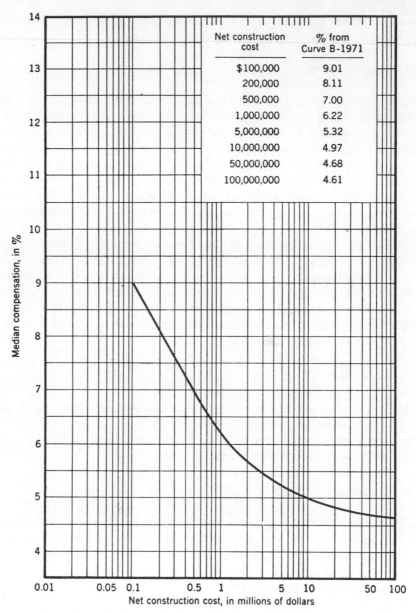

Net construction cost	% from Curve B-1971
$100,000	9.01
200,000	8.11
500,000	7.00
1,000,000	6.22
5,000,000	5.32
10,000,000	4.97
50,000,000	4.68
100,000,000	4.61

Fig. 9.11 Fee curve (average complexity). Copyright: American Society of Civil Engineers, New York, NY, Oct. 1981.[2]

9.11.7   Cost Control

Cost control for the work in progress is usually handled by carefully estimating man-hours per unit of work (e.g., per specification, per drawing). The value of the completed component parts of the work is then established, and as the percentages are worked off, the amount of manhours "earned" is established. Particular care should be taken to realistically evaluate the value of the work completed and of that remain-

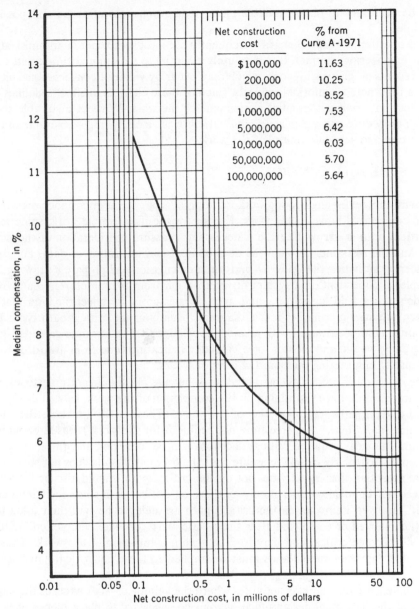

Net construction cost	% from Curve A-1971
$100,000	11.63
200,000	10.25
500,000	8.52
1,000,000	7.53
5,000,000	6.42
10,000,000	6.03
50,000,000	5.70
100,000,000	5.64

Median compensation, in %

Net construction cost, in millions of dollars

Fig. 9.12 Fee curve (above-average complexity). Copyright: American Society of Civil Engineers, New York, NY, Oct. 1981.[2]

ing. It is typical to substantially underestimate the remaining work and thus indicate higher percentage of completion than is actual. By comparing the amount of man-hours earned to the amount of manhours estimated, it is possible to determine broadly whether the work is on schedule and whether the revenue is being realized at the anticipated rate. For deviations from the plan, it is essential to analyze and determine their causes and establish a plan to get back on schedule and within bud-get. Alternatively, where the budget has been substantially overrun or underrun it may be necessary to obtain suitable agreement on changes of scope and possibly schedule.

The question of **scope changes** is frequently the most significant in administration of any engineering contract. It is absolutely essential to maintain accurate and complete records of changes in scope, which may occur by virtue of conversations, phone calls, a brief note, or informal requests. Each of these changes in effect redefines the contract and modifies the initial agreement for the work. Without a suitable scope change procedure and rigorous administration, it is virtually impossible to maintain proper cost and schedule control of the work.

9.12   SCHEDULE CONTROL

Schedule control for engineering work in progress is of critical importance regardless of the contractual basis for the work. Except for the most simple work, some form of **critical path** or **arrow diagram** is necessary to assure consideration of the interfaces and the restraints between the development and utilization of data. For most engineering activities, the ability to perform a calculation and prepare a drawing, for example, is dependent on a receipt of information from manufacturers, data from outside sources, field survey data, and so on. It is essential to reflect the interrelation of these activities in some form of a diagram, usually of the critical path type. The diagram should be arranged with a calendar scale, either an absolute calendar indicating days, weeks, and months, or a calendar indicating time from award of work or release of engineering measured in days or weeks.

The diagram, if properly constructed, will indicate (1) significant milestones, (2) those activities that are restraining on the completion of the work, as well as (3) the several critical paths for completion of the work. (These paths are those that have no or minimal "float," i.e., free time not required for the work.) Based on such a diagram, the early required critical decisions can be identified and those activities monitored particularly closely to assure that the overall schedule does not slip. Noncritical activities, that is, activities not on the critical path, should also be followed, but need not be monitored as closely as those on critical path since their impact is less. If, however, items not on the critical path are delayed beyond their allowable float (i.e., started or completed after their "late start" or "late completion" dates), they will become critical and may in fact govern completion of the work. Thus, a reasonable degree of surveillance must be applied to these noncritical activities as well.

For extensive and complex activities where the volume of work exceeds the capabilities of the individual design office, it may be necessary to place blocks of work with other offices or companies. For such activities, milestone schedules establishing major events and data exchanges taken from the overall critical path schedule are a highly important tool for monitoring this work.

The level of detail to be indicated on the critical path diagrams would depend on the degree of monitoring and control that is desired. It is possible to monitor modest activities with a critical path diagram having perhaps 20 or 30 activities, whereas on very large, multimillion dollar engineering contracts, the critical path diagram may include several hundred activities and may be converted into a computerized schedule that computes float and critical path(s) and generates exception reports, critical item lists, and so forth. Fig. 9.13 shows a typical critical path diagram.

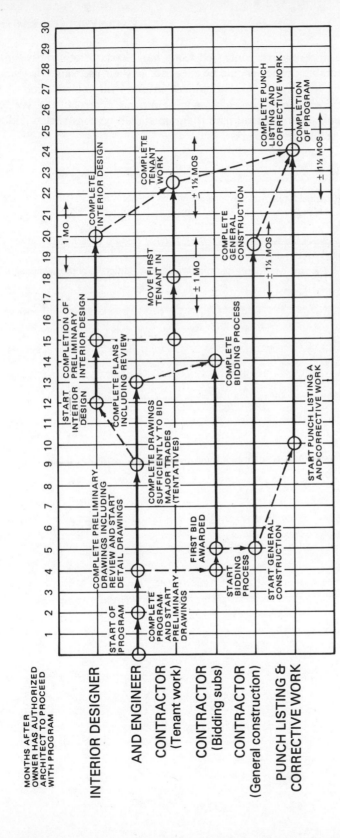

Fig. 9.13 Critical path diagram. From *Professional Construction Management and Project Administration*, second edition, William B. Foxhall, Editor. Copyright 1972, 1976 by the American Institute of Architects and Architectural Record. Reproduced with permission of the American Institute of Architects under permission number 82060. Further reproduction is prohibited. Also, Copyright © 1972 by McGraw-Hill Book Company. Used with the permission of McGraw-Hill Book Company.

Extremely simple work may be controlled using **bar charts** or other less sophisticated tools. Considerable caution should be exercised in their use, since they inherently require that the user maintain continuously in mind the interrelation of the events, and if even modest complications present themselves, control of the work can be lost. In this circumstance, a critical path diagram should be constructed and utilized as the control device for the work.

REFERENCES

1. Wm. B. Foxhall, Professional Construction Management and Project Administration, *Architectural Record,* McGraw-Hill, New York, 1972; and The American Institute of Architects, Washington, D.C.
2. *A Guide for the Engagement of Engineering Services,* Manual No. 45, American Society of Civil Engineers, New York, 1981.
3. H. Malinowsky and J. Richardson, *Science and Engineering,* 3rd edition, Libraries Unlimited, Littleton, Colo., 1980.

SECTION 10
TABLES

Table 10.1 Common Logarithms

N	0	1	2	3	4	5	6	7	8	9
0	....	0000	3010	4771	6021	6990	7782	8451	9031	9542
1	0000	0414	0792	1139	1461	1761	2041	2304	2553	2788
2	3010	3222	3424	3617	3802	3979	4150	4314	4472	4624
3	4771	4914	5051	5185	5315	5441	5563	5682	5798	5911
4	6021	6128	6232	6335	6435	6532	6628	6721	6812	6902
5	6990	7076	7160	7243	7324	7404	7482	7559	7634	7709
6	7782	7853	7924	7993	8062	8129	8195	8261	8325	8388
7	8451	8513	8573	8633	8692	8751	8808	8865	8921	8976
8	9031	9085	9138	9191	9243	9294	9345	9395	9445	9494
9	9542	9590	9638	9685	9731	9777	9823	9868	9912	9956
10	0000	0043	0086	0128	0170	0212	0253	0294	0334	0374
11	0414	0453	0492	0531	0569	0607	0645	0682	0719	0755
12	0792	0828	0864	0899	0934	0969	1004	1038	1072	1106
13	1139	1173	1206	1239	1271	1303	1335	1367	1399	1430
14	1461	1492	1523	1553	1584	1614	1644	1673	1703	1732
15	1761	1790	1818	1847	1875	1903	1931	1959	1987	2014
16	2041	2068	2095	2122	2148	2175	2201	2227	2253	2279
17	2304	2330	2355	2380	2405	2430	2455	2480	2504	2529
18	2553	2577	2601	2625	2648	2672	2695	2718	2742	2765
19	2788	2810	2833	2856	2878	2900	2923	2945	2967	2989
20	3010	3032	3054	3075	3096	3118	3139	3160	3181	3201
21	3222	3243	3263	3284	3304	3324	3345	3365	3385	3404
22	3424	3444	3464	3483	3502	3522	3541	3560	3579	3598
23	3617	3636	3655	3674	3692	3711	3729	3747	3766	3784
24	3802	3820	3838	3856	3874	3892	3909	3927	3945	3962
25	3979	3997	4014	4031	4048	4065	4082	4099	4116	4133
26	4150	4166	4183	4200	4216	4232	4249	4265	4281	4298
27	4314	4330	4346	4362	4378	4393	4409	4425	4440	4456
28	4472	4487	4502	4518	4533	4548	4564	4579	4594	4609
29	4624	4639	4654	4669	4683	4698	4713	4728	4742	4757
30	4771	4786	4800	4814	4829	4843	4857	4871	4886	4900
31	4914	4928	4942	4955	4969	4983	4997	5011	5024	5038
32	5051	5065	5079	5092	5105	5119	5132	5145	5159	5172
33	5185	5198	5211	5224	5237	5250	5263	5276	5289	5302
34	5315	5328	5340	5353	5366	5378	5391	5403	5416	5428
35	5441	5453	5465	5478	5490	5502	5514	5527	5539	5551
36	5563	5575	5587	5599	5611	5623	5635	5647	5658	5670
37	5682	5694	5705	5717	5729	5740	5752	5763	5775	5786
38	5798	5809	5821	5832	5843	5855	5866	5877	5888	5899
39	5911	5922	5933	5944	5955	5966	5977	5988	5999	6010
40	6021	6031	6042	6053	6064	6075	6085	6096	6107	6117
41	6128	6138	6149	6160	6170	6180	6191	6201	6212	6222
42	6232	6243	6253	6263	6274	6284	6294	6304	6314	6325
43	6335	6345	6355	6365	6375	6385	6395	6405	6415	6425
44	6435	6444	6454	6464	6474	6484	6493	6503	6513	6522
45	6532	6542	6551	6561	6571	6580	6590	6599	6609	6618
46	6628	6637	6646	6656	6665	6675	6684	6693	6702	6712
47	6721	6730	6739	6749	6758	6767	6776	6785	6794	6803
48	6812	6821	6830	6839	6848	6857	6866	6875	6884	6893
49	6902	6911	6920	6928	6937	6946	6955	6964	6972	6981
50	6990	6998	7007	7016	7024	7033	7042	7050	7059	7067
N	0	1	2	3	4	5	6	7	8	9

Table 10.1 *Continued*

N	0	1	2	3	4	5	6	7	8	9
50	6990	6998	7007	7016	7024	7033	7042	7050	7059	7067
51	7076	7084	7093	7101	7110	7118	7126	7135	7143	7152
52	7160	7168	7177	7185	7193	7202	7210	7218	7226	7235
53	7243	7251	7259	7267	7275	7284	7292	7300	7308	7316
54	7324	7332	7340	7348	7356	7364	7372	7380	7388	7396
55	7404	7412	7419	7427	7435	7443	7451	7459	7466	7474
56	7482	7490	7497	7505	7513	7520	7528	7536	7543	7551
57	7559	7566	7574	7582	7589	7597	7604	7612	7619	7627
58	7634	7642	7649	7657	7664	7672	7679	7686	7694	7701
59	7709	7716	7723	7731	7738	7745	7752	7760	7767	7774
60	7782	7789	7796	7803	7810	7818	7825	7832	7839	7846
61	7853	7860	7868	7875	7882	7889	7896	7903	7910	7917
62	7924	7931	7938	7945	7952	7959	7966	7973	7980	7987
63	7993	8000	8007	8014	8021	8028	8035	8041	8048	8055
64	8062	8069	8075	8082	8089	8096	8102	8109	8116	8122
65	8129	8136	8142	8149	8156	8162	8169	8176	8182	8189
66	8195	8202	8209	8215	8222	8228	8235	8241	8248	8254
67	8261	8267	8274	8280	8287	8293	8299	8306	8312	8319
68	8325	8331	8338	8344	8351	8357	8363	8370	8376	8382
69	8388	8395	8401	8407	8414	8420	8426	8432	8439	8445
70	8451	8457	8463	8470	8476	8482	8488	8494	8500	8506
71	8513	8519	8525	8531	8537	8543	8549	8555	8561	8567
72	8573	8579	8585	8591	8597	8603	8609	8615	8621	8627
73	8633	8639	8645	8651	8657	8663	8669	8675	8681	8686
74	8692	8698	8704	8710	8716	8722	8727	8733	8739	8745
75	8751	8756	8762	8768	8774	8779	8785	8791	8797	8802
76	8808	8814	8820	8825	8831	8837	8842	8848	8854	8859
77	8865	8871	8876	8882	8887	8893	8899	8904	8910	8915
78	8921	8927	8932	8938	8943	8949	8954	8960	8965	8971
79	8976	8982	8987	8993	8998	9004	9009	9015	9020	9025
80	9031	9036	9042	9047	9053	9058	9063	9069	9074	9079
81	9085	9090	9096	9101	9106	9112	9117	9122	9128	9133
82	9138	9143	9149	9154	9159	9165	9170	9175	9180	9186
83	9191	9196	9201	9206	9212	9217	9222	9227	9232	9238
84	9243	9248	9253	9258	9263	9269	9274	9279	9284	9289
85	9294	9299	9304	9309	9315	9320	9325	9330	9335	9340
86	9345	9350	9355	9360	9365	9370	9375	9380	9385	9390
87	9395	9400	9405	9410	9415	9420	9425	9430	9435	9440
88	9445	9450	9455	9460	9465	9469	9474	9479	9484	9489
89	9494	9499	9504	9509	9513	9518	9523	9528	9533	9538
90	9542	9547	9552	9557	9562	9566	9571	9576	9581	9586
91	9590	9595	9600	9605	9609	9614	9619	9624	9628	9633
92	9638	9643	9647	9652	9657	9661	9666	9671	9675	9680
93	9685	9689	9694	9699	9703	9708	9713	9717	9722	9727
94	9731	9736	9741	9745	9750	9754	9759	9763	9768	9773
95	9777	9782	9786	9791	9795	9800	9805	9809	9814	9818
96	9823	9827	9832	9836	9841	9845	9850	9854	9859	9863
97	9868	9872	9877	9881	9886	9890	9894	9899	9903	9908
98	9912	9917	9921	9926	9930	9934	9939	9943	9948	9952
99	9956	9961	9965	9969	9974	9978	9983	9987	9991	9996
100	0000	0004	0009	0013	0017	0022	0026	0030	0035	0039
N	0	1	2	3	4	5	6	7	8	9

Table 10.2 Degrees to Radians

Degs.	0.0	0.1	0.2	0.3	0.4	0.5	0.6	0.7	0.8	0.9
0	0.0000	0.0017	0.0035	0.0052	0.0070	0.0087	0.0105	0.0122	0.0140	0.0157
1	0.0175	0.0192	0.0209	0.0227	0.0244	0.0262	0.0279	0.0297	0.0314	0.0332
2	0.0349	0.0367	0.0384	0.0401	0.0419	0.0436	0.0454	0.0471	0.0489	0.0506
3	0.0524	0.0541	0.0559	0.0576	0.0593	0.0611	0.0628	0.0646	0.0663	0.0681
4	0.0698	0.0716	0.0733	0.0750	0.0768	0.0785	0.0803	0.0820	0.0838	0.0855
5	0.0873	0.0890	0.0908	0.0925	0.0942	0.0960	0.0977	0.0995	0.1012	0.1030
6	0.1047	0.1065	0.1082	0.1100	0.1117	0.1134	0.1152	0.1169	0.1187	0.1204
7	0.1222	0.1239	0.1257	0.1274	0.1292	0.1309	0.1326	0.1344	0.1361	0.1379
8	0.1396	0.1414	0.1431	0.1449	0.1466	0.1484	0.1501	0.1518	0.1536	0.1553
9	0.1571	0.1588	0.1606	0.1623	0.1641	0.1658	0.1676	0.1693	0.1710	0.1728
10	0.1745	0.1763	0.1780	0.1798	0.1815	0.1833	0.1850	0.1868	0.1885	0.1902
11	0.1920	0.1937	0.1955	0.1972	0.1990	0.2007	0.2025	0.2042	0.2059	0.2077
12	0.2094	0.2112	0.2129	0.2147	0.2164	0.2182	0.2199	0.2217	0.2234	0.2251
13	0.2269	0.2286	0.2304	0.2321	0.2339	0.2356	0.2374	0.2391	0.2409	0.2426
14	0.2443	0.2461	0.2478	0.2496	0.2513	0.2531	0.2548	0.2566	0.2583	0.2601
15	0.2618	0.2635	0.2653	0.2670	0.2688	0.2705	0.2723	0.2740	0.2758	0.2775
16	0.2793	0.2810	0.2827	0.2845	0.2862	0.2880	0.2897	0.2915	0.2932	0.2950
17	0.2967	0.2985	0.3002	0.3019	0.3037	0.3054	0.3072	0.3089	0.3107	0.3124
18	0.3142	0.3159	0.3176	0.3194	0.3211	0.3229	0.3246	0.3264	0.3281	0.3299
19	0.3316	0.3334	0.3351	0.3368	0.3386	0.3403	0.3421	0.3438	0.3456	0.3473
20	0.3491	0.3508	0.3526	0.3543	0.3560	0.3578	0.3595	0.3613	0.3630	0.3648
21	0.3665	0.3683	0.3700	0.3718	0.3735	0.3752	0.3770	0.3787	0.3805	0.3822
22	0.3840	0.3857	0.3875	0.3892	0.3910	0.3927	0.3944	0.3962	0.3979	0.3997
23	0.4014	0.4032	0.4049	0.4067	0.4084	0.4102	0.4119	0.4136	0.4154	0.4171
24	0.4189	0.4206	0.4224	0.4241	0.4259	0.4276	0.4294	0.4311	0.4328	0.4346
25	0.4363	0.4381	0.4398	0.4416	0.4433	0.4451	0.4468	0.4485	0.4503	0.4520
26	0.4538	0.4555	0.4573	0.4590	0.4608	0.4625	0.4643	0.4660	0.4677	0.4695
27	0.4712	0.4730	0.4747	0.4765	0.4782	0.4800	0.4817	0.4835	0.4852	0.4869
28	0.4887	0.4904	0.4922	0.4939	0.4957	0.4974	0.4992	0.5009	0.5027	0.5044
29	0.5061	0.5079	0.5096	0.5114	0.5131	0.5149	0.5166	0.5184	0.5201	0.5219
30	0.5236	0.5253	0.5271	0.5288	0.5306	0.5323	0.5341	0.5358	0.5376	0.5393
31	0.5411	0.5428	0.5445	0.5463	0.5480	0.5498	0.5515	0.5533	0.5550	0.5568
32	0.5585	0.5603	0.5620	0.5637	0.5655	0.5672	0.5690	0.5707	0.5725	0.5742
33	0.5760	0.5777	0.5794	0.5812	0.5829	0.5847	0.5864	0.5882	0.5899	0.5917
34	0.5934	0.5952	0.5969	0.5986	0.6004	0.6021	0.6039	0.6056	0.6074	0.6091
35	0.6109	0.6126	0.6144	0.6161	0.6178	0.6196	0.6213	0.6231	0.6248	0.6266
36	0.6283	0.6301	0.6318	0.6336	0.6353	0.6370	0.6388	0.6405	0.6423	0.6440
37	0.6458	0.6475	0.6493	0.6510	0.6528	0.6545	0.6562	0.6580	0.6597	0.6615
38	0.6632	0.6650	0.6667	0.6685	0.6702	0.6720	0.6737	0.6754	0.6772	0.6789
39	0.6807	0.6824	0.6842	0.6859	0.6877	0.6894	0.6912	0.6929	0.6946	0.6964
40	0.6981	0.6999	0.7016	0.7034	0.7051	0.7069	0.7086	0.7103	0.7121	0.7138
41	0.7156	0.7173	0.7191	0.7208	0.7226	0.7243	0.7261	0.7278	0.7295	0.7313
42	0.7330	0.7348	0.7365	0.7383	0.7400	0.7418	0.7435	0.7453	0.7470	0.7487
43	0.7505	0.7522	0.7540	0.7557	0.7575	0.7592	0.7610	0.7627	0.7645	0.7662
44	0.7679	0.7697	0.7714	0.7732	0.7749	0.7767	0.7784	0.7802	0.7819	0.7837
45	0.7854	0.7871	0.7889	0.7906	0.7924	0.7941	0.7959	0.7976	0.7994	0.8011
	0'	6'	12'	18'	24'	30'	36'	42'	48'	54'

$90° = 1.5708$ radians $\qquad 30° = \dfrac{\pi}{6}, \quad 45° = \dfrac{\pi}{4}, \quad 60° = \dfrac{\pi}{3}, \quad 90° = \dfrac{\pi}{2}$ radians

$180° = 3.1416$ radians $\qquad 120° = \dfrac{2\pi}{3}, \quad 135° = \dfrac{3\pi}{4}, \quad 150° = \dfrac{5\pi}{6}, \quad 180° = \pi$ radians

$270° = 4.7124$ radians $\qquad 210° = \dfrac{7\pi}{6}, \quad 225° = \dfrac{5\pi}{4}, \quad 240° = \dfrac{4\pi}{3}, \quad 270° = \dfrac{3\pi}{2}$ radians

$360° = 6.2832$ radians $\qquad 300° = \dfrac{5\pi}{3}, \quad 315° = \dfrac{7\pi}{4}, \quad 330° = \dfrac{11\pi}{6}, \quad 360° = 2\pi$ radians

Table 10.2 *Continued*

Degs.	0.0	0.1	0.2	0.3	0.4	0.5	0.6	0.7	0.8	0.9
45	0.7854	0.7871	0.7889	0.7906	0.7924	0.7941	0.7959	0.7976	0.7994	0.8011
46	0.8029	0.8046	0.8063	0.8081	0.8098	0.8116	0.8133	0.8151	0.8168	0.8186
47	0.8203	0.8221	0.8238	0.8255	0.8273	0.8290	0.8308	0.8325	0.8343	0.8360
48	0.8378	0.8395	0.8412	0.8430	0.8447	0.8465	0.8482	0.8500	0.8517	0.8535
49	0.8552	0.8570	0.8587	0.8604	0.8622	0.8639	0.8657	0.8674	0.8692	0.8709
50	0.8727	0.8744	0.8762	0.8779	0.8796	0.8814	0.8831	0.8849	0.8866	0.8884
51	0.8901	0.8919	0.8936	0.8954	0.8971	0.8988	0.9006	0.9023	0.9041	0.9058
52	0.9076	0.9093	0.9111	0.9128	0.9146	0.9163	0.9180	0.9198	0.9215	0.9233
53	0.9250	0.9268	0.9285	0.9303	0.9320	0.9338	0.9355	0.9372	0.9390	0.9407
54	0.9425	0.9442	0.9460	0.9477	0.9495	0.9512	0.9529	0.9547	0.9564	0.9582
55	0.9599	0.9617	0.9634	0.9652	0.9669	0.9687	0.9704	0.9721	0.9739	0.9756
56	0.9774	0.9791	0.9809	0.9826	0.9844	0.9861	0.9879	0.9896	0.9913	0.9931
57	0.9948	0.9966	0.9983	1.0001	1.0018	1.0036	1.0053	1.0071	1.0088	1.0105
58	1.0123	1.0140	1.0158	1.0175	1.0193	1.0210	1.0228	1.0245	1.0263	1.0280
59	1.0297	1.0315	1.0332	1.0350	1.0367	1.0385	1.0402	1.0420	1.0437	1.0455
60	1.0472	1.0489	1.0507	1.0524	1.0542	1.0559	1.0577	1.0594	1.0612	1.0629
61	1.0647	1.0664	1.0681	1.0699	1.0716	1.0734	1.0751	1.0769	1.0786	1.0804
62	1.0821	1.0838	1.0856	1.0873	1.0891	1.0908	1.0926	1.0943	1.0961	1.0978
63	1.0996	1.1013	1.1030	1.1048	1.1065	1.1083	1.1100	1.1118	1.1135	1.1153
64	1.1170	1.1188	1.1205	1.1222	1.1240	1.1257	1.1275	1.1292	1.1310	1.1327
65	1.1345	1.1362	1.1380	1.1397	1.1414	1.1432	1.1449	1.1467	1.1484	1.1502
66	1.1519	1.1537	1.1554	1.1572	1.1589	1.1606	1.1624	1.1641	1.1659	1.1676
67	1.1694	1.1711	1.1729	1.1746	1.1764	1.1781	1.1798	1.1816	1.1833	1.1851
68	1.1868	1.1886	1.1903	1.1921	1.1938	1.1956	1.1973	1.1990	1.2008	1.2025
69	1.2043	1.2060	1.2078	1.2095	1.2113	1.2130	1.2147	1.2165	1.2182	1.2200
70	1.2217	1.2235	1.2252	1.2270	1.2287	1.2305	1.2322	1.2339	1.2357	1.2374
71	1.2392	1.2409	1.2427	1.2444	1.2462	1.2479	1.2497	1.2514	1.2531	1.2549
72	1.2566	1.2584	1.2601	1.2619	1.2636	1.2654	1.2671	1.2689	1.2706	1.2723
73	1.2741	1.2758	1.2776	1.2793	1.2811	1.2828	1.2846	1.2863	1.2881	1.2898
74	1.2915	1.2933	1.2950	1.2968	1.2985	1.3003	1.3020	1.3038	1.3055	1.3073
75	1.3090	1.3107	1.3125	1.3142	1.3160	1.3177	1.3195	1.3212	1.3230	1.3247
76	1.3265	1.3282	1.3299	1.3317	1.3334	1.3352	1.3369	1.3387	1.3404	1.3422
77	1.3439	1.3456	1.3474	1.3491	1.3509	1.3526	1.3544	1.3561	1.3579	1.3596
78	1.3614	1.3631	1.3648	1.3666	1.3683	1.3701	1.3718	1.3736	1.3753	1.3771
79	1.3788	1.3806	1.3823	1.3840	1.3858	1.3875	1.3893	1.3910	1.3928	1.3945
80	1.3963	1.3980	1.3998	1.4015	1.4032	1.4050	1.4067	1.4085	1.4102	1.4120
81	1.4137	1.4155	1.4172	1.4190	1.4207	1.4224	1.4242	1.4259	1.4277	1.4294
82	1.4312	1.4329	1.4347	1.4364	1.4382	1.4399	1.4416	1.4434	1.4451	1.4469
83	1.4486	1.4504	1.4521	1.4539	1.4556	1.4573	1.4591	1.4608	1.4626	1.4643
84	1.4661	1.4678	1.4696	1.4713	1.4731	1.4748	1.4765	1.4783	1.4800	1.4818
85	1.4835	1.4853	1.4870	1.4888	1.4905	1.4923	1.4940	1.4957	1.4975	1.4992
86	1.5010	1.5027	1.5045	1.5062	1.5080	1.5097	1.5115	1.5132	1.5149	1.5167
87	1.5184	1.5202	1.5219	1.5237	1.5254	1.5272	1.5289	1.5307	1.5324	1.5341
88	1.5359	1.5376	1.5394	1.5411	1.5429	1.5446	1.5464	1.5481	1.5499	1.5516
89	1.5533	1.5551	1.5568	1.5586	1.5603	1.5621	1.5638	1.5656	1.5673	1.5691
90	1.5708	1.5725	1.5743	1.5760	1.5778	1.5795	1.5813	1.5830	1.5848	1.5865
	0'	6'	12'	18'	24'	30'	36'	42'	48'	54'

$90° = 1.5708$ radians $30° = \dfrac{\pi}{6}$, $45° = \dfrac{\pi}{4}$, $60° = \dfrac{\pi}{3}$, $90° = \dfrac{\pi}{2}$ radians

$180° = 3.1416$ radians $120° = \dfrac{2\pi}{3}$, $135° = \dfrac{3\pi}{4}$, $150° = \dfrac{5\pi}{6}$, $180° = \pi$ radians

$270° = 4.7124$ radians $210° = \dfrac{7\pi}{6}$, $225° = \dfrac{5\pi}{4}$, $240° = \dfrac{4\pi}{3}$, $270° = \dfrac{3\pi}{2}$ radians

$360° = 6.2832$ radians $300° = \dfrac{5\pi}{3}$, $315° = \dfrac{7\pi}{4}$, $330° = \dfrac{11\pi}{6}$, $360° = 2\pi$ radians

Table 10.3(a) Trigonometric Functions
Natural Sines, Cosines and Tangents

0°–14.9°

Degs.	Function	0.0°	0.1°	0.2°	0.3°	0.4°	0.5°	0.6°	0.7°	0.8°	0.9°
0	sin	0.0000	0.0017	0.0035	0.0052	0.0070	0.0087	0.0105	0.0122	0.0140	0 0157
	cos	1.0000	1.0000	1.0000	1.0000	1.0000	1.0000	0 9999	0.9999	0.9999	0.9999
	tan	0.0000	0.0017	0.0035	0.0052	0.0070	0.0087	0.0105	0.0122	0.0140	0.0157
1	sin	0.0175	0.0192	0.0209	0.0227	0.0244	0.0262	0.0279	0.0297	0.0314	0.0332
	cos	0.9998	0.9998	0.9998	0.9997	0.9997	0.9997	0.9996	0.9996	0.9995	0.9995
	tan	0.0175	0.0192	0.0209	0.0227	0.0244	0.0262	0.0279	0.0297	0.0314	0.0332
2	sin	0.0349	0.0366	0.0384	0.0401	0.0419	0.0436	0.0454	0.0471	0.0488	0.0506
	cos	0.9994	0.9993	0.9993	0.9992	0.9991	0.9990	0.9990	0.9989	0.9988	0.9987
	tan	0.0349	0.0367	0.0384	0.0402	0.0419	0.0437	0.0454	0.0472	0.0489	0.0507
3	sin	0.0523	0.0541	0.0558	0.0576	0.0593	0.0610	0.0628	0.0645	0.0663	0.0680
	cos	0.9986	0.9985	0.9984	0.9983	0.9982	0.9981	0.9980	0.9979	0.9978	0.9977
	tan	0.0524	0.0542	0.0559	0.0577	0.0594	0.0612	0.0629	0.0647	⌐.0664	0.0682
4	sin	0.0698	0.0715	0.0732	0.0750	0.0767	0.0785	0.0802	0.0819	0.0837	0.0854
	cos	0.9976	0.9974	0.9973	0.9972	0.9971	0.9969	0.9968	0.9966	0.9965	0.9963
	tan	0.0699	0.0717	0.0734	0.0752	0.0769	0.0787	0.0805	0.0822	0.0840	0.0857
5	sin	0.0872	0.0889	0.0906	0.0924	0.0941	0.0958	0.0976	0.0993	0.1011	0.1028
	cos	0.9962	0.9960	0.9959	0.9957	0.9956	0.9954	0.9952	0.9951	0.9949	0.9947
	tan	0.0875	0.0892	0.0910	0.0928	0.0945	0.0963	0.0981	0.0998	0.1016	0.1033
6	sin	0.1045	0.1063	0.1080	0.1097	0.1115	0.1132	0.1149	0.1167	0.1184	0.1201
	cos	0.9945	0.9943	0.9942	0.9940	0.9938	0.9936	0.9934	0.9932	0.9930	0.9928
	tan	0.1051	0.1069	0.1086	0.1104	0.1122	0.1139	0.1157	0.1175	0.1192	0.1210
7	sin	0.1219	0.1236	0.1253	0.1271	0.1288	0.1305	0.1323	0.1340	0.1357	0.1374
	cos	0.9925	0.9923	0.9921	0.9919	0.9917	0.9914	0.9912	0.9910	0.9907	0.9905
	tan	0.1228	0.1246	0.1263	0.1281	0.1299	0.1317	0.1334	0.1352	0.1370	0.1388
8	sin	0.1392	0.1409	0.1426	0.1444	0.1461	0.1478	0.1495	0.1513	0.1530	0.1547
	cos	0.9903	0.9900	0.9898	0.9895	0.9893	0.9890	0.9888	0.9885	0.9882	0.9880
	tan	0.1405	0.1423	0.1441	0.1459	0.1477	0.1495	0.1512	0.1530	0.1548	0.1566
9	sin	0.1564	0.1582	0.1599	0.1616	0.1633	0.1650	0.1668	0.1685	0.1702	0.1719
	cos	0.9877	0.9874	0.9871	0.9869	0.9866	0.9863	0.9860	0.9857	0.9854	0.9851
	tan	0.1584	0.1602	0.1620	0.1638	0.1655	0.1673	0.1691	0.1709	0.1727	0.1745
10	sin	0.1736	0.1754	0.1771	0.1788	0.1805	0.1822	0.1840	0.1857	0.1874	0.1891
	cos	0.9848	0.9845	0.9842	0.9839	0.9836	0.9833	0.9829	0.9826	0.9823	0.9820
	tan	0.1763	0.1781	0.1799	0.1817	0.1835	0.1853	0.1871	0.1890	0.1908	0.1926
11	sin	0.1908	0.1925	0.1942	0.1959	0.1977	0.1994	0.2011	0.2028	0.2045	0.2062
	cos	0.9816	0.9813	0.9810	0.9806	0.9803	0.9799	0.9796	0.9792	0.9789	0.9785
	tan	0.1944	0.1962	0.1980	0.1998	0.2016	0.2035	0.2053	0.2071	0.2089	0.2107
12	sin	0.2079	0.2096	0.2113	0.2130	0.2147	0.2164	0.2181	0.2198	0.2215	0.2232
	cos	0.9781	0.9778	0.9774	0.9770	0.9767	0.9763	0.9759	0.9755	0.9751	0.9748
	tan	0.2126	0.2144	0.2162	0.2180	0.2199	0.2217	0.2235	0.2254	0.2272	0.2290
13	sin	0.2250	0.2267	0.2284	0.2300	0.2318	0.2334	0.2351	0.2368	0.2385	0.2402
	cos	0.9744	0.9740	0.9736	0.9732	0.9728	0.9724	0.9720	0.9715	0.9711	0.9707
	tan	0.2309	0.2327	0.2345	0.2364	0.2382	0.2401	0.2419	0.2438	0.2456	0.2475
14	sin	0.2419	0.2436	0.2453	0.2470	0.2487	0.2504	0.2521	0.2538	0.2554	0.2571
	cos	0.9703	0.9699	0.9694	0.9690	0.9686	0.9681	0.9677	0.9673	0.9668	0.9664
	tan	0.2493	0.2512	0.2530	0.2549	0.2568	0.2586	0.2605	0.2623	0.2642	0.2661
Degs.	Function	0′	6′	12′	18′	24′	30′	36′	42′	48′	54′

Table 10.3(a) *Continued*

Natural Sines, Cosines and Tangents

15°–29.9°

Degs.	Function	0.0°	0.1°	0.2°	0.3°	0.4°	0.5°	0.6°	0.7°	0.8°	0.9°
15	sin	0.2588	0.2605	0.2622	0.2639	0.2656	0.2672	0.2689	0.2706	0.2723	0.2740
	cos	0.9659	0.9655	0.9650	0.9646	0.9641	0.9636	0.9632	0.9627	0.9622	0.9617
	tan	0.2679	0.2698	0.2717	0.2736	0.2754	0.2773	0.2792	0.2811	0.2830	0.2849
16	sin	0.2756	0.2773	0.2790	0.2807	0.2823	0.2840	0.2857	0.2874	0.2890	0.2907
	cos	0.9613	0.9608	0.9603	0.9598	0.9593	0.9588	0.9583	0.9578	0.9573	0.9568
	tan	0.2867	0.2886	0.2905	0.2924	0.2943	0.2962	0.2981	0.3000	0.3019	0.3038
17	sin	0.2924	0.2940	0.2957	0.2974	0.2990	0.3007	0.3024	0.3040	0.3057	0.3074
	cos	0.9563	0.9558	0.9553	0.9548	0.9542	0.9537	0.9532	0.9527	0.9521	0.9516
	tan	0.3057	0.3076	0.3096	0.3115	0.3134	0.3153	0.3172	0.3191	0.3211	0.3230
18	sin	0.3090	0.3107	0.3123	0.3140	0.3156	0.3173	0.3190	0.3206	0.3223	0.3239
	cos	0.9511	0.9505	0.9500	0.9494	0.9489	0.9483	0.9478	0.9472	0.9466	0.9461
	tan	0.3249	0.3269	0.3288	0.3307	0.3327	0.3346	0.3365	0.3385	0.3404	0.3424
19	sin	0.3256	0.3272	0.3289	0.3305	0.3322	0.3338	0.3355	0.3371	0.3387	0.3404
	cos	0.9455	0.9449	0.9444	0.9438	0.9432	0.9426	0.9421	0.9415	0.9409	0.9403
	tan	0.3443	0.3463	0.3482	0.3502	0.3522	0.3541	0.3561	0.3581	0.3600	0.3620
20	sin	0.3420	0.3437	0.3453	0.3469	0.3486	0.3502	0.3518	0.3535	0.3551	0.3567
	cos	0.9397	0.9391	0.9385	0.9379	0.9373	0.9367	0.9361	0.9354	0.9348	0.9342
	tan	0.3640	0.3659	0.3679	0.3699	0.3719	0.3739	0.3759	0.3779	0.3799	0.3819
21	sin	0.3584	0.3600	0.3616	0.3633	0.3649	0.3665	0.3681	0.3697	0.3714	0.3730
	cos	0.9336	0.9330	0.9323	0.9317	0.9311	0.9304	0.9298	0.9291	0.9285	0.9278
	tan	0.3839	0.3859	0.3879	0.3899	0.3919	0.3939	0.3959	0.3979	0.4000	0.4020
22	sin	0.3746	0.3762	0.3778	0.3795	0.3811	0.3827	0.3843	0.3859	0.3875	0.3891
	cos	0.9272	0.9265	0.9259	0.9252	0.9245	0.9239	0.9232	0.9225	0.9219	0.9212
	tan	0.4040	0.4061	0.4081	0.4101	0.4122	0.4142	0.4163	0.4183	0.4204	0.4224
23	sin	0.3907	0.3923	0.3939	0.3955	0.3971	0.3987	0.4003	0.4019	0.4035	0.4051
	cos	0.9205	0.9198	0.9191	0.9184	0.9178	0.9171	0.9164	0.9157	0.9150	0.9143
	tan	0.4245	0.4265	0.4286	0.4307	0.4327	0.4348	0.4369	0.4390	0.4411	0.4431
24	sin	0.4067	0.4083	0.4099	0.4115	0.4131	0.4147	0.4163	0.4179	0.4195	0.4210
	cos	0.9135	0.9128	0.9121	0.9114	0.9107	0.9100	0.9092	0.9085	0.9078	0.9070
	tan	0.4452	0.4473	0.4494	0.4515	0.4536	0.4557	0.4578	0.4599	0.4621	0.4642
25	sin	0.4226	0.4242	0.4258	0.4274	0.4289	0.4305	0.4321	0.4337	0.4352	0.4368
	cos	0.9063	0.9056	0.9048	0.9041	0.9033	0.9026	0.9018	0.9011	0.9003	0.8996
	tan	0.4663	0.4684	0.4706	0.4727	0.4748	0.4770	0.4791	0.4813	0.4834	0.4856
26	sin	0.4384	0.4399	0.4415	0.4431	0.4446	0.4462	0.4478	0.4493	0.4509	0.4524
	cos	0.8988	0.8980	0.8973	0.8965	0.8957	0.8949	0.8942	0.8934	0.8926	0.8918
	tan	0.4877	0.4899	0.4921	0.4942	0.4964	0.4986	0.5008	0.5029	0.5051	0.5073
27	sin	0.4540	0.4555	0.4571	0.4586	0.4602	0.4617	0.4633	0.4648	0.4664	0.4679
	cos	0.8910	0.8902	0.8894	0.8886	0.8878	0.8870	0.8862	0.8854	0.8846	0.8838
	tan	0.5095	0.5117	0.5139	0.5161	0.5184	0.5206	0.5228	0.5250	0.5272	0.5295
28	sin	0.4695	0.4710	0.4726	0.4741	0.4756	0.4772	0.4787	0.4802	0.4818	0.4833
	cos	0.8829	0.8821	0.8813	0.8805	0.8796	0.8788	0.8780	0.8771	0.8763	0.8755
	tan	0.5317	0.5340	0.5362	0.5384	0.5407	0.5430	0.5452	0.5475	0.5498	0.5520
29	sin	0.4848	0.4863	0.4879	0.4894	0.4909	0.4924	0.4939	0.4955	0.4970	0.4985
	cos	0.8746	0.8738	0.8729	0.8721	0.8712	0.8704	0.8695	0.8686	0.8678	0.8669
	tan	0.5543	0.5566	0.5589	0.5612	0.5635	0.5658	0.5681	0.5704	0.5727	0.5750
Degs.	Function	0′	6′	12′	18′	24′	30′	36′	42′	48′	54′

Table 10.3a   Continued

Natural Sines, Cosines and Tangents

30°–44.9°

Degs.	Function	0.0°	0.1°	0.2°	0.3°	0.4°	0.5°	0.6°	0.7°	0.8°	0.9°
30	sin	0.5000	0.5015	0.5030	0.5045	0.5060	0.5075	0.5090	0.5105	0.5120	0.5135
	cos	0.8660	0.8652	0.8643	0.8634	0.8625	0.8616	0.8607	0.8599	0.8590	0.8581
	tan	0.5774	0.5797	0.5820	0.5844	0.5867	0.5890	0.5914	0.5938	0.5961	0.5985
31	sin	0.5150	0.5165	0.5180	0.5195	0.5210	0.5225	0.5240	0.5255	0.5270	0.5284
	cos	0.8572	0.8563	0.8554	0.8545	0.8536	0.8526	0.8517	0.8508	0.8499	0.8490
	tan	0.6009	0.6032	0.6056	0.6080	0.6104	0.6128	0.6152	0.6176	0.6200	0.6224
32	sin	0.5299	0.5314	0.5329	0.5344	0.5358	0.5373	0.5388	0.5402	0.5417	0.5432
	cos	0.8480	0.8471	0.8462	0.8453	0.8443	0.8434	0.8425	0.8415	0.8406	0.8396
	tan	0.6249	0.6273	0.6297	0.6322	0.6346	0.6371	0.6395	0.6420	0.6445	0.6469
33	sin	0.5446	0.5461	0.5476	0.5490	0.5505	0.5519	0.5534	0.5548	0.5563	0.5577
	cos	0.8387	0.8377	0.8368	0.8358	0.8348	0.8339	0.8329	0.8320	0.8310	0.8300
	tan	0.6494	0.6519	0.6544	0.6569	0.6594	0.6619	0.6644	0.6669	0.6694	0.6720
34	sin	0.5592	0.5606	0.5621	0.5635	0.5650	0.5664	0.5678	0.5693	0.5707	0.5721
	cos	0.8290	0.8281	0.8271	0.8261	0.8251	0.8241	0.8231	0.8221	0.8211	0.8202
	tan	0.6745	0.6771	0.6796	0.6822	0.6847	0.6873	0.6899	0.6924	0.6950	0.6976
35	sin	0.5736	0.5750	0.5764	0.5779	0.5793	0.5807	0.5821	0.5835	0.5850	0.5864
	cos	0.8192	0.8181	0.8171	0.8161	0.8151	0.8141	0.8131	0.8121	0.8111	0.8100
	tan	0.7002	0.7028	0.7054	0.7080	0.7107	0.7133	0.7159	0.7186	0.7212	0.7239
36	sin	0.5878	0.5892	0.5906	0.5920	0.5934	0.5948	0.5962	0.5976	0.5990	0.6004
	cos	0.8090	0.8080	0.8070	0.8059	0.8049	0.8039	0.8028	0.8018	0.8007	0.7997
	tan	0.7265	0.7292	0.7319	0.7346	0.7373	0.7400	0.7427	0.7454	0.7481	0.7508
37	sin	0.6018	0.6032	0.6046	0.6060	0.6074	0.6088	0.6101	0.6115	0.6129	0.6143
	cos	0.7986	0.7976	0.7965	0.7955	0.7944	0.7934	0.7923	0.7912	0.7902	0.7891
	tan	0.7536	0.7563	0.7590	0.7618	0.7646	0.7673	0.7701	0.7729	0.7757	0.7785
38	sin	0.6157	0.6170	0.6184	0.6198	0.6211	0.6225	0.6239	0.6252	0.6266	0.6280
	cos	0.7880	0.7869	0.7859	0.7848	0.7837	0.7826	0.7815	0.7804	0.7793	0.7782
	tan	0.7813	0.7841	0.7869	0.7898	0.7926	0.7954	0.7983	0.8012	0.8040	0.8069
39	sin	0.6293	0.6307	0.6320	0.6334	0.6347	0.6361	0.6374	0.6388	0.6401	0.6414
	cos	0.7771	0.7760	0.7749	0.7738	0.7727	0.7716	0.7705	0.7694	0.7683	0.7672
	tan	0.8098	0.8127	0.8156	0.8185	0.8214	0.8243	0.8273	0.8302	0.8332	0.8361
40	sin	0.6428	0.6441	0.6455	0.6468	0.6481	0.6494	0.6508	0.6521	0.6534	0.6547
	cos	0.7660	0.7649	0.7638	0.7627	0.7615	0.7604	0.7593	0.7581	0.7570	0.7559
	tan	0.8391	0.8421	0.8451	0.8481	0.8511	0.8541	0.8571	0.8601	0.8632	0.8662
41	sin	0.6561	0.6574	0.6587	0.6600	0.6613	0.6626	0.6639	0.6652	0.6665	0.6678
	cos	0.7547	0.7536	0.7524	0.7513	0.7501	0.7490	0.7478	0.7466	0.7455	0.7443
	tan	0.8693	0.8724	0.8754	0.8785	0.8816	0.8847	0.8878	0.8910	0.8941	0.8972
42	sin	0.6691	0.6704	0.6717	0.6730	0.6743	0.6756	0.6769	0.6782	0.6794	0.6807
	cos	0.7431	0.7420	0.7408	0.7396	0.7385	0.7373	0.7361	0.7349	0.7337	0.7325
	tan	0.9004	0.9036	0.9067	0.9099	0.9131	0.9163	0.9195	0.9228	0.9260	0.9293
43	sin	0.6820	0.6833	0.6845	0.6858	0.6871	0.6884	0.6896	0.6909	0.6921	0.6934
	cos	0.7314	0.7302	0.7290	0.7278	0.7266	0.7254	0.7242	0.7230	0.7218	0.7206
	tan	0.9325	0.9358	0.9391	0.9424	0.9457	0.9490	0.9523	0.9556	0.9590	0.9623
44	sin	0.6947	0.6959	0.6972	0.6984	0.6997	0.7009	0.7022	0.7034	0.7046	0.7059
	cos	0.7193	0.7181	0.7169	0.7157	0.7145	0.7133	0.7120	0.7108	0.7096	0.7083
	tan	0.9657	0.9691	0.9725	0.9759	0.9793	0.9827	0.9861	0.9896	0.9930	0.9965
Degs.	Function	0′	6′	12′	18′	24′	30′	36′	42′	48′	54′

Table 10.3a *Continued*

Natural Sines, Cosines and Tangents

Degs.	Function	0.0°	0.1°	0.2°	0.3°	0.4°	0.5°	0.6°	0.7°	0.8°	0.9°
45	sin	0.7071	0.7083	0.7096	0 7108	0 7120	0 7133	0.7145	0.7157	0 7169	0.7181
	cos	0.7071	0.7059	0.7046	0.7034	0 7022	0 7009	0.6997	0.6984	0.6972	0 6959
	tan	1.0000	1.0035	1 0070	1.0105	1.0141	1.0176	1.0212	1.0247	1.0283	1.0319
46	sin	0.7193	0.7206	0.7218	0.7230	0.7242	0 7254	0.7266	0.7278	0.7290	0.7302
	cos	0.6947	0.6934	0.6921	0.6909	0.6896	0.6884	0.6871	0.6858	0.6845	0.6833
	tan	1.0355	1.0392	1.0428	1.0464	1.0501	1.0538	1.0575	1.0612	1.0649	1.0686
47	sin	0.7314	0.7325	0.7337	0.7349	0.7361	0.7373	0.7385	0.7396	0.7408	0.7420
	cos	0.6820	0.6807	0.6794	0.6782	0.6769	0.6756	0.6743	0.6730	0.6717	0.6704
	tan	1.0724	1.0761	1.0799	1.0837	1.0875	1.0913	1.0951	1.0990	1.1028	1.1067
48	sin	0.7431	0.7443	0.7455	0.7466	0.7478	0.7490	0.7501	0.7513	0.7524	0.7536
	cos	0.6691	0.6678	0.6665	0.6652	0.6639	0.6626	0.6613	0.6600	0.6587	0.6574
	tan	1.1106	1.1145	1.1184	1.1224	1.1263	1.1303	1.1343	1.1383	1.1423	1.1463
49	sin	0.7547	0.7559	0.7570	0.7581	0.7593	0.7604	0.7615	0.7627	0.7638	0.7649
	cos	0.6561	0.6547	0.6534	0.6521	0.6508	0.6494	0.6481	0.6468	0.6455	0.6441
	tan	1.1504	1.1544	1.1585	1.1626	1.1667	1.1708	1.1750	1.1792	1.1833	1.1875
50	sin	0.7660	0.7672	0.7683	0.7694	0.7705	0.7716	0.7727	0 7738	0.7749	0.7760
	cos	0.6428	0.6414	0.6401	0.6388	0.6374	0.6361	0.6347	0.6334	0.6320	0.6307
	tan	1.1918	1.1960	1.2002	1.2045	1.2088	1.2131	1.2174	1.2218	1.2261	1.2305
51	sin	0.7771	0.7782	0.7793	0.7804	0.7815	0.7826	0.7837	0.7848	0.7859	0 7869
	cos	0.6293	0.6280	0.6266	0.6252	0.6239	0.6225	0.6211	0.6198	0.6184	0.6170
	tan	1.2349	1.2393	1.2437	1.2482	1.2527	1.2572	1.2617	1.2662	1.2708	1.2753
52	sin	0.7880	0.7891	0.7902	0.7912	0.7923	0.7934	0.7944	0.7955	0 7965	0 7976
	cos	0.6157	0.6143	0.6129	0.6115	0.6101	0.6088	0.6074	0 6060	0 6046	0.6032
	tan	1.2799	1.2846	1.2892	1.2938	1.2985	1.3032	1.3079	1.3127	1.3175	1.3222
53	sin	0.7986	0.7997	0.8007	0.8018	0.8028	0.8039	0.8049	0.8059	0 8070	0.8080
	cos	0.6018	0.6004	0 5990	0.5976	0.5962	0.5948	0.5934	0.5920	0.5906	0.5892
	tan	1.3270	1.3319	1.3367	1.3416	1.3465	1.3514	1.3564	1.3613	1.3663	1.3713
54	sin	0.8090	0.8100	0.8111	0.8121	0.8131	0.8141	0.8151	0.8161	0.8171	0.8181
	cos	0.5878	0.5864	0.5850	0.5835	0.5821	0.5807	0.5793	0.5779	0 5764	0.5750
	tan	1.3764	1.3814	1.3865	1.3916	1.3968	1.4019	1.4071	1.4124	1.4176	1.4229
55	sin	0.8192	0.8202	0.8211	0.8221	0.8231	0.8241	0.8251	0.8261	0 8271	0.8281
	cos	0.5736	0.5721	0.5707	0.5693	0.5678	0.5664	0.5650	0.5635	0.5621	0.5606
	tan	1.4281	1.4335	1.4388	1.4442	1.4496	1.4550	1.4605	1.4659	1.4715	1.4770
56	sin	0.8290	0.8300	0.8310	0.8320	0.8329	0.8339	0.8348	0 8358	0.8368	0.8377
	cos	0.5592	0.5577	0.5563	0.5548	0.5534	0.5519	0.5505	0.5490	0.5476	0.5461
	tan	1.4826	1.4882	1.4938	1.4994	1.5051	1.5108	1.5166	1.5224	1.5282	1.5340
57	sin	0.8387	0.8396	0.8406	0.8415	0.8425	0.8434	0.8443	0.8453	0.8462	0.8471
	cos	0.5446	0.5432	0.5417	0.5402	0.5388	0.5373	0.5358	0.5344	0.5329	0.5314
	tan	1.5399	1.5458	1.5517	1.5577	1.5637	1.5697	1.5757	1.5818	1.5880	1.5941
58	sin	0.8480	0.8490	0.8499	0.8508	0.8517	0.8526	0.8536	0.8545	0.8554	0.8563
	cos	0.5299	0.5284	0.5270	0.5255	0.5240	0.5225	0.5210	0.5195	0.5180	0.5165
	tan	1.6003	1.6066	1.6128	1.6191	1.6255	1.6319	1.6383	1.6447	1.6512	1.6577
59	sin	0.8572	0.8581	0.8590	0.8599	0.8607	0.8616	0.8625	0.8634	0.8643	0.8652
	cos	0.5150	0.5135	0.5120	0.5105	0.5090	0.5075	0.5060	0.5045	0.5030	0.5015
	tan	1.6643	1.6709	1.6775	1.6842	1.6909	1.6977	1.7045	1.7113	1.7182	1.7251
Degs.	Function	0′	6′	12′	18′	24′	30′	36′	42′	48′	54′

Table 10.3a *Continued*
Natural Sines, Cosines and Tangents

60°–74.9°

Degs.	Function	0.0°	0.1°	0.2°	0.3°	0.4°	0.5°	0.6°	0.7°	0.8°	0.9°
60	sin	0.8660	0.8669	0.8678	0.8686	0.8695	0.8704	0.8712	0.8721	0.8729	0.8738
	cos	0.5000	0.4985	0.4970	0.4955	0.4939	0.4924	0.4909	0.4894	0.4879	0.4863
	tan	1.7321	1.7391	1.7461	1.7532	1.7603	1.7675	1.7747	1.7820	1.7893	1.7966
61	sin	0.8746	0.8755	0.8763	0.8771	0.8780	0.8788	0.8796	0.8805	0.8813	0.8821
	cos	0.4848	0.4833	0.4818	0.4802	0.4787	0.4772	0.4756	0.4741	0.4726	0.4710
	tan	1.8040	1.8115	1.8190	1.8265	1.8341	1.8418	1.8495	1.8572	1.8650	1.8728
62	sin	0.8829	0.8838	0.8846	0.8854	0.8862	0.8870	0.8878	0.8886	0.8894	0.8902
	cos	0.4695	0.4679	0.4664	0.4648	0.4633	0.4617	0.4602	0.4586	0.4571	0.4555
	tan	1.8807	1.8887	1.8967	1.9047	1.9128	1.9210	1.9292	1.9375	1.9458	1.9542
63	sin	0.8910	0.8918	0.8926	0.8934	0.8942	0.8949	0.8957	0.8965	0.8973	0.8980
	cos	0.4540	0.4524	0.4509	0.4493	0.4478	0.4462	0.4446	0.4431	0.4415	0.4399
	tan	1.9626	1.9711	1.9797	1.9883	1.9970	2.0057	2.0145	2.0233	2.0323	2.0413
64	sin	0.8988	0.8996	0.9003	0.9011	0.9018	0.9026	0.9033	0.9041	0.9048	0.9056
	cos	0.4384	0.4368	0.4352	0.4337	0.4321	0.4305	0.4289	0.4274	0.4258	0.4242
	tan	2.0503	2.0594	2.0686	2.0778	2.0872	2.0965	2.1060	2.1155	2.1251	2.1348
65	sin	0.9063	0.9070	0.9078	0.9085	0.9092	0.9100	0.9107	0.9114	0.9121	0.9128
	cos	0.4226	0.4210	0.4195	0.4179	0.4163	0.4147	0.4131	0.4115	0.4099	0.4083
	tan	2.1445	2.1543	2.1642	2.1742	2.1842	2.1943	2.2045	2.2148	2.2251	2.2355
66	sin	0.9135	0.9143	0.9150	0.9157	0.9164	0.9171	0.9178	0.9184	0.9191	0.9198
	cos	0.4067	0.4051	0.4035	0.4019	0.4003	0.3987	0.3971	0.3955	0.3939	0.3923
	tan	2.2460	2.2566	2.2673	2.2781	2.2889	2.2998	2.3109	2.3220	2.3332	2.3445
67	sin	0.9205	0.9212	0.9219	0.9225	0.9232	0.9239	0.9245	0.9252	0.9259	0.9265
	cos	0.3907	0.3891	0.3875	0.3859	0.3843	0.3827	0.3811	0.3795	0.3778	0.3762
	tan	2.3559	2.3673	2.3789	2.3906	2.4023	2.4142	2.4262	2.4383	2.4504	2.4627
68	sin	0.9272	0.9278	0.9285	0.9291	0.9298	0.9304	0.9311	0.9317	0.9323	0.9330
	cos	0.3746	0.3730	0.3714	0.3697	0.3681	0.3665	0.3649	0.3633	0.3616	0.3600
	tan	2.4751	2.4876	2.5002	2.5129	2.5257	2.5386	2.5517	2.5649	2.5782	2.5916
69	sin	0.9336	0.9342	0.9348	0.9354	0.9361	0.9367	0.9373	0.9379	0.9385	0.9391
	cos	0.3584	0.3567	0.3551	0.3535	0.3518	0.3502	0.3486	0.3469	0.3453	0.3437
	tan	2.6051	2.6187	2.6325	2.6464	2.6605	2.6746	2.6889	2.7034	2.7179	2.7326
70	sin	0.9397	0.9403	0.9409	0.9415	0.9421	0.9426	0.9432	0.9438	0.9444	0.9449
	cos	0.3420	0.3404	0.3387	0.3371	0.3355	0.3338	0.3322	0.3305	0.3289	0.3272
	tan	2.7475	2.7625	2.7776	2.7929	2.8083	2.8239	2.8397	2.8556	2.8716	2.8878
71	sin	0.9455	0.9461	0.9466	0.9472	0.9478	0.9483	0.9489	0.9494	0.9500	0.9505
	cos	0.3256	0.3239	0.3223	0.3206	0.3190	0.3173	0.3156	0.3140	0.3123	0.3107
	tan	2.9042	2.9208	2.9375	2.9544	2.9714	2.9887	3.0061	3.0237	3.0415	3.0595
72	sin	0.9511	0.9516	0.9521	0.9527	0.9532	0.9537	0.9542	0.9548	0.9553	0.9558
	cos	0.3090	0.3074	0.3057	0.3040	0.3024	0.3007	0.2990	0.2974	0.2957	0.2940
	tan	3.0777	3.0961	3.1146	3.1334	3.1524	3.1716	3.1910	3.2106	3.2305	3.2506
73	sin	0.9563	0.9568	0.9573	0.9578	0.9583	0.9588	0.9593	0.9598	0.9603	0.9608
	cos	0.2924	0.2907	0.2890	0.2874	0.2857	0.2840	0.2823	0.2807	0.2790	0.2773
	tan	3.2709	3.2914	3.3122	3.3332	3.3544	3.3759	3.3977	3.4197	3.4420	3.4646
74	sin	0.9613	0.9617	0.9622	0.9627	0.9632	0.9636	0.9641	0.9646	0.9650	0.9655
	cos	0.2756	0.2740	0.2723	0.2706	0.2689	0.2672	0.2656	0.2639	0.2622	0.2605
	tan	3.4874	3.5105	3.5339	3.5576	3.5816	3.6059	3.6305	3.6554	3.6806	3.7062
Degs.	Function	0'	6'	12'	18'	24'	30'	36'	42'	48'	54'

Table 10.3*a* *Continued*

Natural Sines, Cosines and Tangents

75°–89.9°

Degs.	Function	0.0°	0.1°	0.2°	0.3°	0.4°	0.5°	0.6°	0.7°	0.8°	0.9°
75	sin	0.9659	0.9664	0.9668	0.9673	0.9677	0.9681	0.9686	0.9690	0.9694	0.9699
	cos	0.2588	0.2571	0.2554	0.2538	0.2521	0.2504	0.2487	0.2470	0.2453	0.2436
	tan	3.7321	3.7583	3.7848	3.8118	3.8391	3.8667	3.8947	3.9232	3.9520	3.9812
76	sin	0.9703	0.9707	0.9711	0.9715	0.9720	0.9724	0.9728	0.9732	0.9736	0.9740
	cos	0.2419	0.2402	0.2385	0.2368	0.2351	0.2334	0.2317	0.2300	0.2284	0.2267
	tan	4.0108	4.0408	4.0713	4.1022	4.1335	4.1653	4.1976	4.2303	4.2635	4.2972
77	sin	0.9744	0.9748	0.9751	0.9755	0.9759	0.9763	0.9767	0.9770	0.9774	0.9778
	cos	0.2250	0.2232	0.2215	0.2198	0.2181	0.2164	0.2147	0.2130	0.2113	0.2096
	tan	4.3315	4.3662	4.4015	4.4374	4.4737	4.5107	4.5483	4.5864	4.6252	4.6646
78	sin	0.9781	0.9785	0.9789	0.9792	0.9796	0.9799	0.9803	0.9806	0.9810	0.9813
	cos	0.2079	0.2062	0.2045	0.2028	0.2011	0.1994	0.1977	0.1959	0.1942	0.1925
	tan	4.7046	4.7453	4.7867	4.8288	4.8716	4.9152	4.9594	5.0045	5.0504	5.0970
79	sin	0.9816	0.9820	0.9823	0.9826	0.9829	0.9833	0.9836	0.9839	0.9842	0.9845
	cos	0.1908	0.1891	0.1874	0.1857	0.1840	0.1822	0.1805	0.1788	0.1771	0.1754
	tan	5.1446	5.1929	5.2422	5.2924	5.3135	5.3955	5.4486	5.5026	5.5578	5.6140
80	sin	0.9848	0.9851	0.9854	0.9857	0.9860	0.9863	0.9866	0.9869	0.9871	0.9874
	cos	0.1736	0.1719	0.1702	0.1685	0.1668	0.1650	0.1633	0.1616	0.1599	0.1582
	tan	5.6713	5.7297	5.7894	5.8502	5.9124	5.9758	6.0405	6.1066	6.1742	6.2432
81	sin	0.9877	0.9880	0.9882	0.9885	0.9888	0.9890	0.9893	0.9895	0.9898	0.9900
	cos	0.1564	0.1547	0.1530	0.1513	0.1495	0.1478	0.1461	0.1444	0.1426	0.1409
	tan	6.3138	6.3859	6.4596	6.5350	6.6122	6.6912	6.7720	6.8548	6.9395	7.0264
82	sin	0.9903	0.9905	0.9907	0.9910	0.9912	0.9914	0.9917	0.9919	0.9921	0.9923
	cos	0.1392	0.1374	0.1357	0.1340	0.1323	0.1305	0.1288	0.1271	0.1253	0.1236
	tan	7.1154	7.2066	7.3002	7.3962	7.4947	7.5958	7.6996	7.8062	7.9158	8.0285
83	sin	0.9925	0.9928	0.9930	0.9932	0.9934	0.9936	0.9938	0.9940	0.9942	0.9943
	cos	0.1219	0.1201	0.1184	0.1167	0.1149	0.1132	0.1115	0.1097	0.1080	0.1063
	tan	8.1443	8.2636	8.3863	8.5126	8.6427	8.7769	8.9152	9.0579	9.2052	9.3572
84	sin	0.9945	0.9947	0.9949	0.9951	0.9952	0.9954	0.9956	0.9957	0.9959	0.9960
	cos	0.1045	0.1028	0.1011	0.0993	0.0976	0.0958	0.0941	0.0924	0.0906	0.0889
	tan	9.5144	9.6768	9.8448	10.02	10.20	10.39	10.58	10.78	10.99	11.20
85	sin	0.9962	0.9963	0.9965	0.9966	0.9968	0.9969	0.9971	0.9972	0.9973	0.9974
	cos	0.0872	0.0854	0.0837	0.0819	0.0802	0.0785	0.0767	0.0750	0.0732	0.0715
	tan	11.43	11.66	11.91	12.16	12.43	12.71	13.00	13.30	13.62	13.95
86	sin	0.9976	0.9977	0.9978	0.9979	0.9980	0.9981	0.9982	0.9983	0.9984	0.9985
	cos	0.0698	0.0680	0.0663	0.0645	0.0628	0.0610	0.0593	0.0576	0.0558	0.0541
	tan	14.30	14.67	15.06	15.46	15.89	16.35	16.83	17.34	17.89	18.46
87	sin	0.9986	0.9987	0.9988	0.9989	0.9990	0.9990	0.9991	0.9992	0.9993	0.9993
	cos	0.0523	0.0506	0.0488	0.0471	0.0454	0.0436	0.0419	0.0401	0.0384	0.0366
	tan	19.08	19.74	20.45	21.20	22.02	22.90	23.86	24.90	26.03	27.27
88	sin	0.9994	0.9995	0.9995	0.9996	0.9996	0.9997	0.9997	0.9997	0.9998	0.9998
	cos	0.0349	0.0332	0.0314	0.0297	0.0279	0.0262	0.0244	0.0227	0.0209	0.0192
	tan	28.64	30.14	31.82	33.69	35.80	38.19	40.92	44.07	47.74	52.08
89	sin	0.9998	0.9999	0.9999	0.9999	0.9999	1.000	1.000	1.000	1.000	1.000
	cos	0.0175	0.0157	0.0140	0.0122	0.0105	0.0087	0.0070	0.0052	0.0035	0.0017
	tan	57.29	63.66	71.62	81.85	95.49	114.6	143.2	191.0	286.5	573.0
Degs.	Function	0′	6′	12′	18′	24′	30′	36′	42′	48′	54′

Table 10.3(b)   Hyperbolic Functions

Angle	Function	0.00	0.01	0.02	0.03	0.04	0.05	0.06	0.07	0.08	0.09
0.0	sinh	0.0000	0.0100	0.0200	0.0300	0.0400	0.0500	0.0600	0.0701	0.0801	0.0901
	cosh	1.0000	1.0001	1.0002	1.0005	1.0008	1.0013	1.0018	1.0025	1.0032	1.0041
	tanh	0.0000	0.0100	0.0200	0.0300	0.0400	0.0500	0.0599	0.0699	0.0798	0.0898
0.1	sinh	0.1002	0.1102	0.1203	0.1304	0.1405	0.1506	0.1607	0.1708	0.1810	0.1911
	cosh	1.0050	1.0061	1.0072	1.0085	1.0098	1.0113	1.0128	1.0145	1.0162	1.0181
	tanh	0.0997	0.1096	0.1194	0.1293	0.1391	0.1489	0.1587	0.1684	0.1781	0.1878
0.2	sinh	0.2013	0.2115	0.2218	0.2320	0.2423	0.2526	0.2629	0.2733	0.2837	0.2941
	cosh	1.0201	1.0221	1.0243	1.0266	1.0289	1.0314	1.0340	1.0367	1.0395	1.0423
	tanh	0.1974	0.2070	0.2165	0.2260	0.2355	0.2449	0.2543	0.2636	0.2729	0.2821
0.3	sinh	0.3045	0.3150	0.3255	0.3360	0.3466	0.3572	0.3678	0.3785	0.3892	0.4000
	cosh	1.0453	1.0484	1.0516	1.0549	1.0584	1.0619	1.0655	1.0692	1.0731	1.0770
	tanh	0.2913	0.3004	0.3095	0.3185	0.3275	0.3364	0.3452	0.3540	0.3627	0.3714
0.4	sinh	0.4108	0.4216	0.4325	0.4434	0.4543	0.4653	0.4764	0.4875	0.4986	0.5098
	cosh	1.0811	1.0852	1.0895	1.0939	1.0984	1.1030	1.1077	1.1125	1.1174	1.1225
	tanh	0.3800	0.3885	0.3969	0.4053	0.4136	0.4219	0.4301	0.4382	0.4462	0.4542
0.5	sinh	0.5211	0.5324	0.5438	0.5552	0.5666	0.5782	0.5897	0.6014	0.6131	0.6248
	cosh	1.1276	1.1329	1.1383	1.1438	1.1494	1.1551	1.1609	1.1669	1.1730	1.1792
	tanh	0.4621	0.4700	0.4777	0.4854	0.4930	0.5005	0.5080	0.5154	0.5227	0.5299
0.6	sinh	0.6367	0.6485	0.6605	0.6725	0.6846	0.6967	0.7090	0.7213	0.7336	0.7461
	cosh	1.1855	1.1919	1.1984	1.2051	1.2119	1.2188	1.2258	1.2330	1.2402	1.2476
	tanh	0.5370	0.5441	0.5511	0.5581	0.5649	0.5717	0.5784	0.5850	0.5915	0.5980
0.7	sinh	0.7586	0.7712	0.7838	0.7966	0.8094	0.8223	0.8353	0.8484	0.8615	0.8748
	cosh	1.2552	1.2628	1.2706	1.2785	1.2865	1.2947	1.3030	1.3114	1.3199	1.3286
	tanh	0.6044	0.6107	0.6169	0.6231	0.6292	0.6352	0.6411	0.6469	0.6527	0.6584
0.8	sinh	0.8881	0.9015	0.9150	0.9286	0.9423	0.9561	0.9700	0.9840	0.9981	1.0122
	cosh	1.3374	1.3464	1.3555	1.3647	1.3740	1.3835	1.3932	1.4029	1.4128	1.4229
	tanh	0.6640	0.6696	0.6751	0.6805	0.6858	0.6911	0.6963	0.7014	0.7064	0.7114
0.9	sinh	1.0265	1.0409	1.0554	1.0700	1.0847	1.0995	1.1144	1.1294	1.1446	1.1598
	cosh	1.4331	1.4434	1.4539	1.4645	1.4753	1.4862	1.4973	1.5085	1.5199	1.5314
	tanh	0.7163	0.7211	0.7259	0.7306	0.7352	0.7398	0.7443	0.7487	0.7531	0.7574
1.0	sinh	1.1752	1.1907	1.2063	1.2220	1.2379	1.2539	1.2700	1.2862	1.3025	1.3190
	cosh	1.5431	1.5549	1.5669	1.5790	1.5913	1.6038	1.6164	1.6292	1.6421	1.6552
	tanh	0.7616	0.7658	0.7699	0.7739	0.7779	0.7818	0.7857	0.7895	0.7932	0.7969
1.1	sinh	1.3356	1.3524	1.3693	1.3863	1.4035	1.4208	1.4382	1.4558	1.4735	1.4914
	cosh	1.6685	1.6820	1.6956	1.7093	1.7233	1.7374	1.7517	1.7662	1.7808	1.7956
	tanh	0.8005	0.8041	0.8076	0.8110	0.8144	0.8178	0.8210	0.8243	0.8275	0.8306
1.2	sinh	1.5095	1.5276	1.5460	1.5645	1.5831	1.6019	1.6209	1.6400	1.6593	1.6788
	cosh	1.8107	1.8258	1.8412	1.8568	1.8725	1.8884	1.9045	1.9208	1.9373	1.9540
	tanh	0.8337	0.8367	0.8397	0.8426	0.8455	0.8483	0.8511	0.8538	0.8565	0.8591
1.3	sinh	1.6984	1.7182	1.7381	1.7583	1.7786	1.7991	1.8198	1.8406	1.8617	1.8829
	cosh	1.9709	1.9880	2.0053	2.0228	2.0404	2.0583	2.0764	2.0947	2.1132	2.1320
	tanh	0.8617	0.8643	0.8668	0.8693	0.8717	0.8741	0.8764	0.8787	0.8810	0.8832
1.4	sinh	1.9043	1.9259	1.9477	1.9697	1.9919	2.0143	2.0369	2.0597	2.0827	2.1059
	cosh	2.1509	2.1700	2.1894	2.2090	2.2288	2.2488	2.2691	2.2896	2.3103	2.3312
	tanh	0.8854	0.8875	0.8896	0.8917	0.8937	0.8957	0.8977	0.8996	0.9015	0.9033

Table 10.3(b) *Continued*

1.50–2.99

Angle	Function	0.00	0.01	0.02	0.03	0.04	0.05	0.06	0.07	0.08	0.09
1.5	sinh	2.1293	2.1529	2.1768	2.2008	2.2251	2.2496	2.2743	2.2993	2.3245	2.3499
	cosh	2.3524	2.3738	2.3955	2.4174	2.4395	2.4619	2.4845	2.5074	2.5305	2.5538
	tanh	0.9052	0.9069	c.9087	0.9104	0.9121	0.9138	0.9154	0.9170	c.9186	0.9202
1.6	sinh	2.3756	2.4015	2.4276	2.4540	2.4806	2.5075	2.5346	2.5620	2.5896	2.6175
	cosh	2.5775	2.6013	2.6255	2.6499	2.6746	2.6995	2.7247	2.7502	2.7760	2.8020
	tanh	0.9217	0.9232	0.9246	0.9261	0.9275	0.9289	0.9302	0.9316	0.9329	0.9342
1.7	sinh	2.6456	2.6740	2.7027	2.7317	2.7609	2.7904	2.8202	2.8503	2.8806	2.9112
	cosh	2.8283	2.8549	2.8818	2.9090	2.9364	2.9642	2.9922	3.0206	3.0493	3.0782
	tanh	0.9354	0.9367	0.9379	0.9391	0.9402	0.9414	0.9425	0.9436	0.9447	0.9458
1.8	sinh	2.9422	2.9734	3.0049	3.0367	3.0689	3.1013	3.1340	3.1671	3.2005	3.2341
	cosh	3.1075	3.1371	3.1669	3.1972	3.2277	3.2585	3.2897	3.3212	3.3530	3.3852
	tanh	0.9468	0.9478	0.9488	0.9498	0.9508	0.9518	0.9527	0.9536	0.9545	0.9554
1.9	sinh	3.2682	3.3025	3.3372	3.3722	3.4075	3.4432	3.4792	3.5156	3.5523	3.5894
	cosh	3.4177	3.4506	3.4838	3.5173	3.5512	3.5855	3.6201	3.6551	3.6904	3.7261
	tanh	0.9562	0.9571	0.9579	0.9587	0.9595	0.9603	0.9611	0.9619	0.9626	0.9633
2.0	sinh	3.6269	3.6647	3.7028	3.7414	3.7803	3.8196	3.8593	3.8993	3.9398	3.9806
	cosh	3.7622	3.7987	3.8355	3.8727	3.9103	3.9483	3.9867	4.0255	4.0647	4.1043
	tanh	0.9640	0.9647	0.9654	0.9661	0.9668	0.9674	0.9680	0.9686	0.9693	0.9699
2.1	sinh	4.0219	4.0635	4.1056	4.1480	4.1909	4.2342	4.2779	4.3221	4.3666	4.4117
	cosh	4.1443	4.1847	4.2256	4.2668	4.3085	4.3507	4.3932	4.4362	4.4797	4.5236
	tanh	0.9705	0.9710	0.9716	0.9722	0.9727	0.9732	0.9738	0.9743	0.9748	0.9752
2.2	sinh	4.4571	4.5030	4.5494	4.5962	4.6434	4.6912	4.7394	4.7880	4.8372	4.8868
	cosh	4.5679	4.6127	4.6580	4.7037	4.7499	4.7966	4.8437	4.8914	4.9395	4.9881
	tanh	0.9757	0.9762	0.9767	0.9771	0.9776	0.9780	0.9785	0.9789	0.9793	0.9797
2.3	sinh	4.9370	4.9876	5.0387	5.0903	5.1425	5.1951	5.2483	5.3020	5.3562	5.4109
	cosh	5.0372	5.0868	5.1370	5.1876	5.2388	5.2905	5.3427	5.3954	5.4487	5.5026
	tanh	0.9801	0.9805	0.9809	0.9812	0.9816	0.9820	0.9823	0.9827	0.9830	0.9834
2.4	sinh	5.4662	5.5221	5.5785	5.6354	5.6929	5.7510	5.8097	5.8689	5.9288	5.9892
	cosh	5.5569	5.6119	5.6674	5.7235	5.7801	5.8373	5.8951	5.9535	6.0125	6.0721
	tanh	0.9837	0.9840	0.9843	0.9846	0.9849	0.9852	0.9855	0.9858	0.9861	0.9864
2.5	sinh	6.0502	6.1118	6.1741	6.2369	6.3004	6.3645	6.4293	6.4946	6.5607	6.6274
	cosh	6.1323	6.1931	6.2545	6.3166	6.3793	6.4426	6.5066	6.5712	6.6365	6.7024
	tanh	0.9866	0.9869	0.9871	0.9874	0.9876	0.9879	0.9881	0.9884	0.9886	0.9888
2.6	sinh	6.6947	6.7628	6.8315	6.9009	6.9709	7.0417	7.1132	7.1854	7.2583	7.3319
	cosh	6.7690	6.8363	6.9043	6.9729	7.0423	7.1123	7.1831	7.2546	7.3268	7.3998
	tanh	0.9890	0.9892	0.9895	0.9897	0.9899	0.9901	0.9903	0.9905	0.9906	0.9908
2.7	sinh	7.4063	7.4814	7.5572	7.6338	7.7112	7.7894	7.8683	7.9480	8.0285	8.1098
	cosh	7.4735	7.5479	7.6231	7.6991	7.7758	7.8533	7.9316	8.0106	8.0905	8.1712
	tanh	0.9910	0.9912	0.9914	0.9915	0.9917	0.9919	0.9920	0.9922	0.9923	0.9925
2.8	sinh	8.1919	8.2749	8.3586	8.4432	8.5287	8.6150	8.7021	8.7902	8.8791	8.9689
	cosh	8.2527	8.3351	8.4182	8.5022	8.5871	8.6728	8.7594	8.8469	8.9352	9.0244
	tanh	0.9926	0.9928	0.9929	0.9931	0.9932	0.9933	0.9935	0.9936	0.9937	0.9938
2.9	sinh	9.0596	9.1512	9.2437	9.3371	9.4315	9.5268	9.6231	9.7203	9.8185	9.9177
	cosh	9.1146	9.2056	9.2976	9.3905	9.4844	9.5792	9.6749	9.7716	9.8693	9.9680
	tanh	0.9940	0.9941	0.9942	0.9943	0.9944	0.9945	0.9946	0.9948	c.9949	0.9950

Table **10.3(b)**   *Continued*

Angle	Function	0.00	0.01	0.02	0.03	0.04	0.05	0.06	0.07	0.08	0.09
3.0	sinh	10.018	10.119	10.221	10.324	10.429	10.534	10.640	10.748	10.856	10.966
	cosh	10.068	10.168	10.270	10.373	10.476	10.581	10.687	10.794	10.902	11.011
	tanh	0.9951	0.9952	0.9953	0.9953	0.9954	0.9955	0.9956	0.9957	0.9958	0.9959
3.1	sinh	11.076	11.188	11.301	11.415	11.530	11.647	11.764	11.883	12.003	12.124
	cosh	11.121	11.233	11.345	11.459	11.574	11.689	11.806	11.925	12.044	12.165
	tanh	0.9960	0.9960	0.9961	0.9962	0.9963	0.9963	0.9964	0.9965	0.9966	0.9966
3.2	sinh	12.246	12.369	12.494	12.620	12.747	12.876	13.006	13.137	13.269	13.403
	cosh	12.287	12.410	12.534	12.660	12.786	12.915	13.044	13.175	13.307	13.440
	tanh	0.9967	0.9968	0.9968	0.9969	0.9969	0.9970	0.9971	0.9971	0.9972	0.9972
3.3	sinh	13.538	13.674	13.812	13.951	14.092	14.234	14.377	14.522	14.668	14.816
	cosh	13.575	13.711	13.848	13.987	14.127	14.269	14.412	14.556	14.702	14.850
	tanh	0.9973	0.9973	0.9974	0.9974	0.9975	0.9975	0.9976	0.9976	0.9977	0.9977
3.4	sinh	14.965	15.116	15.268	15.422	15.577	15.734	15.893	16.053	16.215	16.378
	cosh	14.999	15.149	15.301	15.455	15.610	15.766	15.924	16.084	16.245	16.408
	tanh	0.9978	0.9978	0.9979	0.9979	0.9979	0.9980	0.9980	0.9981	0.9981	0.9981
3.5	sinh	16.543	16.709	16.877	17.047	17.219	17.392	17.567	17.744	17.923	18.103
	cosh	16.573	16.739	16.907	17.077	17.248	17.421	17.596	17.772	17.951	18.131
	tanh	0.9982	0.9982	0.9983	0.9983	0.9983	0.9984	0.9984	0.9984	0.9985	0.9985
3.6	sinh	18.285	18.470	18.655	18.843	19.033	19.224	19.418	19.613	19.811	20.010
	cosh	18.313	18.497	18.682	18.870	19.059	19.250	19.444	19.639	19.836	20.035
	tanh	0.9985	0.9985	0.9986	0.9986	0.9986	0.9987	0.9987	0.9987	0.9987	0.9988
3.7	sinh	20.211	20.415	20.620	20.828	21.037	21.249	21.463	21.679	21.897	22.117
	cosh	20.236	20.439	20.644	20.852	21.061	21.272	21.486	21.702	21.919	22.139
	tanh	0.9988	0.9988	0.9988	0.9989	0.9989	0.9989	0.9989	0.9989	0.9990	0.9990
3.8	sinh	22.339	22.564	22.791	23.020	23.252	23.486	23.722	23.961	24.202	24.445
	cosh	22.362	22.586	22.813	23.042	23.273	23.507	23.743	23.982	24.222	24.466
	tanh	0.9990	0.9990	0.9990	0.9991	0.9991	0.9991	0.9991	0.9991	0.9992	0.9992
3.9	sinh	24.691	24.939	25.190	25.444	25.700	25.958	26.219	26.483	26.749	27.018
	cosh	24.711	24.959	25.210	25.463	25.719	25.977	26.238	26.502	26.768	27.037
	tanh	0.9992	0.9992	0.9992	0.9992	0.9992	0.9993	0.9993	0.9993	0.9993	0.9993
4.0	sinh	27.290	27.564	27.842	28.122	28.404	28.690	28.979	29.270	29.564	29.862
	cosh	27.308	27.583	27.860	28.139	28.422	28.707	28.996	29.287	29.581	29.878
	tanh	0.9993	0.9993	0.9994	0.9994	0.9994	0.9994	0.9994	0.9994	0.9994	0.9994
4.1	sinh	30.162	30.465	30.772	31.081	31.393	31.709	32.028	32.350	32.675	33.004
	cosh	30.178	30.482	30.788	31.097	31.409	31.725	32.044	32.365	32.691	33.019
	tanh	0.9995	0.9995	0.9995	0.9995	0.9995	0.9995	0.9995	0.9995	0.9995	0.9995
4.2	sinh	33.336	33.671	34.009	34.351	34.697	35.046	35.398	35.754	36.113	36.476
	cosh	33.351	33.686	34.024	34.366	34.711	35.060	35.412	35.768	36.127	36.490
	tanh	0.9996	0.9996	0.9996	0.9996	0.9996	0.9996	0.9996	0.9996	0.9996	0.9996
4.3	sinh	36.843	37.214	37.588	37.965	38.347	38.733	39.122	39.515	39.913	40.314
	cosh	36.857	37.227	37.601	37.979	38.360	38.746	39.135	39.528	39.925	40.326
	tanh	0.9996	0.9996	0.9997	0.9997	0.9997	0.9997	0.9997	0.9997	0.9997	0.9997
4.4	sinh	40.719	41.129	41.542	41.960	42.382	42.808	43.238	43.673	44.112	44.555
	cosh	40.732	41.141	41.554	41.972	42.393	42.819	43.250	43.684	44.123	44.566
	tanh	0.9997	0.9997	0.9997	0.9997	0.9997	0.9997	0.9997	0.9997	0.9997	0.9998

Table 10.3(*b*)   *Continued*

Angle	Function	0.00	0.01	0.02	0.03	0.04	0.05	0.06	0.07	0.08	0.09
4.5	sinh	45.003	45.455	45.912	46.374	46.840	47.311	47.787	48.267	48.752	49.242
	cosh	45.014	45.466	45.923	46.385	46.851	47.321	47.797	48.277	48.762	49.252
	tanh	0.9998	0.9998	0.9998	0.9998	0.9998	0.9998	0.9998	0.9998	0.9998	0.9998
4.6	sinh	49.737	50.237	50.742	51.252	51.767	52.288	52.813	53.344	53.880	54.422
	cosh	49.747	50.247	50.752	51.262	51.777	52.297	52.823	53.354	53.890	54.431
	tanh	0.9998	0.9998	0.9998	0.9998	0.9998	0.9998	0.9998	0.9998	0.9998	0.9998
4.7	sinh	54.969	55.522	56.080	56.643	57.213	57.788	58.369	58.955	59.548	60.147
	cosh	54.978	55.531	56.089	56.652	57.221	57.796	58.377	58.964	59.556	60.155
	tanh	0.9998	0.9998	0.9998	0.9998	0.9999	0.9999	0.9999	0.9999	0.9999	0.9999
4.8	sinh	60.751	61.362	61.979	62.601	63.231	63.866	64.508	65.157	65.812	66.473
	cosh	60.759	61.370	61.987	62.609	63.239	63.874	64.516	65.164	65.819	66.481
	tanh	0.9999	0.9999	0.9999	0.9999	0.9999	0.9999	0.9999	0.9999	0.9999	0.9999
4.9	sinh	67.141	67.816	68.498	69.186	69.882	70.584	71.293	72.010	72.734	73.465
	cosh	67.149	67.823	68.505	69.193	69.889	70.591	71.300	72.017	72.741	73.472
	tanh	0.9999	0.9999	0.9999	0.9999	0.9999	0.9999	0.9999	0.9999	0.9999	0.9999
5.0	sinh	74.203	74.949	75.702	76.463	77.232	78.008	78.792	79.584	80.384	81.192
	cosh	74.210	74.956	75.709	76.470	77.238	78.014	78.798	79.590	80.390	81.198
	tanh	0.9999	0.9999	0.9999	0.9999	0.9999	0 9999	0.9999	0.9999	0.9999	0.9999
5.1	sinh	82.008	82.832	83.665	84.506	85.355	86.213	87.079	87.955	88.839	89.732
	cosh	82.014	82.838	83.671	84.512	85.361	86.219	87.085	87.960	88.844	89.737
	tanh	0.9999	0.9999	0.9999	0.9999	0.9999	0.9999	0.9999	0.9999	0.9999	0.9999
5.2	sinh	90.633	91.544	92.464	93.394	94.332	95.281	96.238	97.205	98.182	99.169
	cosh	90.639	91.550	92.470	93.399	94.338	95.286	96.243	97.211	98.188	99.174
	tanh	0.9999	0.9999	0.9999	0.9999	0.9999	0.9999	1.0000	1.0000	1.0000	1.0000
5.3	sinh	100.17	101.17	102.19	103.22	104.25	105.30	106.36	107.43	108.51	109.60
	cosh	100.17	101.18	102.19	103.22	104.26	105.31	106.37	107.43	108.51	109.60
	tanh	1.0000	1.0000	1.0000	1.0000	1.0000	1.0000	1.0000	1.0000	1.0000	1.0000
5.4	sinh	110.70	111.81	112.94	114.07	115.22	116.38	117.55	118.73	119.92	121.13
	cosh	110.71	111.82	112.94	114.08	115.22	116.38	117.55	118.73	119.93	121.13
	tanh	1.0000	1.0000	1.0000	1.0000	1.0000	1.0000	1.0000	1.0000	1.0000	1.0000
5.5	sinh	122.34	123.57	124.82	126.07	127.34	128.62	129.91	131.22	132.53	133.87
	cosh	122.35	123.58	124.82	126.07	127.34	128.62	129.91	131.22	132.54	133.87
	tanh	1.0000	1.0000	1.0000	1.0000	1.0000	1.0000	1.0000	1.0000	1.0000	1.0000
5.6	sinh	135.21	136.57	137.94	139.33	140.73	142.14	143.57	145.02	146.47	147.95
	cosh	135.22	136.57	137.95	139.33	140.73	142.15	143.58	145.02	146.48	147.95
	tanh	1.0000	1.0000	1.0000	1.0000	1.0000	1.0000	1.0000	1.0000	1.0000	1.0000
5.7	sinh	149.43	150.93	152.45	153.98	155.53	157.09	158.67	160.27	161.88	163.51
	cosh	149.44	150.94	152.45	153.99	155.53	157.10	158.68	160.27	161.88	163.51
	tanh	1.0000	1.0000	1.0000	1.0000	1.0000	1.0000	1.0000	1.0000	1.0000	1.0000
5.8	sinh	165.15	166.81	168.48	170.18	171.89	173.62	175.36	177.12	178.90	180.70
	cosh	165.15	166.81	168.49	170.18	171.89	173.62	175.36	177.13	178.91	180.70
	tanh	1.0000	1.0000	1.0000	1.0000	1.0000	1.0000	1.0000	1.0000	1.0000	1.0000
5.9	sinh	182.52	184.35	186.20	188.08	189.97	191.88	193.80	195.75	197.72	199.71
	cosh	182.52	184.35	186.21	188.08	189.97	191.88	193.81	195.75	197.72	199.71
	tanh	1.0000	1.0000	1.0000	1.0000	1.0000	1.0000	1.0000	1.0000	1.0000	1.0000

Table 10.4 Values of e^x and e^{-x}

x	Function	0.00	0.01	0.02	0.03	0.04	0.05	0.06	0.07	0.08	0.09
0.0	e^x	1.0000	1.0101	1.0202	1.0305	1.0408	1.0513	1.0618	1.0725	1.0833	1.0942
	e^{-x}	1.0000	0.9900	0.9802	0.9704	0.9608	0.9512	0.9418	0.9324	0.9231	0.9139
0.1	e^x	1.1052	1.1163	1.1275	1.1388	1.1503	1.1618	1.1735	1.1853	1.1972	1.2093
	e^{-x}	0.9048	0.8958	0.8869	0.8781	0.8694	0.8607	0.8521	0.8437	0.8353	0.8270
0.2	e^x	1.2214	1.2337	1.2461	1.2586	1.2712	1.2840	1.2969	1.3100	1.3231	1.3364
	e^{-x}	0.8187	0.8106	0.8025	0.7945	0.7866	0.7788	0.7711	0.7634	0.7558	0.7483
0.3	e^x	1.3499	1.3634	1.3771	1.3910	1.4049	1.4191	1.4333	1.4477	1.4623	1.4770
	e^{-x}	0.7408	0.7334	0.7261	0.7189	0.7118	0.7047	0.6977	0.6907	0.6839	0.6771
0.4	e^x	1.4918	1.5068	1.5220	1.5373	1.5527	1.5683	1.5841	1.6000	1.6161	1.6323
	e^{-x}	0.6703	0.6637	0.6570	0.6505	0.6440	0.6376	0.6313	0.6250	0.6188	0.6126
0.5	e^x	1.6487	1.6653	1.6820	1.6989	1.7160	1.7333	1.7507	1.7683	1.7860	1.8040
	e^{-x}	0.6065	0.6005	0.5945	0.5886	0.5827	0.5769	0.5712	0.5655	0.5599	0.5543
0.6	e^x	1.8221	1.8404	1.8589	1.8776	1.8965	1.9155	1.9348	1.9542	1.9739	1.9939
	e^{-x}	0.5488	0.5434	0.5379	0.5326	0.5273	0.5220	0.5169	0.5117	0.5066	0.5017
0.7	e^x	2.0138	2.0340	2.0544	2.0751	2.0959	2.1170	2.1383	2.1598	2.1815	2.2034
	e^{-x}	0.4966	0.4916	0.4868	0.4819	0.4771	0.4724	0.4677	0.4630	0.4584	0.4538
0.8	e^x	2.2255	2.2479	2.2705	2.2933	2.3164	2.3396	2.3632	2.3869	2.4109	2.4351
	e^{-x}	0.4493	0.4449	0.4404	0.4360	0.4317	0.4274	0.4232	0.4190	0.4148	0.4107
0.9	e^x	2.4596	2.4843	2.5093	2.5345	2.5600	2.5857	2.6117	2.6379	2.6645	2.6912
	e^{-x}	0.4066	0.4025	0.3985	0.3946	0.3906	0.3867	0.3829	0.3791	0.3753	0.3716
1.0	e^x	2.7183	2.7456	2.7732	2.8011	2.8292	2.8577	2.8864	2.9154	2.9447	2.9743
	e^{-x}	0.3679	0.3642	0.3606	0.3570	0.3535	0.3499	0.3465	0.3430	0.3396	0.3362
1.1	e^x	3.0042	3.0344	3.0649	3.0957	3.1268	3.1582	3.1899	3.2220	3.2544	3.2871
	e^{-x}	0.3329	0.3296	0.3263	0.3230	0.3198	0.3166	0.3135	0.3104	0.3073	0.3042
1.2	e^x	3.3201	3.3535	3.3872	3.4212	3.4556	3.4903	3.5254	3.5609	3.5966	3.6328
	e^{-x}	0.3012	0.2982	0.2952	0.2923	0.2894	0.2865	0.2837	0.2808	0.2780	0.2753
1.3	e^x	3.6693	3.7062	3.7434	3.7810	3.8190	3.8574	3.8962	3.9354	3.9749	4.0149
	e^{-x}	0.2725	0.2698	0.2671	0.2645	0.2618	0.2592	0.2567	0.2541	0.2516	0.2491
1.4	e^x	4.0552	4.0960	4.1371	4.1787	4.2207	4.2631	4.3060	4.3492	4.3929	4.4371
	e^{-x}	0.2466	0.2441	0.2417	0.2393	0.2369	0.2346	0.2322	0.2299	0.2276	0.2254
1.5	e^x	4.4817	4.5267	4.5722	4.6182	4.6646	4.7115	4.7588	4.8066	4.8550	4.9037
	e^{-x}	0.2231	0.2209	0.2187	0.2165	0.2144	0.2122	0.2101	0.2080	0.2060	0.2039
1.6	e^x	4.9530	5.0028	5.0531	5.1039	5.1552	5.2070	5.2593	5.3122	5.3656	5.4195
	e^{-x}	0.2019	0.1999	0.1979	0.1959	0.1940	0.1920	0.1901	0.1882	0.1864	0.1845
1.7	e^x	5.4739	5.5290	5.5845	5.6407	5.6973	5.7546	5.8124	5.8709	5.9299	5.9895
	e^{-x}	0.1827	0.1809	0.1791	0.1773	0.1755	0.1738	0.1720	0.1703	0.1686	0.1670
1.8	e^x	6.0496	6.1104	6.1719	6.2339	6.2965	6.3598	6.4237	6.4883	6.5535	6.6194
	e^{-x}	0.1653	0.1637	0.1620	0.1604	0.1588	0.1572	0.1557	0.1541	0.1526	0.1511
1.9	e^x	6.6859	6.7531	6.8210	6.8895	6.9588	7.0287	7.0993	7.1707	7.2427	7.3155
	e^{-x}	0.1496	0.1481	0.1466	0.1451	0.1437	0.1423	0.1409	0.1395	0.1381	0.1367

Table 10.4 *Continued*

2.00–3.99

x	Func-tion	0.00	0.01	0.02	0.03	0.04	0.05	0.06	0.07	0.08	0.09
2.0	e^x	7.3891	7.4633	7.5383	7.6141	7.6906	7.7679	7.8460	7.9248	8.0045	8.0849
	e^{-x}	0.1353	0.1340	0.1327	0.1313	0.1300	0.1287	0.1275	0.1262	0.1249	0.1237
2.1	e^x	8.1662	8.2482	8.3311	8.4149	8.4994	8.5849	8.6711	8.7583	8.8463	8.9352
	e^{-x}	0.1225	0.1212	0.1200	0.1188	0.1177	0.1165	0.1153	0.1142	0.1130	0.1119
2.2	e^x	9.0250	9.1157	9.2073	9.2999	9.3933	9.4877	9.5831	9.6794	9.7767	9.8749
	e^{-x}	0.1108	0.1097	0.1086	0.1075	0.1065	0.1054	0.1044	0.1033	0.1023	0.1013
2.3	e^x	9.9742	10.074	10.176	10.278	10.381	10.486	10.591	10.697	10.805	10.913
	e^{-x}	0.1003	0.0993	0.0983	0.0973	0.0963	0.0954	0.0944	0.0935	0.0926	0.0916
2.4	e^x	11.023	11.134	11.246	11.359	11.473	11.588	11.705	11.822	11.941	12.061
	e^{-x}	0.0907	0.0898	0.0889	0.0880	0.0872	0.0863	0.0854	0.0846	0.0837	0.0829
2.5	e^x	12.182	12.305	12.429	12.554	12.680	12.807	12.936	13.066	13.197	13.330
	e^{-x}	0.0821	0.0813	0.0805	0.0797	0.0789	0.0781	0.0773	0.0765	0.0758	0.0750
2.6	e^x	13.464	13.599	13.736	13.874	14.013	14.154	14.296	14.440	14.585	14.732
	e^{-x}	0.0743	0.0735	0.0728	0.0721	0.0714	0.0707	0.0699	0.0693	0.0686	0.0679
2.7	e^x	14.880	15.029	15.180	15.333	15.487	15.643	15.800	15.959	16.119	16.281
	e^{-x}	0.0672	0.0665	0.0659	0.0652	0.0646	0.0639	0.0633	0.0627	0.0620	0.0614
2.8	e^x	16.445	16.610	16.777	16.945	17.116	17.288	17.462	17.637	17.814	17.993
	e^{-x}	0.0608	0.0602	0.0596	0.0590	0.0584	0.0578	0.0573	0.0567	0.0561	0.0556
2.9	e^x	18.174	18.357	18.541	18.728	18.916	19.106	19.298	19.492	19.688	19.886
	e^{-x}	0.0550	0.0545	0.0539	0.0534	0.0529	0.0523	0.0518	0.0513	0.0508	0.0503
3.0	e^x	20.086	20.287	20.491	20.697	20.905	21.115	21.328	21.542	21.758	21.977
	e^{-x}	0.0498	0.0493	0.0488	0.0483	0.0478	0.0474	0.0469	0.0464	0.0460	0.0455
3.1	e^x	22.198	22.421	22.646	22.874	23.104	23.336	23.571	23.807	24.047	24.288
	e^{-x}	0.0450	0.0446	0.0442	0.0437	0.0433	0.0429	0.0424	0.0420	0.0416	0.0412
3.2	e^x	24.533	24.779	25.028	25.280	25.534	25.790	26.050	26.311	26.576	26.843
	e^{-x}	0.0408	0.0404	0.0400	0.0396	0.0392	0.0388	0.0384	0.0380	0.0376	0.0373
3.3	e^x	27.113	27.385	27.660	27.938	28.219	28.503	28.789	29.079	29.371	29.666
	e^{-x}	0.0369	0.0365	0.0362	0.0358	0.0354	0.0351	0.0347	0.0344	0.0340	0.0337
3.4	e^x	29.964	30.265	30.569	30.877	31.187	31.500	31.817	32.137	32.460	32.786
	e^{-x}	0.0334	0.0330	0.0327	0.0324	0.0321	0.0317	0.0314	0.0311	0.0308	0.0305
3.5	e^x	33.115	33.448	33.784	34.124	34.467	34.813	35.163	35.517	35.874	36.234
	e^{-x}	0.0302	0.0299	0.0296	0.0293	0.0290	0.0287	0.0284	0.0282	0.0279	0.0276
3.6	e^x	36.598	36.966	37.338	37.713	38.092	38.475	38.861	39.252	39.646	40.045
	e^{-x}	0.0273	0.0271	0.0268	0.0265	0.0263	0.0260	0.0257	0.0255	0.0252	0.0250
3.7	e^x	40.447	40.854	41.264	41.679	42.098	42.521	42.948	43.380	43.816	44.256
	e^{-x}	0.0247	0.0245	0.0242	0.0240	0.0238	0.0235	0.0233	0.0231	0.0228	0.0226
3.8	e^x	44.701	45.150	45.604	46.063	46.525	46.993	47.465	47.942	48.424	48.911
	e^{-x}	0.0224	0.0221	0.0219	0.0217	0.0215	0.0213	0.0211	0.0209	0.0207	0.0204
3.9	e^x	49.402	49.899	50.400	50.907	51.419	51.935	52.457	52.985	53.517	54.055
	e^{-x}	0.0202	0.0200	0.0198	0.0196	0.0195	0.0193	0.0191	0.0189	0.0187	0.0185

Table 10.4 *Continued*

x	Func-tion	0.00	0.01	0.02	0.03	0.04	0.05	0.06	0.07	0.08	0.09
4.0	e^x	54.598	55.147	55.701	56.261	56.826	57.397	57.974	58.557	59.145	59.740
	ϵ^{-x}	0.0183	0.0181	0.0180	0.0178	0.0176	0.0174	0.0172	0.0171	0.0169	0.0167
4.1	e^x	60.340	60.947	61.559	62.178	62.803	63.434	64.072	64.715	65.366	66.023
	ϵ^{-x}	0.0166	0.0164	0.0162	0.0161	0.0159	0.0158	0.0156	0.0155	0.0153	0.0151
4.2	e^x	66.686	67.357	68.033	68.717	69.408	70.105	70.810	71.522	72.240	72.966
	ϵ^{-x}	0.0150	0.0148	0.0147	0.0146	0.0144	0.0143	0.0141	0.0140	0.0138	0.0137
4.3	e^x	73.700	74.440	75.189	75.944	76.708	77.478	78.257	79.044	79.838	80.640
	ϵ^{-x}	0.0136	0.0134	0.0133	0.0132	0.0130	0.0129	0.0128	0.0127	0.0125	0.0124
4.4	e^x	81.451	82.269	83.096	83.931	84.775	85.627	86.488	87.357	88.235	89.121
	ϵ^{-x}	0.0123	0.0122	0.0120	0.0119	0.0118	0.0117	0.0116	0.0114	0.0113	0.0112
4.5	e^x	90.017	90.922	91.836	92.759	93.691	94.632	95.583	96.544	97.514	98.494
	ϵ^{-x}	0.0111	0.0110	0.0109	0.0108	0.0107	0.0106	0.0105	0.0104	0.0103	0.0102
4.6	e^x	99.484	100.48	101.49	102.51	103.54	104.58	105.64	106.70	107.77	108.85
	ϵ^{-x}	0.0101	0.0100	0.0099	0.0098	0.0097	0.0096	0.0095	0.0094	0.0093	0.0092
4.7	e^x	109.95	111.05	112.17	113.30	114.43	115.58	116.75	117.92	119.10	120.30
	ϵ^{-x}	0.0091	0.0090	0.0089	0.0088	0.0087	0.0087	0.0086	0.0085	0.0084	0.0083
4.8	e^x	121.51	122.73	123.97	125.21	126.47	127.74	129.02	130.32	131.63	132.95
	ϵ^{-x}	0.0082	0.0081	0.0081	0.0080	0.0079	0.0078	0.0078	0.0077	0.0076	0.0075
4.9	e^x	134.29	135.64	137.00	138.38	139.77	141.17	142.59	144.03	145.47	146.94
	ϵ^{-x}	0.0074	0.0074	0.0073	0.0072	0.0072	0.0071	0.0070	0.0069	0.0069	0.0068
5.0	e^x	148.41	149.90	151.41	152.93	154.47	156.02	157.59	159.17	160.77	162.39
	ϵ^{-x}	0.0067	0.0067	0.0066	0.0065	0.0065	0.0064	0.0063	0.0063	0.0062	0.0062
5.1	ϵ	164.02	165.67	167.34	169.02	170.72	172.43	174.16	175.91	177.68	179.47
	ϵ^{-x}	0.0061	0.0060	0.0060	0.0059	0.0059	0.0058	0.0057	0.0057	0.0056	0.0056
5.2	e^x	181.27	183.09	184.93	186.79	188.67	190.57	192.48	194.42	196.37	198.34
	ϵ^{-x}	0.0055	0.0055	0.0054	0.0054	0.0053	0.0052	0.0052	0.0051	0.0051	0.0050
5.3	e^x	200.34	202.35	204.38	206.44	208.51	210.61	212.72	214.86	217.02	219.20
	ϵ^{-x}	0.0050	0.0049	0.0049	0.0048	0.0048	0.0047	0.0047	0.0047	0.0046	0.0046
5.4	e^x	221.41	223.63	225.88	228.15	230.44	232.76	235.10	237.46	239.85	242.26
	ϵ^{-x}	0.0045	0.0045	0.0044	0.0044	0.0043	0.0043	0.0043	0.0042	0.0042	0.0041
5.5	e^x	244.69	247.15	249.64	252.14	254.68	257.24	259.82	262.43	265.07	267.74
	ϵ^{-x}	0.0041	0.0040	0.0040	0.0040	0.0039	0.0039	0.0038	0.0038	0.0038	0.0037
5.6	e^x	270.43	273.14	275.89	278.66	281.46	284.29	287.15	290.03	292.95	295.89
	ϵ^{-x}	0.0037	0.0037	0.0036	0.0036	0.0036	0.0035	0.0035	0.0034	0.0034	0.0034
5.7	e^x	298.87	301.87	304.90	307.97	311.06	314.19	317.35	320.54	323.76	327.01
	ϵ^{-x}	0.0033	0.0033	0.0033	0.0032	0.0032	0.0032	0.0032	0.0031	0.0031	0.0031
5.8	e^x	330.30	333.62	336.97	340.36	343.78	347.23	350.72	354.25	357.81	361.41
	ϵ^{-x}	0.0030	0.0030	0.0030	0.0029	0.0029	0.0029	0.0029	0.0028	0.0028	0.0028
5.9	e^x	365.04	368.71	372.41	376.15	379.93	383.75	387.61	391.51	395.44	399.41
	ϵ^{-x}	0.0027	0.0027	0.0027	0.0027	0.0026	0.0026	0.0026	0.0026	0.0025	0.0025

Table 10.5 Decimal Equivalents of Fractions

Fractions	Decimals	Fractions	Decimals	Fractions	Decimals	Fractions	Decimals
$\frac{1}{64}$	0.015625	$\frac{17}{64}$	0.265625	$\frac{33}{64}$	0.515625	$\frac{49}{64}$	0.765625
$\frac{1}{32}$	0.03125	$\frac{9}{32}$	0.28125	$\frac{17}{32}$	0.53125	$\frac{25}{32}$	0.78125
$\frac{3}{64}$	0.046875	$\frac{19}{64}$	0.296875	$\frac{35}{64}$	0.546875	$\frac{51}{64}$	0.796875
$\frac{1}{16}$	0.0625	$\frac{5}{16}$	0.3125	$\frac{9}{16}$	0.5625	$\frac{13}{16}$	0.8125
$\frac{5}{64}$	0.078125	$\frac{21}{64}$	0.328125	$\frac{37}{64}$	0.578125	$\frac{53}{64}$	0.828125
$\frac{3}{32}$	0.09375	$\frac{11}{32}$	0.34375	$\frac{19}{32}$	0.59375	$\frac{27}{32}$	0.84375
$\frac{7}{64}$	0.109375	$\frac{23}{64}$	0.359375	$\frac{39}{64}$	0.609375	$\frac{55}{64}$	0.859375
$\frac{1}{8}$	0.125	$\frac{3}{8}$	0.375	$\frac{5}{8}$	0.625	$\frac{7}{8}$	0.875
$\frac{9}{64}$	0.140625	$\frac{25}{64}$	0.390625	$\frac{41}{64}$	0.640625	$\frac{57}{64}$	0.890625
$\frac{5}{32}$	0.15625	$\frac{13}{32}$	0.40625	$\frac{21}{32}$	0.65625	$\frac{29}{32}$	0.90625
$\frac{11}{64}$	0.171875	$\frac{27}{64}$	0.421875	$\frac{43}{64}$	0.671875	$\frac{59}{64}$	0.921875
$\frac{3}{16}$	0.1875	$\frac{7}{16}$	0.4375	$\frac{11}{16}$	0.6875	$\frac{15}{16}$	0.9375
$\frac{13}{64}$	0.203125	$\frac{29}{64}$	0.453125	$\frac{45}{64}$	0.703125	$\frac{61}{64}$	0.953125
$\frac{7}{32}$	0.21875	$\frac{15}{32}$	0.46875	$\frac{23}{32}$	0.71875	$\frac{31}{32}$	0.96875
$\frac{15}{64}$	0.234375	$\frac{31}{64}$	0.484375	$\frac{47}{64}$	0.734375	$\frac{63}{64}$	0.984375
$\frac{1}{4}$	0.25	$\frac{1}{2}$	0.5	$\frac{3}{4}$	0.75	1	1

Table 10.6 Factorials

n	$n! = 1\cdot2\cdot3\cdots n$	$1/n!$	n	$n! = 1\cdot2\cdot3\cdots n$	$1/n!$
1	1	1.	11	$399,168 \times 10^2$	0.250521×10^{-7}
2	2	0.5	12	$479,002 \times 10^3$	$.208768 \times 10^{-8}$
3	6	.166667	13	$622,702 \times 10^4$	$.160590 \times 10^{-9}$
4	24	$.416667 \times 10^{-1}$	14	$871,783 \times 10^5$	$.114707 \times 10^{-10}$
5	120	$.833333 \times 10^{-2}$	15	$130,767 \times 10^7$	$.764716 \times 10^{-12}$
6	720	$.138889 \times 10^{-2}$	16	$209,228 \times 10^8$	$.477948 \times 10^{-13}$
7	5,040	$.198413 \times 10^{-3}$	17	$355,687 \times 10^9$	$.281146 \times 10^{-14}$
8	40,320	$.248016 \times 10^{-4}$	18	$640,237 \times 10^{10}$	$.156192 \times 10^{-15}$
9	362,880	$.275573 \times 10^{-5}$	19	$121,645 \times 10^{12}$	$.822064 \times 10^{-17}$
10	3,628,800	$.275573 \times 10^{-6}$	20	$243,290 \times 10^{13}$	$.411032 \times 10^{-18}$

Table 10.7 Selected Physical Constants

Second	1/86,400 mean solar days
Temperature 0°C, ice point	273.15°K or 491.67°R
Avogadro number	6.02252×10^{23} molecules/mol
Boltzmann constant	1.38054×10^{-23} joules/ K
Elementary charge	1.60210×10^{-19} coulombs
Electron rest mass	9.1091×10^{-28} grams
Faraday constant (electromag)	9.64870×10^{4} coulombs/mol
Gravitational constant	6.670×10^{-8} dyne-cm²/gm²
Ideal gas constant	8.3143 joules/°K-mol
Light speed in vacuum	2.997925×10^{8} meters/second
Planck constant	6.6256×10^{-34} joule-seconds
Standard acceleration of gravity	9.80665 meters/sec²
Standard acceleration of gravity	32.1740 feet/sec²
Standard atmospheric pressure	1013250 dynes/cm²
Stefan-Boltzmann constant	5.6697×10^{-5} ergs/cm²-sec-°K⁴
Volume ideal gas at 0°C, 1 atmos	22413.6 cm³/gram mol

Table 10.8 Solar System Characteristics

Body	Mean dist. from sun 10^6 km	Orbital period sidereal days	Axial period sidereal hr	Equatorial diameter km	Mass 10^{24} kg	Density g/cm³	g m/sec²	Escape velocity km/sec
⸱un	—	—	609 h, 6′	1.392×10^6	1980000	1.39	271	—
⸱ercury	57.85	87.97	—	4800	0.32	5.3	3.33	4.2
⸱nus	108.2	224.70	30 h	12400	4.9	4.95	8.52	10.3
⸱rth	149.6	365.26	23 h, 56′, 4.1″	12756.6	6.0	5.52	9.81	11.2
⸱oon	0.38*	27.32	655 h, 43′, 11″	3478	0.074	3.39	1.62	2.4
⸱ars	227.9	686.98	24 h, 37′, 23″	6783	0.64	3.95	3.77	5.1
⸱piter	778.3	4332.6	9 h, 50′, 30″	142600	1900	1.33	25.1	61
⸱turn	1428	10759	10 h, 14′	119000	570	0.69	10.72	37
⸱ranus	2872	30687	10 h, 49′	51500	87	1.56	8.83	22
⸱eptune	4498	60184	15 h, 40′	49900	103	2.27	11.00	25
⸱uto	5910	90700	16 h	12800	5.6	5	9.1	10

* Distance from earth

Table 10.9   Weights of Materials

Material	Lbs. per cu. ft.	Material	Lbs. per cu. ft.
Air *	0.0809	copper, pure	554
acetylene gas *	0.0733	" cast	549–558
alabaster	168	" wrought	552–558
alcohol	49–57	" wire	555–558
aluminum, pure	168	cork	15.6
" cast	160		
" wire	168	Erbium	297
amber	67	emery	250
ammonia *	0.0482		
antimony	414	Feldspar	158–162
argon *	0.113	flint	162
arsenic	357	fluorine *	0.0920
asbestos	125–175		
asphaltum	69–94	Germanium	341
		german silver	515–535
Barium	234	glass, common	150–175
basalt	180	" flint	180–280
bismuth	609	glucinum	122
boron	159	glycerine	78.6
brass	510–542	gold	1203
brick	100–150	granite	125–187
bromine	196	gravel	90–147
bronze	545–555	gum arabic	90
		gun metal	533
Cadmium	540	gutta percha	61
caesium	117	gypsum	144
calcium	98.6		
carbon	125–144	Hydrogen *	0.00562
" bisulphide	80.6		
" dioxide *	0.124	Ice	55–57
" monoxide *	0.0782	iodine	300
celluloid	90	iridium	1399
cement, loose	72–105	iron, pure	491
" set	168–187	" gray cast	439–445
cerium	437	" white cast	473–482
chalk	119–175	" wrought	487–492
charcoal	17–35	" steel	474–494
chlorine *	0.196	ivory	114
chromium	368		
clay, hard	129–133	Lead	710
" soft	118	leather, dry	54
coal, anthracite	81–106	" greased	64
" " loose	47–58	lime	53–75
" bituminous	78–88	limestone	156–162
" " loose	44–54	lithium	39
" lignite	52	loam	65–88
cobalt	530–563		
coke	62–105	Magnesium	107
" loose	23–32	" carbonate	150
columbium	452	manganese	462
concrete (1 : 2 : 4)	146	marble	157–177
" (1 : 1½ : 3)	139	masonry	100–165
" (1 : 3 : 6)	156	mercury *	849
		mica	165–200
		molybdenum	529

* At 0° Cent. and atmospheric pressure.

Table 10.9 *Continued*

Material	Lbs. per cu. ft.	Material	Lbs. per cu. ft.
mortar, hard.............	103	steel.....................	474–494
muck.....................	40–74	strontium...............	158
mud......................	80–130	sulphur..................	120–130
Naptha..................	53	Talc.....................	168
nickel...................	540–550	tantalum................	1040
nitrogen *...............	0.0782	tar......................	62.4
nitrous oxide *..........	0.0838	tellurium................	389
		thallium.................	739
Oil, cotton-seed.........	60.2	thorium..................	686
" lard................	57.4	tile.....................	113
" linseed.............	58.8	" hollow..............	26–45
" lubricating.........	56.2–57.7	tin......................	455
" petroleum..........	54.8	titanium.................	218
" transformer........	52.6–54.2	trap rock................	187–190
" turpentine..........	54.2	tungsten.................	1174
" whale..............	57.3	turf.....................	20–30
osmium..................	1400		
oxygen *................	0.0895	Uranium.................	1165
Palladium...............	711	Vanadium...............	343
paper....................	44–72		
paraffin..................	54–57	Water, max. dens........	62.4
peat.....................	20–30	" sea...............	64.0–64.3
phosphorus..............	146	wax, bees...............	60.5
pitch....................	67	wood, ash...............	45–47
plaster of Paris..........	144	" bamboo..........	22–25
platinum.................	1336	" beech............	43–56
porcelain.................	143–156	" birch............	32–48
potassium................	53.7	" butternut........	24–28
pumice stone............	23–56	" cedar............	37–38
		" cherry...........	43–56
Quartz..................	165	" chestnut.........	38–40
		" cypress..........	32–37
Resin....................	67	" ebony...........	69–83
rhodium.................	773	" elm.............	35–36
rubber, pure.............	58.0–60.5	" fir..............	34–35
" compound........	106–124	" hemlock.........	25–29
" ebonite...........	74.9–78.0	" hickory..........	53–58
rubidium................	955	" lig. vitæ.........	78–83
ruthenium...............	767	" mahogany........	32–53
		" maple...........	49–50
Salt.....................	129–131	" oak.............	37–56
sand.....................	90–120	" pine............	24–45
sandstone................	124–200	" poplar..........	24–27
selenium.................	300	" red wood........	30–32
shale....................	162	" spruce..........	25–32
silicon...................	131	" walnut..........	38–45
silver....................	660	" willow..........	24–37
slate....................	162–205		
snow, fresh fallen........	5–12	Xenon *.................	0.284
" wet compact.......	15–50		
soapstone................	162–175	Zinc....................	448
sodium..................	60.5	zirconium...............	258
spermaceti..............	59		

* At 0° Cent. and atmospheric pressure.

Table 10.10 Specific Heats

Average values (0° to 100° C. unless otherwise stated) of c in the formula $Q = Mkc (t_2 - t_1)$, c being measured in gram-calories per gram per degree C. or British thermal units per pound per degree F.

Acetylene * (15)	0.383	Ice (−20 to 0)	0.505
air * (−30 to +10)	0.238	iridium	0.0323
air † (−30 to +10)	0.169	iron, cast	0.119
alcohol, ethyl (30)	0.615	iron, wrought	0.115
aluminum	0.226		
ammonia (liq. 0)	1.098	Lead	0.0297
ammonia *	0.520	leather, dry	0.360
ammonia †	0.391		
antimony	0.0504	Marble	0.206
asbestos	0.195	mercury	0.0331
		mica	0.208
Beryllium	0.425	Nickel	0.109
bismuth	0.0297	nitrogen *	0.244
brass (60 Cu, 40 Zn)	0.0917	nitrogen †	0.173
bronze (80 Cu, 20 Sn)	0.0860		
		Oxygen *	0.224
		oxygen †	0.155
Calcium	0.149	osmium	0.0311
carbon, gas	0.315		
carbon, graphite	0.310	Paraffin	0.589
carbon dioxide * (15 to 100)	0.202	petroleum	0.504
carbon dioxide † (15 to 100)	0.168	platinum	0.0319
carbon monoxide *	0.243	porcelain (15 to 950)	0.260
carbon monoxide †	0.173		
cement, Portland	0.271	Quartz (12 to 100)	0.188
chalk	0.220		
chloroform (liq., 30)	0.235		
chloroform (gas, 100 to 200)	0.147	rubber, hard	0.339
chromium	0.111		
clay, dry (20 to 100)	0.220	Selenium (−188 to +18)	0.0680
coal	0.201	silicon	0.175
cobalt	0.103	silver	0.0560
copper	0.0928	steam (100 to 200)	0.480
cork	0.485	steel	0.118
cotton	0.362	sulphur (−188 to +18)	0.137
Gasoline	0.500	Tantalum (58)	0.0360
german silver	0.0945	tin	0.0556
glass	0.180	tungsten	0.0340
glycerine (15 to 50)	0.576	turpentine (0)	0.411
gold	0.0312		
granite (12 to 100)	0.192	Water (15)	1.000
		wood	0.420
		wool	0.393
Hydrogen *	3.41		
hydrogen †	2.42	Zinc	0.0950

* Constant pressure of one atmosphere.   † Constant volume.

Table 10.11   Coefficients of Linear Expansion

Average values (0° to 100° C. unless otherwise stated) of **a** in the formula,
$1 = l_0 (1 + at)$, t being measured in degrees C.

Substance	a × 10⁶	Substance	a × 10⁶
Aluminum (20 to 100)...	23.8	Marble, Rutland blue (15 to 100)...............	15.0
antimony (15 to 101)....	10.9	marble, Georgia gray (20 to 65)................	1.00
Beryllium (20).........	12.2	mercury (− 78 to − 38)..	41.0
bismuth (19 to 101).....	13.4	mica..................	7.60
brass..................	18.7	monel metal (25 to 100)..	14.1
brick..................	9.50		
bronze (80 Cu, 20 Sn) (0 to 800).............	27.0	Nickel (25 to 100).......	12.9
Cadmium..............	31.6	Osmium (40)..........	6.57
calcium (0 to 21)........	25.0		
carbon, diamond (40)...	1.18	Paraffin (0 to 16)	107.
" gas (40)........	5.40	paraffin (16 to 38)	130.
" graphite (40)....	7.86	phosphorous (6 to 44)...	124.
celluloid (20 to 70)	109.	platinum (20)..........	8.93
cobalt (20).............	12.3	porcelain, average......	3.50
copper (25 to 100).......	16.8		
		Quartz, fused..........	0.500
Duralumin, cast (20 to 100)	23.6	Rubber, hard (20 to 60) .	80.0
duralumin, cold rolled (20 to 100)...........	23.7	Selenium (40)..........	36.8
		silicon (40)............	7.63
German silver..........	18.4	silver (20).............	18.8
glass, crown...........	8.97	slate (20)..............	8.00
" flint (50 to 60).....	7.88	solder.................	25.1
gold (16 to 100)........	14.3	sodium (−188 to +17)..	62.2
granite................	8.30	steel, cast.............	13.6
gutta percha...........	198.	sulphur (40)...........	64.1
Ice (− 20 to − 1).......	51.0	Tin (18 to 100)..........	26.9
iridium (− 183 to + 19)..	5.71	tungsten (0 to 500)......	4.60
iron, pure..............	11.9	tungsten (1000 to 2000)..	6.10
" cast (40)	10.6		
" wrought (− 18 to + 100)...........	11.4	Wood, beech (2 to 34)...	2.57
		wood, walnut (2 to 34) ..	6.58
Lead (18 to 100)........	29.4	Zinc (10 to 100)	26.3

Table 10.12 Melting and Boiling Points (at Atmospheric Pressure)

Substance	Melts (°C)	Boils (°C)	Substance	Melts (°C)	Boils (°C)
Acetylene	−81.3	−72.2	Magnesium	651	1110
Alcohol, ethyl	−115	78.3	Manganese	1260	1900
Alcohol, methyl	−97.8	64.7	Mercury	−38.87	356.9
Aluminum	659.7	1800	Molybdenum	2620	3700
Ammonia	−75	−33.5	Neon	−248.7	−245.9
Antimony	630.5	1380	Nickel	1455	2900
Argon	−189.2	−185.7	Nitric oxide	−160.6	−153
Barium	850	1140	Nitrogen	−209.9	−195.8
Beryllium	1350	1500	Oxygen	−218.4	−183
Bismuth	271.3	1450	Ozone	−251.4	−112
Borax	561	—	Palladium	1553	2200
Boron	2300	2550	Paraffin	52.4	—
Brass	950±	—	Phosphorus	44.1	280
Bromine	−7.2	58.8	Platinum	1773.5	4300
Bronze	1000±	—	Potassium	62.3	760
Cadmium	320.9	767	Radium	960	1140
Calcium	810	1170	Radon	−110	—
Carbon	>3500	4200	Rhenium	3000	—
Carbon dioxide	−57	−80	Rhodium	1985	>2500
Carbon monoxide	−207	−191.5	Rubber	100	—
Cerium	640	1400	Rubidium	38.5	700
Cesium	28.5	670	Ruthenium	2450	>2700
Chlorine	−101.6	−34.6	Selenium	220	688
Chromium	1615	2200	Silicon	1420	2600
Cobalt	1480	3000	Silver	960.5	1950
Columbium	1950	2900	Sodium	97.5	880
Copper	1083	2300	Sodium chloride	772	—
Fluorine	−223	−187	Steel, carbon	1400	—
Gallium	29.75	>1600	Steel, stainless	1450	—
German silver	1100±	—	Strontium	800	1150
Germanium	958.5	2700	Sugar	160	—
Glass, flint	1300	—	Sulfur	112.8	444.6
Gold	1063	2600	Tantalum	2850	>4100
Hafnium	1700	>3200	Tellurium	452	1390
Helium	<−272.2	−268.9	Thallium	303.5	1650
Hydrogen	−259.1	−252.7	Thorium	1845	>3000
Indium	155	1450	Tin	231.9	2260
Iodine	113.5	184.3	Titanium	1800	>3000
Iridium	2350	>4800	Tungsten	3370	5900
Iron, pure	1535	3000	Turpentine	—	161
Iron, gray pig	1200	—	Uranium	<1850	—
Iron, white pig	1050	—	Vanadium	1710	3000
Krypton	−169	−151.8	Xenon	−140	−109
Lanthanum	826	1800	Yttrium	1490	2500
Lead	327.4	1620	Zinc	419.5	907
Lithium	186	>1220	Zirconium	1900	>2900

Table 10.13 Physical Constants of Various Fluids[1]

FLUID	FORMULA	MOLECULAR WEIGHT	BOILING POINT (°F AT 14.696 PSIA)	VAPOR PRESSURE @ 70°F (PSIG)	CRITICAL TEMP. (°F)	CRITICAL PRESSURE (PSIA)	SPECIFIC GRAVITY Liquid 60/60°F	SPECIFIC GRAVITY Gas
Acetic Acid	$HC_2H_3O_2$	60.05	245				1.05	
Acetone	C_3H_6O	58.08	133		455	691	0.79	2.01
Air	N_2O_2	28.97	−317		−221	547	0.86‡	1.0
Alcohol, Ethyl	C_2H_6O	46.07	173	2.3†	470	925	0.794	1.59
Alcohol, Methyl	CH_4O	32.04	148	4.63†	463	1174	0.796	1.11
Ammonia	NH_3	17.03	−28	114	270	1636	0.62	0.59
Aniline	C_6H_7N	93.12	365		798	770	1.02	
Argon	A	39.94	−302		−188	705	1.65	1.38
Bromine	Br_2	159.84	138		575		2.93	5.52
Carbon Dioxide	CO_2	44.01	−109	839	88	1072	0.801‡	1.52
Carbon Disulfide	CS_2	76.1	115				1.29	2.63
Carbon Monoxide	CO	28.01	−314		−220	507	0.80	0.97
Carbon Tetrachloride	CCl_4	153.84	170		542	661	1.59	5.31
Chlorine	Cl_2	70.91	−30	85	291	1119	1.42	2.45
Ether	$(C_2H_6)_2O$	74.12	34				0.74	2.55
Fluorine	F_2	38.00	−305	300	−200	809	1.11	1.31
Formaldehyde	H_2CO	30.03	−6				0.82	1.08
Formic Acid	HCO_2H	46.03	214				1.23	
Furfural	$C_6H_4O_2$	96.08	324				1.16	
Glycerine	$C_3H_8O_3$	92.09	554				1.26	
Glycol	$C_2H_6O_2$	62.07	387				1.11	
Helium	He	4.003	−454		−450	33	0.18	0.14
Hydrochloric Acid	HCl	36.47	−115				1.64	
Hydrofluoric Acid	HF	20.01	66	0.9	446		0.92	

Control Valve Handbook, 2nd edition, Fisher © 1977 Reproduced with permission of Fisher Controls International, Inc.

FLUID	FORMULA	MOLECULAR WEIGHT	BOILING POINT (°F AT 14.696 PSIA)	VAPOR PRESSURE @ 70°F (PSIG)	CRITICAL TEMP. (°F)	CRITICAL PRESSURE (PSIA)	SPECIFIC GRAVITY Liquid 60/60°F	SPECIFIC GRAVITY Gas
Hydrogen	H_2	2.016	-422		-400	188	0.07‡	0.07
Hyrogen Chloride	HCl	36.47	-115	613	125	1198	0.86	1.26
Hydrogen Sulfide	H_2S	34.07	-76	252	213	1307	0.79	1.17
Isopropyl Alcohol	C_3H_8O	60.09	180				0.78	2.08
Linseed Oil			538				0.93	
Magnesium Chloride*	$MgCl_2$						1.22	
Mercury	Hg	200.61	670				13.6	6.93
Methyl Bromide	CH_3Br	94.95	38	13	376		1.73	3.27
Methyl Chloride	CH_3Cl	50.49	-11	59	290	969	0.99	1.74
Naphthalene	$C_{10}H_8$	128.16	424				1.14	4.43
Nitric Acid	HNO_3	63.02	187				1.5	
Nitrogen	N_2	28.02	-320		-233	493	0.81‡	0.97
Oxygen	O_2	32	-297		-181	737	1.14‡	1.105
Phosgene	$COCl_2$	98.92	47	10.7	360	823	1.39	3.42
Phosphoric Acid	H_3PO_4	98.00	415				1.83	
Refrigerant 11	CCl_3F	137.38	75	13.4	388	635		5.04
Refrigerant 12	CCl_2F_2	120.93	-22	70.2	234	597		4.2
Refrigerant 13	$CClF_3$	104.47	-115	458.7	84	561		3.82
Refrigerant 21	$CHCl_2F$	102.93	48	8.4	353	750		
Refrigerant 22	$CHClF_2$	86.48	-41	122.5	205	716		
Refrigerant 23	CHF_3	70.02	-119	635	91	691		
Sulfuric Acid	H_2SO_4	98.08	626		316	1145	1.83	
Sulfur Dioxide	SO_2	64.6	14	34.4			1.39	2.21
Turpentine			320				0.87	
Water	H_2O	18.016	212	0.9492†	706	3208	1.00	0.62

†Vapor pressure in psia at 100°F.

‡Density of liquid, gm/ml at normal boiling point.

Table 10.14 Physical Constants of Hydrocarbons[1]

COMPOUND	FORMULA	MOLECULAR WEIGHT	BOILING POINT AT 14.696 psia (°F)	VAPOR PRESSURE AT 100°F (psia)	FREEZING POINT AT 14.696 psia (°F)	CRITICAL CONSTANTS		SPECIFIC GRAVITY at 14.696 psia	
						Critical Temperature (°F)	Critical Pressure (psia)	Liquid,[2,4] 60° F/60° F	Gas at 60° F (Air = 1)[3]
Methane	CH_4	16.043	−258.69	$(5000)^2$	$−296.46^5$	−116.63	667.8	0.3^8	0.5539
Ethane	C_2H_6	30.070	−127.48	$(800)^2$	$−297.89^5$	90.09	707.8	0.3564^7	1.0382
Propane	C_3H_8	44.097	−43.67	190.	$−305.84^5$	206.01	616.3	0.5077^7	1.5225
n-Butane	C_4H_{10}	58.124	31.10	51.6	−217.05	305.65	550.7	0.5844^7	2.0068
Isobutane	C_4H_{10}	58.124	10.90	72.2	−255.29	274.98	529.1	0.5631^7	2.0068
n-Pentane	C_5H_{12}	72.151	96.92	15.570	−201.51	385.7	488.6	0.6310	2.4911
Isopentane	C_5H_{12}	72.151	82.12	20.44	−255.83	369.10	490.4	0.6247	2.4911
Neopentane	C_5H_{12}	72.151	49.10	35.9	2.17	321.13	464.0	0.5967^7	2.4911
n-Hexane	C_6H_{14}	86.178	155.72	4.956	−139.58	453.7	436.9	0.6640	2.9753
2-Methylpentane	C_6H_{14}	86.178	140.47	6.767	−244.63	435.83	436.6	0.6579	2.9753
3-Methylpentane	C_6H_{14}	86.178	145.89	6.098	…	448.3	453.1	0.6689	2.9753
Neohexane	C_6H_{14}	86.178	121.52	9.856	−147.72	420.13	446.8	0.6540	2.9753
2,3-Dimethylbutane	C_6H_{14}	86.178	136.36	7.404	−199.38	440.29	453.5	0.6664	2.9753
n-Heptane	C_7H_{16}	100.205	209.17	1.620	−131.05	512.8	396.8	0.6882	3.4596
2-Methylhexane	C_7H_{16}	100.205	194.09	2.271	−180.89	495.00	396.5	0.6830	3.4596
3-Methylhexane	C_7H_{16}	100.205	197.32	2.130	…	503.78	408.1	0.6917	3.4596
3-Ethylpentane	C_7H_{16}	100.205	200.25	2.012	−181.48	513.48	419.3	0.7028	3.4596
2,2-Dimethylpentane	C_7H_{16}	100.205	174.54	3.492	−190.86	477.23	402.2	0.6782	3.4596
2,4-Dimethylpentane	C_7H_{16}	100.205	176.89	3.292	−182.63	475.95	396.9	0.6773	3.4596
3,3-Dimethylpentane	C_7H_{16}	100.205	186.91	2.773	−210.01	505.85	427.2	0.6976	3.4596
Triptane	C_7H_{16}	100.205	177.58	3.374	−12.82	496.44	428.4	0.6946	3.4596

COMPOUND	FORMULA	MOLECULAR WEIGHT	BOILING POINT AT 14.696 psia (°F)	VAPOR PRESSURE AT 100°F (psia)	FREEZING POINT AT 14.696 psia (°F)	CRITICAL CONSTANTS Critical Temperature (°F)	CRITICAL CONSTANTS Critical Pressure (psia)	SPECIFIC GRAVITY at 14.696 psia Liquid [3,4] 60°F/60°F	SPECIFIC GRAVITY at 14.696 psia Gas at 60°F (Air = 1)[1]
n-Octane	C_8H_{18}	114.232	258.22	0.537	−70.18	564.22	360.6	0.7068	3.9439
Diisobutyl	C_8H_{18}	114.232	228.39	1.101	−132.07	530.44	360.6	0.6979	3.9439
Isooctane	C_8H_{18}	114.232	210.63	1.708	−161.27	519.46	372.4	0.6962	3.9439
n-Nonane	C_9H_{20}	128.259	303.47	0.179	−64.28	610.68	332.	0.7217	4.4282
n-Decane	$C_{10}H_{22}$	142.286	345.48	0.0597	−21.36	652.1	304.	0.7342	4.9125
Cyclopentane	C_5H_{10}	70.135	120.65	9.914	−136.91	461.5	653.8	0.7504	2.4215
Methylcyclopentane	C_6H_{12}	84.162	161.25	4.503	−224.44	499.35	548.9	0.7536	2.9057
Cyclohexane	C_6H_{12}	84.162	177.29	3.264	43.77	536.7	591.	0.7834	2.9057
Methylcyclohexane	C_7H_{14}	98.189	213.68	1.609	−195.87	570.27	503.5	0.7740	3.3900
Ethylene	C_2H_4	28.054	−154.62	...	−272.45[5]	48.58	729.8	0.5220[7]	0.9686
Propene	C_3H_6	42.081	−53.90	226.4	−301.45[5]	196.9	669.	0.6013[7]	1.4529
1-Butene	C_4H_8	56.108	20.75	63.05	−301.63[5]	295.6	583.	0.6271[7]	1.9372
Cis-2-Butene	C_4H_8	56.108	38.69	45.54	−218.06	324.37	610.	0.6100[7]	1.9372
Trans-2-Butene	C_4H_8	56.108	33.58	49.80	−157.96	311.86	595.	0.6004[7]	1.9372
Isobutene	C_4H_8	56.108	19.59	63.40	−220.61	292.55	580.	0.6457	1.9372
1-Pentene	C_5H_{10}	70.135	85.93	19.115	−265.39	376.93	590.	0.658[7]	2.4215
1,2-Butadiene	C_4H_6	54.092	51.53	(20.)[2]	−213.16	(339.)[2]	(653.)[2]	0.6272[7]	1.8676
1,3-Butadiene	C_4H_6	54.092	24.06	(60.)[2]	−164.02	306.	628.	0.6272[7]	1.8676
Isoprene	C_5H_8	68.119	93.30	16.672	−230.74	(412.)[2]	(558.4)[2]	0.6861	2.3519
Acetylene	C_2H_2	26.038	−119.[6]	...	−114.[5]	95.31	890.4	0.615[9]	0.8990
Benzene	C_6H_6	78.114	176.17	3.224	41.96	552.22	710.4	0.8844	2.6969
Toluene	C_7H_8	92.141	231.13	1.032	−138.94	605.55	595.9	0.8718	3.1812
Ethylbenzene	C_8H_{10}	106.168	277.16	0.371	−138.91	651.24	523.5	0.8718	3.6655
o-Xylene	C_8H_{10}	106.168	291.97	0.264	−13.30	675.0	541.4	0.8848	3.6655
m-Xylene	C_8H_{10}	106.168	282.41	0.326	−54.12	651.02	513.6	0.8687	3.6655
p-Xylene	C_8H_{10}	106.168	281.05	0.342	55.86	649.6	509.2	0.8657	3.6655
Styrene	C_8H_8	104.152	293.29	(0.24)[2]	−23.10	706.0	580.	0.9110	3.5959
Isopropylbenzene	C_9H_{12}	120.195	306.34	0.188	−140.82	676.4	465.4	0.8663	4.1498

1. Calculated values.
2. ( ) Estimated values.
3. Air saturated hydrocarbons.
4. Absolute values from weights in vacuum.
5. At saturation pressure (triple point).
6. Sublimation point.
7. Saturation pressure and 60°F.
8. Apparent value for methane at 60°F.
9. Specific gravity. 119°F/60°F (sublimation point).

Source. Control Valve Handbook, 2nd ediitiion, Fisher © 1977. Reproduced with permission of Fisher Controls International, Inc.

Table 10.15 Higher Heating Values

Substance[a]	Btu		
	Per Pound	Per Gallon	Per Foot[3][b]
Acetylene	21,500	—	1,480
Alcohol, ethyl, denatured	11,600	78,900	—
Alcohol, ethyl, pure (0.816)	12,400	84,300	—
Alcohol, methyl (0.798)	9,540	63,700	—
Benzene C_6H_6	18,200	—	3,740
Butane C_4H_{10}	21,300	—	3,390
Carbon, to CO	4,000	—	—
Carbon, to CO_2	14,100	—	—
Carbon disulfide	5,820	62,700	—
Carbon monoxide, to CO_2	4,370	—	323
Charcoal, peat	11,600	—	—
Charcoal, wood	13,500	—	—
Ethane C_2H_6	22,400	—	1,770
Ethylene C_2H_4	21,600	—	1,600
Gas, blast furnace	—	—	90–110
Gas, coal	—	—	630–680
Gas, coke oven	—	—	430–600
Gas, illuminating	—	—	550–600
Gas, natural	—	—	700–2,470
Gas, oil	—	—	450–950
Gas, producer	—	—	110–185
Gas, water, blue	—	—	290–320
Gas, water, carburetted	—	—	400–680
Hexane (liq.) C_6H_{14}	20,700	—	—
Kerosene (0.783)	20,000	131,000	—
Kerosene (0.800)	20,160	136,000	—
Methane CH_4	23,900	—	1,010
Octane (liq.) C_8H_{18}	20,500	—	—
Pentane C_5H_{12}	21,100	—	4,010
Propane C_3H_8	21,700	—	2,520
Straw	5,100–6,700	—	—
Sulfur	4,020	—	—

[a]Numbers in parentheses indicate specify gravity.
[b]At 60°F and atmospheric pressure.

Table 10.16 Thermal Conductivity[a]

Substance	Temp. Range (°C)	$k \times 10^3$	Substance	Temp. Range (°C)	$k \times 10^3$
Air	0	0.0568	Fiber, red	—	1.1
Aluminum	18	480	Fiberglass, blanket	0–120	0.25–0.31[b]
Antimony	0	44.2	Fiberglass, semirigid	0–250	0.23–0.24[b]
Argon	0	0.0389	Fiberglass, semirigid	0–530	0.30–0.23[bc]
Asbestos, paper	—	0.6	Flannel	50	0.035
Bismuth	0	17.7	German silver	0–100	80
Brass	0	204	Glass, crown	—	2.5
Brick, alumina	0–700	2.0	Glass, flint	—	2.0
Brick, building	15–30	1.5	Gold	18	700
Brick, carborundum	100–1000	23	Granite	100	4.5
Brick, fire	0–1300	3.1	Graphite	—	12
Brick, graphite	100–1000	25	Gypsum	—	3.1
Brick, magnesia	100–1000	7.1	Hair	20–155	0.15
Brick, silica	100–1000	2.0	Hair cloth, felt	—	0.042
Cadmium	18	222	Helium	0	0.339
Cambric, varn	—	0.60	Hydrogen	0	0.327
Carbon, gas	100–942	130	Ice	—	3.9
Carbon, graphite	100–914	290	Iron, pure	18	161
Carbon dioxide	0	0.0307	Iron, cast	18	109
Carbon monoxide	0	0.0499	Iron, wrought	18	144
Carborundum	20–100	0.50	Lampblack	100	0.07
Cardboard	—	0.50	Lead	18	83
Cement, portland	0–700	0.17	Leather, cowhide	—	0.42
Chalk	0–100	0.28	Leather, chamois	—	0.15
Charcoal, powd'd	0–100	0.22	Lime	—	0.29
Clinkers, small	0–700	1.1	Linen	—	0.21
Coal	—	0.30	Magnesia	—	0.3
Coke, powdered	0–100	0.44	Magnesium, carb.	100	0.23
Concrete, cinder	—	0.81	Marble	15–30	8.4
Concrete	—	2.2	Mercury	17	19.7
Copper	18	918	Mica	—	0.86
Cotton wool	—	0.043	Nickel	18	142
Cotton batting, loose	—	0.11	Nitrogen	0	0.0524
Cotton batting, packed	—	0.072	Oxygen	0	0.0563
Earth, average	—	4.0	Paper	—	0.31
Eiderdown, loose	—	0.108	Paraffin	—	0.62
Eiderdown, packed	—	0.045	Pasteboard	—	0.45
Feathers	20–155	0.16	Plaster of Paris	20–155	0.42
Felt	21–175	0.22	Plaster, mortar	—	1.3
			Platinum	18–100	170
			Porcelain	165–1055	4.3
			Petroleum	23	0.39

Table 10.16 *Continued*

Substance	Temp. Range (°C)	$k \times 10^3$	Substance	Temp. Range (°C)	$k \times 10^3$
Pumice stone	20–155	0.43	Tin	18	155
Quartz, ‖ to axis	—	30	Water	0	1.4
Quartz, ⊥ to axis	—	160	Water	30	1.6
			Wood, fir, with grain	—	0.30
Rubber, hard	—	0.43			
Rubber	—	0.38	Wood, fir, cross-grain	—	0.09
Sand, dry	20–155	0.86			
Sandstone	—	5.5	Wool, sheep's	20–100	0.14
Sawdust	—	0.14	Wool, mineral	0–175	0.11
Silica, fused	100	2.55	Wool, steel	100	0.20
Silk	50–100	0.13	Woolen, loose wadding	—	0.12
Silver	18	974			
Slate	94	4.8	Woolen, packed wadding	—	0.055
Snow	—	0.60			
Steel	18	115	Zinc	18	265
Terra cotta	100–1000	2.3			

[a]Average values of k in the formula $Q = ckS\theta t/x$. See Subsection 4.7 for descriptions of units.
[b]Btu/hr/ft²/in./°F.
[c]Values decrease with increasing temperature.

Table 10.17 Properties of Elastomers

Property		Natural Rubber	Buna-S	Nitrile	Neoprene	Butyl	Thiokol[1]	Silicone[1]	Hypalon[2]	Viton[2,3,4]	Poly-urethane[4]	Poly-acrylic[3]	Ethylene Propylene[5]
Tensile Strength, Psi (Bar)	Pure Gum	3000 (207)	400 (28)	600 (41)	3500 (241)	3000 (207)	300 (21)	200-450 (14-31)	4000 (276)	...	...	100 (7)	...
	Reinforced	4500 (310)	3000 (207)	4000 (276)	3500 (241)	3000 (207)	1500 (103)	1100 (76)	4400 (303)	2300 (159)	6500 (448)	1800 (124)	2500 (172)
Tear Resistance		Excellent	Poor-Fair	Fair	Good	Good	Fair	Poor-Fair	Excellent	Good	Excellent	Fair	Poor
Abrasion Resistance		Excellent	Good	Good	Excellent	Fair	Poor	Poor	Excellent	Very Good	Excellent	Good	Good
Aging: Sunlight		Poor	Poor	Poor	Excellent	Excellent	Good	Good	Excellent	Excellent	Excellent	Excellent	Excellent
Oxidation		Good	Fair	Fair	Good	Good	Good	Very Good	Very Good	Excellent	Excellent	Excellent	Good
Heat (Max. Temp.)		200°F (93°C)	200°F (93°C)	250°F (121°C)	200°F (93°C)	200°F (93°C)	140°F (60°C)	450°F (232°C)	300°F (149°C)	400°F (204°C)	200°F (93°C)	350°F (177°C)	350°F (177°C)
Static (Shelf)		Good	Good	Good	Very Good	Good	Fair	Good	Good	...	...	Good	Good
Flex Cracking Resistance		Excellent	Good	Good	Excellent	Excellent	Fair	Fair	Excellent	...	Excellent	Good	...
Compression Set Resistance		Good	Good	Very Good	Excellent	Fair	Poor	Good	Poor	Poor	Good	Good	Fair
Solvent Resistance:													
Aliphatic Hydrocarbon		Very Poor	Very Poor	Good	Fair	Poor	Excellent	Poor	Fair	Excellent	Very Good	Good	Poor
Aromatic Hydrocarbon		Very Poor	Very Poor	Fair	Poor	Very Poor	Good	Very Poor	Poor	Very Good	Fair	Poor	Fair
Oxygenated Solvent		Good	Good	Poor	Fair	Good	Fair	Poor	Poor	Good	Poor	Poor	...
Halogenated Solvent		Very Poor	Very Poor	Very Poor	Very Poor	Poor	Poor	Very Poor	Very Poor	...	...	Poor	Poor
Oil Resistance:													
Low Aniline Mineral Oil		Very Poor	Very Poor	Excellent	Fair	Very Poor	Excellent	Poor	Fair	Excellent	...	Excellent	Poor
High Aniline Mineral Oil		Very Poor	Very Poor	Excellent	Good	Very Poor	Excellent	Good	Good	Excellent	...	Excellent	Poor
Synthetic Lubricants		Very Poor	Very Poor	Fair	Very Poor	Poor	Poor	Fair	Poor	...	...	Fair	Poor
Organic Phosphates		Very Poor	Very Poor	Very Poor	Very Poor	Good	Poor	Poor	Poor	Poor	Poor	Poor	Very Good
Gasoline Resistance:													
Aromatic		Very Poor	Very Poor	Good	Poor	Very Poor	Excellent	Poor	Poor	Good	Fair	Fair	Fair
Non-Aromatic		Very Poor	Very Poor	Excellent	Good	Very Poor	Excellent	Good	Fair	Very Good	Good	Poor	Poor
Acid Resistance:													
Diluted (Under 10%)		Good	Good	Good	Fair	Good	Poor	Fair	Good	Excellent	Fair	Poor	Very Good
Concentrated[6]		Fair	Poor	Poor	Fair	Fair	Very Poor	Poor	Good	Very Good	Poor	Poor	Good
Low Temperature Flexibility (Max.)		-65°F (-54°C)	-50°F (-46°C)	-40°F (-40°C)	-40°F (-40°C)	-40°F (-40°C)	-40°F (-40°C)	-100°F (-73°C)	-20°F (-29°C)	-30°F (-34°C)	-40°F (-40°C)	-10°F (-23°C)	-50°F (-45°C)
Permeability to Gases		Fair	Fair	Fair	Very Good	Very Good	Good	Fair	Very Good	Good	Good	Good	Good
Water Resistance		Good	Very Good	Very Good	Fair	Very Good	Fair	Fair	Very Good	Excellent	Fair	Fair	Very Good
Alkali Resistance:													
Diluted (under 10%)		Good	Good	Good	Good	Very Good	Poor	Good	Good	Good	Fair	Poor	Fair
Concentrated		Fair	Fair	Fair	Good	Very Good	Very Poor	Good	Good	Very Good	Poor	Poor	Poor
Resilience		Very Good	Fair	Fair	Very Good	Very Good	Good	Good	Good	Good	Good	Very Poor	Very Good
Elongation (Max.)		700%	500%	500%	500%	700%	400%	300%	300%	425%	625%	200%	500%

1. Trademark of Thiokol Chemical Co.
2. Trademark of E.I. DuPont Co.
3. Do not use with steam.
5. Do not use with petroleum base fluids. Use with ester base non-flammable hydraulic oils and low pressure steam applications to 300°F (149°C).
6. Except for nitric and sulfuric acid.

Source: *Control Valve Handbook*, 2nd edition, Fisher © 1977. Reproduced with permission of Fisher Controls International.

Table 10.18   Density of Gases (at 60°F and 30 in. Hg)

Gas	Molecular Formula	Molecular Weight	Specific Gravity Air = 1.0	Weight lb per cu ft	Volume cu ft per lb
Air	—	28.9	1.000	0.07655	13.063
Oxygen	O_2	32.00	1.105	0.08461	11.819
Hydrogen	H_2	2.02	0.070	0.00533	187.723
Nitrogen (atmospheric)	N_2	28.02	0.972	0.07439	13.443
Carbon Monoxide	CO	28.01	0.967	0.07404	13.506
Carbon Dioxide	CO_2	44.01	1.528	0.1170	8.548
Methane	CH_4	16.04	0.554	0.04243	23.565
Acetylene	C_2H_2	26.04	0.911	0.06971	14.344
Ethylene	C_2H_4	28.05	0.974	0.07456	13.412
Ethane	C_2H_6	30.07	1.049	0.08029	12.455
Sulphur Dioxide	SO_2	64.06	2.264	0.1733	5.770
Hydrogen Sulphide	H_2S	34.08	1.190	0.09109	10.979

From "Fuel Flue Gases," 1941, courtesy American Gas Association

Approximate Percentage Composition of Air

	By Weight	By Volume
Nitrogen	76.8	79.0
Oxygen	23.2	21.0
	94	

Table 10.19   Atmospheric Pressures and Barometer Readings at Different Altitudes (Approximate Values)

Altitude Below or Above Sea Level Feet	Barometer Reading Inches Merc at 32°F	Atmospheric Pressure Lb-Sq In	Equivalent Head of Water (75°) Feet	Boiling Point of Water °F
−1000	31.02	15.2	35.2	213.8
− 500	30.47	15.0	34.7	212.9
0	29.921	14.7	34.0	212.0
+ 500	29.38	14.4	33.4	211.1
+1000	28.86	14.2	32.8	210.2
1500	28.33	13.9	32.2	209.3
2000	27.82	13.7	31.6	208.4
2500	27.31	13.4	31.0	207.4
3000	26.81	13.2	30.5	206.5
3500	26.32	12.9	29.9	205.6
4000	25.84	12.7	29.4	204.7
4500	25.36	12.4	28.8	203.8
5000	24.89	12.2	28.3	202.9
5500	24.43	12.0	27.8	201.9
6000	23.98	11.8	27.3	201.0
6500	23.53	11.5	26.7	200.1
7000	23.09	11.3	26.2	199.2
7500	22.65	11.1	25.7	198.3
8000	22.22	10.9	25.2	197.4
8500	21.80	10.7	24.8	196.5
9000	21.38	10.5	24.3	195.5
9500	20.98	10.3	23.8	194.6
10,000	20.58	10.1	23.4	193.7
15,000	16.88	8.3	19.1	184
20,000	13.75	6.7	15.2	—
30,000	8.88	4.4	10.2	—
40,000	5.54	2.7	6.3	—
50,000	3.44	1.7	3.9	—

Used with permission of Ingersoll-Rand Co. © 1962.

Table 10.20 Orifice Coefficients (Circular)

Coefficients of discharge (c) for circular orifices, with full contractions [*]

Head from center of orifice in feet	Diameters in feet					
	0.02	0.05	0.1	0.2	0.6	1.0
0.5	......	0.627	0.615	0.600	0.592	......
0.8	0.648	0.620	0.610	0.601	0.594	0.591
1.0	0.644	0.617	0.608	0.600	0.595	0.591
1.5	0.637	0.613	0.605	0.600	0.596	0.593
2.0	0.632	0.610	0.604	0.599	0.597	0.595
2.5	0 629	0.608	0.603	0.599	0.598	0.596
3.0	0.627	0.606	0.603	0.599	0.598	0.597
3.5	0.625	0.606	0.602	0.599	0.598	0.596
4.0	0.623	0.605	0.602	0.599	0.597	0.596
6.0	0.618	0.604	0.600	0.598	0.597	0.596
8.0	0.614	0.603	0.600	0.598	0.596	0.596
10.0	0.611	0.601	0.598	0.597	0.596	0.595
20.0	0.601	0.598	0.596	0.596	0.596	0.594
50.0	0.596	0.595	0.594	0.594	0.594	0.593
100.0	0.593	0.592	0.592	0.592	0.592	0.592

Head from center of orifice in feet	Length of side of square in feet					
	0.02	0.05	0.1	0.2	0.6	1.0
0.5	......	0.633	0.619	0.605	0.597	......
0.8	0.652	0.625	0.615	0.605	0.600	0.597
1.0	0.648	0.622	0.613	0.605	0.601	0.599
1.5	0.641	0.619	0.610	0.605	0.602	0.601
2.0	0.637	0.615	0.608	0.605	0.604	0.602
2.5	0.634	0.613	0.607	0.605	0.604	0.602
3.0	0.632	0.612	0.607	0.605	0.604	0.603
3.5	0.630	0.611	0.607	0.605	0.604	0.602
4.0	0.628	0.610	0.606	0.605	0.603	0.602
6.0	0.623	0.609	0.605	0.604	0.603	0.602
8.0	0.619	0.608	0.605	0.604	0.603	0.602
10.0	0.616	0.606	0.604	0.603	0.602	0.601
20.0	0.606	0.603	0.602	0.602	0.601	0.600
50.0	0.602	0.601	0.600	0.600	0.599	0.599
100.0	0.599	0.598	0.598	0.598	0.598	0.598

[*] From Hamilton Smith's Hydraulics.

Table 10.21 Weir Coefficients (Contracted)

Coefficients of discharge (c) for contracted weirs
For use in the Hamilton Smith formula.

Effective head in feet	Length of weir in feet									
	0.66	1	2	3	4	5	7	10	15	19
0.1	0.632	0.639	0.646	0.652	0.653	0.653	0.654	0.655	0.655	0.656
0.2	0.611	0.618	0.626	0.630	0.631	0.631	0.632	0.633	0.634	0.634
0.25	0.605	0.612	0.621	0.624	0.625	0.626	0.627	0.628	0.628	0.629
0.3	0.601	0.608	0.616	0.619	0.621	0.621	0.623	c.624	0.624	0.625
0.4	0.595	0.601	0.609	0.613	0.614	0.615	0.617	0.618	0.619	0.620
0.5	0.590	0.596	0.605	0.608	0.610	0.611	0.613	0.615	0.616	0.617
0.6	0.587	0.593	0.601	0.605	0.607	0.608	0.611	0.613	0.614	0.615
0.8	......	......	0.595	0.600	0.602	0.604	0.607	0.611	0.612	0.613
1.0	......	......	0.590	0.595	0.598	0.601	0.604	0.608	0.610	0.611
1.2	......	......	0.585	0.591	0.594	0.597	0.601	0.605	0.608	0.610
1.4	......	......	0.580	0.587	0.590	0.594	0.598	0.602	0.606	0.609
1.6	......	......	......	0.582	0.587	0.591	0.595	0.600	0.604	0.607

Table 10.22 Weir Coefficients (Suppressed)

Coefficients of discharge (c) for suppressed weirs
For use in the Hamilton Smith formula.

Effective head in feet	Length of weir in feet								
	0.66	2	3	4	5	7	10	15	19
0.1	......	......	......	......	0.659	0.658	0.658	0.657	0.657
0.2	0.656	0.645	0.642	0.641	0.638	0.637	0.637	0.636	0.635
0.25	0.653	0.641	0.638	0.636	0.634	0.633	0.632	0.631	0.630
0.3	0.651	0.639	0.636	0.633	0.631	0.629	0.628	0.627	0.626
0.4	0.650	0.636	0.633	0.630	0.628	0.625	0.623	0.622	0.621
0.5	0.650	0.637	0.633	0.630	0.627	0.624	0.621	0.620	0.619
0.6	0.651	0.638	0.634	0.630	0.627	0.623	0.620	0.619	0.618
0.8	0.656	0.643	0.637	0.633	0.629	0.625	0.621	0.620	0.618
1.0	......	0.648	0.641	0.637	0.633	0.628	0.624	0.621	0.619
1.2	......	......	0.646	0.641	0.636	0.632	0.626	0.623	0.620
1.4	......	......	......	0.644	0.640	0.634	0.629	0.625	0.622
1.6	......	......	......	0.647	0.642	0.637	0.631	0.626	0.623

* From Hamilton Smith's Hydraulics.

515

Table 10.23 Friction Factors (Clean Cast Iron)

Values of friction factor (*f*) for clean cast-iron pipes

Diameter in inches	Velocity in feet per second						
	0.5	1	2	3	6	10	20
1	0.0398	0.0353	0.0317	0.0299	0.0266	0.0244	0.0228
3	0.0354	0.0316	0.0288	0.0273	0.0248	0.0232	0.0218
6	0.0317	0.0289	0.0264	0.0252	0.0231	0.0219	0.0208
9	0.0290	0.0269	0.0247	0.0237	0.0220	0.0209	0.0200
12	0.0268	0.0251	0.0233	0.0224	0.0209	0.0201	0.0192
18	0.0238	0.0224	0.0211	0.0204	0.0193	0.0188	0.0181
24	0.0212	0.0194	0.0193	0.0187	0.0180	0.0176	0.0170
30	0.0194	0.0186	0.0179	0.0175	0.0170	0.0166	0.0161
36	0.0177	0.0172	0.0167	0.0164	0.0160	0.0156	0.0152
48	0.0153	0.0150	0.0147	0.0145	0.0143	0.0141	0.0138
60	0.0137	0.0135	0.0133	0.0132	0.0130	0.0128	0.0125
72	0.0125	0.0124	0.0122	0.0120	0.0118	0.0117	0.0117
96	0.0109	0.0107	0.0106	0.0106	0.0105	0.0104	0.0103

Table 10.24 Friction Factors (Old Cast Iron)

Values of friction factor (*f*) for old cast-iron pipes

Diameter in inches	Velocity in feet per second			
	1	3	6	10
3	0.0608	0.0556	0.0512	0.0488
6	0.0540	0.0468	0.0432	0.0412
9	0.0488	0.0420	0.0400	0.0368
12	0.0432	0.0384	0.0356	0.0336
15	0.0396	0.0348	0.0324	0.0312
18	0.0348	0.0312	0.0292	0.0276
24	0.0304	0.0268	0.0252	0.0240
30	0.0268	0.0244	0.0228	0.0220
36	0.0244	0.0224	0.0208	0.0200
42	0.0232	0.0208	0.0200	0.0192
48	0.0228	0.0204	0.0196	0.0184

Table 10.25 Channel Coefficients (Chezy)

Values of coefficient (*c*) in Chezy Formula

Radius in Feet	Velocity in feet per second						
	1	2	3	4	6	10	15
0.5	96	104	109	112	116	121	124
1.0	109	116	121	124	129	134	138
1.5	117	124	128	132	136	143	147
2.0	123	130	134	137	142	150	155
2.5	128	134	139	142	147	155	....
3.0	132	138	142	145	150	....	....
3.5	135	141	145	149	153	....	....
4.0	137	143	148	151	....	....	....

Table 10.26   Channel Coefficients (Kutter)

Values of coefficients (c) in Kutter's formula

Slope	n	Hydraulic radius r in feet										
		0.2	0.4	0.6	0.8	1.0	1.5	2.0	6.0	10.0	15.0	50.0
0.00005	0.010	87	109	123	133	140	154	164	199	213	220	245
	0.015	52	66	76	83	89	99	107	138	150	159	181
	0.020	35	45	53	59	64	72	80	105	116	125	148
	0.025	26	35	41	45	49	57	62	85	96	104	127
	0.030	22	28	33	37	40	47	51	72	83	90	112
	0.040	15	20	24	27	29	34	38	56	64	71	93
0.0001	0.010	98	118	131	140	147	158	167	196	206	212	227
	0.015	57	72	81	88	93	103	109	134	143	150	166
	0.020	38	50	57	63	67	75	81	102	111	118	134
	0.025	28	38	43	48	51	59	64	84	93	98	114
	0.030	23	30	35	39	42	48	52	72	78	85	100
	0.040	16	22	25	28	31	35	39	54	62	68	83
0.0002	0.010	105	125	137	145	150	162	169	193	202	206	220
	0.015	61	76	84	91	96	105	110	132	140	145	158
	0.020	42	53	60	65	68	76	82	100	108	113	126
	0.025	30	40	45	50	54	60	65	83	90	95	108
	0.030	25	32	37	40	43	49	53	69	77	82	94
	0.040	17	23	26	29	32	36	40	53	60	65	78
0.0004	0.010	110	128	140	148	153	164	171	192	198	203	215
	0.015	64	78	87	93	98	106	112	130	137	142	154
	0.020	43	55	61	67	70	77	83	99	106	110	123
	0.025	32	42	47	51	55	60	65	82	88	92	104
	0.030	26	33	38	41	44	50	54	68	75	80	91
	0.040	18	23	27	30	32	37	40	53	59	63	75
0.001	0.010	113	132	143	150	155	165	172	190	197	201	212
	0.015	66	80	88	94	98	107	112	130	135	141	151
	0.020	45	56	62	68	71	78	84	98	105	109	120
	0.025	33	43	48	52	55	61	65	81	87	91	101
	0.030	27	34	38	42	45	50	54	68	74	78	89
	0.040	18	24	27	30	33	37	40	53	58	61	72
0.01	0.010	114	133	143	151	156	165	172	190	196	200	210
	0.015	67	81	89	95	99	107	113	129	135	140	150
	0.020	46	57	63	68	72	78	84	98	105	108	119
	0.025	34	44	49	52	56	62	65	80	86	90	100
	0.030	27	35	39	43	45	51	55	67	73	77	87
	0.040	19	24	28	30	33	37	40	52	58	61	71

Table 10.27   Channel Coefficients (Bazin)

Values of coefficients (c) in Bazin's Formula *

Hydraulic radius in feet	Coefficient of roughness m					
	0.06	0.16	0.46	0.85	1.30	1.75
0.2	126	96	55	36	25	19
0.3	132	103	63	41	30	23
0.4	134	108	68	46	33	26
0.5	136	112	71	50	36	29
0.75	140	118	80	57	42	34
1.0	142	122	86	62	47	38
1.25	143	125	90	66	51	41
1.5	145	127	94	70	54	44
2.0	146	131	99	76	59	49
2.5	147	133	104	80	63	53
3.0	148	135	106	83	67	57
5.0	150	140	115	93	77	65
10.0	152	144	125	106	91	79
20.0	154	148	133	117	103	92

* From Russell's " Textbook on Hydraulics."

Table 10.28 Tank Capacities (Vertical Cylindrical)

Diameter	Area in Sq. Ft. Cu. Ft. per 1' of Depth	U. S. Gallons per 1' of Depth	Diameter	Area in Sq. Ft. per 1' of Depth	U. S. Gallons per 1' of Depth	Diameter	Area in Sq. Ft. Cu. Ft. per 1' of Depth	U. S. Gallons per 1' of Depth
1'	0.785	5.87	6'	28.27	211.5	28'	615.8	4606.
1' 1"	0.922	6.89	6' 3"	30.68	229.5	28' 6"	637.9	4772.
1' 2"	1.069	8.00	6' 6"	33.18	248.2	29'	660.5	4941.
1' 3"	1.227	9.18	6' 9"	35.78	267.7	29' 6"	683.5	5113.
1' 4"	1.396	10.44	7'	38.48	287.9	30'	706.9	5288.
1' 5"	1.576	11.79	7' 3"	41.28	308.8	31'	754.8	5646.
1' 6"	1.767	13.22	7' 6"	44.18	330.5	32'	804.3	6016.
1' 7"	1.969	14.73	7' 9"	47.17	352.9	33'	855.3	6398.
1' 8"	2.182	16.32	8'	50.27	376.0	34'	907.9	6792.
1' 9"	2.405	17.99	8' 3"	53.46	399.9	35'	962.1	7197.
1' 10"	2.640	19.75	8' 6"	56.75	424.5	36'	1018.	7616.
1' 11"	2.885	21.58	8' 9"	60.13	449.8	37'	1075.	8043.
2'	3.142	23.50	9'	63.62	475.9	38'	1134.	8483.
2' 1"	3.409	25.50	9' 3"	67.20	502.7	39'	1195.	8940.
2' 2"	3.687	27.58	9' 6"	70.88	530.2	40'	1257.	9404.
2' 3"	3.976	29.74	9' 9"	74.66	558.5	41'	1320.	9876.
2' 4"	4.276	31.99	10'	78.54	587.5	42'	1385.	10360.
2' 5"	4.587	34.31	10' 6"	86.59	647.7	43'	1452.	10860.
2' 6"	4.909	36.72	11'	95.03	710.9	44'	1521.	11370.
2' 7"	5.241	39.21	11' 6"	103.9	777.0	45'	1590.	11900.
2' 8"	5.585	41.78	12'	113.1	846.0	46'	1662.	12430.
2' 9"	5.940	44.43	12' 6"	122.7	918.0	47'	1735.	12980.
2' 10"	6.305	47.16	13'	132.7	992.9	48'	1810.	13540.
2' 11"	6.681	49.98	13' 6"	143.1	1071.	49'	1886.	14110.
3'	7.069	52.88	14'	153.9	1152.	50'	1964.	14690.
3' 1"	7.467	55.86	14' 6"	165.1	1235.	52'	2124.	15890.
3' 2"	7.876	58.92	15'	176.7	1322.	54'	2290.	17130.
3' 3"	8.296	62.06	15' 6"	188.7	1412.	56'	2463.	18420.
3' 4"	8.727	65.28	16'	201.1	1504.	58'	2642.	19760.
3' 5"	9.168	68.58	16' 6"	213.8	1600.	60'	2827.	21150.
3' 6"	9.621	71.97	17'	227.0	1698.	62'	3019.	22580.
3' 7"	10.08	75.44	17' 6"	240.5	1799.	64'	3217.	24060.
3' 8"	10.56	78.99	18'	254.5	1904.	66'	3421.	25500.
3' 9"	11.04	82.62	18' 6"	268.8	2011.	68'	3632.	27170.
3' 10"	11.54	86.33	19'	283.5	2121.	70'	3848.	28790.
3' 11"	12.05	90.13	19' 6"	298.6	2234.	72'	4072.	30450.
4'	12.57	94.00	20'	314.2	2350.	74'	4301.	32170.
4' 1"	13.10	97.96	20' 6"	330.1	2469.	76'	4536.	33930.
4' 2"	13.64	102.0	21'	346.4	2591.	78'	4778.	35740.
4' 3"	14.19	106.1	21' 6"	363.1	2716.	80'	5027.	37600.
4' 4"	14.75	110.3	22'	380.1	2844.	82'	5281.	39500.
4' 5"	15.32	114.6	22' 6"	397.6	2974.	84'	5542.	41450.
4' 6"	15.90	119.0	23'	415.5	3108.	86'	5809.	43450.
4' 7"	16.50	123.4	23' 6"	433.7	3245.	88'	6082.	45490.
4' 8"	17.10	128.0	24'	452.4	3384.	90'	6362.	47590.
4' 9"	17.72	132.6	24' 6"	471.4	3527.	92'	6648.	49720.
4' 10"	18.35	137.3	25'	490.9	3672.	94'	6940.	51920.
4' 11"	18.99	142.0	25' 6"	510.7	3820.	96'	7238.	54140.
5'	19.63	146.9	26'	530.9	3972.	98'	7543.	56420.
5' 3"	21.65	161.9	26' 6"	551.5	4126.	100'	7854.	58750.
5' 6"	23.76	177.7	27'	572.6	4283.			
5' 9"	25.97	194.3	27' 6"	59_._	44__			

Table 10.29 Tank Capacities (Horizontal Cylindrical)

Contents of Tanks with Flat Ends when Filled to Various Depths

Contents in U. S. gallons per 1 foot of length.

To ascertain the contents of a tank over one-half full: Let h = depth of unfilled portion. Find from the table the quantity corresponding to a depth h. Subtract this quantity from the contents of a full tank.

Depth of liquid, in inches = h

Diameter of tank inches	Full tank	3"	6"	9"	12"	15"	18"	21"	24"	27"	30"	33"	36"	39"	42"	45"	48"	51"	54"	57"	60"
12"	5.88	1.15	2.94																		
18"	13.22	1.45	3.86	6.61																	
24"	23.50	1.70	4.60	8.05	11.75																
30"	36.72	1.91	5.23	9.27	13.72	18.36															
36"	52.88	2.12	5.79	10.34	15.43	20.85	26.44														
42"	71.97	2.28	6.31	11.31	16.97	23.07	29.47	35.99													
48"	94.01	2.45	6.78	12.20	18.38	25.10	32.20	39.54	47.60												
54"	118.98	2.60	7.22	13.04	19.68	26.95	34.72	42.80	51.08	59.49											
60"	146.89	2.75	7.64	13.82	20.91	28.72	37.06	45.82	54.87	64.11	73.44										
66"	177.73	2.89	8.04	14.56	22.07	30.37	39.28	48.65	58.39	68.41	78.59	88.86									
72"	211.52	3.02	8.42	15.26	23.17	31.92	41.36	51.32	61.71	72.45	83.41	94.54	105.76								
78"	248.24	3.15	8.78	15.94	24.21	33.41	43.34	53.86	64.87	76.27	87.97	99.90	111.97	124.13							
84"	287.90	3.26	9.12	16.57	25.24	34.85	45.24	56.29	67.87	79.91	92.30	104.98	117.85	130.87	143.95						
90"	330.49	3.43	9.46	17.20	26.20	36.21	47.05	58.61	70.75	83.39	96.43	109.81	123.45	137.28	151.23	165.25					
96"	376.02	3.50	9.79	17.80	27.13	37.52	48.81	60.84	73.52	86.73	100.39	114.44	128.79	143.40	158.17	173.06	188.01				
102"	424.50	3.61	10.10	18.37	28.01	39.00	50.49	62.99	76.18	89.94	104.20	118.89	133.92	149.25	164.81	180.53	196.37	212.25			
108"	476.10	3.71	10.39	18.94	28.90	40.03	52.14	65.09	78.74	93.04	107.87	123.17	138.87	154.89	171.19	187.71	204.37	221.14	238.05		
114"	530.25	3.78	10.74	19.49	29.75	41.22	53.73	67.10	81.24	96.05	111.43	127.31	143.63	160.33	177.33	194.60	212.05	229.65	247.37	265.13	
120"	587.54	3.91	10.98	20.02	30.57	42.39	55.26	69.06	83.65	98.95	114.87	131.32	148.25	165.58	183.27	201.24	219.46	237.87	256.43	275.08	293.77

Table 10.30 Contents of Standard Dished Heads When Filled to Various Depths

Contents in U. S. gallons for one head only. This table is only approximate but close enough for practical use.

Radius = Diameter

To ascertain the contents of a head over one-half full: Let h = depth of unfilled portion. Find from the table the quantity corresponding to a depth h. Subtract this quantity from the contents of a full head.

Diameter of head inches	Full head	\multicolumn{20}{c	}{Depth of liquid, in inches = h}
		3"	6"	9"	12"	15"	18"	21"	24"	27"	30"	33"	36"	39"	42"	45"	48"	51"	54"	57"	60"
12"	0.40	0.05	0.20																		
18"	1.36	0.07	0.32	0.68																	
24"	3.22	0.08	0.41	0.95	1.61																
30"	6.30	0.10	0.49	1.18	2.10	3.15															
36"	10.88	0.11	0.56	1.39	2.54	3.92	5.44														
42"	17.28	0.12	0.63	1.59	2.94	4.64	6.57	8.64													
48"	25.79	0.13	0.68	1.75	3.31	5.29	7.62	10.19	12.89												
54"	36.72	0.14	0.74	1.90	3.64	5.91	8.60	11.65	14.95	18.36											
60"	50.37	0.14	0.82	2.07	3.98	6.49	9.54	13.03	16.87	20.96	25.18										
66"	67.04	0.15	0.83	2.19	4.25	6.98	10.35	14.30	18.68	23.42	28.42	33.52									
72"	87.04	0.16	0.88	2.32	4.52	7.47	11.15	15.48	20.38	25.74	31.46	37.43	43.52								
78"	110.66	0.17	0.93	2.44	4.79	7.97	11.94	16.65	22.02	27.97	34.39	41.16	48.20	55.33							
84"	138.22	0.18	0.98	2.59	5.07	8.44	12.69	17.78	23.60	30.11	37.19	44.75	52.67	60.83	69.11						
90"	170.01	0.18	1.00	2.68	5.33	8.91	13.44	18.86	25.12	32.18	39.90	48.22	56.99	66.14	75.52	85.00					
96"	206.32	0.20	1.07	2.83	5.59	9.36	14.14	19.90	26.60	34.17	42.52	51.53	61.13	71.22	81.66	92.34	103.16				
102"	247.46	0.22	1.14	3.01	5.89	9.87	14.92	21.01	28.11	36.18	45.19	54.91	65.31	76.29	87.73	99.56	111.59	123.74			
108"	293.77	0.20	1.13	3.03	6.04	10.21	15.50	21.93	29.47	38.03	47.56	57.07	69.14	81.05	93.53	106.47	119.76	133.26	146.88		
114"	345.51	0.21	1.16	3.12	6.25	10.55	16.06	22.80	30.70	39.73	49.81	60.88	72.85	83.61	99.05	113.07	127.56	142.41	157.51	172.75	
120"	402.27	0.21	1.19	3.23	6.47	10.93	16.68	23.70	31.96	41.43	52.04	63.73	76.40	89.95	104.32	119.39	135.04	151.15	167.62	184.32	201.13

Table 10.31 Pipe Data (Carbon and Alloy Steel—Stainless Steel)

Nominal Pipe Size inches	Outside Diam. Inches	Steel Iron Pipe Size	Steel Sched. No.	Stainless Steel Sched. No.	Wall Thickness (t) Inches	Inside Diameter (d) Inches	Area of Metal Square Inches	Transverse Internal Area (a) Square Inches	Transverse Internal Area (A) Square Feet	Moment of Inertia (I) Inches⁴	Weight Pipe Pounds per foot	Weight Water Pounds per foot of pipe	External Surface Sq. Ft. per foot of pipe	Section Modulus $\left(2\frac{I}{O.D.}\right)$
1/8	0.405	...	...	10S	.049	.307	.0548	.0740	.00051	.00088	.19	.032	.106	.00437
		STD	40	40S	.068	.269	.0720	.0568	.00040	.00106	.24	.025	.106	.00523
		XS	80	80S	.095	.215	.0925	.0364	.00025	.00122	.31	.016	.106	.00602
1/4	0.540	...	...	10S	.065	.410	.0970	.1320	.00091	.00279	.33	.057	.141	.01032
		STD	40	40S	.088	.364	.1250	.1041	.00072	.00331	.42	.045	.141	.01227
		XS	80	80S	.119	.302	.1574	.0716	.00050	.00377	.54	.031	.141	.01395
3/8	0.675	...	...	10S	.065	.545	.1246	.2333	.00162	.00586	.42	.101	.178	.01736
		STD	40	40S	.091	.493	.1670	.1910	.00133	.00729	.57	.083	.178	.02160
		XS	80	80S	.126	.423	.2173	.1405	.00098	.00862	.74	.061	.178	.02554
		...	...	5S	.065	.710	.1583	.3959	.00275	.01197	.54	.172	.220	.02849
		...	...	10S	.083	.674	.1974	.3568	.00248	.01431	.67	.155	.220	.03407
		STD	40	40S	.109	.622	.2503	.3040	.00211	.01709	.85	.132	.220	.04069
1/2	0.840	XS	80	80S	.147	.546	.3200	.2340	.00163	.02008	1.09	.102	.220	.04780
		...	160	...	.187	.466	.3836	.1706	.00118	.02212	1.31	.074	.220	.05267
		XXS	...	...	.294	.252	.5043	.050	.00035	.02424	1.71	.022	.220	.05772
		...	...	5S	.065	.920	.2011	.6648	.00462	.02450	.69	.288	.275	.04667
		...	...	10S	.083	.884	.2521	.6138	.00426	.02969	.86	.266	.275	.05655
		STD	40	40S	.113	.824	.3326	.5330	.00371	.03704	1.13	.231	.275	.07055
3/4	1.050	XS	80	80S	.154	.742	.4335	.4330	.00300	.04479	1.47	.188	.275	.08531
		...	160	...	.219	.612	.5698	.2961	.00206	.05269	1.94	.128	.275	.10036
		XXS	...	...	.308	.434	.7180	.148	.00103	.05792	2.44	.064	.275	.11032
		...	...	5S	.065	1.185	.2553	1.1029	.00766	.04999	.87	.478	.344	.07603
		...	...	10S	.109	1.097	.4130	.9452	.00656	.07569	1.40	.409	.344	.11512
		STD	40	40S	.133	1.049	.4939	.8640	.00600	.08734	1.68	.375	.344	.1328
1	1.315	XS	80	80S	.179	.957	.6388	.7190	.00499	.1056	2.17	.312	.344	.1606
		...	160	...	.250	.815	.8365	.5217	.00362	.1251	2.84	.230	.344	.1903
		XXS	...	...	.358	.599	1.0760	.282	.00196	.1405	3.66	.122	.344	.2136
		...	...	5S	.065	1.530	.3257	1.839	.01277	.1038	1.11	.797	.435	.1250
		...	...	10S	.109	1.442	.4717	1.633	.01134	.1605	1.81	.708	.435	.1934
		STD	40	40S	.140	1.380	.6685	1.495	.01040	.1947	2.27	.649	.435	.2346
1¼	1.660	XS	80	80S	.191	1.278	.8815	1.283	.00891	.2418	3.00	.555	.435	.2913
		...	160	...	.250	1.160	1.1070	1.057	.00734	.2839	3.76	.458	.435	.3421
		XXS	...	...	.382	.896	1.534	.630	.00438	.3411	5.21	.273	.435	.4110
		...	...	5S	.065	1.770	.3747	2.461	.01709	.1579	1.28	1.066	.497	.1662
		...	...	10S	.109	1.682	.6133	2.222	.01543	.2468	2.09	.963	.497	.2598
		STD	40	40S	.145	1.610	.7995	2.036	.01414	.3099	2.72	.882	.497	.3262
1½	1.900	XS	80	80S	.200	1.500	1.068	1.767	.01225	.3912	3.63	.765	.497	.4118
		...	160	...	.281	1.338	1.429	1.406	.00976	.4824	4.86	.608	.497	.5078
		XXS	...	...	.400	1.100	1.885	.950	.00660	.5678	6.41	.42	.497	.5977
		...	...	5S	.065	2.245	.4717	3.958	.02749	.3149	1.61	1.72	.622	.2652
		...	...	10S	.109	2.157	.7760	3.654	.02538	.4992	2.64	1.58	.622	.4204
		STD	40	40S	.154	2.067	1.075	3.355	.02330	.6657	3.65	1.45	.622	.5606
2	2.375	XS	80	80S	.218	1.939	1.477	2.953	.02050	.8679	5.02	1.28	.622	.7309
		...	160	...	.344	1.687	2.190	2.241	.01556	1.162	7.46	.97	.622	.979
		XXS	...	...	.436	1.503	2.656	1.774	.01232	1.311	9.03	.77	.622	1.104
		...	...	5S	.083	2.709	.7280	5.764	.04002	.7100	2.48	2.50	.753	.4939
		...	...	10S	.120	2.635	1.039	5.453	.03787	.9873	3.53	2.36	.753	.6868
		STD	40	40S	.203	2.469	1.704	4.788	.03322	1.530	5.79	2.07	.753	1.064
2½	2.875	XS	80	80S	.276	2.323	2.254	4.238	.02942	1.924	7.66	1.87	.753	1.339
		...	160	...	.375	2.125	2.945	3.546	.02463	2.353	10.01	1.54	.753	1.638
		XXS	...	...	.552	1.771	4.028	2.464	.01710	2.871	13.69	1.07	.753	1.997
		...	...	5S	.083	3.334	.8910	8.730	.06063	1.301	3.03	3.78	.916	.7435
		...	...	10S	.120	3.260	1.274	8.347	.05796	1.822	4.33	3.62	.916	1.041
		STD	40	40S	.216	3.068	2.228	7.393	.05130	3.017	7.58	3.20	.916	1.724
3	3.500	XS	80	80S	.300	2.900	3.016	6.605	.04587	3.894	10.25	2.86	.916	2.225
		...	160	...	.438	2.624	4.205	5.408	.03755	5.032	14.32	2.35	.916	2.876
		XXS	...	...	.600	2.300	5.466	4.155	.02885	5.993	18.58	1.80	.916	3.424

Identification, wall thickness and weights are extracted from ANSI B36.10 and B36.19. The notations STD, XS, and XXS indicate Standard, Extra Strong, and Double Extra Strong pipe respectively.

Transverse internal area values listed in "square feet" also represent volume in cubic feet per foot of pipe length.

521

Table 10.31 Continued

Nominal Pipe Size Inches	Outside Diam. Inches	Identification Steel Iron Pipe Size	Identification Steel Sched. No.	Identification Stainless Steel Sched. No.	Wall Thickness (t) Inches	Inside Diameter (d) Inches	Area of Metal Square Inches	Transverse Internal Area (a) Square Inches	Transverse Internal Area (A) Square Feet	Moment of Inertia (I) Inches^4	Weight Pipe Pounds per foot	Weight Water Pounds per foot of pipe	External Surface Sq. Ft. per foot of pipe	Section Modulus $\left(2\frac{I}{O.D.}\right)$
4	4.500	...	...	5S	.083	4.334	1.152	14.75	.10245	2.810	3.92	6.39	1.178	1.249
		...	...	10S	.120	4.260	1.651	14.25	.09898	3.963	5.61	6.18	1.178	1.761
		STD	40	40S	.237	4.026	3.174	12.73	.08840	7.233	10.79	5.50	1.178	3.214
		XS	80	80S	.337	3.826	4.407	11.50	.07986	9.610	14.98	4.98	1.178	4.271
		...	120	...	.438	3.624	5.595	10.31	.0716	11.65	19.00	4.47	1.178	5.178
		...	160	...	.531	3.438	6.621	9.28	.0645	13.27	22.51	4.02	1.178	5.898
		XXS	...	...	.674	3.152	8.101	7.80	.0542	15.28	27.54	3.38	1.178	6.791
6	6.625	...	...	5S	.109	6.407	2.231	32.24	.2239	11.85	7.60	13.97	1.734	3.576
		...	...	10S	.134	6.357	2.733	31.74	.2204	14.40	9.29	13.75	1.734	4.346
		STD	40	40S	.280	6.065	5.581	28.89	.2006	28.14	18.97	12.51	1.734	8.496
		XS	80	80S	.432	5.761	8.405	26.07	.1810	40.49	28.57	11.29	1.734	12.22
		...	120	...	.562	5.501	10.70	23.77	.1650	49.61	36.39	10.30	1.734	14.98
		...	160	...	.719	5.187	13.32	21.15	.1469	58.97	45.35	9.16	1.734	17.81
		XXS	...	...	.864	4.897	15.64	18.84	.1308	66.33	53.16	8.16	1.734	20.02
8	8.625	...	...	5S	.109	8.407	2.916	55.51	.3855	26.44	9.93	24.06	2.258	6.131
		...	...	10S	.148	8.329	3.941	54.48	.3784	35.41	13.40	23.61	2.258	8.212
		...	20	...	.250	8.125	6.57	51.85	.3601	57.72	22.36	22.47	2.258	13.39
		...	30	...	.277	8.071	7.26	51.16	.3553	63.35	24.70	22.17	2.258	14.69
		STD	40	40S	.322	7.981	8.40	50.03	.3474	72.49	28.55	21.70	2.258	16.81
		...	60	...	.406	7.813	10.48	47.94	.3329	88.73	35.64	20.77	2.258	20.58
		XS	80	80S	.500	7.625	12.76	45.66	.3171	105.7	43.39	19.78	2.258	24.51
		...	100	...	.594	7.437	14.96	43.46	.3018	121.3	50.95	18.83	2.258	28.14
		...	120	...	.719	7.187	17.84	40.59	.2819	140.5	60.71	17.59	2.258	32.58
		...	140	...	.812	7.001	19.93	38.50	.2673	153.7	67.76	16.68	2.258	35.65
		XXS	...	...	.875	6.875	21.30	37.12	.2578	162.0	72.42	16.10	2.258	37.56
		...	160	...	.906	6.813	21.97	36.46	.2532	165.9	74.69	15.80	2.258	38.48
10	10.750	...	...	5S	.134	10.482	4.36	86.29	.5992	63.0	15.19	37.39	2.814	11.71
		...	...	10S	.165	10.420	5.49	85.28	.5922	76.9	18.65	36.95	2.814	14.30
		...	20	...	.250	10.250	8.24	82.52	.5731	113.7	28.04	35.76	2.814	21.15
		...	30	...	.307	10.136	10.07	80.69	.5603	137.4	34.24	34.96	2.814	25.57
		STD	40	40S	.365	10.020	11.90	78.86	.5475	160.7	40.48	34.20	2.814	29.90
		XS	60	80S	.500	9.750	16.10	74.66	.5185	212.0	54.74	32.35	2.814	39.43
		...	80	...	.594	9.562	18.92	71.84	.4989	244.8	64.43	31.13	2.814	45.54
		...	100	...	.719	9.312	22.63	68.13	.4732	286.1	77.03	29.53	2.814	53.22
		...	120	...	.844	9.062	26.24	64.53	.4481	324.2	89.29	27.96	2.814	60.32
		XXS	140	...	1.000	8.750	30.63	60.13	.4176	367.8	104.13	26.06	2.814	68.43
		...	160	...	1.125	8.500	34.02	56.75	.3941	399.3	115.64	24.59	2.814	74.29
12	12.75	...	...	5S	.156	12.438	6.17	121.50	.8438	122.4	20.98	52.65	3.338	19.2
		...	...	10S	.180	12.390	7.11	120.57	.8373	140.4	24.17	52.25	3.338	22.0
		...	20	...	.250	12.250	9.82	117.86	.8185	191.8	33.38	51.07	3.338	30.2
		...	30	...	.330	12.090	12.87	114.80	.7972	248.4	43.77	49.74	3.338	39.0
		STD	...	40S	.375	12.000	14.58	113.10	.7854	279.3	49.56	49.00	3.338	43.8
			40	...	.406	11.938	15.77	111.93	.7773	300.3	53.52	48.50	3.338	47.1
		XS	...	80S	.500	11.750	19.24	108.43	.7528	361.5	65.42	46.92	3.338	56.7
		...	60	...	.562	11.626	21.52	106.16	.7372	400.4	73.15	46.00	3.338	62.8
		...	80	...	.688	11.374	26.03	101.64	.7058	475.1	88.63	44.04	3.338	74.6
		...	100	...	.844	11.062	31.53	96.14	.6677	561.6	107.32	41.66	3.338	88.1
		XXS	120	...	1.000	10.750	36.91	90.76	.6303	641.6	125.49	39.33	3.338	100.7
		...	140	...	1.125	10.500	41.08	86.59	.6013	700.5	139.67	37.52	3.338	109.9
		...	160	...	1.312	10.126	47.14	80.53	.5592	781.1	160.27	34.89	3.338	122.6

Identification, wall thickness and weights are extracted from ANSI B36.10 and B36.19. The notations STD, XS, and XXS indicate Standard, Extra Strong, and Double Extra Strong pipe respectively.

Transverse internal area values listed in "square feet" also represent volume in cubic feet per foot of pipe length.

Table 10.32 Properties of Saturated Steam (Pressure: in. Hg Abs.)

Absolute Pressure (in. Hg)	Temp. (°F)	Specific Volume (ft³/lb)	Absolute Pressure (in. Hg)	Temp. (°F)	Specific Volume (ft³/lb)
0.18	32.00	3306	3.75	123.1	188
0.20	34.56	2997	4.00	125.4	177
0.30	44.96	2039	4.25	127.7	167
0.40	52.64	1553	4.50	129.8	158
0.50	58.80	1256	4.75	131.8	150
0.60	63.96	1057	5.00	133.8	143
0.70	68.40	914	5.50	137.0	131
0.80	72.33	806	6.00	141.0	121
0.90	75.85	721	6.50	144.0	112
1.00	79.03	652	7.00	147.0	105
1.25	85.93	528	8.00	152.0	92
1.50	91.72	445	9.00	157.0	83
1.75	96.72	385	10.00	162.0	75
2.00	101.14	339	12.00	169.0	63
2.25	105.11	304	14.00	176.0	55
2.50	108.71	275	16.00	182.0	48
2.75	112.01	251	18.00	188.0	43
3.00	115.06	232	20.00	192.0	39
3.25	117.90	215	25.00	203.0	32
3.50	120.60	200	29.92	212.0	27

Table 10.33 Properties of Saturated Steam (Temperature)[2,a]

Temp. (°F) t	Press. $\left(\dfrac{\text{lbf}}{\text{in.}^2}\right) p$	Specific Volume Sat. Liquid v_f	Sat. Vapor v_g	Internal Energy Sat. Liquid u_f	Evap. u_{fg}	Sat. Vapor u_g	Enthalpy Sat. Liquid h_f	Evap. h_{fg}	Sat. Vapor h_g	Entropy Sat. Liquid s_f	Evap. s_{fg}	Sat. Vapor s_g
32	.08859	.016022	3305.	−.01	1021.2	1021.2	−.01	1075.4	1075.4	−.00003	2.1870	2.1870
35	.09992	.016021	2948.	2.99	1019.2	1022.2	3.00	1073.7	1076.7	.00607	2.1704	2.1764
40	.12166	.016020	2445.	8.02	1015.8	1023.9	8.02	1070.9	1078.9	.01617	2.1430	2.1592
45	.14748	.016021	2037.	13.04	1012.5	1025.5	13.04	1068.1	1081.1	.02618	2.1162	2.1423
50	.17803	.016024	1704.2	18.06	1009.1	1027.2	18.06	1065.2	1083.3	.03607	2.0899	2.1259
60	.2563	.016035	1206.9	28.08	1002.4	1030.4	28.08	1059.6	1087.7	.05555	2.0388	2.0943
70	.3632	.016051	867.7	38.09	995.6	1033.7	38.09	1054.0	1092.0	.07463	1.9896	2.0642
80	.5073	.016073	632.8	48.08	988.9	1037.0	48.09	1048.3	1096.4	.09332	1.9423	2.0356
90	.6988	.016099	467.7	58.07	982.2	1040.2	58.07	1042.7	1100.7	.11165	1.8966	2.0083
100	.9503	.016130	350.0	68.04	975.4	1043.5	68.05	1037.0	1105.0	.12963	1.8526	1.9822
110	1.2763	.016166	265.1	78.02	968.7	1046.7	78.02	1031.3	1109.3	.14730	1.8101	1.9574
120	1.6945	.016205	203.0	87.99	961.9	1049.9	88.00	1025.5	1113.5	.16465	1.7690	1.9336
130	2.225	.016247	157.17	97.97	955.1	1053.0	97.98	1019.8	1117.8	.18172	1.7292	1.9109
140	2.892	.016293	122.88	107.95	948.2	1056.2	107.96	1014.0	1121.9	.19851	1.6907	1.8892
150	3.722	.016343	96.99	117.95	941.3	1059.3	117.96	1008.1	1126.1	.21503	1.6533	1.8684
160	4.745	.016395	77.23	127.94	934.4	1062.3	127.96	1002.2	1130.1	.23130	1.6171	1.8484
170	5.996	.016450	62.02	137.95	927.4	1065.4	137.97	996.2	1134.2	.24732	1.5819	1.8293
180	7.515	.016509	50.20	147.97	920.4	1068.3	147.99	990.2	1138.2	.26311	1.5478	1.8109
190	9.343	.016570	40.95	158.00	913.3	1071.3	158.03	984.1	1142.1	.27866	1.5146	1.7932
200	11.529	.016634	33.63	168.04	906.2	1074.2	168.07	977.9	1145.9	.29400	1.4822	1.7762

Table 10.33 *Continued*

Temp. (°F) t	Press. $\left(\dfrac{\text{lbf}}{\text{in.}^2}\right) p$	Specific Volume Sat. Liquid v_f	Specific Volume Sat. Vapor v_g	Internal Energy Sat. Liquid u_f	Internal Energy Evap. u_{fg}	Internal Energy Sat. Vapor u_g	Enthalpy Sat. Liquid h_f	Enthalpy Evap. h_{fg}	Enthalpy Sat. Vapor h_g	Entropy Sat. Liquid s_f	Entropy Evap. s_{fg}	Entropy Sat. Vapor s_g
210	14.125	.016702	27.82	178.10	898.9	1077.0	178.14	971.6	1149.7	.30913	1.4508	1.7599
212	14.698	.016716	26.80	180.11	897.5	1077.6	180.16	970.3	1150.5	.31213	1.4446	1.7567
220	17.188	.016772	23.15	188.17	891.7	1079.8	188.22	965.3	1153.5	.32406	1.4201	1.7441
230	20.78	.016845	19.386	198.26	884.3	1082.6	198.32	958.8	1157.1	.33880	1.3901	1.7289
240	24.97	.016922	16.327	208.36	876.9	1085.3	208.44	952.3	1160.7	.35335	1.3609	1.7143
250	29.82	.017001	13.826	218.49	869.4	1087.9	218.59	945.6	1164.2	.36772	1.3324	1.7001
260	35.42	.017084	11.768	228.64	861.8	1090.5	228.76	938.8	1167.6	.38193	1.3044	1.6864
270	41.85	.017170	10.066	238.82	854.1	1093.0	238.95	932.0	1170.9	.39597	1.2771	1.6731
280	49.18	.017259	8.650	249.02	846.3	1095.4	249.18	924.9	1174.1	.40986	1.2504	1.6602
290	57.53	.017352	7.467	259.25	838.5	1097.7	259.44	917.8	1177.2	.42360	1.2241	1.6477
300	66.98	.017448	6.472	269.52	830.5	1100.0	269.73	910.4	1180.2	.43720	1.1984	1.6356
310	77.64	.017548	5.632	279.81	822.3	1102.1	280.06	903.0	1183.0	.45067	1.1731	1.6238
320	89.60	.017652	4.919	290.14	814.1	1104.2	290.43	895.3	1185.8	.46400	1.1483	1.6123
330	103.00	.017760	4.312	300.51	805.7	1106.2	300.84	887.5	1188.4	.47722	1.1238	1.6010
340	117.93	.017872	3.792	310.91	797.1	1108.0	311.30	879.5	1190.8	.49031	1.0997	1.5901
350	134.53	.017988	3.346	321.35	788.4	1109.8	321.80	871.3	1193.1	.50329	1.0760	1.5793
360	152.92	.018108	2.961	331.84	779.6	1111.4	332.35	862.9	1195.2	.51617	1.0526	1.5688
370	173.23	.018233	2.628	342.37	770.6	1112.9	342.96	854.2	1197.2	.52894	1.0295	1.5585
380	195.60	.018363	2.339	352.95	761.4	1114.3	353.62	845.4	1199.0	.54163	1.0067	1.5483
390	220.2	.018498	2.087	363.58	752.0	1115.6	364.34	836.2	1200.6	.55422	.9841	1.5383
400	247.1	.018638	1.8661	374.27	742.4	1116.6	375.12	826.8	1202.0	.56672	.9617	1.5284
410	276.5	.018784	1.6726	385.01	732.6	1117.6	385.97	817.2	1203.1	.57916	.9395	1.5187
420	308.5	.018936	1.5024	395.81	722.5	1118.3	396.89	807.2	1204.1	.59152	.9175	1.5091
430	343.3	.019094	1.3521	406.68	712.2	1118.9	407.89	796.9	1204.8	.60381	.8957	1.4995
440	381.2	.019260	1.2192	417.62	701.7	1119.3	418.98	786.3	1205.3	.61605	.8740	1.4900
450	422.1	.019433	1.1011	428.6	690.9	1119.5	430.2	775.4	1205.6	.6282	.8523	1.4806
460	466.3	.019614	.9961	439.7	679.8	1119.6	441.4	764.1	1205.5	.6404	.8308	1.4712
470	514.1	.019803	.9025	450.9	668.4	1119.4	452.8	752.4	1205.2	.6525	.8093	1.4618
480	565.5	.020002	.8187	462.2	656.7	1118.9	464.3	740.3	1204.6	.6646	.7878	1.4524
490	620.7	.020211	.7436	473.6	644.7	1118.3	475.9	727.8	1203.7	.6767	.7663	1.4430
500	680.0	.02043	.6761	485.1	632.3	1117.4	487.7	714.8	1202.5	.6888	.7448	1.4335
520	811.4	.02091	.5605	508.5	606.2	1114.8	511.7	687.3	1198.9	.7130	.7015	1.4145
540	961.5	.02145	.4658	532.6	578.4	1111.0	536.4	657.5	1193.8	.7374	.6576	1.3950
560	1131.8	.02207	.3877	557.4	548.4	1105.8	562.0	625.0	1187.0	.7620	.6129	1.3749
580	1324.3	.02278	.3225	583.1	515.9	1098.9	588.6	589.3	1178.0	.7872	.5668	1.3540
600	1541.0	.02363	.2677	609.9	480.1	1090.0	616.7	549.7	1166.4	.8130	.5187	1.3317
620	1784.4	.02465	.2209	638.3	440.2	1078.2	646.4	505.0	1151.4	.8398	.4677	1.3075
640	2057.1	.02593	.1805	668.7	394.5	1063.2	678.6	453.4	1131.9	.8681	.4122	1.2803
660	2362.	.02767	.14459	702.3	340.0	1042.3	714.4	391.1	1105.5	.8990	.3493	1.2483
680	2705.	.03032	.11127	741.7	269.3	1011.0	756.9	309.8	1066.7	.9350	.2718	1.2068
700	3090.	.03666	.07438	801.7	145.9	947.7	822.7	167.5	990.2	.9902	.1444	1.1346
705.44	3204.	.05053	.05053	872.6	0	872.6	902.5	0	902.5	1.0580	0	1.0580

[a]Symbols used: h, specific enthalpy, Btu per lb; p, pressure, lbf per in.^{2}; s, specific entropy, Btu per lb, degrees Rankine; t, thermodynamic temperature, degrees Fahrenheit; u, specific internal energy, Btu per lb; and v, specific volume, ft^3 per lb.

Note the use of the following subscripts: f, refers to a property of liquid in equilibrium with vapor; g, refers to a property of vapor in equilibrium with liquid; i, refers to a property of solid in equilibrium with vapor; fg, refers to a change by evaporation; and ig, refers to a change by sublimation.

Table 10.34 Properties of Saturated Steam (Pressure)[2,a]

Press. $\left(\dfrac{lbf}{in.^2}\right)$ p	Temp. (°F) t	Specific Volume Sat. Liquid v_f	Sat. Vapor v_g	Internal Energy Sat. Liquid u_f	Evap. u_{fg}	Sat. Vapor u_g	Enthalpy Sat. Liquid h_f	Evap. h_{fg}	Sat. Vapor h_g	Entropy Sat. Liquid s_f	Evap. s_{fg}	Sat. Vapor s_g
.08866	32.02	.016022	3302.	.00	1021.2	1021.2	.01	1075.4	1075.4	.00000	2.1869	2.1869
.10	35.02	.016021	2946.	·3.02	1019.2	1022.2	3.02	1073.7	1076.7	.00612	2.1702	2.1764
.20	53.15	.016027	1526.	21.22	1007.0	1028.2	21.22	1063.5	1084.7	.04225	2.0736	2.1158
.30	64.46	.016041	1039.	32.55	999.4	1031.9	32.56	1057.1	1089.6	.06411	2.0166	2.0807
.40	72.84	.016056	792.0	40.94	993.7	1034.7	40.94	1052.3	1093.3	.07998	1.9760	2.0559
.50	79.56	.016071	641.5	47.64	989.2	1036.9	47.65	1048.6	1096.2	.09250	1.9443	2.0368
.60	85.19	.016086	540.0	53.26	985.4	1038.7	53.27	1045.4	1098.6	.10287	1.9184	2.0213
.70	90.05	.016099	466.9	58.12	982.1	1040.3	58.12	1042.6	1100.7	.11174	1.8964	2.0081
.80	94.35	.016112	411.7	62.41	979.2	1041.7	62.41	1040.2	1102.6	.11951	1.8773	1.9968
.90	98.20	.016124	368.4	66.25	976.6	1042.9	66.25	1038.0	1104.3	.12642	1.8604	1.9868
1.0	101.70	.016136	333.6	69.74	974.3	1044.0	69.74	1036.0	1105.8	.13266	1.8453	1.9779
1.5	115.65	.016187	227.7	83.65	964.8	1048.5	83.65	1028.0	1111.7	.15714	1.7867	1.9438
2.0	126.04	.016230	173.7	94.02	957.8	1051.8	94.02	1022.1	1116.1	.17499	1.7448	1.9198
3.0	141.43	.016300	118.7	109.38	947.2	1056.6	109.39	1013.1	1122.5	.20089	1.6852	1.8861
4.0	152.93	.016358	90.64	120.88	939.3	1060.2	120.89	1006.4	1127.3	.21983	1.6426	1.8624
5.0	162.21	.016407	73.53	130.15	932.9	1063.0	130.17	1000.9	1131.0	.23486	1.6093	1.8441
6.0	170.03	.016451	61.98	137.98	927.4	1065.4	138.00	996.2	1134.2	.24736	1.5819	1.8292
7.0	176.82	.016490	53.65	144.78	922.6	1067.4	144.80	992.1	1136.9	.25811	1.5585	1.8167
8.0	182.84	.016526	47.35	150.81	918.4	1069.2	150.84	988.4	1139.3	.26754	1.5383	1.8058
9.0	188.26	.016559	42.41	156.25	914.5	1070.8	156.27	985.1	1141.4	.27596	1.5203	1.7963
10	193.19	.016590	38.42	161.20	911.0	1072.2	161.23	982.1	1143.3	.28358	1.5041	1.7877
14.696	211.99	.016715	26.80	180.10	897.5	1077.6	180.15	970.4	1150.5	.31212	1.4446	1.7567
15	213.03	.016723	26.29	181.14	896.8	1077.9	181.19	969.7	1150.9	.31367	1.4414	1.7551
20	227.96	.016830	20.09	196.19	885.8	1082.0	196.26	960.1	1156.4	.33580	1.3962	1.7320
25	240.08	.016922	16.306	208.44	876.9	1085.3	208.52	952.2	1160.7	.35345	1.3607	1.7142
30	250.34	.017004	13.748	218.84	869.2	1088.0	218.93	945.4	1164.3	.36821	1.3314	1.6996
35	259.30	.017078	11.900	227.93	862.4	1090.3	228.04	939.3	1167.4	.38093	1.3064	1.6873
40	267.26	.017146	10.501	236.03	856.2	1092.3	236.16	933.8	1170.0	.39214	1.2845	1.6767
45	274.46	.017209	9.403	243.37	850.7	1094.0	243.51	928.8	1172.3	.40218	1.2651	1.6673
50	281.03	.017269	8.518	250.08	845.5	1095.6	250.24	924.2	1174.4	.41129	1.2476	1.6589
55	287.10	.017325	7.789	256.28	840.8	1097.0	256.46	919.9	1176.3	.41963	1.2317	1.6513
60	292.73	.017378	7.177	262.06	836.3	1098.3	262.25	915.8	1178.0	.42733	1.2170	1.6444
65	298.00	.017429	6.657	267.46	832.1	1099.5	267.67	911.9	1179.6	.43450	1.2035	1.6380
70	302.96	.017478	6.209	272.56	828.1	1100.6	272.79	908.3	1181.0	.44120	1.1909	1.6321
75	307.63	.017524	5.818	277.37	824.3	1101.6	277.61	904.8	1182.4	.44749	1.1790	1.6265
80	312.07	.017570	5.474	281.95	820.6	1102.6	282.21	901.4	1183.6	.45344	1.1679	1.6214
85	316.29	.017613	5.170	286.30	817.1	1103.5	286.58	898.2	1184.8	.45907	1.1574	1.6165
90	320.31	.017655	4.898	290.46	813.8	1104.3	290.76	895.1	1185.9	.46442	1.1475	1.6119
95	324.16	.017696	4.654	294.45	810.6	1105.0	294.76	892.1	1186.9	.46952	1.1380	1.6076
100	327.86	.017736	4.434	298.28	807.5	1105.8	298.61	889.2	1187.8	.47439	1.1290	1.6034
110	334.82	.017813	4.051	305.52	801.6	1107.1	305.88	883.7	1189.6	.48355	1.1122	1.5957
120	341.30	.017886	3.730	312.27	796.0	1108.3	312.67	878.5	1191.1	.49201	1.0966	1.5886
130	347.37	.017957	3.457	318.61	790.7	1109.4	319.04	873.5	1192.5	.49989	1.0822	1.5821
140	353.08	.018024	3.221	324.58	785.7	1110.3	325.05	868.7	1193.8	.50727	1.0688	1.5761
150	358.48	.018089	3.016	330.24	781.0	1111.2	330.75	864.2	1194.9	.51422	1.0562	1.5704
160	363.60	.018152	2.836	335.63	776.4	1112.0	336.16	859.8	1196.0	.52078	1.0443	1.5651
170	368.47	.018214	2.676	340.76	772.0	1112.7	341.33	855.6	1196.9	.52700	1.0330	1.5600
180	373.13	.018273	2.533	345.68	767.7	1113.4	346.29	851.5	1197.8	.53292	1.0223	1.5553
190	377.59	.018331	2.405	350.39	763.6	1114.0	351.04	847.5	1198.6	.53857	1.0122	1.5507
200	381.86	.018387	2.289	354.9	759.6	1114.6	355.6	843.7	1199.3	.5440	1.0025	1.5464

Table 10.34 *Continued*

Press. $\left(\dfrac{\text{lbf}}{\text{in.}^2}\right)$ p	Temp. (°F) t	Specific Volume Sat. Liquid v_f	Specific Volume Sat. Vapor v_g	Internal Energy Sat. Liquid u_f	Internal Energy Evap. u_{fg}	Internal Energy Sat. Vapor u_g	Enthalpy Sat. Liquid h_f	Enthalpy Evap. h_{fg}	Enthalpy Sat. Vapor h_g	Entropy Sat. Liquid s_f	Entropy Evap. s_{fg}	Entropy Sat. Vapor s_g
250	401.04	.018653	1.8448	375.4	741.4	1116.7	376.2	825.8	1202.1	.5680	.9594	1.5274
300	417.43	.018896	1.5442	393.0	725.1	1118.2	394.1	809.8	1203.9	.5883	.9232	1.5115
350	431.82	.019124	1.3267	408.7	710.3	1119.0	409.9	795.0	1204.9	.6060	.8917	1.4978
400	444.70	.019340	1.1620	422.8	696.7	1119.5	424.2	781.2	1205.5	.6218	.8638	1.4856
450	456.39	.019547	1.0326	435.7	683.9	1119.6	437.4	768.2	1205.6	.6360	.8385	1.4746
500	467.13	.019748	.9283	447.7	671.7	1119.4	449.5	755.8	1205.3	.6490	.8154	1.4645
600	486.33	.02013	.7702	469.4	649.1	1118.6	471.7	732.4	1204.1	.6723	.7742	1.4464
700	503.23	.02051	.6558	488.9	628.2	1117.0	491.5	710.5	1202.0	.6927	.7378	1.4305
800	518.36	.02087	.5691	506.6	608.4	1115.0	509.7	689.6	1199.3	.7110	.7050	1.4160
900	532.12	.02123	.5009	523.0	589.6	1112.6	526.6	669.5	1196.0	.7277	.6750	1.4027
1000	544.75	.02159	.4459	538.4	571.5	1109.9	542.4	650.0	1192.4	.7432	.6471	1.3903
1100	556.45	.02195	.4005	552.9	553.9	1106.8	557.4	631.0	1188.3	.7576	.6209	1.3786
1200	567.37	.02232	.3623	566.7	536.8	1103.5	571.7	612.3	1183.9	.7712	.5961	1.3673
1300	577.60	.02269	.3297	579.9	519.9	1099.8	585.4	593.8	1179.2	.7841	.5724	1.3565
1400	587.25	.02307	.3016	592.7	503.3	1096.0	598.6	575.5	1174.1	.7964	.5497	1.3461
1500	596.39	.02346	.2769	605.0	486.9	1091.8	611.5	557.2	1168.7	.8082	.5276	1.3359
1750	617.31	.02450	.2268	634.4	445.9	1080.2	642.3	511.4	1153.7	.8361	.4748	1.3109
2000	636.00	.02565	.18813	662.4	404.2	1066.6	671.9	464.4	1136.3	.8623	.4238	1.2861
2250	652.90	.02698	.15692	689.9	360.7	1050.6	701.1	414.8	1115.9	.8876	.3728	1.2604
2500	668.31	.02860	.13059	717.7	313.4	1031.0	730.9	360.5	1091.4	.9131	.3196	1.2327
2750	682.46	.03077	.10717	747.3	258.6	1005.9	763.0	297.4	1060.4	.9401	.2604	1.2005
3000	695.52	.03431	.08404	783.4	185.4	968.8	802.5	213.0	1015.5	.9732	.1843	1.1575
3200	705.27	.04805	.05444	862.1	25.6	887.7	890.6	29.3	919.9	1.0478	.0252	1.0730
3203.6	705.44	.05053	.05053	872.6	0	872.6	902.5	0	902.5	1.0580	0	1.0580

[a]See footnotes to Table 10.33.

Table 10.35 Properties of Superheated Steam²,ᵃ

Absolute Pressure (lb/in.²) Sat. Temp. (°F)		Temperature (°F)										
		200°	300°	400°	500°	600°	700°	800°	900°	1000°	1100°	1200°
1 (101.70)	v	392.5	452.3	511.9	571.5	631.1	690.7	750.3	809.9	869.5	929.0	988.6
	h	1150.1	1195.7	1241.8	1288.5	1336.1	1384.5	1433.7	1483.8	1534.8	1586.8	1639.6
	s	2.0508	2.1150	2.1720	2.2235	2.2706	2.3142	2.3550	2.3932	2.4294	2.4638	2.4967
5 (162.21)	v	78.15	90.24	102.24	114.20	126.15	138.08	150.01	161.94	173.86	185.78	197.70
	h	1148.6	1194.8	1241.2	1288.2	1335.8	1384.3	1433.5	1483.7	1534.7	1586.7	1639.5
	s	1.8715	1.9367	1.9941	2.0458	2.0930	2.1367	2.1775	2.2158	2.2520	2.2864	2.3192
10 (193.19)	v	38.85	44.99	51.03	57.04	63.03	69.01	74.98	80.95	86.91	92.88	98.84
	h	1146.6	1193.7	1240.5	1287.7	1335.5	1384.0	1433.3	1483.5	1534.6	1586.6	1639.4
	s	1.7927	1.8592	1.9171	1.9690	2.0164	2.0601	2.1009	2.1393	2.1755	2.2099	2.2428
14.696 (211.99)	v		30.52	34.67	38.77	42.86	46.93	51.00	55.07	59.13	63.19	67.25
	h		1192.6	1239.9	1287.3	1335.2	1383.8	1433.1	1483.4	1534.5	1586.4	1639.3
	s		1.8157	1.8741	1.9263	1.9737	2.0175	2.0584	2.0967	2.1330	2.1674	2.2003
20 (227.96)	v		22.36	25.43	28.46	31.47	34.47	37.46	40.45	43.44	46.42	49.41
	h		1191.5	1239.2	1286.8	1334.8	1383.5	1432.9	1483.2	1534.3	1586.3	1639.2
	s		1.7805	1.8395	1.8919	1.9395	1.9834	2.0243	2.0627	2.0989	2.1334	2.1663
40 (267.26)	v		11.038	12.623	14.164	15.685	17.196	18.701	20.202	21.700	23.20	24.69
	h		1186.8	1236.4	1284.9	1333.4	1382.4	1432.1	1482.5	1533.8	1585.9	1638.9
	s		1.6993	1.7606	1.8140	1.8621	1.9063	1.9474	1.9859	2.0223	2.0568	2.0897
60 (292.73)	v		7.260	8.353	9.399	10.425	11.440	12.448	13.452	14.454	15.454	16.452
	h		1181.9	1233.5	1283.0	1332.1	1381.4	1431.2	1481.8	1533.2	1585.4	1638.5
	s		1.6496	1.7134	1.7678	1.8165	1.8609	1.9022	1.9408	1.9773	2.0119	2.0448
80 (312.07)	v			6.217	7.017	7.794	8.561	9.321	10.078	10.831	11.583	12.333
	h			1230.6	1281.1	1330.7	1380.3	1430.4	1481.2	1532.6	1584.9	1638.1
	s			1.6790	1.7346	1.7838	1.8285	1.8700	1.9087	1.9453	1.9799	2.0130

Table 10.35 Continued

Absolute Pressure (lb/in.²) Sat. Temp. (°F)		200°	300°	400°	500°	600°	700°	800°	900°	1000°	1100°	1200°
100 (327.86)	v			4.934	5.587	6.216	6.834	7.445	8.053	8.657	9.260	9.861
	h			1227.5	1279.1	1329.3	1379.2	1429.6	1480.5	1532.1	1584.5	1637.7
	s			1.6517	1.7085	1.7582	1.8033	1.8449	1.8838	1.9204	1.9551	1.9882
120 (341.30)	v			4.079	4.633	5.164	5.682	6.195	6.703	7.208	7.711	8.213
	h			1224.4	1277.1	1327.8	1378.2	1428.7	1479.8	1531.5	1584.0	1637.3
	s			1.6288	1.6868	1.7371	1.7825	1.8243	1.8633	1.9000	1.9348	1.9679
140 (353.08)	v			3.466	3.952	4.412	4.860	5.301	5.739	6.173	6.605	7.036
	h			1221.2	1275.1	1326.4	1377.1	1427.9	1479.1	1531.0	1583.6	1636.9
	s			1.6088	1.6682	1.7191	1.7648	1.8068	1.8459	1.8827	1.9176	1.9507
160 (363.60)	v			3.007	3.440	3.848	4.243	4.631	5.015	5.397	5.776	6.154
	h			1217.8	1273.0	1325.0	1376.0	1427.0	1478.4	1530.4	1583.1	1636.5
	s			1.5911	1.6518	1.7034	1.7494	1.7916	1.8308	1.8677	1.9026	1.9358
180 (373.13)	v			2.648	3.042	3.409	3.763	4.110	4.453	4.793	5.131	5.467
	h			1214.4	1270.9	1323.5	1374.9	1426.2	1477.7	1529.8	1582.6	1636.1
	s			1.5749	1.6372	1.6893	1.7357	1.7781	1.8175	1.8545	1.8894	1.9227
200 (381.86)	v			2.361	2.724	3.058	3.379	3.693	4.003	4.310	4.615	4.918
	h			1210.8	1268.8	1322.1	1373.8	1425.3	1477.1	1529.3	1582.2	1635.7
	s			1.5600	1.6239	1.6767	1.7234	1.7660	1.8055	1.8425	1.8776	1.9109
220 (389.94)	v			2.125	2.463	2.771	3.065	3.352	3.635	3.914	4.192	4.468
	h			1207.2	1266.6	1320.6	1372.7	1424.5	1476.4	1528.7	1581.7	1635.4
	s			1.5461	1.6116	1.6651	1.7122	1.7549	1.7946	1.8318	1.8668	1.9002
240 (397.46)	v			1.9283	2.246	2.531	2.803	3.068	3.328	3.585	3.840	4.094
	h			1203.3	1264.4	1319.1	1371.6	1423.6	1475.7	1528.2	1581.2	1635.0
	s			1.5329	1.6002	1.6545	1.7018	1.7448	1.7846	1.8219	1.8570	1.8904

Temperature (°F)

Pressure (Sat. Temp)									
260 (404.51)	v	2.061	2.329	2.582	2.827	3.068	3.306	3.542	3.777
	h	1262.1	1317.6	1370.5	1422.7	1475.0	1527.6	1580.8	1634.6
	s	1.5896	1.6446	1.6923	1.7355	1.7754	1.8128	1.8480	1.8814
280 (411.15)	v	1.9033	2.155	2.392	2.621	2.846	3.067	3.287	3.505
	h	1259.9	1316.0	1369.4	1421.9	1474.3	1527.0	1580.3	1634.2
	s	1.5796	1.6353	1.6834	1.7268	1.7669	1.8043	1.8396	1.8731
300 (417.43)	v	1.7662	2.004	2.227	2.442	2.653	2.860	3.066	3.270
	h	1257.5	1314.5	1368.3	1421.0	1473.6	1526.5	1579.8	1633.8
	s	1.5701	1.6266	1.6751	1.7187	1.7589	1.7964	1.8317	1.8653
350 (431.82)	v	1.4913	1.7025	1.8975	2.085	2.267	2.446	2.624	2.799
	h	1251.5	1310.6	1365.4	1418.8	1471.8	1525.0	1578.6	1632.8
	s	1.5482	1.6068	1.6562	1.7004	1.7409	1.7787	1.8142	1.8478
400 (444.70)	v	1.2843	1.4760	1.6503	1.8163	1.9776	2.136	2.292	2.446
	h	1245.2	1306.6	1362.5	1416.6	1470.1	1523.6	1577.4	1631.8
	s	1.5282	1.5892	1.6397	1.6844	1.7252	1.7632	1.7989	1.8327
450 (456.39)	v	1.1226	1.2996	1.4580	1.6077	1.7524	1.8941	2.034	2.172
	h	1238.5	1302.5	1359.6	1414.4	1468.3	1522.2	1576.3	1630.8
	s	1.5097	1.5732	1.6248	1.6701	1.7113	1.7495	1.7853	1.8192
500 (467.13)	v	.9924	1.1583	1.3040	1.4407	1.5723	1.7008	1.8271	1.9518
	h	1231.5	1298.3	1356.7	1412.1	1466.5	1520.7	1575.1	1629.8
	s	1.4923	1.5585	1.6112	1.6571	1.6987	1.7371	1.7731	1.8072
550 (477.07)	v	.8850	1.0424	1.1779	1.3040	1.4249	1.5426	1.6581	1.7720
	h	1224.1	1293.9	1353.7	1409.8	1464.7	1519.3	1573.9	1628.8
	s	1.4755	1.5448	1.5987	1.6452	1.6872	1.7259	1.7620	1.7962
600 (486.33)	v	.7947	.9456	1.0727	1.1900	1.3021	1.4108	1.5173	1.6222
	h	1216.2	1289.5	1350.6	1407.6	1462.9	1517.8	1572.7	1627.8
	s	1.4592	1.5320	1.5872	1.6343	1.6766	1.7155	1.7519	1.7861
700 (503.23)	v		.7929	.9073	1.0109	1.1089	1.2036	1.2960	1.3868
	h		1280.2	1344.4	1402.9	1459.3	1514.9	1570.2	1625.8
	s		1.5081	1.5661	1.6145	1.6576	1.6970	1.7337	1.7682

Table 10.35 *Continued*

Absolute Pressure (lb/in²) Sat. Temp. (°F)		200°	300°	400°	500°	600°	700°	800°	900°	1000°	1100°	1200°
								Temperature (°F)				
800 (518.36)	v					.6776	.7829	.8764	.9640	1.0482	1.1300	1.2102
	h					1270.4	1338.0	1398.2	1455.6	1511.9	1567.8	1623.8
	s					1.4861	1.5471	1.5969	1.6408	1.6807	1.7178	1.7526
900 (532.12)	v					.5871	.6859	.7717	.8513	.9273	1.0009	1.0729
	h					1260.0	1331.4	1393.4	1451.9	1508.9	1565.4	1621.7
	s					1.4652	1.5297	1.5810	1.6257	1.6662	1.7036	1.7386
1000 (544.75)	v					.5140	.6080	.6878	.7610	.8305	.8976	.9630
	h					1248.8	1324.6	1388.5	1448.1	1505.9	1562.9	1619.7
	s					1.4450	1.5135	1.5664	1.6120	1.6530	1.6908	1.7261
1100 (556.45)	v					.4532	.5441	.6190	.6871	.7513	.8131	.8731
	h					1236.7	1317.5	1383.5	1444.3	1502.8	1560.4	1617.6
	s					1.4252	1.4982	1.5529	1.5993	1.6409	1.6790	1.7146
1200 (567.37)	v					.4017	.4906	.5617	.6255	.6583	.7426	.7982
	h					1223.6	1310.2	1378.4	1440.4	1499.7	1557.9	1615.5
	s					1.4054	1.4837	1.5402	1.5876	1.6297	1.6682	1.7040
1400 (587.25)	v					.3175	.4059	.4713	.5285	.5815	.6319	.6805
	h					1193.1	1294.8	1367.9	1432.5	1493.5	1552.8	1611.4
	s					1.3642	1.4562	1.5168	1.5661	1.6094	1.6487	1.6851
1600 (605.06)	v						.3415	.4032	.4557	.5036	.5488	.5921
	h						1278.1	1357.0	1424.4	1487.1	1547.7	1607.1
	s						1.4299	1.4953	1.5468	1.5913	1.6315	1.6684
1800 (621.21)	v						.2905	.3500	.3989	.4430	.4842	.5235
	h						1259.9	1345.7	1416.1	1480.7	1542.5	1602.9
	s						1.4042	1.4753	1.5291	1.5749	1.6159	1.6534

2000	v	.2487	.3071	.3534	.3945	.4325	.4685
(636.00)	h	1239.8	1333.8	1407.6	1474.1	1537.2	1598.6
	s	1.3782	1.4562	1.5126	1.5598	1.6017	1.6398
2500	v	.16839	.2291	.2712	.3069	.3393	.3696
(668.31)	h	1176.6	1301.7	1385.4	1457.2	1523.8	1587.7
	s	1.3073	1.4112	1.4752	1.5262	1.5704	1.6101
3000	v	.09771	.17572	.2160	.2485	.2772	.3036
(695.52)	h	1058.1	1265.2	1361.7	1439.6	1510.1	1576.6
	s	1.1944	1.3675	1.4414	1.4967	1.5434	1.5848
3206.2[b]	v		.1591	.1980	.2290	.2564	.2816
(705.34)	h		1250.6	1353.4	1434.8	1507.3	1575.1
	s		1.3506	1.4293	1.4872	1.5351	1.5773

[a]See footnotes to Table 10.33.

Table 10.36 Metric Prefixes and Symbols

Multiplication Factor		Prefix	Symbol
1 000 000 000 000 000 000	= 10^{18}	exa	E
1 000 000 000 000 000	= 10^{15}	peta	P
1 000 000 000 000	= 10^{12}	tera	T
1 000 000 000	= 10^{9}	giga	G
1 000 000	= 10^{6}	mega	M
1 000	= 10^{3}	kilo	k
100	= 10^{2}	hecto*	h
10	= 10^{1}	deka*	da
0.1	= 10^{-1}	deci *	d
0.01	= 10^{-2}	centi*	c
0.001	= 10^{-3}	milli	m
0.000 001	= 10^{-6}	micro	μ
0.000 000 001	= 10^{-9}	nano	n
0.000 000 000 001	= 10^{-12}	pico	p
0.000 000 000 000 001	= 10^{-15}	femto	f
0.000 000 000 000 000 001	= 10^{-18}	atto	a

*Avoid usage, if possible.

Table 10.37 Conversion Factors[a]

Multiply	By	To Obtain
Abamperes	10	amperes
Abamperes	2.99796×10^{10}	statamperes
Abampere-turns	12.566	gilberts
Abcoulombs	10	coulombs (abs.)
Abcoulombs	2.99796×10^{10}	statcoulombs
Abcoulombs/kg	30577	statcoulombs/dyne
Abfarads	1×10^{9}	farads (abs.)
Abfarads	8.98776×10^{20}	statfarads
Abhenrys	1×10^{-9}	henrys (abs.)
Abhenrys	1.11263×10^{-21}	stathenrys
Abohms	1×10^{-9}	ohms (abs.)
Abohms	1.11263×10^{-21}	statohms
Abvolts	3.33560×10^{-11}	statvolts
Abvolts	1×10^{-8}	volts (abs.)
Abvolts/centimeters	2.540005×10^{-8}	volts (abs.)/inch
Acre feet	43,560	cubic feet
Acre feet	1233	cubic meters
Acre feet	325,850	gallons (U.S.)
Acres	43,560	square feet
Acres	4047	square meters
Ampere-hours (abs.)	3600	coulombs (abs.)
Amperes (abs.)	0.1	abamperes
Amperes (abs.)	1.036×10^{-5}	faradays/second
Amperes (abs.)	2.9980×10^{9}	statamperes
Angstrom units	1×10^{-8}	centimeters
Angstrom units	3.937×10^{-9}	inches
Ares	1076	square feet
Ares	100	square meters

Table 10.37 *Continued*

Multiply	by	to obtain
Assay tons	29.17	grams
Astronomical units	1.495×10^8	kilometers
Astronomical units	9.290×10^7	miles
Atmospheres	1.0133	bars
Atmospheres	1.01325×10^6	dynes/square centimeter
Atmospheres	10,333	kilograms/square meter
Atmospheres	14.696	pounds/square inch
Avograms	1.66036×10^{-24}	grams
Bags, cement	94	pounds of cement
Barleycorns (British)	⅓	inches
Barleycorns (British)	8.467×10^{-3}	meters
Barrels (British, dry)	5.780	cubic feet
Barrels (British, dry)	0.1637	cubic meters
Barrels (British, dry)	36	gallons (British)
Barrels, cement	170.6	kilograms
Barrels, cement	376	pounds of cement
Barrels, cranberry	3.371	cubic feet
Barrels, cranberry	0.09547	cubic meters
Barrels, oil	5.615	cubic feet
Barrels, oil	0.1590	cubic meters
Barrels, oil	42	gallons (U.S.)
Barrels (U.S., dry)	4.083	cubic feet
Barrels (U.S., dry)	0.11562	cubic meters
Barrels (U.S., liquid)	4.211	cubic feet
Barrels (U.S., liquid)	0.1192	cubic meters
Barrels (U.S., liquid)	31.5	gallons (U.S.)
Bars	10^6	dynes/square centimeter
Bars	10,197	kilograms/square meter
Bars	14.50	pounds/square inch
Baryes	10^{-6}	bars
Board feet	¹⁄₁₂	cubic feet
Boiler horsepower	33,475	Btu (mean)/hour
Boiler horsepower	34.5	pounds of water evaporated from and at 212°F (per hour)
Bolts (U.S., cloth)	120	linear feet
Bolts (U.S., cloth)	36.58	meters
Bougie decimales	1	candles (int.)
Btu (mean)	251.98	calories, gram (mean)
Btu (mean)	0.55556	centigrade heat units (chu)
Btu (mean)	778.2	foot-pounds
Btu (mean)	1054.8	joules (abs.)
Btu (mean)	107.565	kilogram-meters
Btu (mean)	6.876×10^{-5}	pounds of carbon to CO_2
Btu (mean)	0.29305	watt-hours
Btu (mean)/pound	0.5556	calories, gram (mean)/gram
Btu (mean)/pound/°F	1	calories, gram/gram/°C
Btu (mean)/hour (feet²)°F	4.882	kilogram-calorie/hr (m²)°C

Table 10.37 *Continued*

Multiply	by	to obtain
Btu (mean)/hour (feet²) °F	1.3562×10^{-4}	gram-calorie/second (cm²) °C
Btu (mean)/hour (feet²)°F	3.94×10^{-4}	horsepower/(ft²)°F
Btu (mean)/hour (feet²)°F	5.682×10^{-4}	watts/(cm²)°C
Btu (mean)/hour (feet²)°F	2.035×10^{-3}	watts/(in.²)°C
Btu (mean)/(hour)(feet²)(°F/ inch)	3.4448×10^{-4}	calories, gram (15°C)/ (sec)(cm²)(°C/cm)
Btu (mean)/(hour)(feet²)(°F/ inch)	1	chu/(hr)(ft²)(°C/in.)
Btu (mean)/(hour)(feet²)(°F/ inch)	1.442×10^{-3}	joules (abs.)/(sec)(cm²)(°C/ cm)
Btu (mean)/(hour)(feet²)(°F/ inch)	1.442×10^{-3}	watts/(cm²)(°C/cm)
Btu (60°F)	1054.6	joules (abs.)
Buckets (British, dry)	4	gallons (British)
Bushels (British)	1.03205	bushels (U.S.)
Bushels (British)	0.03637	cubic meters
Bushels (British)	1.2843	cubic feet
Bushels (U.S.)	1.2444	cubic feet
Bushels (U.S.)	0.035239	cubic meters
Bushels (U.S.)	4	pecks (U.S.)
Butts (British)	20.2285	cubic feet
Butts (British)	126	gallons (British)
Cable lengths	720	feet
Cable lengths	219.46	meters
Calories, gram (mean)	3.9685×10^{-3}	Btu (mean)
Calories, gram (mean)	0.001459	cubic feet atmospheres
Calories, gram (mean)	3.0874	foot-pounds
Calories, gram (mean)	4.186	joules (abs.)
Calories, gram (mean)	0.42685	kilogram-meters
Calories, gram (mean)	0.0011628	watt-hours
Calories (thermochemical)	0.999346	calories (Int. Steam Tables)
Calories, gram (mean)/gram	1.8	Btu (mean)/pound
Candle power (spherical)	12.566	lumens
Candles (int.)	0.104	carcel units
Candles (int.)	1.11	Hefner units
Candles (int.)	1	lumens (int.)/steradian
Candles (int.)/square centimeter	2919	foot-lamberts
Candles (int.)/square centimeter	3.1416	lamberts
Candles (int.)/square foot	3.1416	foot-lamberts
Candles (int.)/square foot	3.382×10^{-3}	lamberts
Candles (int.)/square inch	452.4	foot-lamberts
Candles (int.)/square inch	0.4870	lamberts
Carats (metric)	3.0865	grains
Carats (metric)	0.2	grams
Centals	100	pounds
Centigrade heat units (chu)	1.8	Btu
Centigrade heat units (chu)	453.6	calories, gram (15°C)

Table 10.37 *Continued*

Multiply	By	To Obtain
Centigrade heat units (chu)	1897.8	joules (abs.)
Centimeters	0.0328083	feet (U.S.)
Centimeters	0.3937	inches (U.S.)
Centipoises	3.60	kilograms/meter-hour
Centipoises	10^{-3}	kilograms/meter-second
Centipoises	2.42	pounds/foot-hour
Centipoises	6.72×10^{-4}	pounds/foot-second
Centipoises	2.089×10^{-5}	pound force second/ft^2
Chains (engineers' or Ramden's)	100	feet
Chains (engineers' or Ramden's)	30.48	meters
Chains (surveyors' or Gunter's)	66	feet
Chains (surveyors' or Gunter's)	20.12	meters
Chaldrons (British)	32	bushels (British)
Chaldrons (U.S.)	36	bushels (U.S.)
Cheval-vapours	0.9863	horsepower
Cheval-vapours	735.5	watts (abs.)
Cheval-vapours heures	2.648×10^6	joules (abs.)
Chu/(hr)(ft^2)($^\circ$C/in.)	1	Btu/(hr)(ft^2)($^\circ$F/in.)
Circular inches	0.7854	square inches
Circular millimeters	7.854×10^{-7}	square meters
Circular mils	7.854×10^{-7}	square inches
Circumferences	360	degrees
Circumferences	400	grades
Cloves	8	pounds
Coombs (British)	4	bushels (British)
Cords	8	cord feet
Cords	128	cubic feet
Cords	3.625	cubic meters
Coulombs (abs.)	0.1	abcoulombs
Coulombs (abs.)	6.281×10^{18}	electronic charges
Coulombs (abs.)	2.998×10^9	statcoulombs
Cubic centimeters	3.531445×10^{-5}	cubic feet (U.S.)
Cubic centimeters	2.6417×10^{-4}	gallons (U.S.)
Cubic centimeters	0.033814	ounces (U.S., fluid)
Cubic feet (British)	0.9999916	cubic feet (U.S.)
Cubic feet (U.S.)	28317.016	cubic centimeters
Cubic feet (U.S.)	1728	cubic inches
Cubic feet (U.S.)	7.48052	gallons (U.S.)
Cubic feet (U.S.)	28.31625	liters
Cubic foot-atmospheres	2.7203	Btu (mean)
Cubic foot-atmospheres	680.74	calories, gram (mean)
Cubic foot-atmospheres	2116	foot-pounds
Cubic foot-atmospheres	2869	joules (abs.)
Cubic foot-atmospheres	292.6	kilogram-meters
Cubic foot-atmospheres	7.968×10^{-4}	kilowatt-hours
Cubic feet of common brick	120	pounds
Cubic feet of water (60°F)	62.37	pounds
Cubic feet/second	1.9834	acre-feet/day

Table 10.37 *Continued*

Multiply	By	To Obtain
Cubic feet/second	448.83	gallons/minute
Cubic feet/second	0.64632	million gallons/day
Cubic inches (U.S.)	16.387162	cubic centimeters
Cubic inches (U.S.)	1.0000084	cubic inches (British)
Cubic inches (U.S.)	0.55411	ounces (U.S., fluid)
Cubic meters	8.1074×10^{-4}	acre-feet
Cubic meters	8.387	barrels (U.S., liquid)
Cubic meters	35.314	cubic feet (U.S.)
Cubic meters	61023	cubic inches (U.S.)
Cubic meters	1.308	cubic yards (U.S.)
Cubic meters	264.17	gallons (U.S.)
Cubic meters	999.973	liters
Cubic yards (British)	0.76455	cubic meters
Cubic yards (British)	0.9999916	cubic yards (U.S.)
Cubic yards (U.S.)	27	cubic feet (U.S.)
Cubic yards (U.S.)	0.76456	cubic meters
Cubic yards of sand	2700	pounds
Cubits	45.720	centimeters
Cubits	1.5	feet
Days (sidereal)	86,164	seconds (mean solar)
Debye units (dipole moment)	10^{18}	electrostatic units
Drachms (British, fluid)	3.5516×10^{-6}	cubic meters
Drachms (British, fluid)	0.125	ounces (British, fluid)
Drams (troy)	2.1943	drams (avoirdupois)
Drams (troy)	60	grains
Drams (troy)	3.8879351	grams
Drams (troy)	0.125	ounces (troy)
Drams (avoirdupois)	1.771845	grams
Drams (avoirdupois)	0.0625	ounces (avoirdupois)
Drams (avoirdupois)	0.00390625	pounds (avoirdupois)
Drams (U.S., fluid)	3.6967×10^{-6}	cubic meters
Drams (U.S., fluid)	0.125	ounces (fluid)
Dynes	0.00101972	grams
Dynes	2.24809×10^{-6}	pounds
Dyne-centimeters (torque)	7.3756×10^{-8}	pound-feet
Dynes/centimeter	1	ergs/square centimeter
Dynes/square centimeter	9.8692×10^{-7}	atmospheres
Dynes/square centimeter	0.01020	kilograms/square meter
Dynes/square centimeter	0.1	newtons/square meter
Dynes/square centimeter	1.450×10^{-5}	pounds/square inch
Electromagnetic cgs units of magnetic permeability	1.1128×10^{-21}	electrostatic cgs units of magnetic permeability
Electromagnetic cgs units of mass resistance	9.9948×10^{-6}	ohms (int.)-meter-gram
Electromagnetic ft-lb-sec units of magnetic permeability	0.0010764	electromagnetic cgs units of magnetic permeability
Electromagnetic ft-lb-sec units of magnetic permeability	1.03382×10^{-18}	electrostatic cgs units of magnetic permeability
Electronic charges	1.5921×10^{-19}	coulombs (abs.)

Table 10.37 *Continued*

Multiply	By	To Obtain
Electron-volts	1.6020×10^{-12}	ergs
Electron-volts	1.0737×10^{-9}	mass units
Electron-volts	0.07386	Rydberg units of energy
Electrostatic cgs units of Hall effect	2.6962×10^{31}	electromagnetic cgs units of Hall effect
Electrostatic ft-lb-sec units of charge	1.1952×10^{-6}	coulombs (abs.)
Electrostatic ft-lb-sec units of magnetic permeability	929.03	electrostatic cgs units of magnetic permeability
Ells	114.30	centimeters
Ells	45	inches
Ems, pica (printing)	0.42333	centimeters
Ems, pica (printing)	⅙	inches
Ergs	9.4805×10^{-11}	Btu (mean)
Ergs	2.3889×10^{-8}	calories, gram (mean)
Ergs	7.3756×10^{-8}	foot-pounds
Ergs	10^{-7}	joules (abs.)
Ergs	1.01972×10^{-8}	kilogram-meters
Faradays	26.80	ampere-hours
Faradays	96,500	coulombs (abs.)
Faradays/second	96,500	amperes (abs.)
Farads (abs.)	10^{-9}	abfarads
Farads (abs.)	8.9877×10^{11}	statfarads
Fathoms	6	feet
Feet (U.S.)	30.4801	centimeters
Feet (U.S.)	1.0000028	feet (British)
Feet (U.S.)	0.30480	meters
Feet (U.S.)	1.893939×10^{-4}	miles (statute)
Feet of air (1 atmosphere, 60°F)	5.30×10^{-4}	pounds/square inch
Feet of water at 39.2°F	304.79	kilograms/square meter
Feet of water at 39.2°F	0.43352	pounds/square inch
Feet/second	1.0973	kilometers/hour
Feet/second	0.68182	miles/hour
Feet/second/second	1.0973	kilometers/hour/second
Firkins (British)	9	gallons (British)
Firkins (U.S.)	9	gallons (U.S.)
Foot-poundals	3.9951×10^{-5}	Btu (mean)
Foot-poundals	0.0421420	joules (abs.)
Foot-pounds	0.0012854	Btu (mean)
Foot-pounds	0.32389	calories, gram (mean)
Foot-pounds	1.13558×10^{7}	ergs
Foot-pounds	32.174	foot-poundals
Foot-pounds	1.35582	joules (abs.)
Foot-pounds	0.138255	kilogram-meters
Foot-pounds	0.013381	liter-atmospheres
Foot-pounds	3.7662×10^{-4}	watt-hours (abs.)
Foot-pounds/second	4.6275	Btu (mean)/hour
Foot-pounds/second	0.0018182	horsepower
Foot-pounds/second	1.35582	watts (abs.)

Table 10.37 *Continued*

Multiply	By	To Obtain
Furlongs	660	feet
Furlongs	201.17	meters
Furlongs	0.125	miles
Gallons (British)	4516.086	cubic centimeters
Gallons (British)	1.20094	gallons (U.S.)
Gallons (British)	10	pounds (avoirdupois) of water at 62°F
Gallons (U.S.)	3785.434	cubic centimeters
Gallons (U.S.)	0.13368	cubic feet (U.S.)
Gallons (U.S.)	231	cubic inches
Gallons (U.S.)	3.78533	liters
Gallons (U.S.)	128	ounces (U.S., fluid)
Gausses (abs.)	3.3358×10^{-4}	electrostatic cgs units of magnetic flux density
Gausses (abs.)	0.99966	gausses (int.)
Gausses (abs.)	1	maxwells (abs.)/square centimeters
Gausses (abs.)	6.4516	maxwells (abs.)/square inch
Gausses (abs.)	1	lines/square centimeter
Gilberts (abs.)	0.07958	abampere-turns
Gilberts (abs.)	0.7958	ampere-turns
Gilberts (abs.)	2.998×10^{10}	electrostatic cgs units of magnetomotive force
Gills (British)	5	ounces (British, fluid)
Gills (U.S.)	32	drams (fluid)
Grains	0.036571	drams (avoirdupois)
Grains	1/7000	pounds (avoirdupois)
Gram-centimeters	9.2967×10^{-3}	Btu (mean)
Gram-centimeters	2.3427×10^{-5}	calories, gram (mean)
Gram-centimeters	7.2330×10^{-5}	foot-pounds
Gram-centimeters	9.8067×10^{-5}	joules (abs.)
Gram-centimeters	2.7241×10^{-8}	watt-hours
Gram-centimeters/second	9.80665×10^{-5}	watts (abs.)
Grams-centimeters² (moment of inertia)	2.37305×10^{-6}	pounds-feet²
Grams-centimeters² (moment of inertia)	3.4172×10^{-4}	pounds-inch²
Grams/cubic centimeter	62.428	pounds/cubic foot
Grams/cubic centimeter	8.3454	pounds/gallon (U.S.)
Grams/cubic meter	0.43700	grains/cubic foot
Grams/liter	58.417	grains/gallon (U.S.)
Grams/liter	9.99973×10^{-4}	grams/cubic centimeter
Grams/liter	1000	parts per million (ppm)
Grams/liter	0.06243	pounds/cubic foot
Grams/square centimeter	0.0142234	pounds/square inch
Hands	4	inches
Hemispheres	0.5	spheres
Hemispheres	4	spherical right angles
Hemispheres	6.2832	steradians

538

Table 10.37 *Continued*

Multiply	By	To Obtain
Henrys (abs.)	10^9	abhenrys
Henrys (abs.)	1.1126×10^{-12}	stathenrys
Hogsheads (British)	63	gallons (British)
Hogsheads (U.S.)	8.422	cubic feet
Hogsheads (U.S.)	0.2385	cubic meters
Hogsheads (U.S.)	63	gallons (U.S.)
Horsepower	2545.08	Btu (mean)/hour
Horsepower	550	foot-pounds/second
Horsepower	0.74570	kilowatts
Horsepower, electrical	1.0004	horsepower
Horsepower (metric)	0.98632	horsepower
Horsepower-hours	2545	Btu (mean)
Horsepower-hours	1.98×10^6	foot-pounds
Horsepower-hours	2.737×10^5	kilogram-meters
Horsepower-hours	0.7457	kilowatt-hours (abs.)
Hundredweights (long)	112	pounds
Inches (British)	2.540	centimeters
Inches (British)	0.9999972	inches (U.S.)
Inches (U.S.)	2.54000508	centimeters
Inches of mercury at 32°F	345.31	kilograms/square meter
Inches of mercury at 32°F	0.4912	pounds/square inch
Joules (abs.)	10^7	ergs
Joules (abs.)	0.73756	foot-pounds
Joules (abs.)	0.101972	kilogram-meters
Joule-seconds	1.5258×10^{33}	quanta
Karats (1 of gold to 24 of mixture)	41.667	milligrams/gram
Kilderkins (British)	18	gallons (British)
Kilograms	980,665	dynes
Kilograms	15,432.4	grains
Kilograms	35.2740	ounces (avoirdupois)
Kilograms	70.931	poundals
Kilograms	2.20462	pounds
Kilograms	9.84207×10^{-4}	tons (long)
Kilograms	0.001	tons (metric)
Kilograms	0.0011023	tons (short)
Kilogram-meters	0.0092967	Btu (mean)
Kilogram-meters	2.3427	calories, gram (mean)
Kilogram-meters	9.80665×10^7	ergs
Kilogram-meters	7.2330	foot-pounds
Kilogram-meters	3.6529×10^{-6}	horsepower-hours
Kilogram-meters	9.80665	joules (abs.)
Kilogram-meters	2.52407×10^{-6}	kilowatt-hours (abs.)
Kilogram-meters	9.579×10^{-6}	pounds water evap. at 212°F
Kilogram-meters	6.392×10^{-7}	pounds carbon to CO_2
Kilograms/square meter	9.6784×10^{-5}	atmospheres
Kilograms/square meter	98.0665	dynes/square centimeters
Kilograms/square meter	3.281×10^{-3}	feet of water at 39.2°F
Kilograms/square meter	0.1	grams/square centimeters
Kilograms/square meter	2.896×10^{-3}	inches of mercury at 32°F

Table 10.37 *Continued*

Multiply	By	To Obtain
Kilograms/square meter	0.07356	mm of mercury at 0°C
Kilograms/square meter	0.2048	pounds/square foot
Kilograms/square meter	0.00142234	pounds/square inch
Kilowatt-hours (abs.)	3413	Btu (mean)
Kilowatt-hours (abs.)	2.6552×10^6	foot-pounds
Kilowatt-hours (abs.)	3.6709×10^5	kilogram-meters
Kilowatts (abs.)	1.341	horsepower
Knots	1	miles (nautical)/hour
Leagues (nautical)	3	miles (nautical)
Leagues (statute)	3	miles (statute)
Light years	63,274	astronomical units
Light years	9.4599×10^{12}	kilometers
Light years	5.8781×10^{12}	miles
Lignes (Paris lines)	½₂	pouces (Paris inches)
Links (Gunter's)	0.01	chains (Gunter's)
Links (Gunter's)	0.66	feet
Links (Ramden's)	1	feet
Links (Ramden's)	0.01	chains (Ramden's)
Liter-atmospheres (normal)	0.096064	Btu (mean)
Liter-atmospheres (normal)	24.206	calories, gram (mean)
Liter-atmospheres (normal)	74.735	foot-pounds
Liter-atmospheres (normal)	2.815×10^{-5}	kilowatt-hours
Liter-atmospheres (normal)	101.33 .	joules (abs.)
Liter-atmospheres (normal)	10.33	kilogram-meters
Liters	1000.028	cubic centimeters
Liters	0.035316	cubic feet
Liters	0.26417762	gallons (U.S.)
Liters	1.0566828	quarts (U.S., liquid)
Lumens	0.07958	candle-power (spherical)
Lumens	0.00147	watts of maximum visibility radiation
Lumens/square centimeters	1	lamberts
Lumens/square centimeters/ steradian	3.1416	lamberts
Lumens/square foot	1	foot-candles
Lumens/square foot	10.764	lumens/square meter
Lumens/square feet/steradian	3.3816	millilamberts
Lumens/square meter	0.09290	foot-candles or lumens/ square foot
Lumens/square meter	10^{-4}	phots
Lux	0.09290	foot-candles
Lux	1	lumens/square meter
Lux	10^{-4}	phots
Meter-candles	1	lumens/square meter
Meters	10^{10}	angstrom units
Meters	3.280833	feet (U.S.)
Meters	39.37	inches
Meters	6.2137×10^{-4}	miles (statute)
Meters	1.09361	yards (U.S.)

Table 10.37 *Continued*

Multiply	By	To Obtain
Meters/second	2.23693	miles/hour
Microns	1×10^{-4}	centimeters
Miles (British)	1.6093425	kilometers
Miles (int. nautical)	1.852	kilometers
Miles (U.S. statute)	5280	feet
Miles (U.S. statute)	1609	meters
Miles (U.S. statute)	1760	yards
Miles/hour	44.7041	centimeters/second
Miles/hour	1.4667	feet/second
Miles/hour	0.86839	knots
Miles/hour/second	1.4667	feet/second/second
Milligrams/assay ton	1	ounces (troy)/ton (short)
Mils	0.001	inches
Mils	25.40	microns
Miner's inches (Colorado)	0.02604	cubic feet/second
Miner's inches (Ariz., Calif., Mont., and Ore.)	0.025	cubic feet/second
Miner's inches (Ida., Kan., Neb., Nev., N. Mex., N. Dak., S. Dak., and Utah)	0.020	cubic feet/second
Minims (British)	0.05919	cubic centimeters
Minims (U.S.)	0.06161	cubic centimeters
Months (mean calendar)	30.4202	days
Months (mean calendar)	730.1	hours
Months (mean calendar)	43,805	minutes
Months (mean calendar)	2.6283×10^6	seconds
Newtons	10^5	dynes
Newtons	0.10197	kilograms
Newtons	0.22481	pounds
Noggins (British)	⅟₃₂	gallons (British)
Oersteds (abs.)	1	electromagnetic cgs units of magnetizing force
Oersteds (abs.)	2.9978×10^{10}	electrostatic cgs units of magnetizing force
Ohms (abs.)	10^9	abohms
Ohms (abs.)	1.1126×10^{-12}	statohms
Ounces (avoirdupois)	16	drams (avoirdupois)
Ounces (avoirdupois)	28.349527	grams
Ounces (avoirdupois)	0.9114583	ounces (troy)
Ounces (avoirdupois)	⅟₁₆	pounds (avoirdupois)
Ounces (troy)	31.103481	grams
Ounces (troy)	1.09714	ounces (avoirdupois)
Ounces (troy)	⅟₁₂	pounds (troy)
Ounces (U.S., fluid)	29.5737	cubic centimeters
Ounces (U.S., fluid)	1.80469	cubic inches
Ounces (U.S., fluid)	8	drams (fluid)
Ounces (U.S., fluid)	⅟₁₂₈	gallons (U.S.)
Paces	30	inches
Palms (British)	3	inches

Table 10.37 *Continued*

Multiply	By	To Obtain
Parsecs	3.084×10^{16}	meters
Parsecs	3.260	light years
Parts per million (ppm)	0.058417	grains/gallon (U.S.)
Pecks (British)	0.25	bushels (British)
Pecks (U.S.)	0.25	bushels (U.S.)
Pennyweights	24	grains
Pennyweights	1.555174	grams
Pennyweights	0.05	ounces (troy)
Perches (masonry)	24.75	cubic feet
Phots	929.0	foot-candles
Picas (printers')	⅙	inches
Pieds (French feet)	0.3249	meters
Pints (U.S., dry)	33.6003	cubic inches
Pints (U.S., liquid)	473.179	cubic centimeters
Pints (U.S., liquid)	16	ounces (U.S., fluid)
Planck's constant	6.6256×10^{-27}	erg-seconds
Pottles (British)	0.5	gallons (British)
Pouces (Paris inches)	0.02707	meters
Pouces (Paris inches)	0.08333	pieds (Paris feet)
Poundals	14.0981	grams
Poundals	0.031081	pounds
Pound-feet (torque)	1.3558×10^{7}	dyne-centimeters
Pounds (avoirdupois)	7000	grains
Pounds (avoirdupois)	453.5924	grams
Pounds (avoirdupois)	1.2152778	pounds (troy)
Pounds (troy)	373.2418	grams
Pounds of carbon to CO_2	14,544	Btu (mean)
Pounds of water evaporated at 212°F	970.3	Btu
Pounds/square foot	4.88241	kilograms/square meter
Pounds/square inch	0.068046	atmospheres
Pounds/square inch	70.307	grams/square centimeter
Pounds/square inch	703.07	kilograms/square meter
Pounds/square inch	51.715	millimeters of mercury at 0°C
Proof (U.S.)	0.5	percent alcohol by volume
Puncheons (British)	70	gallons (British)
Quadrants	90	degrees
Quarts (U.S., liquid)	0.033420	cubic feet
Quarts (U.S., liquid)	32	ounces (U.S., fluid)
Quarts (U.S., liquid)	0.832674	quarts (British)
Quintals (metric)	100	kilograms
Quintals (long)	112	pounds
Quintals (short)	100	pounds
Quires	24	sheets
Radians	57.29578	degrees
Reams	500	sheets
Register tons (British)	100	cubic feet

Table 10.37 *Continued*

Multiply	By	To Obtain
Revolutions/minute	0.10472	radians/second
Revolutions/minute^2	0.0017453	radians/second/second
Reyns	6.8948×10^6	centipoises
Rods	16.5	feet
Rods	5.0292	meters
Rods	3.125×10^{-3}	miles
Roods (British)	0.25	acres
Scruples	⅓	drams (troy)
Scruples	20	grains
Sections	1	square miles
Slugs	32.174	pounds
Space, entire (solid angle)	12.566	steradians
Spans	9	inches
Square centimeters	1.07639×10^{-3}	square feet (U.S.)
Square centimeters	0.15499969	square inches (U.S.)
Square centimeter-square centimeter (moment of area)	0.024025	square inch-square inch
Square chains (Gunter's)	0.1	acres
Square chains (Gunter's)	404.7	square meters
Square chains (Ramden's)	0.22956	acres
Square chains (Ramden's)	10,000	square feet
Square feet (British)	0.092903	square meters
Square feet (U.S.)	929.0341	square centimeters
Square foot-square foot (moment of area)	20,736	square inch-square inch
Square inches (U.S.)	6.4516258	square centimeters
Square inches (U.S.)	7.71605×10^{-4}	square yards
Square kilometers	0.3861006	square miles (U.S.)
Square links (Gunter's)	10^{-5}	acres (U.S.)
Square links (Gunter's)	0.04047	square meters
Square meters	2.471×10^{-4}	acres (U.S.)
Square meters	10.76387	square feet (U.S.)
Square meters	1550	square inches
Square meters	3.8610×10^{-7}	square miles (statute)
Square meters	1.196	square yards (U.S.)
Square miles	640	acres
Square miles	2.78784×10^7	square feet
Square miles	2.590×10^6	square meters
Square rods	272.3	square feet
Square yards (U.S.)	1296	square inches
Square yards (U.S.)	0.8361	square meters
Statamperes	3.33560×10^{-10}	amperes (abs.)
Statcoulombs	3.33560×10^{-10}	coulombs (abs.)
Statcoulombs/kilogram	1.0197×10^{-6}	statcoulombs/dyne
Statfarads	1.11263×10^{-12}	farads (abs.)
Stathenrys	8.98776×10^{11}	henrys (abs.)
Statohms	8.98776×10^{11}	ohms (abs.)
Statvolts	299.796	volts (abs.)

Table 10.37 *Continued*

Multiply	By	To Obtain
Statvolts/inch	118.05	volts (abs.)/centimeter
Statwebers	2.99796×10^{10}	electromagnetic cgs units of magnetic flux
Statwebers	1	electrostatic cgs units of magnetic flux
Stones (British)	6.350	kilograms
Stones (British)	14	pounds
Toises (French)	6	pieds (Paris feet)
Tons (long)	1016	kilograms
Tons (long)	2240	pounds
Tons (metric)	1000	kilograms
Tons (metric)	2204.6	pounds
Tons (short)	907.2	kilograms
Tons (short)	2000	pounds
Townships (U.S.)	23,040	acres
Townships (U.S.)	36	square miles
Tuns	252	gallons
Volts (abs.)	10^8	abvolts
Volts (abs.)	3.336×10^{-3}	statvolts
Watts (abs.)	3.41304	Btu (mean)/hour
Watts (abs.)	0.01433	calories, kilogram (mean)/minute
Watts (abs.)	10^7	ergs/second
Watts (abs.)	0.7376	foot-pounds/second
Watts (abs.)	0.0013405	horsepower (electrical)
Watts (abs.)	0.10197	kilogram-meters/second
Watts/(cm^2)($^\circ$C/cm)	693.6	Btu/(hr)(ft^2)($^\circ$F/in.)
Wavelength of the red line of cadmium	6.43847×10^{-7}	meters
Webers	10^3	electromagnetic cgs units
Webers	3.336×10^{-3}	electrostatic cgs units
Webers	10^8	lines
Webers	10^8	maxwells
Webers	3.336×10^{-3}	statwebers
Yards	3	feet
Yards	0.91440	meters
Years (tropical, mean solar)	365.2422	days (mean solar)
Years (tropical, mean solar)	8765.8128	hours (mean solar)
Years (tropical, mean solar)	3.155693×10^7	seconds (mean solar)
Years (tropical, mean solar)	1.00273780	years (sidereal)
Years (sidereal)	365.2564	days (mean solar)
Years (sidereal)	366.2564	days (sidereal)

[a]Unless otherwise stated, pounds are U.S. avoirdupois, feet are U.S., and seconds are mean solar.

Table 10.38 Impurities in Water

Source: Reprinted from *The Permutit Water and Waste Treatment Data Book*, © The Permutit Co., Inc.

U. S. Systems of Expressing Impurities

1 grain per gallon	= 1 grain calcium carbonate (CaCO₃) per U.S. gallon of water
1 part per million	= 1 part calcium carbonate (CaCO₃) per 1,000,000 parts of water
1 part per hundred thousand	= 1 part calcium carbonate (CaCO₃) per 100,000 parts of water

Foreign Systems of Expressing Impurities

1 English degree (or °Clark)	= 1 gram calcium carbonate (CaCO₃) per British Imperial gal of water
1 French degree	= 1 part calcium carbonate (CaCO₃) per 100,000 parts of water
1 German degree	= 1 part calcium oxide (CaO) per 100,000 parts of water

Conversions

	Parts $CaCO_3$ per Million (ppm)	Parts $CaCO_3$ per Hundred Thousand (Pts./100,000)	Grains $CaCO_3$ per U.S. Gallon (gpg)	English Degrees (° Clark)	French Degrees (° French)	German Degrees (° German)	Milliequivalents per Liter or Equivalents per Million
1 Part per million	1.	0.1	0.0583	0.07	0.1	0.0560	0.020
1 Part per hundred thousand	10.0	1.	0.583	0.7	1.	0.560	0.20
1 Grain per U. S. gallon	17.1	1.71	1.	1.2	1.71	0.958	0.343
1 English or Clark degree	14.3	1.43	0.833	1.	1.43	0.800	0.286
1 French degree	10.	1.	0.583	0.7	1.	0.560	0.20
1 German degree	17.9	1.79	1.04	1.24	1.79	1.	0.357
1 Milliequivalent per liter or 1 Equivalent per million	50.	5.	2.92	3.50		2.80	1.

545

Table 10.39 Water Analysis

Conversions

	Parts per Million (ppm)	Milligrams per Liter (mg/l)	Grams per Liter (g/l)	Parts per Hundred Thousand (Pts./100,000)	Grains per U.S. Gallon (gr/U.S. gal)	Grains per British Imperial Gallon	Kilograins per Cubic Foot (kgr/ft³)
1 Part per million	1.	1.	0.001	0.1	0.0583	0.07	0.0004
1 Milligram per liter	1.	1.	0.001	0.1	0.0583	0.07	0.0004
1 Gram per liter	1000.	1000.	1.	100.	58.3	70.	0.436
1 Part per hundred thousand	10.	10.	0.01	1.	0.583	0.7	0.00436
1 Grain per U.S. gallon	17.1	17.1	0.017	1.71	1.	1.2	0.0075
1 Grain per British Imperial gallon	14.3	14.3	0.014	1.43	0.833	1.	0.0062
1 Kilograin per cubic foot	2294.	2294.	2.294	229.4	134.	161.	1.

NOTE: In practice, water analysis samples are measured by volume, not by weight, and corrections for variations in specific gravity are practically never made. Therefore, parts per million are assumed to be the same as milligrams per liter and hence the above relationships are, for practical purposes, true.

Equivalents

Water analyses may also be expressed as:

(1) Equivalents per million (epm) $= \dfrac{\text{No. of ppm of substance present}}{\text{Equivalent weight of substance}}$

(2) Milliequivalents per liter (meq/l) $=$ Equivalents per million

(3) Parts per million expressed as $CaCO_3$ $=$ No. of ppm $CaCO_3$ equivalent to No. of ppm of substance present

(4) Fiftieths of equivalents per million (epm/50) $= \dfrac{\text{No. of ppm of substance present} \times 50}{\text{Equivalent weight of substance}}$

NOTES: Numerically (1) and (2) are equal.
Numerically (3) and (4) are equal.

Source: Reprinted from *The Permutit Water and Waste Treatment Data Book*. © The Permutit Co., Inc.

Table 10.40 Sodium Zeolite Reactions[a]

Softening

$$Na_2Z + \left.\begin{matrix} Ca \\ Mg \end{matrix}\right\} \left\{\begin{matrix} (HCO_3)_2 \\ SO_4 \\ Cl_2 \end{matrix}\right. \quad = \quad \left.\begin{matrix} Ca \\ Mg \end{matrix}\right\} Z \quad + \left\{\begin{matrix} 2NaHCO_3 \\ Na_2SO_4 \\ 2NaCl \end{matrix}\right.$$

$$\text{Sodium Zeolite} + \left.\begin{matrix} \text{Calcium} \\ \text{and/or} \\ \text{Magnesium} \end{matrix}\right\} \left\{\begin{matrix} \text{Bicarbonates,} \\ \text{Sulfates,} \\ \text{and/or Chlorides} \end{matrix}\right. = \left.\begin{matrix} \text{Calcium} \\ \text{and/or} \\ \text{Magnesium} \end{matrix}\right\} \text{Zeolite} + \left\{\begin{matrix} \text{Sodium bicarbonate,} \\ \text{Sodium sulfate,} \\ \text{and/or} \\ \text{Sodium chloride} \end{matrix}\right.$$

   (insoluble)   (soluble)   (insoluble)   (soluble)

Regeneration

$$\left.\begin{matrix} Ca \\ Mg \end{matrix}\right\} Z \quad + \; 2NaCl \quad = \quad Na_2Z \quad + \left.\begin{matrix} Ca \\ Mg \end{matrix}\right\} Cl_2$$

$$\left.\begin{matrix} \text{Calcium} \\ \text{and/or} \\ \text{Magnesium} \end{matrix}\right\} \text{Zeolite} + \text{Sodium Chloride} = \text{Sodium Zeolite} + \left.\begin{matrix} \text{Calcium} \\ \text{and/or} \\ \text{Magnesium} \end{matrix}\right\} \text{Chlorides}$$

   (insoluble)   (soluble)   (insoluble)   (soluble)

COMPENSATED HARDNESS: The hardness of a water for softening by the zeolite process should be compensated when:
1. The total hardness (T.H.) is over 400 ppm as $CaCO_3$, or
2. The sodium salts (Na) are over 100 ppm as $CaCO_3$.

Calculate compensated hardness as follows:

$$\text{Compensated Hardness (ppm)} = \text{Total Hardness (ppm)} \times \frac{9000}{9000 - \text{Total Cations (ppm)}}.$$

Express compensated hardness as:
1. Next higher tenth of a grain up to 5.0 grains per gallon.
2. Next higher half of a grain from 5.0 to 10.0 grains per gallon.
3. Next higher grain above 10.0 grains per gallon.

SALT CONSUMPTION: The salt consumption with the zeolite water softener ranges between 0.37 and 0.45 lb of salt per 1000 grains of hardness, expressed as calcium carbonate, removed. This range is due to two factors: (1) the composition of the water and (2) the operating exchange value at which the zeolite is to be worked. The lower salt consumptions may be attained with waters that are not excessively hard nor high in sodium salts and where the zeolite is not worked at its maximum capacity.

Source: Reprinted from *The Permutit Water and Waste Treatment Data Book.* © The Permutit Co., Inc.
[a] The symbol Z represents Zeolite radical.

Table 10.41 Salt and Brine Equivalents

1 U.S. gallon of saturated salt brine weighs 10 lb and contains 2.48 lb salt.
1 Cubic foot of saturated salt brine weighs 75 lb and contains 18.5 lb salt.
Pounds of salt $\times$ 0.405 = gallons of saturated salt brine.
1 Cubic foot of dry salt weighs from slightly less than 50 lb to somewhat over 70 lb.
(Evaporated salts are near the lower limits; rock salts are near the upper limits.)

Source: Reprinted from *The Permutit Water and Waste Treatment Data Book,* © The Permutit Co., Inc.

Table 10.42 Hydrogen Cation Exchanger Reactions[a]

Reactions with Bicarbonates

$$\begin{Bmatrix} Ca \\ Mg \\ Na_2 \end{Bmatrix}(HCO_3)_2 \;+\; H_2Z \;=\; \begin{Bmatrix} Ca \\ Mg \\ Na_2 \end{Bmatrix}Z \;+\; 2H_2O \;+\; 2CO_2$$

Calcium, Magnesium, and/or Sodium Bicarbonate + Hydrogen cation exchanger = Calcium, Magnesium, and/or Sodium Cation Exchanger + Water + Carbon dioxide

(soluble) (insoluble) (insoluble) (soluble gas)

Reactions with Sulfates or Chlorides

$$\begin{Bmatrix} Ca \\ Mg \\ Na_2 \end{Bmatrix}\begin{Bmatrix} SO_4 \\ Cl_2 \end{Bmatrix} \;+\; H_2Z \;=\; \begin{Bmatrix} Ca \\ Mg \\ Na_2 \end{Bmatrix}Z \;+\; \begin{Bmatrix} H_2SO_4 \\ 2HCl \end{Bmatrix}$$

Calcium, Magnesium, and/or Sodium Sulfates and/or Chlorides + Hydrogen Cation Exchanger = Calcium, Magnesium, and/or Sodium Cation Exchanger + Sulfuric and/or Hydrochloric acids

(soluble) (insoluble) (insoluble) (soluble)

Regeneration Reactions

$$\left.\begin{matrix} Ca \\ Mg \\ Na_2 \end{matrix}\right\} Z \; + \; H_2SO_4 \; = \; H_2Z \; + \; \left.\begin{matrix} Ca \\ Mg \\ Na_2 \end{matrix}\right\} SO_4$$

Calcium, Magnesium, and/or Sodium Cation Exchanger	+	Sulfuric acid	=	Hydrogen Cation Exchanger	+	Calcium, Magnesium, and/or Sodium Sulfates
(insoluble)		(soluble)		(insoluble)		(soluble)

Source: Reprinted from *The Permutit Water and Waste Treatment Data Book*, © The Permutit Co., Inc.
ᵃThe symbol Z represents Hydrogen Cation Exchanger radical.

Table 10.43 Weakly Basic Anion Exchanger Reactions[a]

Reactions with Sulfuric and Hydrochloric Acids

$$\text{D} + \begin{Bmatrix} H_2SO_4 \\ 2HCl \end{Bmatrix} = \text{D}\begin{Bmatrix} \cdot H_2SO_4 \\ \cdot 2HCl \end{Bmatrix}$$

Weakly Basic Anion Exchanger + Sulfuric and/or Hydrochloric acids = Weakly Basic Anion Exchanger Hydrosulfate and/or Hydrochloride

(insoluble)   (soluble)   (insoluble)

Regeneration Reactions

$$\text{D}\begin{Bmatrix} \cdot H_2SO_4 \\ \cdot 2HCl \end{Bmatrix} + Na_2CO_3 = \text{D} + Na_2\begin{Bmatrix} SO_4 \\ Cl_2 \end{Bmatrix} + H_2O + CO_2$$

Weakly Basic Anion Exchanger Hydrosulfate and/or Hydrochloride + Soda ash = Weakly Basic Anion Exchanger + Sodium Sulfate and/or Chloride + Water + Carbon dioxide

(insoluble)   (soluble)   (insoluble)   (soluble)   (soluble gas)

Source: Reprinted from *The Permutit Water and Waste Treatment Data Book,* © The Permutit Co., Inc.

[a] The symbol D represents the Anion Exchanger radical.

Table 10.44 Selected Copy Editor and Proofreader Marks

Mark	Meaning	Marked in Manuscript	Set in Type
—	Set in italics (ital)	Goethe's <u>Faust</u>	Goethe's *Faust*
=	Set in small capitals (sc, esc)	<u>d</u>–glucose	D-glucose
≡	Set in capitals (cap)	<u>new york</u>	New York
/	Set in lowercase (lc)	the $\cancel{N}$uclear $\cancel{A}$ge	the nuclear age
⌇	Set in boldface (bf)	A · B	**A · B**
⌐	Move left; flush left (fl, fl left)	⌐ where <u>t</u> = time.	where t = time.
⌐	Move right; flush right (fl right)	(25a) ⎯⎯⎯⎯⎯⎯⌐	(25a)
¶	Paragraph indent (para indent)	¶ The ethologist D. Morris gave his young chimpanzee Congo a pencil and studied his "drawing activity."	The ethologist D. Morris gave his young chimpanzee Congo a pencil and studied his "drawing activity." Even the first scratchings were not random:
↶	Run in	Even the first scratchings were not random: ¶ [Congo] carried in him the germ ... of visual patterns." The	"[Congo] carried in him the germ ... of visual patterns."
¶	New line, paragraph indent		
⌐	New line, flush left	chimpanzee passed through several stages characteristic of early child art.	The chimpanzee passed through several stages characteristic of early child art.
⊔	Move down	x<u>y</u>	*zy*
⊓	Move up	x<u>y</u>	*zy*
⌉⌐	Center (ctr), display	⌉ x + y ⌐	$x + y$
⊓⊔	Transpose (tr)	⌐Smith,⌐ A. B.⌐	A. B. Smith
∼	Transpose (tr)	cent$\cap$re	center
⌗	Insert full space (#)	A.⌗B.⌗Smith et$\cancel{a}$l.	A. B. Smith et al.
⌒	Close up	non⌒uniform	nonuniform
( )	Close up	1. Purpose ( ) 2. Methods 3. Results	1. Purpose 2. Methods 3. Results
⌒	Bring closer	In ~~the~~ two cases	In two cases

Table 10.44 *Continued*

Mark	Meaning	Marked in Manuscript	Set in Type
ℓ	Delete	right–hand hand page	right-hand page
⌐	Delete	right–haͅnd page	right-hand page
(stet)	Restore deleted copy (stet)	right–hand page	right-hand page
∧	Insert copy	right⌄page = *hand*	right-hand page
(en)	Set en fraction	½ (en)	½
(em)	Set em fraction	½ (em)	½
(shill)	Set shilling fraction (shill, sh)	½ (shill)	1/2
⊙	Set a period	J. Am⊙Chem. Soc.	J. Am. Chem. Soc.
⋏	Set a comma	a⋏, a⋏, . . . ⋏a⋏	$a_1, a_2, \ldots, a_n$
⊙	Set a colon	New York⊙ Wiley, 1973	New York: Wiley, 1973
⋏	Set a semicolon	a = 2⋏ b = 5	$a = 2; b = 5$
⋎	Set a prime	A⋎	A'
⋎⋎	Set quotation marks	⋎Bromine⋎ comes from the French	"Bromine" comes from the French
=	Set hyphen	high≠frequency radiation	high-frequency radiation
⊜ ⋏=	Insert hyphen	un⊜ionized calcium	un-ionized calcium

Mark in Margin	Mark in Proof	Meaning
cap /	are capitalized in ɇnglish titles;	Capital letter
ℓc /	only the first word and Ᵽroper	Lowercase letter
tr /	nouns, in French and Italaiͪn ti-	Transpose
⋏/	tles; and all nouns⋏in German.	Insert comma
ℯ/	Setting the voͫlume number of	Delete and close up
✕/	ⓐperiodical in distinctive type	Fix broken letter
ℒ/	(italics or bold⌿face) and page	Delete hyphen, close up
ℳ	numbers in Ᵽrdinary type makes	Invert letter
⋎	"Vol.⋏and "p." unnecessary.	Insert quotes
=/	The year of publication is en⋏	Insert hyphen
e/	closed in parenthesis only when	Substitute e for i
tr/	it⌐is⌐actually⌐parenthetical	Transpose
eq #	information, as when, for ⋎ex-	Equalize spacing
⊏/	⊏ ample, the volume and page of	Move left
stet /	a periodical are cited and the	Let type stand
w.f /	year is⊙added for the convenience	Change to right font
¶ /	of the reader. ⌐The following	Begin new paragraph
:/	illustrates a bibliographic item/	Change period to colon
sc/	GAY, CHARLES M., and Charles	Set in small caps
ital /	DE VAN FAWCETT, *Mechanical and*	Set in itals
for/	*Electrical Equipment of Buildings,*	Change *of* to *for*
rom/	John Wiley & Sons, New York,	Set in roman
⌒/ # /	second edition, p. 229 ⋏1945.	Close up; insert space

Proofreader's marks.

Table 10.45 Expectation of Life and Expected Deaths, by Race, Age, and Sex: 1978

AGE IN 1978 (years)	EXPECTATION OF LIFE IN YEARS				
	Total	White		Black and other	
		Male	Female	Male	Female
At birth	73.3	70.2	77.8	65.0	73.6
1	73.3	70.1	77.6	65.5	74.0
2	72.4	69.2	76.7	64.6	73.1
3	71.5	68.3	75.7	63.7	72.1
4	70.5	67.3	74.8	62.8	71.2
5	69.5	66.3	73.8	61.8	70.2
6	68.6	65.4	72.8	60.8	69.3
7	67.6	64.4	71.8	59.9	68.3
8	66.6	63.4	70.8	58.9	67.3
9	65.6	62.4	69.9	57.9	66.3
10	64.6	61.5	68.9	57.0	65.4
11	63.7	60.5	67.9	56.0	64.4
12	62.7	59.5	66.9	55.0	63.4
13	61.7	58.5	65.9	54.0	62.4
14	60.7	57.5	64.9	53.1	61.4
15	59.7	56.6	64.0	52.1	60.4
16	58.8	55.6	63.0	51.1	59.5
17	57.8	54.7	62.0	50.2	58.5
18	56.9	53.8	61.1	49.3	57.5
19	56.0	52.9	60.1	48.3	56.6
20	55.0	52.0	59.1	47.4	55.6
21	54.1	51.1	58.2	46.5	54.7
22	53.2	50.2	57.2	45.7	53.7
23	52.2	49.3	56.2	44.8	52.8
24	51.3	48.4	55.3	43.9	51.8
25	50.4	47.5	54.3	43.1	50.9
26	49.5	46.5	53.3	42.2	49.9
27	48.5	45.6	52.4	41.4	49.0
28	47.6	44.7	51.4	40.5	48.1
29	46.6	43.8	50.4	39.7	47.1
30	45.7	42.8	49.5	38.8	46.2
31	44.8	41.9	48.5	37.9	45.2
32	43.8	41.0	47.5	37.1	44.3
33	42.9	40.0	46.6	36.2	43.4
34	41.9	39.1	45.6	35.4	42.5
35	41.0	38.2	44.6	34.5	41.5
36	40.1	37.2	43.7	33.7	40.6
37	39.1	36.3	42.7	32.9	39.7
38	38.2	35.4	41.8	32.0	38.8
39	37.3	34.5	40.8	31.2	37.9
40	36.4	33.6	39.9	30.4	37.0
41	35.5	32.6	38.9	29.6	36.1
42	34.6	31.7	38.0	28.9	35.3
43	33.7	30.8	37.1	28.1	34.4
44	32.8	29.9	36.1	27.3	33.5
45	31.9	29.1	35.2	26.5	32.7
46	31.0	28.2	34.3	25.8	31.8
47	30.1	27.3	33.4	25.0	31.0
48	29.3	26.5	32.5	24.3	30.2
49	28.4	25.6	31.6	23.5	29.3
50	27.6	24.8	30.7	22.8	28.5
51	26.7	24.0	29.8	22.1	27.8
52	25.9	23.2	29.0	21.4	27.0
53	25.1	22.4	28.1	20.8	26.2
54	24.3	21.6	27.3	20.1	25.5
55	23.5	20.8	26.4	19.5	24.7
56	22.7	20.1	25.6	18.9	24.0
57	22.0	19.3	24.7	18.3	23.3
58	21.2	18.6	23.9	17.7	22.5
59	20.5	17.9	23.1	17.1	21.8
60	19.7	17.2	22.3	16.5	21.2
61	19.0	16.5	21.5	16.0	20.5
62	18.3	15.8	20.7	15.5	19.9
63	17.6	15.2	19.9	15.1	19.3
64	17.0	14.6	19.2	14.6	18.6
65	16.3	14.0	18.4	14.1	18.0
70	13.1	11.1	14.8	11.6	14.8
75	10.4	8.6	11.5	9.8	12.5
80	8.1	6.7	8.8	8.8	11.5
85 and over	6.4	5.3	6.7	7.8	9.9

Source: U.S. National Center for Health Statistics, unpublished data.

553

Table 10.46 The Number of Each Day of the Year

Day of Mo.	Jan.	Feb.	Mar.	Apr.	May	Jun.	Jul.	Aug.	Sep.	Oct.	Nov.	Dec.
1	1	32	60	91	121	152	182	213	244	274	305	335
2	2	33	61	92	122	153	183	214	245	275	306	336
3	3	34	62	93	123	154	184	215	246	276	307	337
4	4	35	63	94	124	155	185	216	247	277	308	338
5	5	36	64	95	125	156	186	217	248	278	309	339
6	6	37	65	96	126	157	187	218	249	279	310	340
7	7	38	66	97	127	158	188	219	250	280	311	341
8	8	39	67	98	128	159	189	220	251	281	312	342
9	9	40	68	99	129	160	190	221	252	282	313	343
10	10	41	69	100	130	161	191	222	253	283	314	344
11	11	42	70	101	131	162	192	223	254	284	315	345
12	12	43	71	102	132	163	193	224	255	285	316	346
13	13	44	72	103	133	164	194	225	256	286	317	347
14	14	45	73	104	134	165	195	226	257	287	318	348
15	15	46	74	105	135	166	196	227	258	288	319	349
16	16	47	75	106	136	167	197	228	259	289	320	350
17	17	48	76	107	137	168	198	229	260	290	321	351
18	18	49	77	108	138	169	199	230	261	291	322	352
19	19	50	78	109	139	170	200	231	262	292	323	353
20	20	51	79	110	140	171	201	232	263	293	324	354
21	21	52	80	111	141	172	202	233	264	294	325	355
22	22	53	81	112	142	173	203	234	265	295	326	356
23	23	54	82	113	143	174	204	235	266	296	327	357
24	24	55	83	114	144	175	205	236	267	297	328	358
25	25	56	84	115	145	176	206	237	268	298	329	359
26	26	57	85	116	146	177	207	238	269	299	330	360
27	27	58	86	117	147	178	208	239	270	300	331	361
28	28	59	87	118	148	179	209	240	271	301	332	362
29	29	a	88	119	149	180	210	241	272	302	333	363
30	30		89	120	150	181	211	242	273	303	334	364
31	31		90		151		212	243		304		365

aIn leap years, after February 28, add 1 to the tabulated number.

Table 10.47　Latin Terms

ca.	*circa*	about
cf.	*confer*	compare
e. g.	*exempli gratia*	for example
et al.	*et alibi*	and elsewhere
	et alii or aliae	and others
etc.	*et cetera*	and others, and so forth
fl.	*floruit*	flourished
	fluidus	fluid
i. e.	*id est*	that is
ibid.	*ibidem*	the same place
inf.	*infra*	below
in loc. cit.	*in loco citato*	in the place cited
op. cit.	*opere citato*	in the work cited
pass.	*passim*	here and there
per	*per*	through, by means of
q. v.	*quod vide*	which see
sc.	*scilicet*	namely
sic	*sic*	thus
sup.	*supra*	above
viz.	*videlicet*	namely
v. or vs.	*versus*	against

Table 10.48　Greek Alphabet

A	α	Alpha	N	ν	Nu
B	β	Beta	Ξ	ξ	Xi
Γ	γ	Gamma	O	o	Omicron
Δ	δ	Delta	Π	π	Pi
E	ε	Epsilon	P	ρ	Rho
Z	ζ	Zeta	Σ	σ	Sigma
H	η	Eta	T	τ	Tau
Θ	θ	Theta	Υ	υ	Upsilon
I	ι	Iota	Φ	φ	Phi
K	κ	Kappa	X	χ	Chi
Λ	λ	Lambda	Ψ	ψ	Psi
M	μ	Mu	Ω	ω	Omega

Table 10.49 Standard Electrode (Oxidation) Potentials of Some Electrodes in Aqueous Solution at 25°C

Electrode Reaction	Standard Potential (volts)	Electrode Reaction	Standard Potential (volts)
$Li = Li^+ + \epsilon$	3.045	$Ag + I^- = AgI + \epsilon$	0.151
$K = K^+ + \epsilon$	2.925	$Sn = Sn^{2+} + 2\epsilon$	0.136
$Na = Na^+ + \epsilon$	2.714	$Pb = Pb^{2+} + 2\epsilon$	0.036
$Mg = Mg^{2+} + 2\epsilon$	2.37	$H_2 = 2H^+ + 2\epsilon$	0.000
$Al = Al^{3+} + 3\epsilon$	1.66	$Cu = Cu^{2+} + 2\epsilon$	−0.337
$Zn = Zn^{2+} + 2\epsilon$	0.763	$2OH^- = H_2O + \frac{1}{2}O_2 + \epsilon$	−0.401
$Fe = Fe^{2+} + 2\epsilon$	0.440	$2Hg = Hg^{2+} + 2\epsilon$	−0.789
$Cd = Cd^{2+} + 2\epsilon$	0.403	$Ag = Ag^+ + \epsilon$	−0.799
$Ni = Ni^{2+} + 2\epsilon$	0.250	$2Cl^- = Cl_2 + 2\epsilon$	−1.360

Source: Handbook of Physics, E. U. Condon and Hugh Odishaw. © McGraw-Hill. Used with the permission of McGraw-Hill Book Co.

Table 10.50 Organization Abbreviations

AAR	Association of American Railroads
AAAS	American Association for the Advancement of Science
AASHO	American Association of State Highway Officials
ACI	American Concrete Institute
ACS	American Chemical Society
AFBMA	Anti-Friction Bearing Manufacturers Association
AGA	American Gas Association
AGI	American Geological Institute
AGMA	American Gear Manufacturers Association
AIA	American Institute of Architects
AIChE	American Institute of Chemical Engineers
AIME	American Institute of Mining Engineering
AIM&M	American Institute of Mining & Metallurgy
AISC	American Institute of Steel Construction
AISI	American Iron and Steel Institute
AITC	American Institute of Timber Construction
ANSI	American National Standards Institute
ANS	American Nuclear Society
API	American Petroleum Institute
ASA	Acoustical Society of America
ASCE	American Society of Civil Engineers
ASCET	American Society of Certified Engineering Technicians
ASCT	American Society of Computing Technicians
ASEC	American Standard Elevator Code
ASHRAE	American Society of Heating, Refrigerating and Air Conditioning Engineers

Table 10.50 *Continued*

ASM	American Society for Metals
ASME	American Society of Mechanical Engineers
ASQC	American Society for Quality Control
ASTM	American Society for Testing and Materials
AWS	American Welding Society
AWWA	American Water Works Association
CEC	Consulting Engineers Council
CEMA	Conveyor Equipment Manufacturers Association
CGA	Compressed Gas Association
DOE	Department of Energy
EEI	Edison Electric Institute
EIA	Electronic Industries Association
EOHCI	Electric Overhead Crane Institute
FM	Factory Mutual
FPC	Federal Power Commission
FSPT	Federation of Societies for Paint Technology
GSA	Geological Society of America
HI	Hydraulic Institute
ICC	Interstate Commerce Commission
IEC	International Electrotechnical Commission
IEEE	Institute of Electrical and Electronic Engineers
IES	Illuminating Engineering Society
IPCEA	Insulated Power Cable Engineers Association
ISA	Instrument Society of America
ISO	International Organization for Standardization
ITE	Institute of Traffic Engineers
JAN	Joint Army–Navy
JANAF	Joint Army–Navy–Airforce
MPTA	Mechanical Power Transmission Association
MTS	Marine Technology Society
NACE	National Association of Corrosion Engineers
NAM	National Association of Manufacturers
NBS	National Bureau of Standards
NCARB	National Council of Architectural Registration Boards
NDHA	National District Heating Association
NEC	National Electrical Code
NECA	National Electrical Contractors Association
NEMA	National Electrical Manufacturers Association
NFPA	National Fire Protection Association
NFPA	National Fluid Power Association
NRC	Nuclear Regulatory Commission
NSPE	National Society of Professional Engineers
RWMA	Resistance Welder Manufacturers Association
SAE	Society of Automotive Engineers
SEIA	Solar Energy Industries Association
TAPPI	Technical Association of the Pulp and Paper Industry
TEMA	Tubular Exchanger Manufacturers Association
UL	Underwriters' Laboratories
VMA	Valve Manufacturers Association

Table 10.51   Periodic Table of the Elements

GROUP
1A

Key

atomic number → 29   63.54 ← atomic weight; parentheses indicates longest lived isotope

symbol → Cu

Copper ← name

oxidation states +1 and +2 → 1,2

(Ar)3d¹⁰4s¹ ← electron structure same as Ar, plus 10e⁻ in 3d and 1e⁻ in 4s orbitals

1A	2A	3B	4B	5B	6B	7B	8		
1 1.00797 **H** Hydrogen ±1 1s¹									
3 6.939 **Li** Lithium 1 1s²2s¹	4 9.0122 **Be** Beryllium 2 1s²2s²								
11 22.9898 **Na** Sodium 1 (Ne)3s¹	12 24.312 **Mg** Magnesium 2 (Ne)3s²								
19 39.102 **K** Potassium 1 (Ar)4s¹	20 40.08 **Ca** Calcium 2 (Ar)4s²	21 44.956 **Sc** Scandium 3 (Ar)3d¹4s²	22 47.90 **Ti** Titanium 2,3,4 (Ar)3d²4s²	23 50.942 **V** Vanadium 2,3,4,5 (Ar)3d³4s²	24 51.996 **Cr** Chromium 2,3,6 (Ar)3d⁵4s¹	25 54.938 **Mn** Manganese 2,3,4,6,7 (Ar)3d⁵4s²	26 55.847 **Fe** Iron 2,3,4,6 (Ar)3d⁶4s²	27 58.933 **Co** Cobalt 2,3,4 (Ar)3d⁷4s²	
37 85.47 **Rb** Rubidium 1 (Kr)5s¹	38 87.62 **Sr** Strontium 2 (Kr)5s²	39 88.905 **Y** Yttrium 3 (Kr)4d¹5s²	40 91.22 **Zr** Zirconium 4 (Kr)4d²5s²	41 92.906 **Nb** Niobium 3,5 (Kr)4d⁴5s¹	42 95.94 **Mo** Molybdenum 3,5,6 (Kr)4d⁵5s¹	43 (98) **Tc** Technetium 2,4,7 (Kr)4d⁵5s²	44 101.07 **Ru** Ruthenium 2,3,4,6,8 (Kr)4d⁷5s¹	45 102.905 **Rh** Rhodium 2,3,4,6 (Kr)4d⁸5s¹	
55 132.905 **Cs** Cesium 1 (Xe)6s¹	56 137.34 **Ba** Barium 2 (Xe)6s²	57 138.91 **La** Lanthanum 3 (Xe)5d¹6s²	72 178.49 **HF** Hafnium 4 (Xe)4f¹⁴5d²6s²	73 180.948 **Ta** Tantalum 5 (Xe)4f¹⁴5d³6s²	74 183.85 **W** Tungsten 2,4,5,6 (Xe)4f¹⁴5d⁴6s²	75 186.2 **Re** Rhenium −1,3,4,6,7 (Xe)4f¹⁴5d⁵6s²	76 190.2 **Os** Osmium 2,3,4,6,8 (Xe)4f¹⁴5d⁶6s²	77 192.2 **Ir** Iridium 2,3,4,6 (Xe)4f¹⁴5d⁷6s²	
87 (223) **Fr** Francium 1 (Rn)7s¹	88 (226) **Ra** Radium 2 (Rn)7s²	89 (227) **Ac** Actinium 3 (Rn)6d¹7s²							

58 140.12 **Ce** Cerium 3,4 (Xe)4f²5d⁰6s²	59 140.907 **Pr** Praseodymium 3,4 (Xe)4f³5d⁰6s²	60 144.24 **Nd** Neodymium 3,4 (Xe)4f⁴5d⁰6s²	61 (147) **Pm** Promethium 3 (Xe)4f⁵5d⁰6s²	62 150.35 **Sm** Samarium 2,3 (Xe)4f⁶5d⁰6s²	63 151.96 **Eu** Europium 2,3 (Xe)4f⁷5d⁰6s²
90 232.038 **Th** Thorium 3,4 (Rn)6d²7s²	91 (231) **Pa** Protoactinium 4,5 (Rn)5f²6d¹7s²	92 238.03 **U** Uranium 3,4,5,6 (Rn)5f³6d¹7s²	93 (237) **Np** Neptunium 3,4,5,6 (Rn)5f⁴6d¹7s²	94 (244) **Pu** Plutonium 3,4,5,6 (Rn)5f⁶6d¹7s²	95 (243) **Am** Americium 3,4,5,6 (Rn)5f⁷6d⁰7s²

Basis: ¹²C = 12.0000.

					NOBLE GASES
					2 4.0026 **He** Helium 0 $1s^2$

3A	4A	5A	6A	7A	
5 10.811 **B** Boron 3 $1s^22s^22p^1$	6 12.0112 **C** Carbon ±4,2 $1s^22s^22p^2$	7 14.0067 **N** Nitrogen −3,2,3,4,5 $1s^22s^22p^3$	8 15.9994 **O** Oxygen −2 $1s^22s^22p^4$	9 18.9984 **F** Fluorine −1 $1s^22s^22p^5$	10 20.183 **Ne** Neon 0 $1s^22s^22p^6$
13 26.9815 **Al** Aluminum 3 $(Ne)3s^23p^1$	14 28.086 **Si** Silicon 4 $(Ne)3s^23p^2$	15 30.9738 **P** Phosphorus ±3,4,5 $(Ne)3s^23p^3$	16 32.064 **S** Sulfur −2,4,6 $(Ne)3s^23p^4$	17 35.453 **Cl** Chlorine ±1,3,5,7 $(Ne)3s^23p^5$	18 39.948 **Ar** Argon 0 $(Ne)3s^23p^6$

1B	2B	3A	4A	5A	6A	7A	

1B	2B						
28 58.71 **Ni** Nickel 2,4,6 $(Ar)3d^84s^2$	29 63.54 **Cu** Copper 1,2 $(Ar)3d^{10}4s^1$	30 65.37 **Zn** Zinc 2 $(Ar)3d^{10}4s^2$	31 69.72 **Ga** Gallium 3 $(Ar)3d^{10}4s^24p^1$	32 72.59 **Ge** Germanium 2,4 $(Ar)3d^{10}4s^24p^2$	33 74.922 **As** Arsenic ±3,5 $(Ar)3d^{10}4s^24p^3$	34 78.96 **Se** Selenium −2,4,6 $(Ar)3d^{10}4s^24p^4$	35 79.909 **Br** Bromine ±1,5 $(Ar)3d^{10}4s^24p^5$ · 36 83.80 **Kr** Krypton 0 $(Ar)3d^{10}4s^24p^6$
46 106.4 **Pd** Palladium 2,4,6 $(Kr)4d^{10}5s^0$	47 107.870 **Ag** Silver 1 $(Kr)4d^{10}5s^1$	48 112.40 **Cd** Cadmium 2 $(Kr)4d^{10}5s^2$	49 114.82 **In** Indium 3 $(Kr)4d^{10}5s^25p^1$	50 118.69 **Sn** Tin 2,4 $(Kr)4d^{10}5s^25p^2$	51 121.75 **Sb** Antimony ±3,5 $(Kr)4d^{10}5s^25p^3$	52 127.60 **Te** Tellurium −2,4,6 $(Kr)4d^{10}5s^25p^4$	53 126.904 **I** Iodine ±1,5,7 $(Kr)4d^{10}5s^25p^5$ · 54 131.30 **Xe** Xenon 0(+4 +6) $(Kr)4d^{10}5s^25p^6$
78 195.09 **Pt** Platinum 2,4,6 $(Xe)4f^{14}5d^96s^1$	79 196.967 **Au** Gold 1,3 $(Xe)4f^{14}5d^{10}6s^1$	80 200.59 **Hg** Mercury 2 $(Xe)4f^{14}5d^{10}6s^2$	81 204.37 **Tl** Thallium 1,3 $(Xe)4f^{14}5d^{10}6s^26p^1$	82 207.19 **Pb** Lead 2,4 $(Xe)4f^{14}5d^{10}6s^26p^2$	83 208.980 **Bi** Bismuth ±3,5 $(Xe)4f^{14}5d^{10}6s^26p^3$	84 (210) **Po** Polonium −2,4,6 $(Xe)4f^{14}5d^{10}6s^26p^4$	85 (210) **At** Astatine ±1,5 $(Xe)4f^{14}5d^{10}6s^26p^5$ · 86 (222) **Rn** Radon 0(+6) $(Xe)4f^{14}5d^{10}6s^26p^6$

LANTHANIDES

64 157.25 **Gd** Gadolinium 3 $(Xe)4f^75d^16s^2$	65 158.924 **Tb** Terbium 3,4 $(Xe)4f^85d^16s^2$	66 162.50 **Dy** Dysprosium 3 $(Xe)4f^{10}5d^06s^2$	67 164.930 **Ho** Holmium 3 $(Xe)4f^{11}5d^06s^2$	68 167.26 **Er** Erbium 3 $(Xe)4f^{12}5d^06s^2$	69 168.934 **Tm** Thulium 2,3 $(Xe)4f^{13}5d^06s^2$	70 173.04 **Yb** Ytterbium 2,3 $(Xe)4f^{14}5d^06s^2$	71 174.97 **Lu** Lutetium 3 $(Xe)4f^{14}5d^16s^2$
96 (245) **Cm** Curium 3 $(Rn)5f^76d^17s^2$	97 247 **Bk** Berkelium 3,4 $(Rn)5f^86d^17s^2$	98 (251) **Cf** Californium 3 $(Rn)5f^{10}6d^07s^2$	99 (252) **Es** Einsteinium (3) $(Rn)5f^{11}6d^07s^2$	100 (253) **Fm** Fermium (3) $(Rn)5f^{12}6d^07s^2$	101 (256) **Md** Mendelevium (3) $(Rn)5f^{13}6d^07s^2$	102 (253) **No** Nobelium (3) $(Rn)5f^{14}6d^07s^2$	103 (257) **Lw** Lawrencium (3) $(Rn)5f^{14}6d^17s^2$

ACTINIDES

REFERENCES

1. *Control Valve Handbook,* 2nd edition, Fisher Controls International, Inc., Marshalltown, Iowa, 1977.
2. J. H. Keenan, F. G. Keyes, P. G. Hill, and J. G. Moore, *Steam Tables,* John Wiley & Sons, New York, 1969.

INDEX